Fuzzy Systems Engineering

THE WILEY BICENTENNIAL–KNOWLEDGE FOR GENERATIONS

Each generation has its unique needs and aspirations. When Charles Wiley first opened his small printing shop in lower Manhattan in 1807, it was a generation of boundless potential searching for an identity. And we were there, helping to define a new American literary tradition. Over half a century later, in the midst of the Second Industrial Revolution, it was a generation focused on building the future. Once again, we were there, supplying the critical scientific, technical, and engineering knowledge that helped frame the world. Throughout the 20th Century, and into the new millennium, nations began to reach out beyond their own borders and a new international community was born. Wiley was there, expanding its operations around the world to enable a global exchange of ideas, opinions, and know-how.

For 200 years, Wiley has been an integral part of each generation's journey, enabling the flow of information and understanding necessary to meet their needs and fulfill their aspirations. Today, bold new technologies are changing the way we live and learn. Wiley will be there, providing you the must-have knowledge you need to imagine new worlds, new possibilities, and new opportunities.

Generations come and go, but you can always count on Wiley to provide you the knowledge you need, when and where you need it!

WILLIAM J. PESCE
PRESIDENT AND CHIEF EXECUTIVE OFFICER

PETER BOOTH WILEY
CHAIRMAN OF THE BOARD

Fuzzy Systems Engineering
Toward Human-Centric Computing

Witold Pedrycz
Department of Electrical & Computer Engineering
University of Alberta, Edmonton, Canada
and
Systems Research Institute,
Polish Academy of Sciences Warsaw, Poland

Fernando Gomide
Faculty of Electrical & Computer Engineering
Department of Computer Engineering & Automation
State University of Campinas, Campinas, Brazil

IEEE PRESS

WILEY-INTERSCIENCE
A JOHN WILEY & SONS, INC., PUBLICATION

Copyright © 2007 by John Wiley & Sons, Inc. All rights reserved

Published by John Wiley & Sons, Inc., Hoboken, New Jersey
Published simultaneously in Canada

No part of this publication may be reproduced, stored in a retrieval system, or transmitted in any form or by any means, electronic, mechanical, photocopying, recording, scanning, or otherwise, except as permitted under Section 107 or 108 of the 1976 United States Copyright Act, without either the prior written permission of the Publisher, or authorization through payment of the appropriate per-copy fee to the Copyright Clearance Center, Inc., 222 Rosewood Drive, Danvers, MA 01923, (978) 750-8400, fax (978) 750-4470, or on the web at www.copyright.com. Requests to the Publisher for permission should be addressed to the Permissions Department, John Wiley & Sons, Inc., 111 River Street, Hoboken, NJ 07030, (201) 748-6011, fax (201) 748-6008, or online at http://www.wiley.com/go/permission.

Limit of Liability/Disclaimer of Warranty: While the publisher and author have used their best efforts in preparing this book, they make no representations or warranties with respect to the accuracy or completeness of the contents of this book and specifically disclaim any implied warranties of merchantability or fitness for a particular purpose. No warranty may be created or extended by sales representatives or written sales materials. The advice and strategies contained herein may not be suitable for your situation. You should consult with a professional where appropriate. Neither the publisher nor author shall be liable for any loss of profit or any other commercial damages, including but not limited to special, incidental, consequential, or other damages.

For general information on our other products and services or for technical support, please contact our Customer Care Department within the United States at (800) 762-2974, outside the United States at (317) 572-3993 or fax (317) 572-4002.

Wiley also publishes its books in a variety of electronic formats. Some content that appears in print may not be available in electronic formats. For more information about Wiley products, visit our web site at www.wiley.com.

Wiley Bicentennial Logo: Richard J. Pacifico

Library of Congress Cataloging-in-Publication Data:

Pedrycz, Witold, 1953-
  Fuzzy systems engineering : toward human-centric computing/by Witold Pedrycz and Fernando Gomide.
    p. cm.
  ISBN 978-0-471-78857-7 (cloth)
 1. Soft computing. 2. Fuzzy systems. I. Gomide, Fernando. II. Title.
  QA76.9.S63P44 2007
  006.3--dc22   2007001711

Printed in the United States of America
10 9 8 7 6 5 4 3 2 1

To Ewa, Thais, Adam, Tiago, Barbara, Flavia, Arthur, Ari, and Maria de Lourdes

Contents

Preface   xvii

1 Introduction   1

   1.1 Digital communities and a fundamental quest for human-centric systems   1
   1.2 A historical overview: towards a non-Aristotelian perspective of the world   3
   1.3 Granular computing   5
        1.3.1 Sets and interval analysis   7
        1.3.2 The role of fuzzy sets: a perspective of information granules   8
        1.3.3 Rough sets   13
        1.3.4 Shadowed sets   15
   1.4 Quantifying information granularity: generality versus specificity   16
   1.5 Computational intelligence   16
   1.6 Granular computing and computational intelligence   17
   1.7 Conclusions   18
      Exercises and problems   19
      Historical notes   20
      References   25

2 Notions and Concepts of Fuzzy Sets   27

   2.1 Sets and fuzzy sets: a departure from the principle of dichotomy   27
   2.2 Interpretation of fuzzy sets   31
   2.3 Membership functions and their motivation   33
        2.3.1 Triangular membership functions   34
        2.3.2 Trapezoidal membership functions   35
        2.3.3 Γ-membership functions   36
        2.3.4 S-membership functions   36
        2.3.5 Gaussian membership functions   37
        2.3.6 Exponential-like membership functions   37
   2.4 Fuzzy numbers and intervals   39
   2.5 Linguistic variables   40
   2.6 Conclusions   42

viii Contents

      Exercises and problems   43
      Historical notes   43
      References   44

3 Characterization of Fuzzy Sets   45

3.1 A generic characterization of fuzzy sets: some fundamental descriptors   45

    3.1.1 Normality   46
    3.1.2 Normalization   46
    3.1.3 Support   47
    3.1.4 Core   47
    3.1.5 α-Cut   47
    3.1.6 Convexity   48
    3.1.7 Cardinality   49

3.2 Equality and inclusion relationships in fuzzy sets   50

    3.2.1 Equality   50
    3.2.2 Inclusion   50

3.3 Energy and entropy measures of fuzziness   52

    3.3.1 Energy measure of fuzziness   52
    3.3.2 Entropy measure of fuzziness   54

3.4 Specificity of fuzzy sets   54
3.5 Geometric interpretation of sets and fuzzy sets   56
3.6 Granulation of information   57
3.7 Characterization of the families of fuzzy sets   59

    3.7.1 Frame of cognition   59
    3.7.2 Coverage   59
    3.7.3 Semantic soundness   60
    3.7.4 Main characteristics of the frames of cognition   61

3.8 Fuzzy sets, sets and the representation theorem   62
3.9 Conclusions   64
      Exercises and problems   64
      Historical notes   65
      References   65

4 The Design of Fuzzy Sets   67

4.1 Semantics of fuzzy sets: some general observations   67
4.2 Fuzzy set as a descriptor of feasible solutions   69
4.3 Fuzzy set as a descriptor of the notion of typicality   71
4.4 Membership functions in the visualization of preferences of solutions   72
4.5 Nonlinear transformation of fuzzy sets   73

4.6 Vertical and horizontal schemes of membership estimation 76
4.7 Saaty's priority method of pairwise membership function estimation 78
4.8 Fuzzy sets as granular representatives of numeric data 81
4.9 From numeric data to fuzzy sets 86
4.10 Fuzzy equalization 93
4.11 Linguistic approximation 95
4.12 Design guidelines for the construction of fuzzy sets 95
4.13 Conclusions 97
Exercises and problems 97
Historical notes 99
References 99

5 Operations and Aggregations of Fuzzy Sets 101

5.1 Standard operations on sets and fuzzy sets 101
5.2 Generic requirements for operations on fuzzy sets 105
5.3 Triangular norms 105
    5.3.1 Defining t-norms 105
    5.3.2 Constructors of t-norms 108
5.4 Triangular conorms 112
    5.4.1 Defining t-conorms 112
    5.4.2 Constructors of t-conorms 115
5.5 Triangular norms as a general category of logical operators 118
5.6 Aggregation operations 120
    5.6.1 Averaging operations 121
    5.6.2 Ordered weighted averaging operations 123
    5.6.3 Uninorms and nullnorms 123
    5.6.4 Symmetric sums 128
    5.6.5 Compensatory operations 129
5.7 Fuzzy measure and integral 130
5.8 Negations 134
5.9 Conclusions 135
Historical notes 135
Exercises and problems 136
References 137

6 Fuzzy Relations 139

6.1 The concept of relations 139
6.2 Fuzzy relations 141

x   Contents

    6.3   Properties of the fuzzy relations   142

           6.3.1   Domain and codomain of fuzzy relations   142
           6.3.2   Representation of fuzzy relations   143
           6.3.3   Equality of fuzzy relations   143
           6.3.4   Inclusion of fuzzy relations   143

    6.4   Operations on fuzzy relations   143

           6.4.1   Union of fuzzy relations   144
           6.4.2   Intersection of fuzzy relations   144
           6.4.3   Complement of fuzzy relations   144
           6.4.4   Transpose of fuzzy relations   144

    6.5   Cartesian product, projections, and cylindrical extension of fuzzy sets   145

           6.5.1   Cartesian product   145
           6.5.2   Projection of fuzzy relations   145
           6.5.3   Cylindrical extension   146

    6.6   Reconstruction of fuzzy relations   149
    6.7   Binary fuzzy relations   150

           6.7.1   Transitive closure   152
           6.7.2   Equivalence and similarity relations   153
           6.7.3   Compatibility and proximity relations   155

    6.8   Conclusions   155
         Exercises and problems   155
         Historical notes   156
         References   156

7 Transformations of Fuzzy Sets   **157**

    7.1   The extension principle   157
    7.2   Compositions of fuzzy relations   161

           7.2.1   Sup-t composition   162
           7.2.2   Inf-s composition   165
           7.2.3   Inf-φ composition   167

    7.3   Fuzzy relational equations   168

           7.3.1   Solutions to the estimation problem   170
           7.3.2   Fuzzy relational system   172
           7.3.3   Relation-relation fuzzy equations   172
           7.3.4   Multi-input, single-output fuzzy relational equations   173
           7.3.5   Solution of the estimation problem for equations with inf-s composition   174
           7.3.6   Solution of the inverse problem   175
           7.3.7   Relation–relation fuzzy equations   176

Contents xi

      7.3.8 Multi-input, single-output fuzzy relational equations   177
      7.3.9 Solvability conditions for maximal solutions   178

7.4 Associative memories   179

      7.4.1 Sup-t fuzzy associative memories   179
      7.4.2 Inf-s fuzzy associative memories   181

7.5 Fuzzy numbers and fuzzy arithmetic   181

      7.5.1 Algebraic operations on fuzzy numbers   181
      7.5.2 Computing with fuzzy numbers   183
      7.5.3 Interval arithmetic and α-cuts   183
      7.5.4 Fuzzy arithmetic and the extension principle   185
      7.5.5 Computing with triangular fuzzy numbers   187

7.6 Conclusions   191
Exercises and problems   191
Historical notes   192
References   193

8 Generalizations and Extensions of Fuzzy Sets   195

8.1 Fuzzy sets of higher order   195
8.2 Rough fuzzy sets and fuzzy rough sets   197
8.3 Interval-valued fuzzy sets   200
8.4 Type-2 fuzzy sets   201
8.5 Shadowed sets as a three-valued logic characterization of fuzzy sets   204

      8.5.1 Defining shadowed sets   204
      8.5.2 The development of shadowed sets   206

8.6 Probability and fuzzy sets   211
8.7 Probability of fuzzy events   213
8.8 Conclusions   216
Exercises and problems   217
Historical notes   218
References   218

9 Interoperability Aspects of Fuzzy Sets   220

9.1 Fuzzy set and its family of α-cuts   220
9.2 Fuzzy sets and their interfacing with the external world   226

      9.2.1 Encoding mechanisms   228
      9.2.2 Decoding mechanisms   229

9.3 Encoding and decoding as an optimization problem of vector quantization   231

    9.3.1 Fuzzy scalar quantization   231
    9.3.2 Forming the mechanisms of the fuzzy quantization: beyond a winner-takes-all scheme   234
    9.3.3 Coding and decoding with the use of fuzzy codebooks   235

9.4 Decoding of a fuzzy set through a family of fuzzy sets   238

    9.4.1 Possibility and necessity measures in the encoding of fuzzy data   238
    9.4.2 The design of the decoder of fuzzy data   240

9.5 Taxonomy of data in structure description with shadowed sets   242

    9.5.1 Core data structure   244
    9.5.2 Shadowed data structure   244
    9.5.3 Uncertain data structure   244

9.6 Conclusions   248
Exercises and problems   248
Historical notes   249
References   250

10 Fuzzy Modeling: Principles and Methodology   252

10.1 The architectural blueprint of fuzzy models   252
10.2 Key phases of the development and use of fuzzy models   254
10.3 Main categories of fuzzy models: an overview   256

    10.3.1 Tabular fuzzy models   256
    10.3.2 Rule-based systems   257
    10.3.3 Fuzzy relational models and associative memories   258
    10.3.4 Fuzzy decision trees   260
    10.3.5 Fuzzy neural networks   260
    10.3.6 Network of fuzzy processing units   261

10.4 Verification and validation of fuzzy models   265

    10.4.1 Verification of fuzzy models   265
    10.4.2 Training, validation, and testing data in the development of fuzzy models   268
    10.4.3 Validation of fuzzy models   269

10.5 Conclusions   270
Exercises and Problems   271

Contents **xiii**

    Historical notes   273
    References   274

11 Rule-Based Fuzzy Models   276

11.1 Fuzzy rules as a vehicle of knowledge representation   276
11.2 General categories of fuzzy rules and their semantics   277

    11.2.1 Certainty-qualified rules   278
    11.2.2 Gradual rules   278
    11.2.3 Functional fuzzy rules   278

11.3 Syntax of fuzzy rules   279
11.4 Basic functional modules: rule base, database, and inference scheme   280

    11.4.1 Input interface   281
    11.4.2 Rule base   283
    11.4.3 Main types of rule bases   290
    11.4.4 Data base   297
    11.4.5 Fuzzy inference   298

11.5 Types of rule-based systems and architectures   302

    11.5.1 Linguistic fuzzy models   303
    11.5.2 Functional (local) fuzzy models   312
    11.5.3 Gradual fuzzy models   316

11.6 Approximation properties of fuzzy rule-based models   318
11.7 Development of rule-based systems   318

    11.7.1 Expert-based development   319
    11.7.2 Data-driven development   320

11.8 Parameter estimation procedure for functional rule-based systems   324
11.9 Design issues of rule-based systems – consistency, completeness, and the curse of dimensionality   326

    11.9.1 Completeness of rules   327
    11.9.2 Consistency of rules   327

11.10 The curse of dimensionality in rule-based systems   330
11.11 Development scheme of fuzzy rule-based models   330
11.12 Conclusions   331
    Exercises and problems   331
    Historical notes   332
    References   333

12 From Logic Expressions to Fuzzy Logic Networks   335

12.1 Introduction   335
12.2 Main categories of fuzzy neurons   337

    12.2.1 Aggregative neurons   337
    12.2.2 Referential (reference) neurons   342

12.3 Uninorm-based fuzzy neurons   345

    12.3.1 Main classes of unineurons   345
    12.3.2 Properties and characteristics of the unineurons   347

12.4 Architectures of logic networks   349

    12.4.1 Logic processor in the processing of fuzzy logic functions: a canonical realization   351
    12.4.2 Fuzzy neural networks with feedback loops   353

12.5 The development mechanisms of the fuzzy neural networks   354

    12.5.1 The key design phases   355
    12.5.2 Gradient-based learning schemes for the networks   356

12.6 Interpretation of the fuzzy neural networks   360

    12.6.1 Retention of the most significant connections   361
    12.6.2 Conversion of the fuzzy network to the Boolean version   362

12.7 From fuzzy logic networks to Boolean functions and their minimization through learning   365
12.8 Interfacing the fuzzy neural network   366
12.9 Interpretation aspects—a refinement of induced rule-based system   367
12.10 Reconciliation of perception of information granules and granular mappings   371

    12.10.1 Reconciliation of perception of information granule   371
    12.10.2 The optimization process   372
    12.10.3 An application of the perception mechanism to fuzzy rule-based systems   374
    12.10.4 Reconciliation of granular mappings   374

12.11 Conclusions   378
Exercises and problems   379
Historical notes   381
References   382

13 Fuzzy Systems and Computational Intelligence — 383

- 13.1 Computational intelligence   384
- 13.2 Recurrent neurofuzzy systems   386
    - 13.2.1 Recurrent neural fuzzy network model   386
    - 13.2.2 Learning algorithm   389
- 13.3 Genetic fuzzy systems   394
- 13.4 Coevolutionary hierarchical genetic fuzzy system   396
- 13.5 Hierarchical collaborative relations   398
    - 13.5.1 Fitness evaluation   400
    - 13.5.2 Pruning algorithm   401
- 13.6 Evolving fuzzy systems   407
    - 13.6.1 Functional fuzzy model   408
    - 13.6.2 Evolving participatory learning algorithm   409
- 13.7 Conclusions   415
- Exercises and problems   415
- Historical notes   416
- References   417

14 Granular Models and Human-Centric Computing — 419

- 14.1 The cluster-based representation of the input–output mappings   420
- 14.2 Context-based clustering in the development of granular models   423
- 14.3 Granular neuron as a generic processing element in granular networks   427
- 14.4 Architecture of granular models based on conditional fuzzy clustering   429
- 14.5 Refinements of granular models   431
    - 14.5.1 Bias of granular neurons   431
    - 14.5.2 Refinement of the contexts   432
- 14.6 Incremental granular models   433
    - 14.6.1 The principle of incremental fuzzy model and its design and architecture   434
- 14.7 Human-centric fuzzy clustering   439
    - 14.7.1 Fuzzy clustering with partial supervision   440
    - 14.7.2 The development of the human-centric clusters   443
    - 14.7.3 Proximity-based fuzzy clustering   446
    - 14.7.4 Interaction aspects of sources of information in the P-FCM   449

14.8 Participatory Learning in fuzzy clustering   450
14.9 Conclusions   457
Exercises and problems   458
Historical notes   459
References   459

15 Emerging Trends in Fuzzy Systems   461

15.1 Relational ontology in information retrieval   461

15.1.1 Fuzzy relational ontological model   462
15.1.2 Information retrieval model and structure   463
15.1.3 Documents representation   463
15.1.4 Query representation   464
15.1.5 Information retrieval with relational ontological model   464

15.2 Multiagent fuzzy systems   469

15.2.1 Agents and multiagents   469
15.2.2 Electricity market   470
15.2.3 Genetic fuzzy system   473

15.3 Distributed fuzzy control   482

15.3.1 Resource allocation   482
15.3.2 Control systems and economy   483
15.3.3 Fuzzy market-based control   484

15.4 Conclusions   490
Exercises and problems   491
Historical notes   491
References   491

APPENDIX A Mathematical Prerequisites   494

APPENDIX B Neurocomputing   502

APPENDIX C Biologically Inspired Optimization   513

Index   525

Preface

Over 40 years have already passed since the inception of fuzzy sets. During this period we have witnessed a truly impressive wealth of theoretical developments and conceptual pursuits, emergence of novel methodologies, algorithmic environments, and a variety of applications. Contemporary technologies in the areas such as information storage and retrieval, web search, image processing and understanding, control, pattern recognition, bioinformatics and computational biology, e-markets, autonomous navigation, and guidance have benefited considerably from the developments in fuzzy sets. What becomes equally important is that we have accumulated a body of knowledge, developed sound design practices, and gained a comprehensive insight into the role of the technology of fuzzy sets in system design and analysis.

With the existing affluence and apparent diversity of the landscape of intelligent systems, fuzzy sets exhibit an important and unique position by forming a unified framework supporting various facets of human-centric computing. Given the current trends in the information technology, it becomes apparent that the increasing level of intelligence, autonomy, and required flexibility comes hand in hand with the increased human centricity of resulting systems. This manifests at the end level when the delivered systems are expected to exhibit flexibility, significant communication abilities, user awareness, and a substantial level of adaptive behavior.

The human-centric facet of processing (or human centricity, briefly) supported by the use of fuzzy sets is concerned with (a) user-friendly nature of the resulting systems (manifesting though a high level of context awareness, realization of relevance feedback, etc.), (b) forming a sound trade-off between accuracy and transparency (interpretability), and (c) incorporation of designer-friendly mechanisms of system development facilitating an efficient aggregation of various sources of available information being present at several quite different levels of abstraction (say, a highly seamless integration of domain knowledge and numeric experimental data).

LEITMOTIV

The fundamental objective of this book is to offer a comprehensive, systematic, fully updated, and self-contained treatise of fuzzy sets that will be of vital interest to a broad audience of students, researchers, and practitioners. Our ultimate goal is to offer solid conceptual fundamentals, a carefully selected collection of design methodologies, a wealth of development guidelines, and pertinent, carefully selected illustrative material. The book constitutes a departure from the conventional approach to fuzzy systems engineering used to date. We explicitly cover concepts, design methodologies, and algorithms inherently coupled with interpretation, analysis, and underlying engineering knowledge. This *holistic* view of the discipline is helpful in stressing the role of fuzzy sets as a fundamental component of computational intelligence (CI) and human-centric systems.

FOCAL POINTS

There are several focal points that make this book highly unique and relevant from the perspective of the key fundamentals and practice of fuzzy sets

- *Systematic exposure of the concepts, design methodology, and detailed algorithms*: Overall, we adhere to the top-down strategy starting with the concepts and motivating arguments and afterward proceeding with the detailed design that materializes in some specific algorithms.
- *A wealth of illustrative material*: All concepts covered are illustrated with a series of small, numeric examples to make the material more readable, motivating, and appealing.
- *Self-containment of the material*: No specific prerequisites are required (standard calculus, linear algebra, probability, and logic are deemed to be fully sufficient).
- *More advanced concepts explained in great detail and augmented by pertinent illustrative material*: Appendices offer a concise and focused coverage of the subjects of neural networks and biologically inspired optimization.
- *Down-to-earth exposure of the material*: Although we maintain a required level of formalism and necessary mathematical rigor, our ultimate goal is to present the material in the way it emphasizes its applied side so that the reader becomes fully aware of direct applicability and limitations of the presented concepts, algorithms, and modeling techniques.
- *Auxiliary editorial features*: Historical and bibliographical notes included in each chapter help the reader view the developments of fuzzy sets in a broader perspective. Each chapter comes with a suite of well-balanced exercises and problems.

CHAPTER SUMMARY

The following offers a concise summary of the topics covered in each chapter and underlines the essential aspects of each of them.

In Chapter 1—*Introduction*—we introduce the reader to the subject, highlight several motivating factors, elaborate on the origin of fuzzy sets, and cast them in a certain historical perspective. Similarly, it is shown what role fuzzy sets play vis-á-vis existing technologies. This discussion helps emphasize the enabling role of the technology of fuzzy sets as well as highlight its key role in human-centric systems when addressing the acute and widely spread problem of the semantic gap. Fuzzy sets are also linked to the ideas of granular computing treated as a generalized conceptual and algorithmic environment. The fundamental areas of Artificial Intelligence and CI are concisely discussed and contrasted in the context of the development of intelligent systems. Some illustrative examples and case studies are covered.

Chapter 2—*Notions and Concepts of Fuzzy Sets*—serves as a coherent and systematic introduction of the fundamental concept of a fuzzy set with focus on the ideas of partial membership conveyed by membership functions, underlying rationale, examples, and most commonly encountered categories (classes) of membership functions. The ideas of fuzzy quantities and numbers are also introduced.

In Chapter 3—*Characterization of Fuzzy Sets*—major properties of membership functions are studied along with their interpretation. Discussed are geometric features of fuzzy sets to help underline the differences between sets and fuzzy sets in terms of operations defined therein, and in the sequel some global characterization of fuzzy sets (expressed through energy, granularity, etc.) is offered. The properties of families of fuzzy sets defined in the same space are presented. We emphasize here the semantics of information granules represented as fuzzy sets.

Chapter 4—*The Design of Fuzzy Sets*—elaborates on the development of fuzzy sets (membership functions) by emphasizing their syntax and semantics and linking those concepts with the user- and data-driven mechanisms of elicitation of membership functions. Some typical mechanisms supporting the construction of fuzzy sets are presented and contrasted. A great deal of attention is paid to fuzzy clustering that is regarded as one of the dominant technologies of information granulation.

In Chapter 5—*Operations and Aggregations of Fuzzy Sets*—we are concerned with operations (union, intersection, complement) on fuzzy sets. The presentation addresses the issues of formal requirements, interpretations, and realizations and possible parametric adjustments. Covered are triangular norms (t-norms and t-conorms) along with their conceptual and computing refinements such as, for example, ordinal sums, uninorms, and nullnorms. Outlined are the aspects of semantics conveyed by such logic operators and their possible parametric refinements invoked by available experimental data.

The predominant concept presented in Chapter 6—*Fuzzy Relations*—is relations. Fuzzy relations are fundamental concepts expanding the idea of fuzzy sets to a multivariable case. The very concept is introduced and illustrated. The closely linked ideas of Cartesian products, projections, and cylindric extensions of fuzzy relations are discussed.

In Chapter 7—*Transformations of Fuzzy Sets*—we introduce an idea of mappings of fuzzy sets between spaces and elaborate on its realization in the case of functions (extension principle) and relations (relational calculus). In particular, discussed are the principles of fuzzy arithmetic.

In Chapter 8—*Generalizations and Extensions of Fuzzy Sets*—various concepts and ideas that augment fuzzy sets and discussed. In this chapter, fuzzy sets are discussed in the framework of granular computing involving various formalisms of information granulation including interval analysis and rough sets. Through an extended contrastive analysis, we are able to emphasize the role played by fuzzy sets. The generalizations in the form of type-2 and order-2 fuzzy sets are covered. The concept of shadowed sets is presented as a vehicle of a more qualitative (three-valued) interpretation of fuzzy sets.

The orthogonality of fuzzy sets and probability is underlined and illustrated. Furthermore, hybrid constructs resulting through a joint treatment of fuzzy sets and probability, such as fuzzy probabilities, are discussed.

Chapter 9—*Interoperability Aspects of Fuzzy Sets*—is concerned with various dependencies between fuzzy sets and other environments of granular computing and numeric settings. In the latter case, the ideas of encoding and decoding (referred to as a fuzzification and defuzzification mechanism, respectively) are introduced and studied in detail. Both scalar and vector cases are investigated. The linkages between fuzzy sets and sets are revealed and articulated in the language of α-cuts.

Chapter 10—*Fuzzy Modeling: Principles and Methodology*—offers an in-depth discussion on the principles and underlying methodology of fuzzy modeling, their design objectives (accuracy, interpretability, etc.), an overall design process, and pertinent verification and validation procedures.

In Chapter 11—*Rule-based Fuzzy Models*—we concentrate on a class of models that play a dominant and highly noticeable role in fuzzy modeling. We introduce the main concepts and underlying terminology, classes of architectures, and discuss a variety of design processes. The mechanisms of structural and parametric learning with examples of the ensuing optimization vehicles are discussed as well.

The focal point of Chapter 12—*From Logic Expressions to Fuzzy Logic Networks*—is a category of fuzzy systems exhibiting logic-driven semantics and significant parametric flexibility. Different classes of logic neurons are introduced and afterward exploited as generic building components in the formation of highly heterogeneous logic networks. The underlying interpretability issues are raised and investigated in great detail.

In Chapter 13—*Fuzzy Systems and Computational Intelligence*—fuzzy systems are discussed vis-à-vis the research agenda and main concerns of CI. The synergistic linkages between fuzzy sets and other leading technologies of CI such as neural networks and evolutionary methods are discussed. Several representative examples are studied including recurrent neurofuzzy systems.

Human centricity of fuzzy systems is studied in Chapter 14—*Granular Models and Human-Centric Computing*—This chapter serves as a carefully organized compendium of human-centric architectures in the areas of data analysis, clustering, and granular modeling. It involves a general methodological discussion and formulates a series of guidelines. We highlight an important and active role of fuzzy sets in learning processes.

Chapter 15—*Emerging Trends in Fuzzy Systems*—ventures into several emerging and already promising areas of further developments of fuzzy sets with emphasis placed on their applied side. In particular, this concerns examples of relational ontology, information retrieval, and multiagent systems.

To make the material highly self-contained, we have included three appendices. The first one is a concise summary of the most useful and commonly encountered ideas and concepts of linear algebra and unconstrained and constrained optimization. The two others offer a brief view of the essence of neurocomputing and biologically inspired optimization that plays a vital role in the development of fuzzy systems and various constructs of CI, in general.

READERSHIP

Given the content of the book and an arrangement of the material, it will appeal to at least three large communities of readers:

Senior undergraduate students: The key objective is to present fuzzy sets as a coherent enabling technology that offers a unique and highly functional environment for building human-centric systems in numerous subject areas to which the students have been already exposed during the earlier years of their undergraduate programs. The pedagogy of the resulting course may succinctly highlight the capabilities fuzzy sets can offer as a coherent analysis platform augmenting, formalizing, and expanding the existing detailed subject knowledge. The fundamental design practices supported by fuzzy sets build upon the existing body of design knowledge being now substantially enriched by fuzzy sets. The book can be also used to deliver a standalone one-term course on fuzzy sets. Depending upon the objectives of the instructor, the material could be structured to emphasize the fundamentals of fuzzy sets or concentrate on their applied facet including modeling, classification, and data analysis.

Graduate students: The book supports the need of a broad audience of graduate students in engineering and science. Given this audience, we anticipate that the detailed presentation of the fundamentals of fuzzy sets (along with their necessary mathematical details) and the comprehensive design principles would be equally appealing to them. Again, we envision that this might involve students working in the realm of fuzzy sets or pursuing advanced research in other disciplines.

Researchers and practitioners: The organization and coverage of the material will appeal to all those who are already familiar with fuzzy sets and are interested in exploring further advancements in the area. The readers can benefit from a thorough, in-depth, and critical assessment of the current state of the art of the area. Along with the presentation of novel pursuits within the realm of the well-established domains of fuzzy sets, the book embarks on a number of emerging areas of fuzzy sets. For those who are looking for a brief yet highly informative introduction to fuzzy sets, the core of the book brings solid exposure to the area. The holistic view of the discipline embracing the fundamentals with the practice of fuzzy sets could greatly appeal to those interested in pursuing the applied side of fuzzy sets.

Throughout the book we emphasize the role of fuzzy sets as an *enabling* technology whose impact, contributions, and methodology stretch far beyond any specific community and research area. Taking this into account, a substantial interest arises from a vast array of disciplines like engineering, computer science, business, medicine, bioinformatics, computational biology, and so on.

THE ROADMAP

The book is intended to serve the needs of a broad audience by covering a wealth of territory of the discipline of fuzzy sets. Depending on the needs, several possible routes can be projected:

- A one-semester undergraduate course could cover Chapters 1–4 (possibly excluding Sections 4.5, 4.8, and 4.10), Chapter 5–11.
- A one-term graduate course could be composed of Chapters 1–12 with some selective choice of content of Chapters 13–15.
- For some specialized, short-term courses one could consider covering Chapters 1–11.

INSTRUCTOR RESOURCES

Instructors will be provided with the following classroom-ready electronic resources:

- Viewgraphs to be used in class. They aim to be customized when used in more specialized presentations or short courses.
- Solution manual with graphics presenting answers to selected problems.
- Sample assignments and examinations.

Although we strived for the delivery of a flawless material, we are aware that some typos may be inevitable. Some concepts could have been presented differently. Some algorithms could have been outlined in a more readable manner. Some interesting generalizations could have been included. We greatly appreciate your comments; please drop us a line (pedrycz@ee.ualberta.ca or gomide@dca.fee.unicamp.br).

While working on this book, we enjoyed generous support from the Natural Sciences and Engineering Research Council of Canada (NSERC) and Canada Research Chair (CRC) program—W. Pedrycz, and the Brazilian National Research Council (CNPq)—F. Gomide.

We fully acknowledge assistance of Marcos Eduardo Ribeiro do Valle Mesquita who offered us a number of valuable and constructive comments.

We owe our thanks to the dedicated and friendly people at John Wiley: George Telecki, Rachel Witmer and S. Bhuvaneshwari who did a superb job in copy-editing of the manuscript. We greatly benefited from their encouragement and continuous professional assistance.

WITOLD PEDRYCZ
University of Alberta, Edmonton, Canada &
Systems
Science Institute Polish Academy of Sciences, Warsaw, Poland

FERNANDO GOMIDE
UNICAMP, Campinas, Brazil
January 2007

Chapter 1

Introduction

We live in the world of digital technology that surrounds us and without which we can barely function. There are myriads of examples (which we take for granted) in which computers bring a wealth of services. Computers constitute an omnipresent fabric of the society (Vasilakos and Pedrycz, 2006). As once succinctly captured by Weiser (1991), "the most profound technologies are those that disappear. They weave themselves into the fabric of everyday life until they are indistinguishable from it."

There is an ongoing challenge of building intelligent systems whose functionality could make them predominantly human centric. Human centricity is one of the driving forces of ubiquitous and pervasive computing. Although there are interesting developments along this line, there is a still a long way to go. Some important milestones have been achieved, yet a lot of challenges lie ahead.

In this chapter, we investigate some fundamental features of human centricity of intelligent systems and in this context raise a need for comprehensive studies in information granulation and fuzzy sets, in particular.

1.1 DIGITAL COMMUNITIES AND A FUNDAMENTAL QUEST FOR HUMAN-CENTRIC SYSTEMS

Problem solving, design, and creative thinking—these are all endeavors in which we are inherently faced with conflicting requirements, incomplete information, numerous constraints, and finally collections of alternative solutions. All of these lead us to situations in which we have to effectively manage enormous amounts of heterogeneous data, deal with conflicting or missing evidence, and arrive at meaningful conclusions being aware of the confidence associated with our findings.

In spite of ever growing complexity of the problems, we somewhat manage to develop solutions. Both in analysis and in design (synthesis), we follow the key principles of abstraction and decomposition that help us handle a phenomenon of complexity and arrive at meaningful solutions. In essence, the effective use of abstraction means that instead of being buried in a flood of details and mountains

---

Fuzzy Systems Engineering: Toward Human-Centric Computing, by Witold Pedrycz and Fernando Gomide
Copyright © 2007 John Wiley & Sons, Inc.

of data, we establish certain, perhaps most suitable conceptual perspective and set up a framework in which the problems could be tackled. Granularity of the problem representation is a fundamental manifestation of the principle of abstraction. The decomposition is a meaningful and commonly used strategy in which on the basis of some prudently established granularity we solve the problem by isolating its loosely connected subproblems and handling them on an individual basis.

Computing systems that are around us in so visible abundance operate on completely different principles of binary (Boolean logic), numeric information and solutions, and predefined models of the world of two-valued logic and human information processing. It becomes apparent that we are concerned with two conceptually distinct worlds. To make them work together and take full advantage of the computing faculties, we need a well-developed interface through which both worlds could talk to each other. This is the key rationale behind the emergence of human-centric systems and human-centric computing (HC^2). The primary objective of the HC^2 is to make computers adjust to people by being more *natural* and *intuitive* to use and seamlessly integrated within the existing environment. Various pursuits along the line of e-society include intelligent housing, ambient intelligence (Vasilakos and Pedrycz, 2006) and ubiquitous computing, semantic web, e-health, e-commerce and manufacturing, sensor networks, intelligent data analysis, and wearable hardware. All of these are concrete examples of the general tendency existing in the development of HC^2 systems. Referring to the general architectural framework as portrayed in Figure 1.1, we easily note that in such endeavors a middleware of the semantic layer plays a crucial role in securing all necessary efficient interaction and communication between various sources of data and groups of users coming with their diversified needs and objectives. In the development of HC^2 systems, we are ultimately faced with an omnipresent challenge known as a semantic gap. To alleviate its consequences, we have to focus on how to reconcile and interpret detailed numeric information with the qualitative, descriptive, and usually linguistic input coming from the user. For instance, in the design of a typical HC^2 system, such

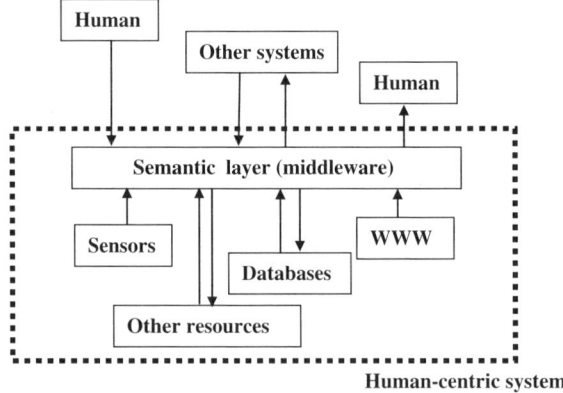

Figure 1.1 An overall architecture of human-centric systems; note a critical role of the semantic layer linking the layers of computing and humans together.

Table 1.1 Selected Examples of Human-Centric Systems and their Underlying Objectives.

Area	Key objectives, existing trends, and solutions
Intelligent data analysis	Effective explanatory analysis, delivery of findings at the level of information granules, and effective mechanisms of summarization.
System modeling	Building transparent models that could be easily interpreted and whose outcomes are readily understood. Models should help the user justify decisions being taken.
Adaptive hypermedia	Personalization of hypermedia to meet needs of individual users, development of specialized web services, building collaborative filtering, recommendation, content-based filtering, personalization of web engines, and so on.
e-commerce	Expressing preferences of customers formulated at different levels of specificity (granularity).
Intelligent interfaces	Face expression, emotion recognition and tracking, formation and use of face-related features.

as a personalized digital photo album, we encounter a lot of detailed numeric data (pixels of images) and have to accommodate a significant and highly descriptive user's input that comes in the form of some relevance feedback. The context awareness and personalization invoke numerous collaborative aspects of processing involving various sources of data and information (including those available directly from the users). The user-based processing capability is an important aspect of HC^2 systems that has to be taken into account in any design considerations.

The crux of the semantic layer lies in the formation and usage of entities that are easily perceived and processed by humans. The difficulty is that the world of numeric processing has to interact with humans who are quite resistant to the explicit use of numbers and uncomfortable to process them. We operate at the higher level of abstraction, and this essential design perspective has to be embraced by human-centric systems through their underlying functionality.

Let us offer a sample of examples in which human centricity plays a pivotal role (Table 1.1) (Frias-Martinez et al., 2005; Perkowitz and Etzioni, 2000; Spott and Nauck, 2006). Most of them heavily rely on the idea of an effective relevance feedback that needs to be implemented in an efficient manner.

1.2 A HISTORICAL OVERVIEW: TOWARDS A NON-ARISTOTELIAN PERSPECTIVE OF THE WORLD

From the brief investigations covered above, it becomes apparent that in the realization of the quest for human centricity of systems, the leitmotiv of many investigations is in building effective mechanisms of communication including various schemes of relevance feedback. Given that human processing is carried out at some level of abstraction, a concept of information granules and information granulation plays a visible role. The question of dichotomy offered by some formal frameworks of information granules has to be revisited as well.

The concept of dichotomy becomes profoundly imprinted into our education, philosophy, and many branches of science, management, and engineering. Although the formalism and vocabulary of Boolean concepts being effective in handling various discrimination processes involving binary quantification (yes–no, true–false) has been with us from the very beginning of our education, it becomes evident that this limited, two-valued view at world is painfully simplified and in many circumstances lacks rapport with the reality. In real world, there is nothing like black–white, good–bad, and so on. All of us recognize that the notion of dichotomy is quite simple and does not look realistic. Concepts do not possess sharp boundaries. Definitions are not binary unless they tackle very simple concepts (say odd–even numbers). Let us allude here to the observation made by Russell (1923)

> "... the law of excluded middle is true when precise symbols are employed, but it is not true when symbols are vague, as, in fact, all symbols are."

In reality, we use terms whose complexities are far higher and which depart from the principle of dichotomy. Consider the notions used in everyday life such as *warm* weather, *low* inflation, *long* delay, and so on. How could you define them if you were to draw a single line? Is 25°C *warm*? Is 24.9°C *warm*? Or is 24.95°C *warm* as well? Likewise in any image: Could you draw a single line to discriminate between objects such as sky, land, trees, and lake. Evidently, as illustrated in Figure 1.2, identifying boundaries delineating the objects in this way is a fairly futile task and in many cases produces pretty much meaningless results. Objects in images do not exhibit clear and unique boundaries (the location of the horizon line is not obvious at all) (Fig. 1.2(a)). Experimental data do not come in well-formed and distinct clusters; there are always some points in-between (Fig. 1.2(b)).

One might argue that these are concepts that are used in everyday language and, therefore, they need not possess any substantial level of formalism. Yet, one has to admit that the concepts that do not adhere to the principle of dichotomy are also

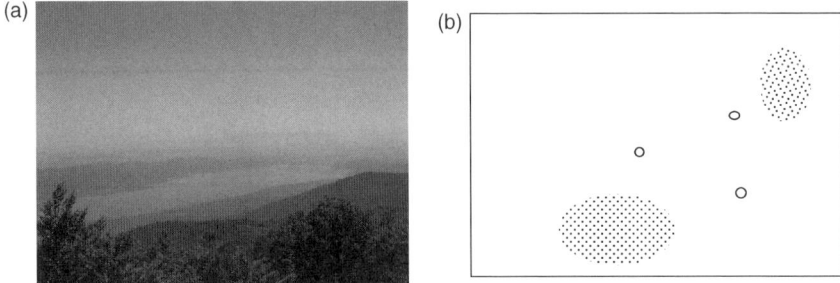

Figure 1.2 Objects, as we perceive and describe them, do not exhibit sharp boundaries. Such boundaries implementing a principle of dichotomy exhibit limitations. Practically, they may not exist at all: (a) images and (b) experimental data.

visible in science, mathematics, and engineering. For instance, we often carry out a linear approximation of nonlinear function and make a quantifying statement that such linearization is valid in some *small* neighborhood of the linearization point. Under these circumstances, the principle of dichotomy does not offer too much.

The principle of dichotomy, or as we say an Aristotelian perspective at the description of the world, has been subject to a continuous challenge predominantly from the standpoint of philosophy and logic. Let us recall some of the most notable developments that have led to the revolutionary paradigm shift. Indisputably, the concept of a three-valued and multivalued logic put forward by Jan Lukasiewicz and then pursued by others, including Emil Post, is one of the earliest and the most prominent logical attempts made toward the direction of abandoning the supremacy of the principle of dichotomy. As noted by Lukasiewicz (1920, 1930,) the question of the suitability or relevance of two-valued logic in evaluating the truth of propositions was posed in the context of those statements that allude to the future. "Tomorrow will rain." Is this statement true? If we can answer this question, this means that we have already predetermined the future. We start to sense that this two-valued model, no matter how convincing it could be, is conceptually limited if not wrong. The non-Aristotelian view of the world was vividly promoted by Korzybski (1933). Although the concept of the three-valued logic was revolutionary in 1920s, we somewhat quietly endorsed it over the passage of time. For instance, in database engineering, a certain entry may be two-valued (yes–no), but the third option of "unknown" is equally possible—here we simply indicate that no value of this entry has been provided.

1.3 GRANULAR COMPUTING

Information granules permeate human endeavors (Zadeh, 1973, 1979, 1996, 1997, 2005; Pedrycz, 2001; Bargiela and Pedrycz, 2003). No matter what problem is taken into consideration, we usually cast it into a certain conceptual framework of basic entities, which we regard to be of relevance to the problem formulation and problem solving. This becomes a framework in which we formulate generic concepts adhering to some level of abstraction, carry out processing, and communicate the results to the external environment. Consider, for instance, image processing. In spite of the continuous progress in the area, a human being assumes a dominant and very much uncontested position when it comes to understanding and interpreting images. Surely, we do not focus our attention on individual pixels and process them as such but group them together into semantically meaningful constructs—familiar objects we deal with in everyday life. Such objects involve regions that consist of pixels or categories of pixels drawn together because of their proximity in the image, similar texture, color, and so on. This remarkable and unchallenged ability of humans dwells on our effortless ability to construct information granules, manipulate them, and arrive at sound conclusions. As another example, consider a collection of time series. From our perspective, we can describe them in a semiqualitative manner by pointing at specific regions of such signals. Specialists can effortlessly interpret electrocardiograms (ECG signals). They distinguish some segments of such signals and interpret their combinations. Experts can interpret temporal readings of sensors and assess the status of the monitored system. Again, in all these situations, the individual samples of the signals are not the focal point of the analysis and the ensuing signal interpretation. We always granulate all phenomena (no matter if they are originally discrete or analog in their nature). Time is another important variable that is subjected to granulation. We use seconds, minutes, days, months, and years. Depending on which specific problem we have in mind and who the user is, the size of information granules (time intervals) could vary quite dramatically. To the high-level management, time intervals of quarters of year or a few years could be meaningful temporal information granules on the basis of which one develops any predictive model. For those in charge of everyday operation of a dispatching power plant, minutes and hours could form a viable scale of time granulation. For the designer of high-speed integrated circuits and digital systems, the temporal information granules concern nanoseconds, microseconds, and perhaps seconds. Even such commonly encountered and simple examples are convincing enough to lead us to ascertain that (a) information granules are the key components of knowledge representation and processing, (b) the level of granularity of information granules (their size, to be more descriptive) becomes crucial to the problem description and an overall strategy of problem solving, and (c) there is no universal level of granularity of information; the size of granules is problem oriented and user dependent.

What has been said so far touched a qualitative aspect of the problem. The challenge is to develop a computing framework within which all these representation and processing endeavors could be formally realized. The common platform emerging within this context comes under the name of granular computing. In essence, it is an emerging paradigm of information processing. Although we have already noticed a number of important conceptual and computational constructs built in the domain of system modeling, machine learning, image processing, pattern recognition, and data compression in which various abstractions (and ensuing information granules) came into existence, granular computing becomes innovative and intellectually proactive in several fundamental ways:

- It identifies the essential commonalities between the surprisingly diversified problems and technologies used there, which could be cast into a unified framework we usually refer to as a granular world. This is a fully operational processing entity that interacts with the external world (that could be another granular or numeric world) by collecting necessary granular information and returning the outcomes of the granular computing.
- With the emergence of the unified framework of granular processing, we get a better grasp as to the role of interaction between various formalisms and visualize a way in which they communicate.
- It brings together the existing formalisms of set theory (interval analysis), fuzzy sets, rough sets, and so on under the same roof by clearly visualizing that in spite of their visibly distinct underpinnings (and ensuing processing), they exhibit some fundamental commonalities. In this sense, granular computing

establishes a stimulating environment of synergy between the individual approaches.
- By building upon the commonalities of the existing formal approaches, granular computing helps build heterogeneous and multifaceted models of processing of information granules by clearly recognizing the orthogonal nature of some of the existing and well-established frameworks (say, probability theory coming with its probability density functions and fuzzy sets with their membership functions).
- Granular computing fully acknowledges a notion of variable granularity whose range could cover detailed numeric entities and very abstract and general information granules. It looks at the aspects of compatibility of such information granules and ensuing communication mechanisms of the granular worlds.
- Interestingly, the inception of information granules is highly motivated. We do not form information granules without reason. Information granules are an evident realization of the fundamental paradigm of abstraction.

Granular computing forms a unified conceptual and computing platform. Yet, it directly benefits from the already existing and well-established concepts of information granules formed in the setting of set theory, fuzzy sets, rough sets and others. Let us now take a quick look at the fundamental technologies of information granulation and contrast their key features.

1.3.1 Sets and Interval Analysis

Sets are fundamental concepts of mathematics and science. Referring to the classic notes, set is described as "any multiplicity, which can be thought of as one... *any totality of definite elements, which can be bound up into a whole by means of a law*" or being more descriptive "*...any collection into a whole M of definite and separate objects m of our intuition or our thought*" (Cantor, 1883, 1895). Likewise, interval analysis ultimately dwells upon a concept of sets, which in this case are collections of elements in the line of reals, say $[a,b]$, $[c,d]$,... and so on. Multidimensional constructs are built upon Cartesian products of numeric intervals and give rise to computing with hyperboxes. Going back to the history, computing with intervals is intimately linked with the world of digital technology. One of the first papers in this area was published in 1956 by Warmus. Some other early research was done by Sunaga and Moore (1966). This was followed by a wave of research in so-called interval mathematics or interval calculus. Conceptually, sets (intervals) are rooted in a two-valued logic with their fundamental predicate of membership ($\in$). Here holds an important isomorphism between the structure of two-valued logic endowed with its truth values (false–true) and set theory with sets being fully described by their characteristic functions. The interval analysis is a cornerstone of reliable computing, which in turn is ultimately associated with digital computing in which any variable is associated with a finite accuracy (implied by the fixed number of bits used to represent numbers). This limited accuracy gives rise to a certain pattern of propagation of

Table 1.2 Arithmetic Operations on Numeric Intervals A and B.

Algebraic operation	Result
Addition	$[a+c, b+d]$
Subtraction	$[a-d, b-c]$
Multiplication	$[\min(ac, ad, bc, bd), \max(ac, ad, bc, bd)]$
Division	$\left[\min\left(\frac{a}{c}, \frac{a}{d}, \frac{b}{c}, \frac{b}{d}\right), \max\left(\frac{a}{c}, \frac{a}{d}, \frac{b}{c}, \frac{b}{d}\right)\right]$ assumption: the interval $[c, d]$ does not contain 0

error of computing. For instance, addition of two intervals $[a, b]$ and $[c, d]$ leads to a broader interval in the form $[a+c, b+d]$ (Hansen, 1975; Jaulin et al., 2001; Moore, 1966). Here, the accumulation of uncertainty (or equivalently the decreased granularity of the result) depends upon the specific algebraic operation completed for given intervals. Table 1.2 summarizes four algebraic operations realized on numeric intervals A $= [a, b]$ and B $= [c, d]$.

Interestingly, intervals distributed uniformly in a certain space are at the center of any mechanism of analog-to-digital conversion; the higher the number of bits, the finer the intervals and the higher their number. The well-known fundamental relationship states that with n bits we can build a collection of 2^n intervals of width $(b-a)/2^n$ for the original range of numeric values in $[a, b]$. Intervals offer a straightforward mechanism of abstraction: all elements lying within a certain interval become indistinguishable and therefore are treated as identical. In addition to algebraic manipulation, the area of interval mathematics embraces a wealth of far more advanced and practically relevant processing including differentiation, integral calculus, as well as interval-valued optimization.

1.3.2 The Role of Fuzzy Sets: A Perspective of Information Granules

Fuzzy sets offer an important and unique feature of describing information granules whose contributing elements may belong to varying degrees of membership (belongingness). This helps us describe the concepts that are commonly encountered in real world. The notions, such as *low* income, *high* inflation, *small* approximation error, and many others, are examples of concepts to which the yes–no quantification does not apply or becomes quite artificial and restrictive. We are cognizant that there is no way of quantifying the Boolean boundaries as there are a lot of elements whose membership to the concept is only partial and quite different from 0 and 1.

The binary view of the world supported by set theory and two-valued logic has been vigorously challenged by philosophy and logic. The revolutionary step in logic was made by Lukasiewicz with his introduction of three and afterward multivalued logic (Lukasiewic, 1930, 1970). It took 'however' more decades to dwell on the ideas of the non-Aristotelian view of the world before fuzzy sets were introduced. This happened in 1965 with the publication of the seminal paper on fuzzy sets by Zadeh (1965). Refer also to other influential papers by Zadeh (1979, 1996, 1997, 1999, 2005). The concept of fuzzy set is surprisingly simple and elegant. Fuzzy set A captures its elements by assigning them to it with some varying degrees of membership. A so-called membership function is a vehicle that quantifies different degrees of membership. The higher the degree of membership $A(x)$, the stronger is the level of belongingness of this element to A (Gottwald, 2005; Zimmermann, 1996).

The obvious yet striking difference between sets (intervals) and fuzzy sets lies in the notion of partial membership supported by fuzzy sets. In fuzzy sets, we discriminate between elements that are "typical" to the concept and those of borderline character. Information granules such as *high* speed, *warm* weather, *fast* car are examples of information granules falling under this category and can be conveniently represented by fuzzy sets. As we cannot specify a single, well-defined element that forms a solid border between full belongingness and full exclusion, fuzzy sets offer an appealing alternative and a practical solution to this problem. Fuzzy sets with their smooth transition boundaries form an ideal vehicle to capture the notion of partial membership. In this sense, information granules formalized in the language of fuzzy sets support a vast array of human-centric pursuits. They are predisposed to play a vital role when interfacing human to intelligent systems.

In problem formulation and problem solving, fuzzy sets emerge in two fundamentally different ways.

Explicit. Here, they typically pertain to some generic and fairly basic concepts we use in our communication and description of reality. There is a vast amount of examples as such concepts being commonly used every day, say *short* waiting time, *large* dataset, *low* inflation, *high* speed, *long* delay, and so on. All of them are quite simple as we can easily capture their meaning. We can easily identify a universe of discourse over which such variable are defined. For instance, this could be time, number of records, velocity, and alike.

Implicit. Here we are concerned with more complex and inherently multifaceted concepts and notions where fuzzy sets could be incorporated into the formal description and quantification of such problems, yet not in so instantaneous manner. Some examples could include concepts such as "*preferred* car," "*stability* of the control system," "*high performance* computing architecture," "*good convergence* of the learning scheme," "*strong* economy," and so on. All of these notions incorporate some components that could be quantified with the use of fuzzy sets, yet this translation is not that completely straightforward and immediate as it happens for the category of the explicit usage of fuzzy sets. For instance, the concept of "*preferred* car" is evidently multifaceted and may involve a number of essential descriptors that when put together are really reflective of the notion we have in mind. For instance, we may involve a number of qualities such as speed, economy, reliability, depreciation, maintainability, and alike. Interestingly, each of these features could be easily rephrased in simpler terms and through this process at some level of this refinement phase, we may arrive at fuzzy sets that start to manifest themselves in an explicit manner.

10 Chapter 1 Introduction

Table 1.3 Examples of Concepts Whose Description and Processing Invoke the Use of Fuzzy Sets and Granular Computing.

---

p. 65: *small* random errors in the measurement vector...
p. 70: The success of the method depends on whether the first initial guess is already *close enough* to the global minimum...
p. 72: Hence, the convergence region of a numerical optimizer will be *large*
(van der Heijden et al., 2004).
p. 162: Comparison between bipolar and MOS technology (a part of the table)

	bipolar	MOS
integration	low	very high
power	high	low
cost	low	low

(Katz and Borriello, 2005).
p. 50: validation costs are *high* for *critical systems*
p. 660: ...A *high* value for fan-in means that X is *highly coupled* to the rest of the design and changes to X will have extensive knock-on effect. A *high* value for fan-out suggests that the overall complexity of X may be *high* because of the complexity of control logic needed to coordinate the called components.
...Generally, the *larger* the size of the code of a component, the more *complex* and error-prone the component is likely to be...
...The *higher* the value of the Fog index, the more difficult the document is to understand
(Sommerville, 2007).

As we stressed, the omnipresence of fuzzy sets is surprising. Even going over any textbook or research monograph, not mentioning newspapers and magazines, we encounter a great deal of fuzzy sets coming in their implicit or explicit format. Table 1.3 offers a handful of selected examples.

From the optimization standpoint, the properties of continuity and commonly encountered differentiability of the membership functions become a genuine asset. We may easily envision situations where those information granules incorporated as a part of the neurofuzzy system are subject to optimization—hence the differentiability of their membership functions becomes critical relevance. What becomes equally important is the fact that fuzzy sets bridge numeric and symbolic concepts. On one hand, fuzzy set can be treated as some symbol. We can regard it as a single conceptual entity by assigning to it some symbol, say L (for *low*). In the sequel, it could be processed as a purely symbolic entity. On the other hand, a fuzzy set comes with a numeric membership function and these membership grades could be processed in a numeric fashion.

Fuzzy sets can be viewed from several fundamentally different standpoints. Here we emphasize the four of them that play a fundamental role in processing and knowledge representation.

As an Enabling Processing Technology of Some Universal Character and of Profound Human-Centric Character

Fuzzy sets build upon the existing information technologies by forming a user-centric interface using which one could communicate essential design knowledge thus guiding problem solving and making it more efficient. For instance, in signal processing and image processing we might incorporate a collection of rules capturing specific design knowledge about filter development in a certain area. Say, "if the level of noise is *high*, consider using a *large* window of averaging." In control engineering, we may incorporate some domain knowledge about the specific control objectives. For instance, "if the constraint of fuel consumption is *very important*, consider settings of a PID controller producing *low* overshoot." Some other examples of highly representative human-centric systems concern those involving (a) construction and usage of relevance feedback in retrieval, organization, and summarization of video and images, (b) queries formulated in natural languages, and (c) summarization of results coming as an outcome of some query.

Second, there are unique areas of applications in which fuzzy sets form a methodological backbone and deliver the required algorithmic setting. This concerns fuzzy modeling in which we start with collections of information granules (typically realized as fuzzy sets) and construct a model as a web of links (associations) between them. This approach is radically different from the numeric, function-based models encountered in "standard" system modeling. Fuzzy modeling emphasizes an augmented agenda in comparison with the one stressed in numeric models. Whereas we are still concerned with the accuracy of the resulting model, its interpretability and transparency become of equal and sometimes even higher relevance.

It is worth stressing that fuzzy sets provide an additional conceptual and algorithmic layer to the existing and well-established areas. For instance, there are profound contributions of fuzzy sets to pattern recognition. In this case, fuzzy sets build upon the well-established technology of feature selection, classification, and clustering.

Fuzzy sets are an ultimate mechanism of communication between humans and computing environment. The essence of this interaction is illustrated in Figure 1.3(a). Any input is translated in terms of fuzzy sets and thus made comprehensible at the level of the computing system. Likewise, we see a similar role of fuzzy sets when communicating the results of detailed processing, retrieval, and alike. Depending upon application and the established mode of interaction, the communication layer may involve a substantial deal of processing of fuzzy sets. Quite often, we combine the mechanisms of communication and represent them in a form of a single module (Fig. 1.3(b)). This architectural representations stress the human-centricity aspect of the developed systems.

As an Efficient Computing Framework of Global Character

Rather than processing individual elements, say a single numeric datum, an encapsulation of a significant number of the individual elements that is realized in the form of some fuzzy sets, offers immediate benefits of joint and orchestrated processing.

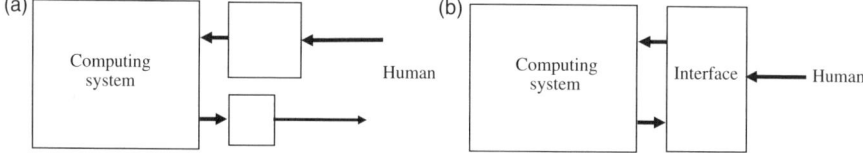

Figure 1.3 Fuzzy sets in the realization of communication mechanisms (a) both at the user end and at the computing system side, (b) a unified representation of input and output mechanisms of communication in the form of the interface, which could also embrace a certain machinery of processing at the level of fuzzy sets.

Instead of looking at the individual number, we embrace a more general point of view and process a entire collection of elements represented now in the form of a single fuzzy set. This effect of a *collective* handling of individual elements is seen very profoundly in the so-called fuzzy arithmetic. The basic constructs here are fuzzy numbers. In contrast to single numeric quantities (real numbers), fuzzy numbers represent collections of numbers where each of them belongs to the concept (fuzzy number) to some degree. These constructs are then subject to processing, say addition, subtraction, multiplication, division, and so on. Noticeable is the fact that by processing fuzzy numbers we are in fact handling a significant number of individual elements at the same time. Fuzzy numbers and fuzzy arithmetic provide an interesting advantage over interval arithmetic (viz. arithmetic in which we are concerned with intervals—sets of numeric values). Intervals come with abrupt boundaries as elements can belong to or are excluded from the given set. This means, for example, that any gradient-based techniques of optimization invoked when computing solutions become very limited: the derivative is equal to zero with an exception at the point where the abrupt boundary is located.

Fuzzy Sets as a Vehicle of Raising and Quantifying Awareness About Granularity of Outcomes

Fuzzy sets form the results of granular computing. As such they convey a global view at the elements of the universe of discourse over which they are constructed. When visualized, the values of the membership function describe a suitability of the individual points as compatible (preferred) with the solution. In this sense, fuzzy sets serve as a useful visualization vehicle: when displayed, the user could gain an overall view of the character of solution (regarded as a fuzzy set) and make a final choice. Note that this is very much in line with the idea of the human-centricity: We present the user with all possible results however do not put any pressure as to the commitment of selecting a certain numeric solution.

Fuzzy Sets as a Mechanism Realizing a Principle of the Least Commitment

As the computing realized in the setting of granular computing returns a fuzzy set as its result, it could be effectively used to realize a principle of the least commitment.

1.3 Granular Computing

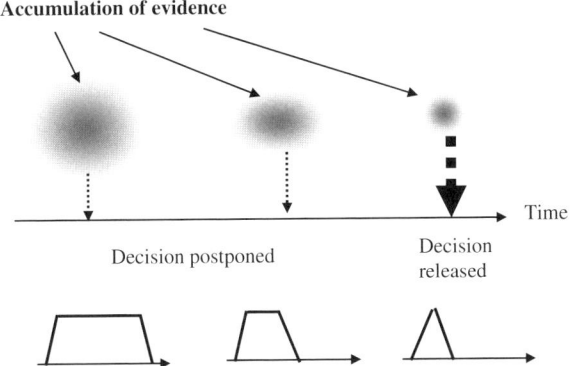

Figure 1.4 An essence of the principle of the least commitment; the decision is postponed until the phase where there is enough evidence accumulated and the granularity of the result becomes specific enough. Also examples of fuzzy sets formed at successive phases of processing that become more specific along with the increased level of evidence are shown.

The crux of this principle is to use fuzzy set as a mechanism of making us cognizant of the quality of obtained result. Consider a fuzzy set being a result of computing in some problem of multiphase decision making. The fuzzy set is defined over various alternatives and associates with them the corresponding degrees of preference, see Figure 1.4. If there are several alternatives with very similar degrees of membership, this serves as a clear indicator of uncertainty or hesitation as to the making of a decision. In other words, in light of the form of the generated fuzzy set, we do not intend to commit ourselves to making any decision (selection of one of the alternatives) at this time. Our intent would be to postpone decision and collect more evidence. For instance, this could involve further collecting of data, soliciting expert opinion, and alike. Based on this evidence, we could continue with computing and evaluate the form of the resulting fuzzy set. It could well be that the collected evidence has resulted in more specific fuzzy set of decisions on the basis of which we could either still postpone decision and keep collecting more evidence or proceed with decision making. Thus, the principle of the least commitment offers us an interesting and useful guideline as to the mechanism of decision making versus evidence collection.

1.3.3 Rough Sets

The description of information granules completed with the aid of some vocabulary is usually imprecise. Intuitively, such description may lead to some approximations called lower and upper bounds. This is the essence of rough sets introduced by Pawlak (1982; 1991); refer also to Skowron (1989) and Polkowski and Skowron (1998). Interesting generalizations, conceptual insights, and algorithmic investigations are offered in a series of fundamental papers by Pawlak and Skowron (2007a,b,c).

14   Chapter 1   Introduction

To explain the concept of rough sets and show what they are to offer in terms of representing information granules, we use an illustrative example. Consider a description of environmental conditions expressed in terms of temperature and pressure. For each of these factors, we fix several ranges of possible values where each of such ranges comes with some interpretation such as "values below," "values in-between," "values above," and so on. By admitting such selected ranges in both variables, we construct a grid of concepts formed in the Cartesian product of the spaces of temperature and pressure, refer to Figure 1.5. In more descriptive terms, this grid forms a vocabulary of generic terms using which we would like to describe all new information granules.

Now let us consider that the environmental conditions monitored over some time have resulted in some values of temperature and pressure ranging in-between some lower and upper bound as illustrated in Figure 1.5. Denote this result by X. When describing it in terms of the elements of the vocabulary, we end up with a collection of elements that are fully included in X. They form a lower bound of description of X when being completed in presence of the given vocabulary. Likewise, we may

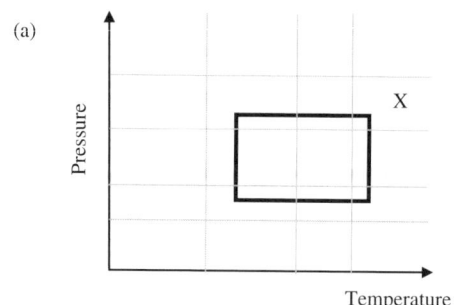

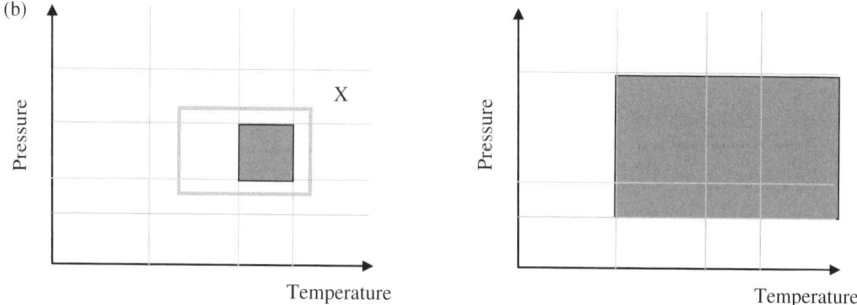

Figure 1.5 A collection of vocabularies and their use in the problem description. Environmental conditions X result in some interval of possible values (a). In the sequel, this gives rise to the concept of a rough set with the roughness of the description being captured by the lower and upper bounds (approximations) as illustrated in (b).

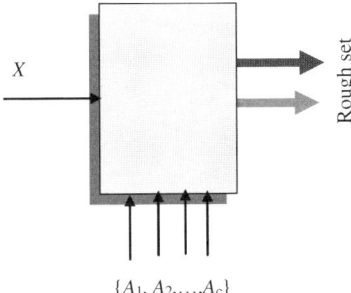

Figure 1.6 Rough set as a result of describing X in terms of some fixed vocabulary of information granules $\{A_1, A_2, \ldots, A_c\}$.

identify elements of the vocabulary that have a nonempty overlap with X and in this sense constitute an upper bound of the description of the given environmental conditions. Along with the vocabulary, the description forms a certain rough set.

As succinctly visualized in Figure 1.6, we are concerned with a description of a given concept X realized in the language of a certain collection (vocabulary) of rather generic and simple terms $A_1, A_2, \ldots, A_c$. The lower and upper boundaries (approximation) are reflective of the resulting imprecision caused by the conceptual incompatibilities between the concept itself and the existing vocabulary.

It is interesting to note that the vocabulary used in the above construct could comprise information granules being expressed in terms of any other formalism, say fuzzy sets. Quite often we can encounter constructs like rough fuzzy sets and fuzzy rough sets in which both fuzzy sets and rough sets are put together (Dubois and Prade, 1990).

1.3.4 Shadowed Sets

Fuzzy sets are associated with the collections of numeric membership grades. Shadowed sets (Pedrycz, 1998; 2005) are based upon fuzzy sets by forming a more general and highly synthetic view at the numeric concept of membership. Using shadowed sets, we quantify numeric membership values into three categories: complete belongingness, complete exclusion, and unknown (which could be also conveniently referred to as do not know condition or a *shadow*). A graphic illustration of a shadowed set along with the principles of sets and fuzzy sets is schematically shown in Figure 1.7. This helps us contrast these three fundamental constructs of information granules.

In a nutshell, shadowed sets can be regarded as a general and far more concise representation of a fuzzy set that could be of particular interest when dealing with further computing (in which case we could come up with substantial reduction of the overall processing effort).

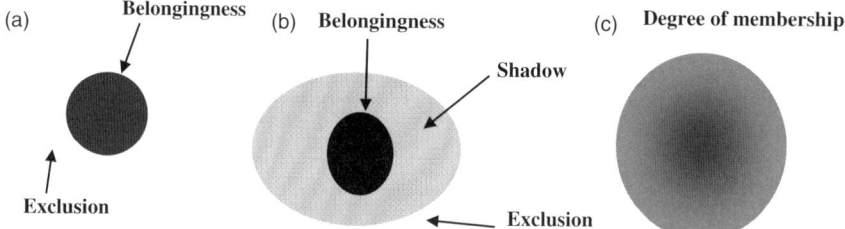

Figure 1.7 A schematic view at sets (a), shadowed sets (b), and fuzzy sets (c). Shadowed sets reveal interesting linkages between fuzzy sets and sets.

1.4 QUANTIFYING INFORMATION GRANULARITY: GENERALITY VERSUS SPECIFICITY

The notion of granularity itself and a level of specificity/generality seem to be highly intuitive: We can easily sense what is more detailed and specific and what looks more abstract and general. Formally, we can easily quantify granularity of information granule by counting its number of elements. The more the elements are located in the information granule, the lower its granularity (and the higher the generality). In this limit, a single element exhibits the highest level of granularity (specificity). In the case of sets, this will be the cardinality (number of elements) or the length of the interval or a similar measure expressing a count of the elements. In case of fuzzy sets, we usually use a so-called sigma count that is produced by summing up the membership grades of the elements belonging to the fuzzy set under consideration. For rough sets, we may consider the cardinality of their lower or upper approximations.

1.5 COMPUTATIONAL INTELLIGENCE

Emerged in the early 1990s (Bezdek, 1992; Pedrycz, 1997), Computational intelligence (CI) offers a unique and interesting opportunity to narrow down the acute semantic gap we encounter when building HC^2 systems. The contributing technologies of CI (in particular, neural networks, granular computing, and evolutionary optimization) along with their research thrusts are complementary to a high degree. This has triggered a great deal of synergy, which in turn has made the CI a highly cohesive conceptual and algorithmic platform exhibiting significant modifiability (adaptability) and supporting mechanisms of context-awareness, human-centricity, and user-friendliness. In this highly symbiotic CI environment, each of the technologies listed above plays an important role. For instance, through the use of fuzzy sets, detailed numeric data may be arranged into meaningful and tangible information granules. Information granulation allows for the incorporation of a users' prior domain knowledge and preferences, as well as facilitates the management of uncertainty. Neurocomputing delivers a rich diversity of learning techniques and flexible neural or neuro-fuzzy architectures. Evolutionary methods help cope with structural optimization and are often essential in the design of complex systems. CI benefits from this both in terms of the overall methodology of problem understanding and problem solving, as well as the ensuing system architectures. Again as strongly advocated in the literature, CI addresses the very nature of human problem solving, namely, problem modularization, dealing, for example, with numerous conflicting criteria. The recently developed ideas and practices of granular computing promote a general top-down design approach: knowledge tidbits are collected, afterward analyzed, refined, and used as a *blueprint* (backbone) of the ensuing detailed architecture. Neurocomputing, on the contrary, supports the bottom-up design approach: here one starts from "clouds" of data and attempts to reveal and describe some common regularities (e.g., trends) and encapsulate them in the form of specific models. The omnipresent tendency in the development of HC^2 systems lies in its *multistrategy* and *multifaceted* approach. It is strongly manifested in various architectures, different design (learning) techniques, and more advanced user-friendly interfaces. In this sense, CI becomes an ideal methodological, development, and experimental platform for HC^2 systems.

1.6 GRANULAR COMPUTING AND COMPUTATIONAL INTELLIGENCE

Granular computing seamlessly integrates with architectures of CI. Given the fact that information granules help set up the most suitable perspective when dealing with the problem, collecting data (that could be of heterogeneous character), carrying out processing, and releasing the results (in a formal acceptable to the environment), the general architecture is shown in Figure 1.8.

Although the communication layers are supported by granular computing, the underlying processing is a domain of neurocomputing, while the overall optimization of the architecture is supported by the machinery of evolutionary computing. There are different levels of synergy; for instance, one could regard the overall architecture as a neurofuzzy system. In this case, the interface delivers a unified setting where various sources of data are effortlessly combined and presented to the neural network, which constitutes the core of the processing layer. In many cases, the architecture could have somewhat blurred delineation between the communication layers and the processing core, in particular, when information granules become an integral part of the basic processing elements. A typical example here comes when we are concerned with a granular neuron—a construct in which the connections are treated as information granules, say fuzzy sets (and then we may refer to it as a fuzzy neuron) or rough sets (which gives rise to the concept of rough neurons).

As discussed earlier, information granules help us cast the problem into some perspective. This becomes visible in case of neural networks. To cope with huge masses of data, we could granulate them (which naturally reduce their number and

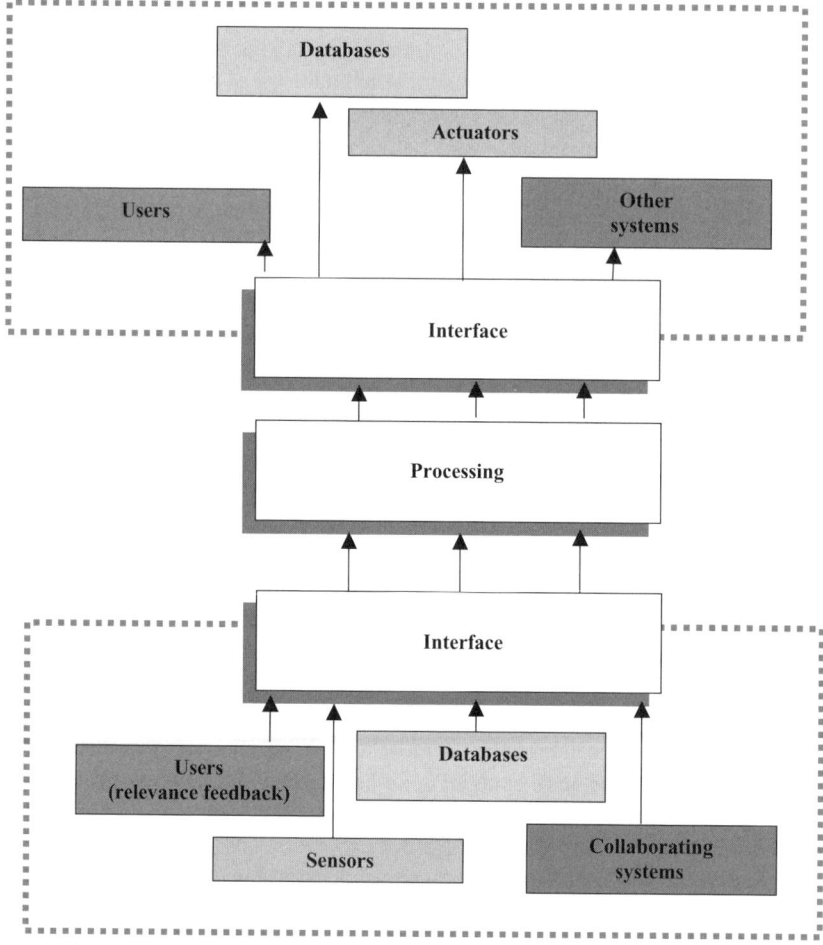

Figure 1.8 The layered architecture of systems of Computational Intelligence with the functional layers of communication (interfacing) with the environment and the processing core.

dimensionality) and treat those as meaningful aggregates and components of the learning set.

1.7 CONCLUSIONS

Human centricity becomes a feature that is of growing interest, especially when dealing with the development of more sophisticated and intelligent systems. Whereas there is a remarkably diversified spectrum of possible applications and ensuing realizations, in all of them, we can identify some commonalities and a visible role of information granules and information granulation. The chapter offers some introduction to fuzzy sets and brings a number of motivating comments as far as their methodology and applied side are concerned. Similarly, we looked at fuzzy sets being an important component of granular computing. We also clarified a concept of CI and pointed at the role of fuzzy sets within this framework.

EXERCISES AND PROBLEMS

1. Consider a certain concept A defined in the space of two variables (attributes) x_1 and x_2 whose geometric representation is shown below. We would like to describe it by means of some Cartesian products of intervals. It becomes evident that such characterization cannot be perfect. How would you define lower and upper bounds of the description of the concept so that its "roughness" becomes as small as possible? Justify your construction of the bounds.

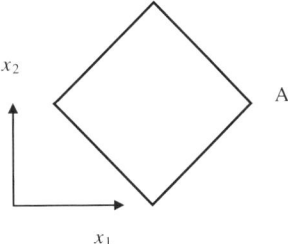

2. Pick up some textbooks, newspapers, and magazines and identify terms (concepts) that could be formalized as fuzzy sets. Justify your choice. Suggest possible models of membership functions and link them with the semantics of the concepts being described in this manner.

3. Discuss some additional functionality in commonly encountered computer systems that could be beneficial in making them highly user centric or could be useful in enhancing their user centricity.

4. Identify some concepts in which fuzzy sets could be used in explicit and implicit manner.

5. Unleash your imagination and suggest some functionality of future computing systems in which human centricity could play an important role.

6. For the differentiable membership functions, we could evaluate their sensitivity by determining the absolute value of derivative of the membership function. Discuss the sensitivity of piecewise linear membership functions (triangular fuzzy sets), parabolic membership functions, and Gaussian membership functions. They are described by the following membership functions:

(a) Triangular $\quad A(x) = \begin{cases} \dfrac{x-a}{m-a} & \text{if } x \in [a, m] \\ 1 - \dfrac{x-m}{b-m} & \text{if } x \in [m, b] \\ 0, & \text{otherwise} \end{cases}$

where $a < m < b$

(b) Parabolic $\quad A(x) = \begin{cases} 1 - \left(\dfrac{x}{a}\right)^2 & \text{if } x \in [-a, a] \\ 0, & \text{otherwise} \end{cases}$

(c) Gaussian $\quad A(x) = \exp(-(x-m)^2/\sigma^2)$

7. You are about to buy a new car. The info sticker you see on the windshield of the vehicle in the dealer's exhibition area tells you about economy "22 mpg in city and 35 mpg on highway." How could you interpret this information? Would you be dissatisfied after buying this vehicle and learning that it makes 20 mpg in city driving? Suggest models of fuzzy sets capturing the semantics of the concept of economy of a vehicle; be realistic. While dealing with cars, also suggest some other concepts that directly lead to the emergence of fuzzy sets that could serve as the meaningful descriptors of the concepts.

HISTORICAL NOTES

While the inception of fuzzy sets has to be attributed to 1965 paper by Zadeh (Zadeh, 1965), we have indicated that their conceptual and philosophical roots are dated back to the beginning of 20th century where the most influential and prominent ideas of three-valued and multivalued logic came into existence (Łukasiewicz, 1920, 1930, 1970). The philosophical underpinnings of non-Aristotelian view at the world were laid down by Korzybski (1933). The Aristotelian view of the world was challenged by Black in his 1938 study entitled "Vagueness: an exercise in logical analysis." The others include Klaua and Post.

Jan Łukasiewicz

Jan Łukasiewicz (1878–1956) is known as a founder of three-valued and multivalued logics. After studies of law at the University of Lvov (Poland), his interests were focused on philosophy in which he received his Ph.D. in 1902. While at the University of Lvov, in 1907–1908, he offered the first Polish course in mathematical logic. During the WW I in 1915, he moved to Warsaw where he occupied one of the two chairs of philosophy at the Warsaw University. In 1946, not accepting the new political system set up in Poland under the Soviet occupation, he moved to Dublin, Ireland. Łukasiewicz's Polish notation (known as reverse Polish notation or postfix notation) of 1920 was an inspiration behind the idea of the *recursive* *stack*, a last-in, first-out computer memory store. The reverse Polish notation is used in Hewlett Packard calculators and postscript language.

Alfred Korzybski (1879–1950) has contributed to the area of general semantics and the fundamentals of non-Aristotelian systems. He studied in Warsaw University of Technology, Germany and Italy. Then, he volunteered in the Russian army and was sent to Canada and USA as an artillery expert. His book entitled *Science and Sanity: An Introduction to Non-Aristotelian Systems and General Semantics* and published in 1933 has become a landmark in studies of general semantics. Here, it is worth to recall Korzybski's note from this book that succinctly highlights the shortcomings of the Aristotelian perspective.

> *..in analyzing the Aristotelian codification, I had to deal with the two-valued, "either-or" type of orientation. In living, many issues are not so sharp, and therefore a system that posits the general sharpness of "either-or" and so objectifies "kind," is unduly limited; it must be revised and more flexible in terms of "degree"…*

The developments of interval calculus emerged with inception of the era of digital computing and the paper by J. Warmus was one of the first publications in this realm. It is interesting to follow a general way in which the computing with such information granules is carried out (Fig. 1.9).

Fuzzy sets came into existence when the fundamental paper of **L. A. Zadeh** was published in *Information and Control* (Fig. 1.10). Fuzzy sets departed from the principle of dichotomy by admitting a notion of partial membership (degree of membership defined in the unit interval). Fuzzy sets offered a rich conceptual and algorithmic setting in which granular information could be handled. Furthermore, they provide a highly effective vehicle to express and quantify general principles of modeling and human-centric systems, for example, the principle of incompatibility coined by Zadeh (1973).

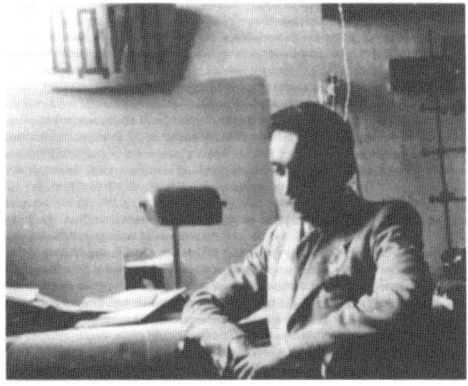

BULLETIN DE L'ACADÉMIE
POLONAISE DES SCIENCES
Cl. III — Vol. IV, No. 5, 1956

MATHEMATICS

Calculus of Approximations

by

M. WARMUS

Presented by H. STEINHAUS on February 6, 1956

This paper presents a theory which lays down the foundations for numerical computations and makes it possible to formulate properly many numerical problems.

By the approximate number $[a, A]$ we shall indicate the interval $[a, A]$, i. e. the set of all real numbers x that satisfy the inequality $a < x < A$. The approximate number $[B-b,\ B+b]$ can also be denoted by $\overset{b}{B}$. Thus, the approximate number $[a, A]$ can be expressed in the form $\overset{\frac{1}{2}(A-a)}{\frac{1}{2}(A+a)}$. We shall omit initial zeros in the upper part, if they lie to the left of the last digit of the lower one. For example, we shall write $\overset{02}{3.1416}$ instead of $\overset{0.00002}{3.1416}$.

We say that the approximate numbers α and β are equal and we write $\alpha = \beta$ if, and only if, they are two identical intervals. Hence, we have $[a, A] = [b, B]$ if, and only if, $a = b$ and $A = B$, and similarly $\overset{a}{A} = \overset{b}{B}$ if, and only if, $A = B$ and $a = b$.

We say that the approximate number β approximates the approximate number α and we write $\alpha \Rightarrow \beta$ or $\beta \Leftarrow \alpha$, if, and only if, the interval β includes the interval α. Thus, we have $[a, A] \Rightarrow [b, B]$ if, and only if, $a > b$ and $A < B$, and similarly $\overset{a}{A} \Rightarrow \overset{b}{B}$ if, and only if, $b - a > |B - A|$. It is easy to prove that the approximations $\alpha \Rightarrow \beta$ and $\beta \Rightarrow \gamma$ imply $\alpha \Rightarrow \gamma$.

The relation $\alpha \Rightarrow \beta$ is a partial ordering of the set of all approximate numbers.

In practical computations it is convenient to use the following two rules:

the rounding-off rule: $\overset{a}{A+c} \Rightarrow \overset{a+|c|}{A}$;

the extending rule: $\overset{a}{A} \Rightarrow \overset{b}{A}$ if $b > a$.

Figure 1.9 The first page of the paper by Warmus in which he outlined the concept of computing with numeric intervals.

As the complexity of a system increases, our ability to make precise and yet significant statements about its behavior diminishes until a threshold is reached beyond which precision and significance (or relevance) become almost mutually exclusive characteristics

The theory of rough sets established by Z. Pawlak (Fig. 1.11) opened another successful avenue of investigations of information granules whose description realized in the setting of a certain vocabulary leads to the concept of roughness of description (which itself manifests through lower and upper boundaries or approximations).

Losfi Zadeh during his Student years in Tehran in the early 1940s (the large Russian sign ODIN which means "alone," was his early proclamation of Independence).

Fuzzy Sets*

L. A. ZADEH

*Department of Electrical Engineering and Electronics Research Laboratory,
University of California, Berkeley, California*

A fuzzy set is a class of objects with a continuum of grades of membership. Such a set is characterized by a membership (characteristic) function which assigns to each object a grade of membership ranging between zero and one. The notions of inclusion, union, intersection, complement, relation, convexity, etc., are extended to such sets, and various properties of these notions in the context of fuzzy sets are established. In particular, a separation theorem for convex fuzzy sets is proved without requiring that the fuzzy sets be disjoint.

I. INTRODUCTION

More often than not, the classes of objects encountered in the real physical world do not have precisely defined criteria of membership. For example, the class of animals clearly includes dogs, horses, birds, etc. as its members, and clearly excludes such objects as rocks, fluids, plants, etc. However, such objects as starfish, bacteria, etc. have an ambiguous status with respect to the class of animals. The same kind of ambiguity arises in the case of a number such as 10 in relation to the "class" of all real numbers which are much greater than 1.

Clearly, the "class of all real numbers which are much greater than 1," or "the class of beautiful women," or "the class of tall men," do not constitute classes or sets in the usual mathematical sense of these terms. Yet, the fact remains that such imprecisely defined "classes" play an important role in human thinking, particularly in the domains of pattern recognition, communication of information, and abstraction.

The purpose of this note is to explore in a preliminary way some of the basic properties and implications of a concept which may be of use in dealing with "classes" of the type cited above. The concept in question is that of a *fuzzy set*,[1] that is, a "class" with a continuum of grades of membership. As will be seen in the sequel, the notion of a fuzzy set provides a convenient point of departure for the construction of a conceptual framework which parallels in many respects the framework used in the case of ordinary sets, but is more general than the latter and, potentially, may prove to have a much wider scope of applicability, particularly in the fields of pattern classification and information processing. Essentially, such a framework provides a natural way of dealing with problems in which the source of imprecision is the absence of sharply defined criteria of class membership rather than the presence of random variables.

We begin the discussion of fuzzy sets with several basic definitions.

II. DEFINITIONS

Let X be a space of points (objects), with a generic element of X denoted by x. Thus, $X = \{x\}$.

---

* This work was supported in part by the Joint Services Electronics Program (U.S. Army, U.S. Navy and U.S. Air Force) under Grant No. AF-AFOSR-139-64 and by the National Science Foundation under Grant GP-2413.

[1] An application of this concept to the formulation of a class of problems in pattern classification is described in RAND Memorandum RM-4307-PR, "Abstraction and Pattern Classification," by R. Bellman, R. Kalaba and L. A. Zadeh, October, 1964.

Figure 1.10 The first page of the Zadeh's seminal paper on fuzzy sets.

24 Chapter 1 Introduction

> # ROUGH SETS
> *Theoretical Aspects
> of Reasoning about Data*
>
> by
>
> ZDZISŁAW PAWLAK
> *Institute of Computer Science,
> Warsaw University of Technology*
>
> KLUWER ACADEMIC PUBLISHERS
> DORDRECHT / BOSTON / LONDON

Figure 1.11 Dealing with information with unclear boundaries—an emergence of rough sets.

Zdzisław Pawlak (1926–2006) was born in Lodz, 130 km south–west from Warsaw, Poland. He studied in Lodz University of Technology and Warsaw University of Technology. He has contributed to the number of disciplines of computer science and was one of the pioneers of computing. In 1961, he was on a research team that constructed one of the first computers in Poland named UMC 1. He proposed and investigated parenthesis-free languages, a generalization of reverse Polish notation introduced by Jan Lukasiewicz. While working at the Institute of Mathematics, in 1965 he introduced the foundations for modeling DNA what has come to be known as molecular computing. In 1968, he proposed a new formal model of a computing machine. In 1970s, he introduced knowledge representation systems. The early 1980s saw the inception of rough sets with the seminal papers published in the International Journal of Computer Information Systems. The most comprehensive coverage of this subject was presented in his book entitled "*Rough Sets. Theoretical Aspects of Reasoning about Data*" published in 1991.

REFERENCES

Bargiela, A., Pedrycz, W. *Granular Computing: An Introduction*, Kluwer Academic Publishers, Dordercht, 2003.

Bezdek, J. C. On the relationship between neural networks, pattern recognition and intelligence, *Int. J. Approx. Reason.* 6(2), 1992, 85–107.

Cantor, G. *Grundlagen Einer Allgemeinen Mannigfaltigkeitslehre*, Teubner, Leipzig, 1883.

Cantor, G. Beitraage zur Begraundung der transfniten Mengenlehre. *Math. Ann.* 46, 1895, 207–246.

Dubois, D., Prade, H., Rough fuzzy sets and fuzzy rough sets, *Int. J. Gen. Syst.* 17(2–3), 1990, 191–209.

Frias-Martinez E., et al., Modeling human behavior in user-adaptive systems: recent advances using soft computing techniques, *Expert Syst. Appl.* 29, 2005, 320–329.

Gottwald, S. Mathematical fuzzy logic as a tool for the treatment of vague information, *Inf. Sci.* 172(1–2), 2005, 41–71.

Hansen, E. A generalized interval arithmetic, *Lecture Notes in Computer Science*, vol. 29, Springer Verlag, Berlin 1975, pp. 7–18.

van der Heijden F., et al., *Classification, Parameter Estimation and State Estimation*, John Wiley & Sons, Ltd, Chichester, 2004.

Jaulin, L., Kieffer, M., Didrit, O., Walter, E. *Applied Interval Analysis*, Springer, London, 2001.

Katz, R. H., Borriello, G. *Contemporary Logic Design*, 2nd ed., Prentice Hall, Upper Saddle River, NJ, 2005.

Klir, G. J., Yuan, B. *Fuzzy Sets and Fuzzy Logic: Theory and Applications*, Prentice-Hall, Inc., Upper Saddle River, NJ, USA, 1995.

Korzybski, A. *Science and Sanity: An Introduction to Non-Aristotelian Systems and General Semantics*, 3rd ed, The International Non-Aristotelian Library Publishing Co., Lakeville, CT, 1933.

Lin, T. Y. Data mining and machine oriented modeling: a granular computing approach, *J. Appl. Intell.* 13(2), 2000, 113–124.

Łukasiewicz, J., O logice trójwartościowej, *Ruch Filoz.* 5, 1920, 170.

Łukasiewicz, J. Philosophische Bemerkungen zu mehrwertigen Systemen des Aussagenkalk, *C. R. Soc. Sci. Lett. Varsovie* 23, 1930, 51–77.

Łukasiewicz, J. Studies in logic and the foundations of mathematics, in: L. Borkowski (ed.), *Selected Works*, North-Holland, Amsterdam, 1970.

Moore, R. *Interval Analysis*, Prentice Hall, Englewood Cliffs, NJ, 1966.

Pawlak, Z. Rough sets, *Int. J. Comput. Inform. Sci.* 11, 1982, 341–356.

Pawlak, Z. *Rough Sets. Theoretical Aspects of Reasoning About Data*, Kluwer Academic Publishers, Dordercht, 1991.

Pawlak, Z., Skowron, A. Rudiments of rough sets, *Inf. Sci.* 177(1), 2007a, 3–27.

Pawlak, Z., Skowron, A. Rough sets: some extensions, *Inf. Sci.* 177(1), 2007b, 28–40.

Pawlak, Z., Skowron, A. Rough sets and Boolean reasoning, *Inf. Sci.* 177(1), 2007c, 41–73.

Pedrycz, W. *Computational Intelligence: An Introduction*, CRC Press, Boca Raton, FL, 1997.

Pedrycz, W. Shadowed sets: representing and processing fuzzy sets, *IEEE Trans. Syst. Man Cy. B* 28, 1998, 103–109.

Pedrycz W. (ed.), *Granular Computing: An Emerging Paradigm*, Physica-Verlag, Heidelberg, 2001.

Pedrycz, W. *Knowledge-Based Clustering*, John Wiley & Sons, Inc., Hoboken, NJ, 2005.

Perkowitz, M., Etzioni, O. Adaptive web sites, *Commun. ACM* 43(8), 2000, 152–158.

Polkowski, L., Skowron A. (eds.), *Rough Sets in Knowledge Discovery*, Physica-Verlag, Heidelberg, 1998.

Skowron, A. Rough decision problems in information systems, *Bull. Acad. Polonaise Sci. (Tech)* 37, 1989, 59–66.

Sommerville, I. *Software Engineering*, 8th ed., Addison-Wesley, Harlow, 2007.

Spott, M., Nauck, D. Towards the automation of intelligent data analysis, *Appl. Soft Comput.* 6, 2006, 348–356.

Warmus, M. Calculus of approximations, *Bull. Acad. Polonaise Sci.* 4(5), 1956, 253–259.

Weiser, M. The computer of the twenty-first century, *Sci. Am.* 163(3), 1991, 94–104.

Vasilakos, A., Pedrycz W. (eds.), *Ambient Intelligence, Wireless Networking, and Ubiquitous Computing*, Artech House, Boston, MA, 2006.

Zadeh, L. A. Fuzzy sets, *Inf. Control* 8, 1965, 338–353.

Zadeh, L. A. Outline of a new approach to the analysis of complex system and decision process, *IEEE Trans. Syst. Man Cyb.* 3, 1973, 28–44.

Zadeh, L. A. Fuzzy sets and information granularity, in: M. M. Gupta, R. K. Ragade, R. R. Yager (eds.), *Advances in Fuzzy Set Theory and Applications*, North Holland, Amsterdam, 1979, pp. 3–18.

Zadeh, L. A. Fuzzy logic = Computing with words, *IEEE Trans. Fuzzy Syst.* 4, 1996, 103–111.

Zadeh, L. A. Toward a theory of fuzzy information granulation and its centrality in human reasoning and fuzzy logic, *Fuzzy Sets Syst.* 90, 1997, 111–117.

Zadeh, L. A. From computing with numbers to computing with words-from manipulation of measurements to manipulation of perceptions, *IEEE Trans. Circ. Syst.* 45, 1999, 105–119.

Zadeh, L. A. Toward a generalized theory of uncertainty (GTU)—an outline, *Inf. Sci.* 172, 2005, 1–40.

Zimmermann, H. J. *Fuzzy Set Theory and Its Applications*, 3rd ed., Kluwer Academic Publishers, Norwell, MA, 1996.

Chapter 2

Notions and Concepts of Fuzzy Sets

In this chapter, we introduce the fundamental concepts of fuzzy sets. We focus on the underlying idea of partial membership being conveniently quantified through membership functions. We present the basic rationale and then move on to the detailed quantification of fuzzy sets by discussing the most commonly encountered classes of membership functions and presenting their semantics. Furthermore, some fundamental interpretations of fuzzy sets are given. In the sequel, we refine the concept of fuzzy sets by introducing an idea of fuzzy numbers along with their key operational aspects.

2.1 SETS AND FUZZY SETS: A DEPARTURE FROM THE PRINCIPLE OF DICHOTOMY

Conceptually and algorithmically, fuzzy sets constitute one of the most fundamental and influential notions in science and engineering. The notion of fuzzy set is highly intuitive and transparent as it captures what really becomes an essence of a way in which a real world is perceived and described. We encounter categories of objects whose belongingness to a given category (concept) is always a matter of degree. There are numerous examples in which we find elements whose allocation to the concept we want to define can be satisfied to some degree. One may eventually claim that continuity of transition from full belongingness and full exclusion is the major and ultimate feature of the physical world and natural systems. For instance, we may qualify an indoor environment as *comfortable* when its temperature is kept *around* $20°C$. If we observe a value of $19.5°C$, it is very likely that we still feel quite *comfortable*. The same holds if we encounter $20.5°C$—humans usually do not discriminate changes in temperature within the range of $1°C$. A value of $20°C$ would be fully compatible with the concept of *comfortable* temperature, yet $0°C$ or $30°C$ would not. In these two cases, as well as for temperatures close to these two values,

---

Fuzzy Systems Engineering: Toward Human-Centric Computing, by Witold Pedrycz and Fernando Gomide
Copyright © 2007 John Wiley & Sons, Inc.

we would describe them as being *cold* and *warm*, respectively. We could question whether the temperature of 25°C is viewed as *warm* or *comfortable* or, similarly, if 15°C is *comfortable* or *cold*. Intuitively, we know that 25°C is somehow between *comfortable* and *warm*, whereas 15°C is between *comfortable* and *cold*. The value 25°C is partially compatible with the term *comfortable* and *warm*, and somewhat compatible or, depending on observer's perception, incompatible with the term *cold* temperature. Similarly, we may say that 15°C is partially compatible with the comfortable and cold temperature, and slightly compatible or incompatible with the warm temperature. In spite of this highly intuitive categorization of environment temperatures into the three classes, namely *cold*, *comfortable*, and *warm*, we note that the transition between the classes is not instantaneous and sharp. Simply when moving across the range of temperatures, these values become gradually perceived as *cold*, *comfortable*, or *warm*. Similar phenomenon happens when we are dealing with the concept of height of people. An individual of height of 1 m is *short*, whereas a person of 1.90 m is perceived to be *tall*. Again the question is, what is the range of height values that could qualify a person to be *tall*? Does a height of 1.85 m discriminate between *tall* and *short* individuals? Or maybe 1.86 m would be the right choice? Asking these questions, we sense that they do not make too much sense. We realize that the nature of these concepts is such that we cannot use a single number—a transition between the notion of tall and short is not abrupt in any way. Hence, we cannot assign a single number that does the job. This sends a clear message: The concept of dichotomy does not apply when defining even simple concepts. The illustration of dichotomy is included in Figure 2.1 (a). In contrast, defining a concept by not confining ourselves to the dichotomy is illustrated in Figure 2.1(b).

Fuzzy sets and the corresponding membership functions form a viable and mathematically sound setting. When talking about heights of Europeans, we may refer to real numbers within the interval [0, 3] to represent a universe of heights that range in between 0 and 3 m. This universe of discourse is suitable for describing the concept of *tall* people.

Let us denote by **X** a universe of discourse (space) of all elements. The universe can be either continuous or discrete. For instance, the closed interval [0, 3] constitutes a continuous and bounded universe, whereas the set $\mathbf{N} = \{0, 1, 2, \ldots\}$ of natural numbers is discrete, and countable, but there are no bounds.

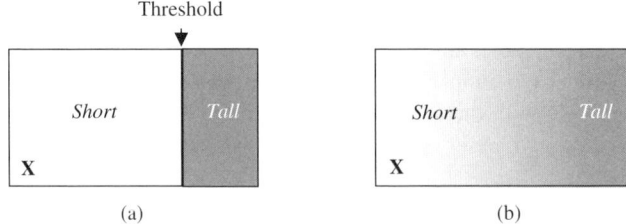

Figure 2.1 Contrasting a concept of a set and the principle of dichotomy itself versus a relaxation of the concept of complete inclusion and exclusion.

2.1 Sets and Fuzzy Sets: a Departure from the Principle of Dichotomy

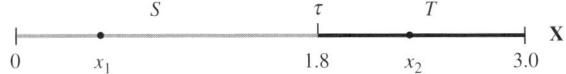

Figure 2.2 Set as a collection of values in intervals.

Consider the universe of discourse $\mathbf{X} = [0, 3]$ and the collection S of values in $\mathbf{X}$ that are less than a threshold value τ in $\mathbf{X}$, for example, $\tau = 1.8$. Consider the sets $S = \{x \in \mathbf{X} | 0 \leq x \leq 1.8\}$ and $T = \{x \in \mathbf{X} | 1.8 \leq x \leq 3.0\}$ (Fig. 2.2). Each set is a class whose members are elements of the universe that satisfy the same property. This set is equivalent to a list of elements of the universe that are members of the set.

Given a value in $\mathbf{X}$, the process of dichotomization imposes a binary, all or none classification decision: either accept or reject the value as belonging to a given collection. For instance, consider the set S shown in Figure 2.2. Clearly, the point x_1 belongs to S whereas x_2 does not, that is, $x_1 \in S$ and $x_2 \notin S$. Similarly, for the set T we have $x_1 \notin T$, and $x_2 \in T$. If we denote the accept decision by 1 and the reject decision by 0, for short, we may express the classification decision of $x \in \mathbf{X}$ through a characteristic function as follows:

$$S(x) = \begin{cases} 1, & \text{if } x \in S \\ 0, & \text{if } x \notin S \end{cases} \qquad T(x) = \begin{cases} 1, & \text{if } x \in T \\ 0, & \text{if } x \notin T \end{cases}$$

Figure 2.3 illustrates sets S and T with the use of their characteristic functions. Because a characteristic function fully characterizes a set, it is synonymous of the notion of set.

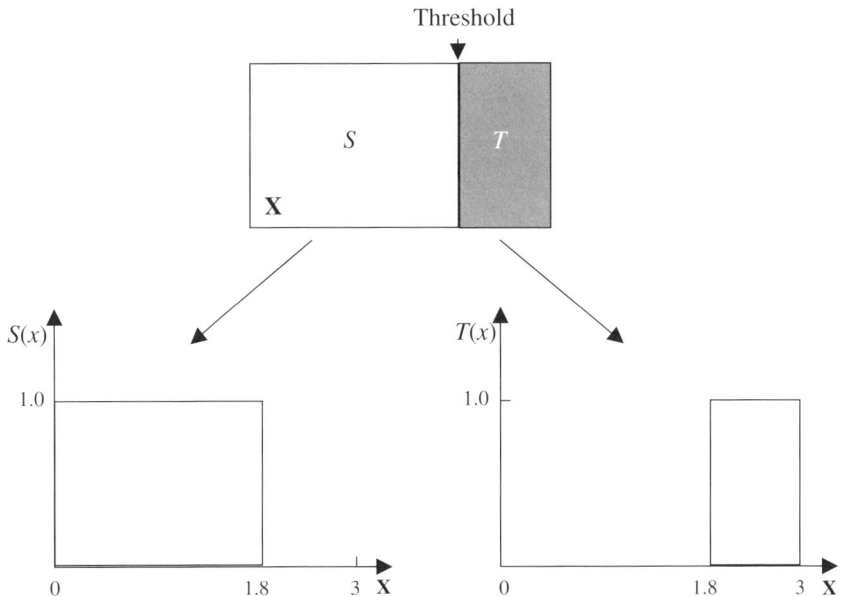

Figure 2.3 Sets and their corresponding characteristic functions.

In general, a characteristic function of set *A* defined on **X** assumes the following form:

$$A(x) = \begin{cases} 1, & \text{if } x \in A \\ 0, & \text{if } x \notin A \end{cases} \qquad (2.1)$$

The empty set Ø has a characteristic function that is identically equal to zero, $Ø(x) = 0$ for all *x* in **X**. The universe **X** itself comes with the characteristic function that is identically equal to one, $X(x) = 1$ for all *x* in **X**. Also, a singleton $A = \{a\}$, a set with only a single element, has a characteristic function such that $A(x) = 1$ if $x = a$ and $A(x) = 0$ otherwise.

Characteristic functions $A: \mathbf{X} \to \{0,1\}$ induce a constraint with well-defined boundaries on the elements of the universe **X** that can be assigned to a set *A*. The fundamental idea of fuzzy set is to relax this requirement by admitting intermediate values of class membership. Therefore, we may assign intermediate values between 0 and 1 to quantify our perception on how compatible these values are with the class with 0 meaning incompatibility (complete exclusion) and 1 compatibility (complete membership). Membership values thus express the degrees to which each element of a universe is compatible with the properties distinctive to the class. Intermediate membership values means that no natural threshold exists and that elements of a universe can be a member of a class and at the same time belong to other classes with different degrees. Gradual, less strict membership degrees are the essence of fuzzy sets.

Formally, a fuzzy set *A* is described by a membership function mapping the elements of a universe **X** to the unit interval [0,1] (Zadeh, 1965):

$$A : \mathbf{X} \to [0, 1] \qquad (2.2)$$

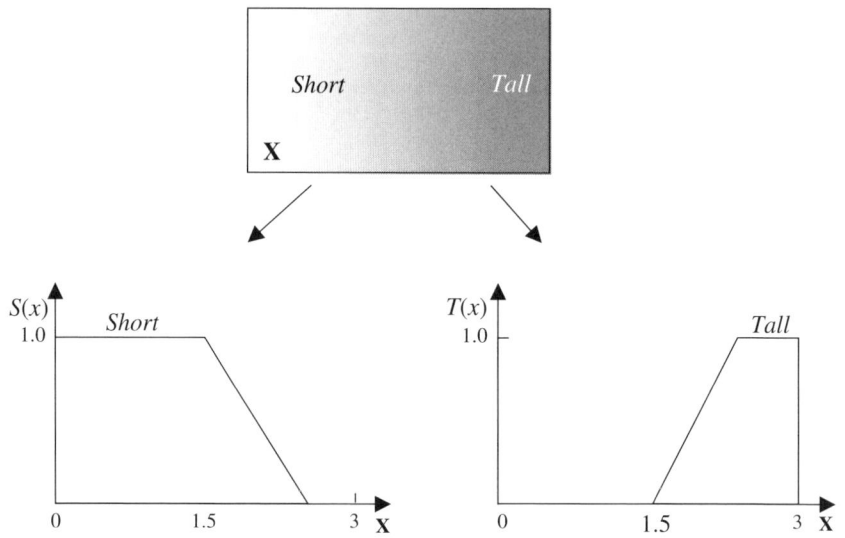

Figure 2.4 Fuzzy sets and their membership functions.

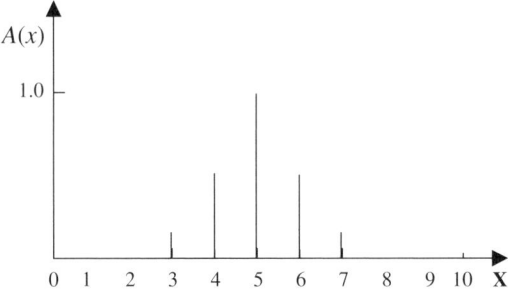

Figure 2.5 Fuzzy set A in a discrete universe.

The membership functions are therefore synonymous of fuzzy sets. In a nutshell, membership functions generalize characteristic functions in the same way as fuzzy sets generalize sets; refer to Figure 2.4.

Fuzzy sets can also be viewed as a set of ordered pairs of the form $\{x, A(x)\}$, where x is an element of $\mathbf{X}$ and $A(x)$ denotes its corresponding degree of membership. For a finite universe of discourse $\mathbf{X} = \{x_1, x_2, \ldots, x_n\}$, A can be represented by an n-dimensional vector $A = [a_1, a_2, \ldots, a_n]$ with $a_i = A(x_i)$. Figure 2.5 illustrates a fuzzy set whose membership function captures the concept of integer *around* 5. Here $n = 10$ and expressing the integer quantity around 5 in the finite universe $x = \{0, 1, 2, \ldots, 10\}$ $A = [0, 0, 0, 0.2, 0.5, 1.0, 0.5, 0.2, 0, 0, 0]$. An equivalent notation of A is $A = \{0/1, 0/2, 0.2/3, 0.5/4, 1.0/5, 0.5/6, 0.2/7, \ldots, 0/10\}$.

The choice of the unit interval for the values of membership degrees could be a matter of convenience. The choice of the detailed membership values, say $A(4) = 0.5865$, is not crucial; in describing membership grades we are predominantly after reflecting an order of the elements in A in terms of their belongingness to the fuzzy set (Dubois and Prade, 1979).

Being more descriptive, we may view fuzzy sets as elastic constraints imposed on the elements of a universe. As emphasized before, fuzzy sets deal primarily with the concept of elasticity, graduality, or absence of sharply defined boundaries. In contrast, when dealing with sets we are concerned with rigid boundaries, lack of graded belongingness, and sharp binary boundaries. Gradual membership means that no natural boundary exists and that some elements of the universe of discourse can, contrary to sets, coexist (belong) in different fuzzy sets with different degrees of membership. For instance, as shown in Figure 2.6, $x_1 = 1.5$ is compatible with the concept of *short* and $x_2 = 1.8$ belongs to the category of *tall* people (when assuming the model of sets), but x_1 is 0.8 *short* and 0.2 *tall* and x_2 is 0.6 *short* and 0.6 *tall* when assuming a perspective of fuzzy sets.

2.2 INTERPRETATION OF FUZZY SETS

In fuzzy theory, the concept of fuzziness comes with a precise meaning. Fuzziness primarily means lack of precise boundaries of a collection of objects and, as such, it is

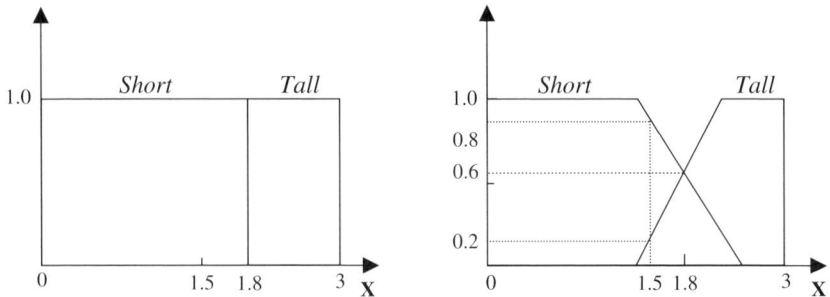

Figure 2.6 The concept of Boolean (two-valued) membership in characteristic functions and gradual membership represented by membership functions (fuzzy sets).

an evident manifestation of imprecision and a particular type of uncertainty. Let us make some comments regarding this topic.

First, it is worth indicating that fuzziness is both conceptually and formally different from the fundamental concept of probability. In general, it is difficult to foresee the result of tossing a fair coin once it is impossible to know if either head or tail will occur for certain. We may, at most, say that there is 50% chance to have head or tail occur, but as soon as the coin falls, uncertainty vanishes. But, in the case of the height of a person, say, *tall* imprecision remains. Formally, fuzzy sets are membership functions, which are mappings from some given universe of discourse to the unit interval as in (2.2). In contrast, probability is a set function, a mapping whose universe is a set of subsets of a domain.

Second, there are differences between fuzziness, generality, and ambiguity. A notion is general when it applies to a multiplicity of objects and keeps only a common essential property. An ambiguous notion stands for several unrelated objects. Therefore, from this point of view, fuzziness does not mean generality nor ambiguity, and applications of fuzzy sets exclude these categories. Fuzzy set theory assumes that the universe is well defined and has its elements assigned to the classes by means of a numerical scale.

Applications of fuzzy set to areas such as data analysis, reasoning under uncertainty, and decision-making suggest different interpretations of membership grades in terms of similarity, uncertainty, and preference (Dubois and Prade, 1997, 1998). From the similarity point of view, $A(x)$ means the degree of compatibility of an element $x \in \mathbf{X}$ with representative elements of A. This is the primary and most intuitive interpretation of a fuzzy set, one that is particularly suitable for data analysis. An example is the case where we question how to qualify an environment as *comfortable* when we know that the current temperature is 25°C. As discussed at the beginning of this chapter, such quantification is a matter of degree. For instance, assuming a universe of discourse $\mathbf{X} = [0, 40]$ and choosing 20°C as the representative of *comfortable* temperature, we note, in Figure 2.7, that the degree at which 25°C is comfortable is 0.2. In the example, we have adopted piecewise linearly decreasing functions of the distance between temperature values and the representative value 20°C to determine the corresponding membership degree.

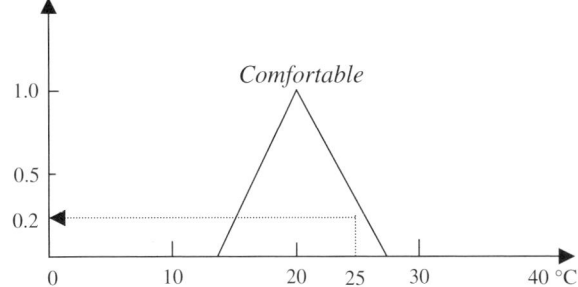

Figure 2.7 Membership function for a fuzzy set of *comfortable* temperature.

Now let us assume that the values of a variable x are located within the support of a fuzzy set A. Then given a value v of **X**, $A(v)$ expresses a possibility that $x = v$ given that x is in A is all that is known about. In this situation, the membership degree of a given tentative value v to the class A reflects the degree of plausibility that this value is the same as x. This idea reflects a type of uncertainty because if the membership degree is high, our confidence about the value of x may still be low, but if the degree is low, then the tentative value may be rejected as an plausible candidate. The variable labeled by the class A is uncontrollable. This allows assignment of fuzzy sets to the possibility distributions as presented in possibility theory (Zadeh, 1978). For instance, suppose someone said that he felt comfortable when watching a soccer game. In this situation, the membership degree of a given tentative temperature value, say 25°C, reflects the degree of plausibility that this value of temperature is the same as the one when he felt comfortable. Note that the temperature value felt is unknown, but there is no question if it did occur or not. Possibility is whether an event may occur and with what degree. On the contrary, probability is about whether an event will occur.

Finally, let us assume that A reflects a preference on the values of a variable x in **X**. For instance, x can be a decision variable and fuzzy set A be an elastic constraint characterizing feasible values and decision-maker preferences. In this case, $A(v)$ denotes the grade of preference in favor of v as the value of x. This is the interpretation that prevails in fuzzy optimization and decision analysis. For instance, we may be interested in finding a comfortable value of temperature. The membership degree of a candidate temperature value v reflects our degree of satisfaction with the particular temperature value chosen. In this situation, the value of the variable is controllable in the sense that the value being adopted depends on our choice.

2.3 MEMBERSHIP FUNCTIONS AND THEIR MOTIVATION

Formally speaking, any function $A : \mathbf{X} \to [0, 1]$ could be qualified to serve as a membership function describing the corresponding fuzzy set. In practice, the form of the membership functions should be reflective of the problem at hand for which we construct fuzzy sets. They should mirror our perception of the concept to be represented and used in problem solving, the level of detail we intend to capture, and the context in which the fuzzy set are going to be used. It is also essential to assess the type of a fuzzy set from the standpoint of its suitability when handling the ensuing optimization procedures. Given these criteria in mind, we elaborate on the most commonly used categories of membership functions. All of them are defined in the universe of real numbers, which is $\mathbf{X} = \mathbf{R}$.

2.3.1 Triangular Membership Functions

They are described by their piecewise linear segments described in the form

$$A(x, a, m, b) = \begin{cases} 0, & \text{if } x \leq a \\ \dfrac{x-a}{m-a}, & \text{if } x \in [a, m) \\ \dfrac{b-x}{b-m}, & \text{if } x \in [m, b] \\ 0, & \text{if } x \geq b \end{cases}$$

Using the more concise notation, the above expression can be written in the form $A(x, a, m, b) = \max\{\min[(x-a)/(m-a), (b-x)/(b-m)], 0\}$. Also refer to Figure 2.8. The meaning of the parameters is straightforward: m denotes a modal (typical) value of the fuzzy set whereas a and b are the lower and upper bounds, respectively. They could be sought as the extreme elements of the universe of

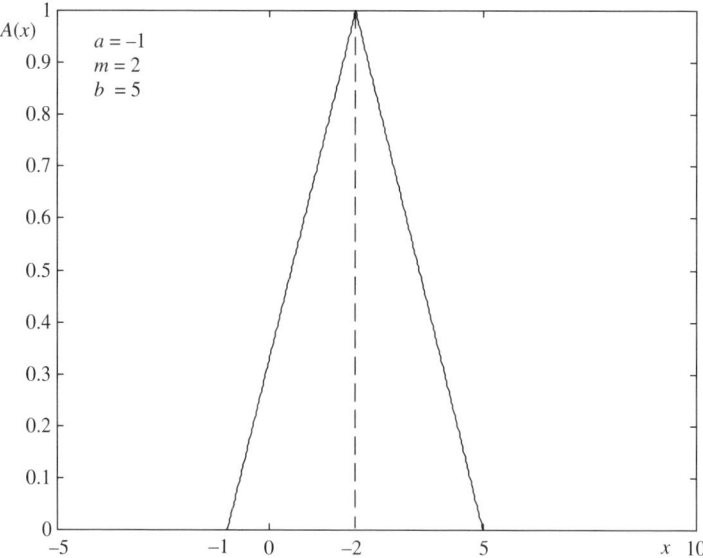

Figure 2.8 Triangular membership function.

discourse that delineate the elements belonging to A with nonzero membership degrees.

Triangular fuzzy sets (membership functions) are the simplest possible models of grades of membership as they are fully defined by only three parameters. As mentioned, the semantics is evident as the fuzzy sets are expressed on the basis of knowledge of the spreads of the concepts and their typical values. The linear change in the membership grades is the simplest possible model of membership one could think of. Taking the derivative of the triangular membership function, which could be sought as a measure of sensitivity of A, $\frac{\partial A}{\partial x}$ we conclude that its sensitivity is constant for each of the linear segments of the fuzzy set.

2.3.2 Trapezoidal Membership Functions

They are piecewise linear function characterized by four parameters, a, m, n, and b, each of which defines one of the four linear parts of the membership function, as illustrated in Figure 2.9. They assume the following form:

$$A(x) = \begin{cases} 0, & \text{if } x < a \\ \dfrac{x-a}{m-a}, & \text{if } x \in [a, m) \\ 1, & \text{if } x \in [m, n) \\ \dfrac{b-x}{b-n}, & \text{if } x \in [n, b] \\ 0, & \text{if } x > b \end{cases}$$

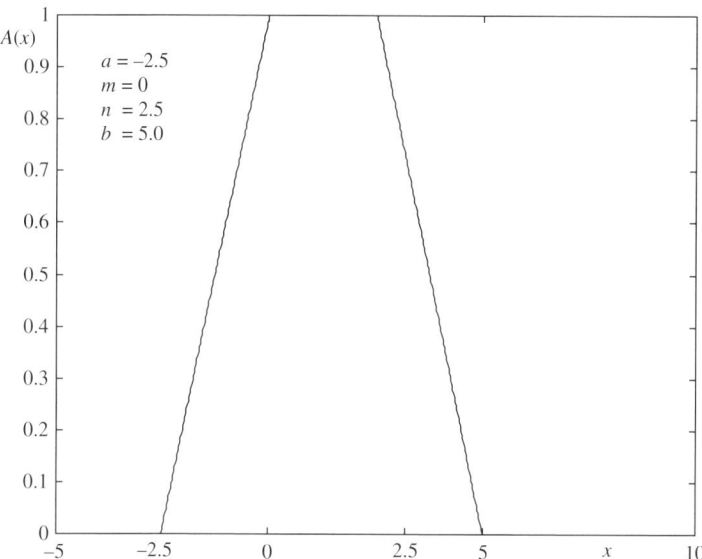

Figure 2.9 Trapezoidal membership functions.

Using an equivalent notation, we can rewrite A as follows:

$$A(x,a,m,n,b) = \max\{\min[(x-a)/(m-a), 1, (b-x)/(b-n)], 0\}$$

2.3.3 Γ-Membership Functions

They are expressed in the following form:

$$A(x) = \begin{cases} 0, & \text{if } x \leq a \\ 1 - e^{-k(x-a)^2}, & \text{if } x > a \end{cases} \quad \text{or} \quad A(x) = \begin{cases} 0, & \text{if } x \leq a \\ \dfrac{k(x-a)^2}{1 + k(x-a)^2}, & \text{if } x > a \end{cases}$$

where $k > 0$, as illustrated in Figure 2.10.

2.3.4 S-Membership Functions

These functions are of the following form:

$$A(x) = \begin{cases} 0, & \text{if } x \leq a \\ 2\left(\dfrac{x-a}{b-a}\right)^2, & \text{if } x \in [a,m] \\ 1 - 2\left(\dfrac{x-b}{b-a}\right)^2, & \text{if } x \in [m,b] \\ 1, & \text{if } x > b \end{cases}$$

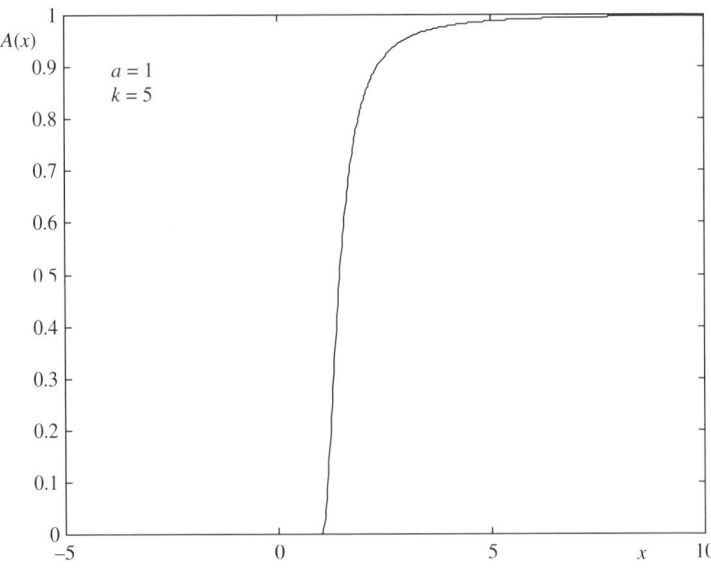

Figure 2.10 Γ-membership function.

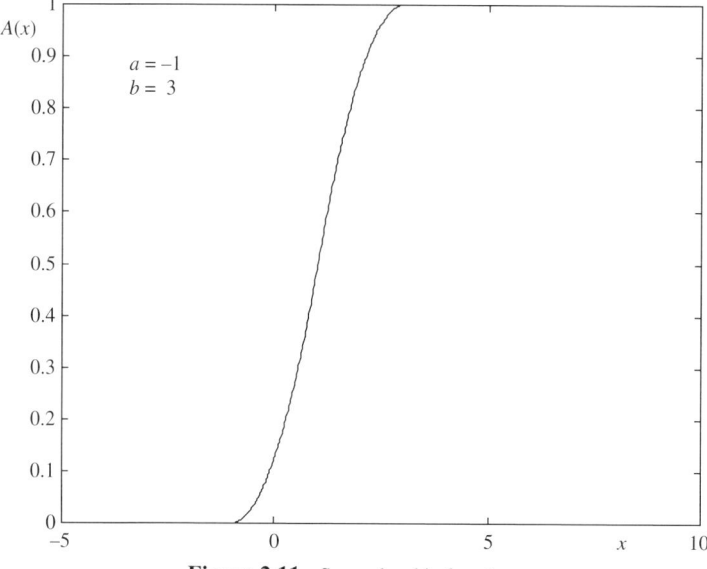

Figure 2.11 S-membership function.

The point $m = (a+b)/2$ is the crossover point of the S-function, shown in Figure 2.11.

2.3.5 Gaussian Membership Functions

These membership functions are described by the following relationship:

$$A(x, m, \sigma) = \exp\left(-\frac{(x-m)^2}{\sigma^2}\right)$$

An example of the membership function is shown in Figure 2.12. Gaussian membership functions have two important parameters. The modal value m represents the typical element of A, whereas σ denotes a spread of A. Higher values of σ correspond to larger spreads of the fuzzy sets.

2.3.6 Exponential-Like Membership Functions

They are described in the following form:

$$A(x) = \frac{1}{1 + k(x-m)^2}, \quad k > 0$$

See Figure 2.13. The spread of the exponential-like membership function increases as the values of k get lower.

38 Chapter 2 Notions and Concepts of Fuzzy Sets

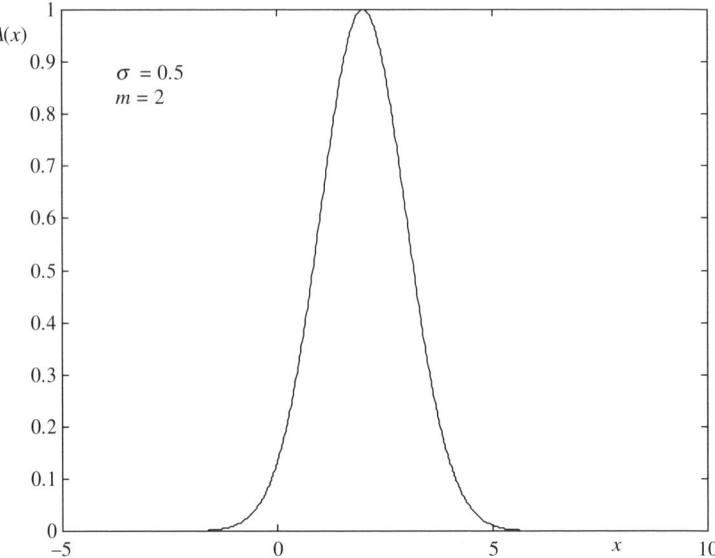

Figure 2.12 Gaussian membership functions.

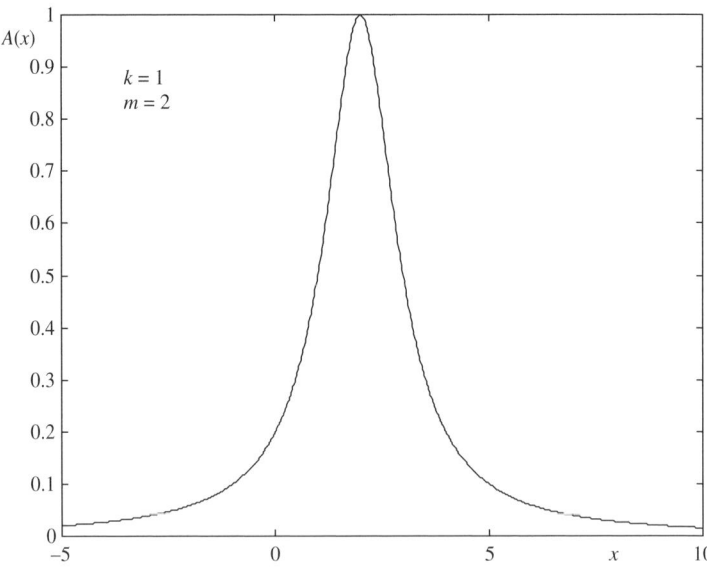

Figure 2.13 An example of the exponential-like membership function.

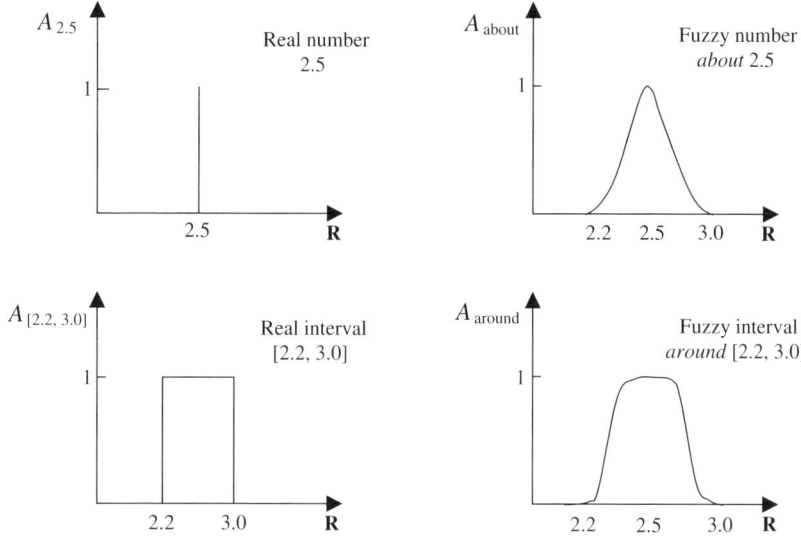

Figure 2.14 Examples of quantities and fuzzy quantities.

2.4 FUZZY NUMBERS AND INTERVALS

In practice, the exact values of the parameters of models are not so common. Normally, uncertainty and imprecision arise due to the lack of knowledge and incomplete information reflected in system structure, parameters, inputs, and possible bounds.

Fuzzy numbers and intervals model imprecise quantities and capture our innate conception of approximate numbers, such as about five and around 10, and intervals such as below 100, around two and three, and above 10. Fuzzy quantities are intended to model our intuitive notions of approximate numbers and intervals as a generalization of numbers and intervals, as Figure 2.14 suggests. In general, fuzzy quantities summarize numerical data by means of linguistically labeled fuzzy sets whose universe is **R**, the set of real numbers. For instance, if a value of a real variable is certain, say $x = 2.5$, then we can represent it as a certain quantity, a singleton whose characteristic function is $A_{2.5}(x) = 1$ if $x = 2.5$ and $A_{2.5}(0) = 0$ otherwise, as shown in Figure 2.14. In this situation, the quantity has both precise value and precise meaning. If we are uncertain of the value of the variable, but certain about its bounds, then the quantity is uncertain and can be represented, for instance, by the closed interval [2.2, 3.0] a set whose characteristic function is $A_{[2.2, 3.0]}(x) = 1$ if $x \in [2.2, 3.0]$ and $A_{[2.2, 3.0]}(x) = 0$ otherwise. Here the variable is characterized by an imprecise value, but its meaning is precise. When bounds are also not sharply defined, the quantities become fuzzy numbers or intervals, respectively, as Figure 2.14 shows. In these cases, both fuzzy numbers and intervals are also quantities with precise meaning, but with imprecise values.

40 Chapter 2 Notions and Concepts of Fuzzy Sets

To be meaningful, we expect that the membership functions of fuzzy quantities should posses certain properties (Nguyen and Walker, 1999). For instance, we expect that all the sets formed by the elements of the universe for which membership degrees are different from zero always be bounded sets because numbers and intervals, even imprecise ones, are bounded. Also, there must exist at least an element of the universe for which the membership degree is unitary because numbers and intervals have precise meanings and when represented by sets they must have characteristic functions that value one for the corresponding numbers and intervals. This must also be the case with fuzzy quantities once real numbers and intervals are particular instances of fuzzy numbers and intervals (Klir and Yuan, 1995). Moreover, membership functions cannot be multimodal because we would otherwise face a conflict to assign a meaning to the description of fuzzy numbers or intervals, as Figure 2.15 indicates.

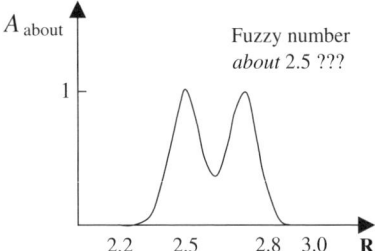

Figure 2.15 Bimodal membership function.

Fuzzy quantities are essential in many applications including time-series modeling, optimization, decision-making, control, and approximate reasoning. For instance, time-series models produce forecasts by extrapolating the historical behavior of the values of a variable. Time-series data are historical data in chronological order. In general, it is possible to use nonlinear or linear functions to extrapolate a series of observations. One alternative that is often considered in practice is to assume a higher-order polynomial for which appropriate values for its parameters must be derived from the available, often imprecise and noisy, observations. Under these circumstances, fuzzy time-series models are of value because they are built assuming that the parameters of the model are fuzzy numbers and hence convey more information. The same issue appears in fuzzy optimization models, as we rarely know coefficients of the objective function and constraints. In most practical cases model parameters can only be the rough estimates because of the lack of information, data, or cost to obtain their values. Although in classical optimization models uncertain data are replaced by their average surrogates, fuzzy optimization models allow the use of subjective and approximate (granular) data. This increases robustness of decisions, increases model credibility, and reduces costs of data processing.

2.5 LINGUISTIC VARIABLES

One can often deal with variables describing phenomena of physical or human systems assuming a finite, quite small number of descriptors.

2.5 Linguistic Variables

We often describe observations about a phenomenon by characterizing its states that we naturally translate in terms of the idea of a variable. For instance, we may refer to an environment through words such as comfortable, sunny, and neat. In particular, we can qualify the environment condition through the variable temperature with values chosen in a range such as the interval $\mathbf{X} = [0, 40]$. Alternatively, temperature could be qualified using labels such as cold, comfortable, and warm. A precise numerical value such as 20°C seems simpler to characterize the environment than the ill-defined term comfortable. But the linguistic label comfortable is a choice of one out of three values, whereas 20°C is a choice out of many. The statement could be strengthened if the underlying meaning of comfortable is conceived as about 20°C. Although the numerical quantity 20°C can be visualized as a point in a set, the linguistic temperature value *comfortable* can be viewed as a collection of temperature values in a bounded region centered in 20°C. The label *comfortable* can, therefore, be regarded as a form of information summarization, namely; granulation, because it serves to approximate a characterization of ill-defined or complex phenomena (Zadeh, 1975). In these circumstances, fuzzy sets provide a way to map a finite term set to a linguistic scale whose values are fuzzy sets. In general, it is difficult to find incontestable thresholds, such as 15°C and 30°C for instance, which allows us to assign $cold = [0,15]$, $comfortable = [15,30]$, and $warm = [30,40]$. *Cold, comfortable*, and *warm* are fuzzy sets instead of single numbers or sets (intervals). As fuzzy sets concern the representation of collections with unclear boundaries by means of membership functions taking values in an ordered set of membership values, they provide a means to interface numerical and linguistic quantities, a way to link computing with words and granular computing.

In contrast to the idea of numeric variables as being commonly used, the notion of linguistic variable can be regarded as a variable whose values are fuzzy sets. In general, linguistic variables may assume values consisting of words or sentences expressed in a certain language (Zadeh, 1999). Formally, a linguistic variable is characterized by a quintuple $\langle X, T(X), \mathbf{X}, G, M \rangle$ where its components are as follows:

X—the name of the variable,

$T(X)$—a term set of X whose elements are labels L of linguistic values of X,

G—a grammar that generates the labels of X,

M—a semantic rule that assigns to each label $L \in T(X)$ a meaning whose realization is a fuzzy set on the universe $\mathbf{X}$ whose base variable is x.

EXAMPLE 2.1

Let us consider the linguistic variable of temperature. Here, the linguistic variable is formalized by explicitly identifying all the components of the formal definition:

$X = $ temperature, $\mathbf{X} = [0, 40]$

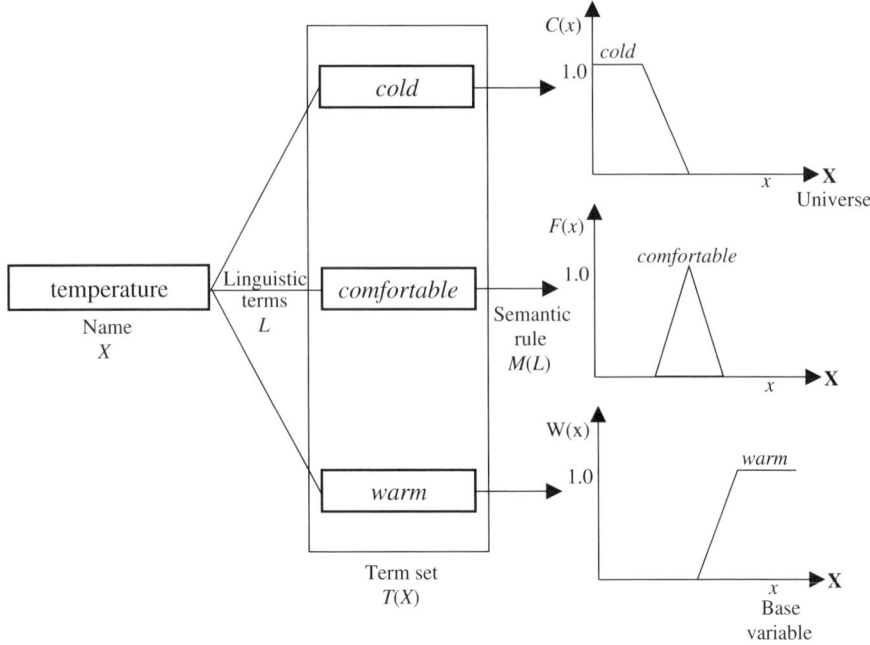

Figure 2.16 An example of the linguistic variable temperature.

T(temperature) = {*cold, comfortable, warm*}

$M(cold) \rightarrow C$, $M(comfortable) \rightarrow F$ and $M(warm) \rightarrow W$, where C, F, and W are fuzzy sets whose membership functions are illustrated in Figure 2.16.

The notion of the linguistic variable plays a major role in applications of fuzzy sets. In fuzzy logic and approximate reasoning, truth values can be viewed as linguistic variables whose truth values form the term set as, for example, *true, very true, false, more or less true*, and the like.

2.6 CONCLUSIONS

Fuzzy set is a concept that extends the notion of a set by assigning to each element of a reference set, the universe, a value representing its degree of membership in the fuzzy set. Membership values correspond to the degree an element is compatible or similar to typical elements of the class associated with the fuzzy set. Elements may belong in a fuzzy set to a lesser or greater degree as quantified by lower or higher membership values. Usually membership degrees are values in the unit interval, but generally it can be any partially ordered set. Therefore, fuzzy sets abolish dichotomy and provide flexibility needed to match real-world requirements. Fuzzy sets capture important characteristics of different application contexts. These provide distinct semantics and interpretations, namely similarity, preference, and uncertainty. Fuzzy sets are also essential to interface words with computing carried out at the level of linguistic variables and fuzzy numbers. Linguistic variable is a key notion to construct fuzzy systems in application areas such as data analysis, pattern recognition, classification, approximate reasoning, fuzzy control, fuzzy optimization, and decision analysis.

EXERCISES AND PROBLEMS

1. There is an interesting problem posed by Borel (1950) that could now be conveniently handled in the setting of fuzzy sets:

   One seed does not constitute a pile nor two or three. From the other side, everybody will agree that 100 million seeds constitute a pile. What therefore is the appropriate limit?

   Given this description, suggest a membership function of the concept of pile. What type of membership function would you consider in this problem?

2. Consider two situations: (a) The number of expected people to ride on a bus on a certain day. (b) The number of people who could ride in a bus at any time. Both situations describe an uncertain scenario. Which of these two situations involves randomness? Which one involves fuzziness? What is the nature of fuzziness: similarity, possibility, or preference?

3. We are interested in describing the state of an environment by quantifying temperature as *very cold, cold, comfortable, warm*, and *hot*. Choose an appropriate universe of discourse. Represent state values using (a) sets and (b) fuzzy sets.

4. Suppose that the allowed speed values in a city streets range between 0 and 60 km/h. Describe the speed values such as *low, medium*, and *high* using sets and fuzzy sets. Would this description be adequate also for highways? Justify the answer.

5. Given is the fuzzy set A with the membership function

$$A(x) = \begin{cases} x - 4, & \text{if } 4 \leq x \leq 5 \\ -x + 6, & \text{if } 5 < x \leq 6 \\ 0, & \text{otherwise} \end{cases}$$

   (a) Plot the membership function and identify its type.

   (b) What type of linguistic label could be associated with the concept conveyed by A?

6. Fuzzy set is a precise theory to describe imprecision. Elaborate on this statement and justify your opinion.

HISTORICAL NOTES

A different notion of fuzziness from the one originally introduced by Zadeh (1965) was suggested by Sugeno (1974). Assume an element x of universe $\mathbf{X}$. Given any subset A of $\mathbf{X}$, a set function $g_x(A)$ assigns a value in [0, 1] to specify the degree of fuzziness of the claim that $x \in A$. The value quantifies a measure of certainty if x is in A or is excluded from A. The function g_x is such that $g_x(\phi) = 0, g_x(\mathbf{X}) = 1$, and if $A, B \subseteq X$ then $A \subseteq B \Rightarrow g_x(A) \leq g_x(B)$. It has been shown (Klir and Yuan, 1995) that g_x, called fuzzy measure by Sugeno, encompasses other measures of uncertainty, notably the probability and possibility measures.

Kosko (1992, 1997) argues that fuzziness is a type of deterministic uncertainty that emerges as a consequence of simultaneous membership of an element in a set and a complement of this set. This is a contradiction in set theory because the intersection of a set and its complement is the empty set. The reason is that set theory assumes dichotomy. Partial membership of fuzzy set theory avoids such a contradiction once it admits an element of a universe to belong to different classes with distinct grades. Therefore, fuzzy set theory does not require an assumption about empty intersection of a fuzzy set and its complement.

REFERENCES

Borel, E. *Probabilité e Certitude*, Press Universite de France, Paris, 1950.

Dubois, D., Prade, H. Outline of fuzzy set theory: an introduction, in: M. M. Gupta, R. K. Ragade, R. R. Yager (eds.), *Advances in Fuzzy Set Theory and Applications*, North Holland, Amsterdam, 1979, pp. 27–39.

Dubois, D., Prade, H. The three semantics of fuzzy sets, *Fuzzy Set Syst.*, **2**, 1997, 141–150.

Dubois, D., Prade, H. An introduction to fuzzy sets, *Clin. Chim. Acta* **70**, 1998, 3–29.

Klir, G., Yuan, B. *Fuzzy Sets and Fuzzy Logic: Theory and Applications*, Prentice-Hall, Upper Saddle River, NJ, 1995.

Kosko, B. *Neural Networks and Fuzzy Systems*, Prentice-Hall International, Englewood Cliffs, NJ, 1992.

Kosko, B. *Fuzzy Engineering: A Dynamical Systems Approach to Machine Intelligence*, Prentice-Hall International, Upper Saddle River, NJ, 1997.

Nguyen, H., Walker, E. *A First Course in Fuzzy Logic*, Chapman & Hall CRC Press, Boca Raton, FL, 1999.

Sugeno, M. *Theory of fuzzy integrals and its applications*, Ph.D. dissertation, Tokyo Institute of Technology, Tokyo, Japan, 1974.

Zadeh, L. A. Fuzzy sets, *Inf. Cont.* **8**, 1965, 338–353.

Zadeh, L. A. The concept of linguistic variable and its application to approximate reasoning (part I), *Inf. Sci.* **8**, 1975, 301–357.

Zadeh, L. A. Fuzzy sets as a basis for a theory of possibility, *Fuzzy Set Syst.* **3**, 1978, 3–28.

Zadeh, L. A. From computing with numbers to computing with words: from manipulation of measurements to manipulation of perceptions, *IEEE Trans. Circ. Syst.* **45**, 1999, 105–119.

Chapter 3

Characterization of Fuzzy Sets

Fuzzy sets are fully characterized by their membership functions. Hence, properties of fuzzy sets originate directly from the properties of the membership functions. In particular, geometric features of fuzzy sets help visualize and underline similarities and differences between sets and fuzzy sets. This chapter covers major properties of membership functions, presents pertinent characterizations of fuzzy sets, and offers their various interpretations. Furthermore, we emphasize the role of semantics of fuzzy sets and discuss the position of fuzzy sets in granular computing. The characteristics of individual fuzzy sets and their families are expressed in terms of the fundamental concepts of specificity, energy measure of fuzziness, and granularity. Furthermore, we elaborate on some fundamental properties of families of fuzzy sets being regarded *en block* and used in this way in numerous constructs of fuzzy models.

3.1 A GENERIC CHARACTERIZATION OF FUZZY SETS: SOME FUNDAMENTAL DESCRIPTORS

In principle, any function $A : \mathbf{X} \to [0, 1]$ becomes potentially eligible to represent the membership function of fuzzy set A. In practice, however, the type and shape of membership functions should fully reflect the nature of the underlying phenomenon we are interested to model. We require that fuzzy sets should be semantically sound, which implies that the selection of membership functions needs to be guided by the character of the application and the nature of the problem we intend to solve.

Given the enormous diversity of potentially useful (viz. semantically sound) membership functions, there are certain common characteristics (descriptors) that are conceptually and operationally qualified to capture the essence of the granular

---

Fuzzy Systems Engineering: Toward Human-Centric Computing, by Witold Pedrycz and Fernando Gomide
Copyright © 2007 John Wiley & Sons, Inc.

46 Chapter 3 Characterization of Fuzzy Sets

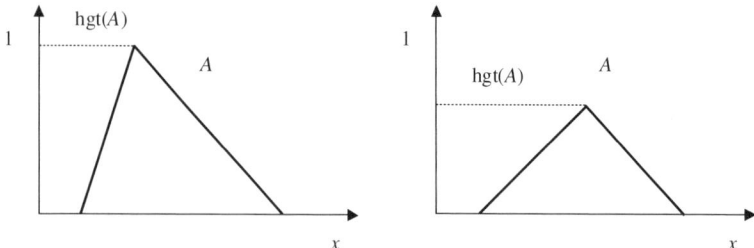

Figure 3.1 Examples of normal and subnormal fuzzy sets.

constructs represented in terms of fuzzy sets. In what follows, we provide a list of the descriptors commonly encountered in practice.

3.1.1 Normality

We say that the fuzzy set A is *normal* if its membership function attains 1, that is,

$$\sup_{x \in \mathbf{X}} A(x) = 1 \tag{3.1}$$

If this property does not hold, we call the fuzzy set *subnormal*. An illustration of the corresponding fuzzy set is shown in Figure 3.1. The supremum (sup) in the above expression is also referred to as the height of the fuzzy set A, $\mathrm{hgt}(A) = \sup_{x \in \mathbf{X}} A(x)$.

The normality of A has a simple interpretation: By determining the height of the fuzzy set, we identify an element with the highest membership degree. The value of the height being equal to one states that there is at least one element in $\mathbf{X}$ whose typicality with respect to A is the highest one and which could be sought as fully compatible with the semantic category presented by A. A subnormal fuzzy set whose height is lower than 1, that is, $\mathrm{hgt}(A) < 1$, means that the degree of typicality of elements in this fuzzy set is somewhat lower (weaker) and we cannot identify any element in $\mathbf{X}$ that is fully compatible with the underlying concept. Generally, while forming a fuzzy set we expect its normality (otherwise why would such a fuzzy set for which there are no typical elements come into existence in the first place?).

3.1.2 Normalization

The normalization operation, Norm(A), is a transformation mechanism that is used to convert a subnormal nonempty fuzzy set A into its normal counterpart. This is done by dividing the original membership function by the height of this fuzzy set, that is,

$$\mathrm{Norm}(A) = \frac{A(x)}{\mathrm{hgt}(A)} \tag{3.2}$$

Although the height describes the global property of the membership grades, the following notions offer an interesting characterization of the elements of $\mathbf{X}$ vis-à-vis their membership degrees.

3.1.3 Support

The support of a fuzzy set A, denoted by Supp(A), is a set of all elements of **X** with nonzero membership degrees in A

$$\text{Supp}(A) = \{x \in \mathbf{X} | A(x) > 0\} \tag{3.3}$$

In other words, support identifies all elements of **X** that exhibit some association with the fuzzy set under consideration (by being allocated to A with nonzero membership degrees).

3.1.4 Core

The core of a fuzzy set A, Core(A), is a set of all elements of the universe that are typical to A, that is, they come with membership grades equal to 1,

$$\text{Core}(A) = \{x \in \mathbf{X} | A(x) = 1\} \tag{3.4}$$

The support and core are related in the sense that they identify and collect elements belonging to the fuzzy set yet at two different levels of membership. Given the character of the core and support, we note that all elements of the core of A are subsumed by the elements of the support of this fuzzy set. Note that both support and core are sets and not fuzzy sets (Fig. 3.2). We refer to them as the set-based characterizations of fuzzy sets.

Although core and support are somewhat extreme (in the sense that they identify the elements of A that exhibit the strongest and the weakest linkages with A), we may also be interested in characterizing sets of elements that come with some intermediate membership degrees. A notion of a so-called α-cut offers here an interesting insight into the nature of fuzzy sets.

3.1.5 α-Cut

The α-cut of a fuzzy set A, denoted by A_α, is a set consisting of the elements of the universe whose membership values are equal to or exceed a certain threshold level α where $\alpha \in [0, 1]$. Formally speaking, we have $A_\alpha = \{x \in \mathbf{X} | A(x) \geq \alpha\}$. A strong

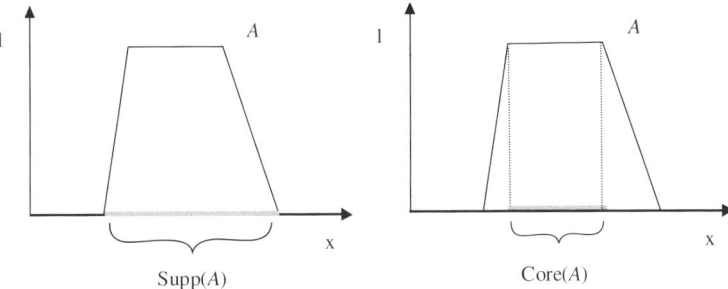

Figure 3.2 Support and core of A.

48 Chapter 3 Characterization of Fuzzy Sets

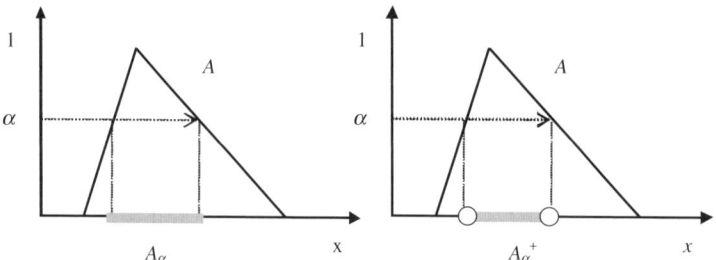

Figure 3.3 Examples of α-cut and strong α-cut.

α-cut differs from the α-cut in the sense that it identifies all elements in **X** for which we have the following equality: $A_\alpha^+ = \{x \in \mathbf{X} | A(x) > \alpha\}$. An illustration of the concept of the α-cut and strong α-cut is presented in Figure 3.3. Both support and core are limit cases of α-cuts and strong α-cuts. For $\alpha = 0$ and the strong α-cut, we arrive at the concept of the support of A. The threshold $\alpha = 1$ means that the corresponding α-cut is the core of A.

3.1.6 Convexity

We say that a fuzzy set is convex if its membership function satisfies the following condition:

for all $x_1, x_2 \in \mathbf{X}$ and all $\lambda \in [0,1]$:

$$A[\lambda x_1 + (1 - \lambda)x_2] \geq \min [A(x_1), A(x_2)] \tag{3.5}$$

The above relationship states that whenever we choose a point x on a line segment between x_1 and x_2, the point $(x, A(x))$ is always located above or on the line passing through the two points $(x_1, A(x_1))$ and $(x_2, A(x_2))$; refer to Figure 3.4. Note that the membership function is not a convex function in the traditional sense (Klir and Yuan, 1995).

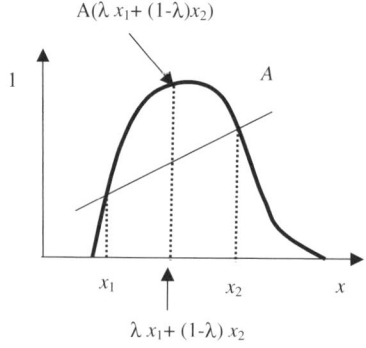

Figure 3.4 An example of a convex fuzzy set A.

3.1 A Generic Characterization of Fuzzy Sets: Some Fundamental Descriptors

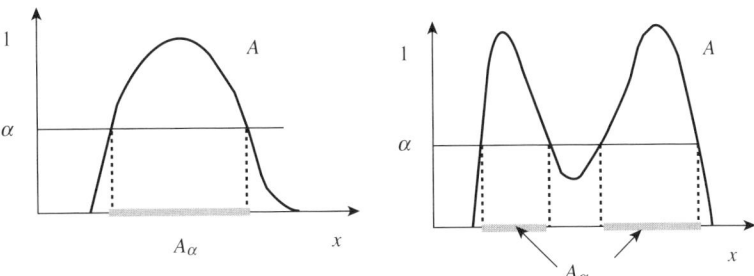

Figure 3.5 Examples of convex and nonconvex fuzzy sets.

Let us recall that a set S is convex if, for all $x_1, x_2 \in S$, then $x = \lambda x_1 + (1-\lambda)x_2 \in S$ for all $\lambda \in [0,1]$. In other words, convexity means that any line segment identified by any two points in S is also contained in S. For instance, intervals of real numbers are convex sets. Therefore, if a fuzzy set is convex, then all of its α-cuts are convex, and conversely, if a fuzzy set has all its α-cuts convex, then it is a convex fuzzy set; refer to Figure 3.5. Thus, we may say that a fuzzy set is convex if and only if all its α-cuts are convex (intervals).

Fuzzy sets can be characterized by counting their elements and bringing a single numeric quantity as a meaningful descriptor of this count. While in the case of sets, this sounds convincing, here we have to take into account different membership grades. In the simplest form, this counting comes under the name of cardinality.

3.1.7 Cardinality

Given a fuzzy set A defined in a finite or countable universe $\mathbf{X}$, its cardinality, denoted by $\text{Card}(A)$, is expressed as the following sum:

$$\text{Card}(A) = \sum_{x \in \mathbf{X}} A(x) \qquad (3.6)$$

or alternatively as the following integral:

$$\text{Card}(A) = \int_{\mathbf{X}} A(x) dx \qquad (3.7)$$

(we assume that the integral shown above does make sense). The cardinality produces a count of the number of elements in the given fuzzy set. As there are different degrees of membership, the use of the sum here makes sense as we keep adding contributions coming from the individual elements of this fuzzy set. Note that in the case of sets, we count the number of elements belonging to the corresponding sets. We also use the alternative notation of $\text{Card}(A) = |A|$ and refer to it as a sigma count (σ-count).

The cardinality of fuzzy sets is explicitly associated with the concept of granularity of information granules realized in this manner. More descriptively, the more the elements of A we encounter, the higher the level of abstraction supported by A and the lower the granularity of the construct. Higher values of cardinality come with the higher level of abstraction (generalization) and the lower values of granularity (specificity).

EXAMPLE 3.1

Consider fuzzy sets $A = [1.0, 0.6, 0.8, 0.1]$, $B = [0.1, 0.8, 1.0, 0.1]$, and $C = [0.6, 0.9, 1.0, 1.0]$ defined in the same space. We can order them in a linear fashion by computing their cardinalities. Here we obtain Card(A) = 2.5, Card(B) = 2.0, and Card(C) = 3.5. In terms of the levels of abstraction, C is the most general, A lies in-between, and B is the least general.

So far we have discussed properties of a single fuzzy set. The operations to be studied look into the characterizations of relationships between two fuzzy sets.

3.2 EQUALITY AND INCLUSION RELATIONSHIPS IN FUZZY SETS

We investigate two essential relationships between two fuzzy sets defined in the same space that offer a useful insight into their fundamental dependencies. When defining these notions, bear in mind that they build upon the well-known definitions encountered in set theory.

3.2.1 Equality

We say that two fuzzy sets A and B defined in the same universe **X** are equal if and only if their membership functions are identical, meaning that

$$A(x) = B(x), \quad \forall x \in \mathbf{X} \tag{8.8}$$

3.2.2 Inclusion

Fuzzy set A is a subset of B (A is included in B), denoted by $A \subseteq B$, if and only if every element of A also is an element of B. This property expressed in terms of membership degrees means that the following inequality is satisfied:

$$A(x) \leq B(x), \quad \forall x \in \mathbf{X} \tag{3.9}$$

An illustration of these two relationships in the case of sets is shown in Figure 3.6. In order to satisfy the relationship of inclusion, we require that the characteristic functions adhere to (3.9) for all elements of **X**. If the inclusion is not satisfied even for a single point of **X**, the inclusion property does not hold.

If A and B are fuzzy sets in **X**, we have adopted the same definition of inclusion as being available in set theory.

Interestingly, the definitions of equality and inclusion exhibit an obvious dichotomy as the property of equality (or inclusion) is satisfied or is not satisfied. Although this quantification could be acceptable in the case of sets, fuzzy sets require more

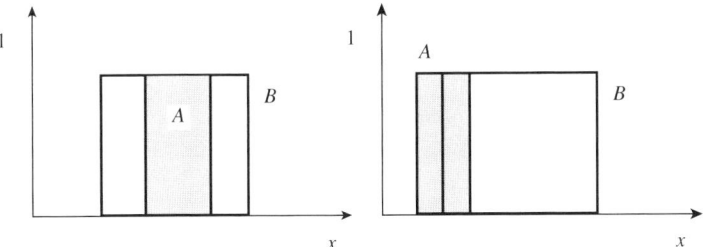

Figure 3.6 Set inclusion: (a) $A \subset B$ and (b) inclusion not satisfied as $A \not\subset B$.

attention in this regard given that the membership degrees are involved in expressing the corresponding definitions.

The approach being envisioned here takes into consideration the degrees of membership and sets up a conjecture that any comparison of membership values should rather return a degree of equality or inclusion. For a given element of finite universe **X**, let us introduce the following degree of inclusion of $A(x)$ in $B(x)$ and denote it by $A(x) \Rightarrow B(x)$ ($\Rightarrow$ is the symbol of implication; the operation of implication itself will be discussed in detail later on; we do not need these details for the time being):

$$A(x) \Rightarrow B(x) = \begin{cases} 1, \text{if} & A(x) \leq B(x) \\ 1 - A(x) + B(x) & \text{otherwise} \end{cases} \quad (3.10)$$

If $A(x)$ and $B(x)$ are confined to 0 and 1 as in the case of sets, we come up with the standard definition of Boolean inclusion being used in set theory. Computing (3.10) for all elements of **X**, we introduce a degree of inclusion of A in B, denoted by $\| A \subset B \|$, to be in the form

$$\| A \subset B \| = \frac{1}{\text{Card}(\mathbf{X})} \int_\mathbf{X} (A(x) \Rightarrow B(x)) dx \quad (3.11)$$

We characterize the equality of A and B, $\| A = B \|$, using the following expression:

$$\| A = B \| = \frac{1}{\text{Card}(\mathbf{X})} \int_\mathbf{X} [\min((A(x) \Rightarrow B(x)), (B(x) \Rightarrow A(x)))] dx \quad (3.12)$$

Again this definition is appealing as it results as a direct consequence of the inclusion relationships that have to be satisfied with respect to the inclusion of A in B and B in A.

EXAMPLE 3.2

Let us consider two fuzzy sets A and B described by the Gaussian and triangular membership functions. Recall that Gaussian membership function is described as $(\exp(-(x-m)^2/\sigma^2))$, where the modal value and spread are denoted by m and s, respectively. The triangular fuzzy set

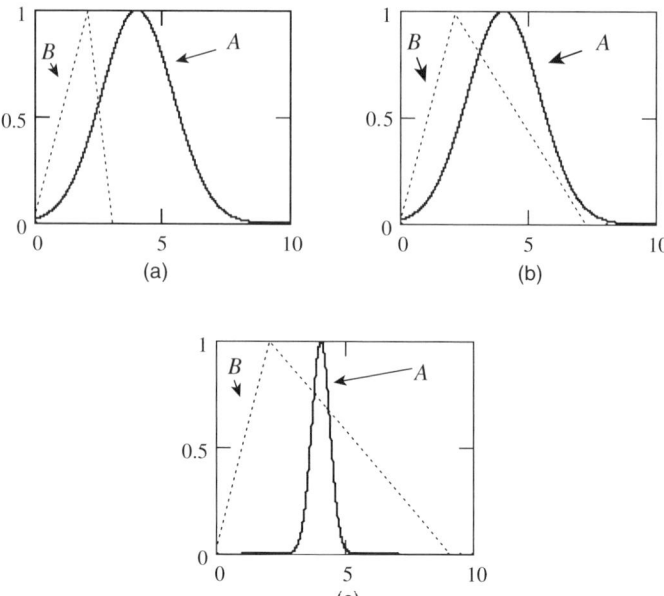

Figure 3.7 Examples of fuzzy sets A and B along with their degrees of inclusion: (a) $a = 0, n = 2$, $b = 3, m = 4, s = 2$, $\| A = B \| = 0.637$; (b) $b = 7$, $\| A = B \| = 0.864$; (c) $a = 0, n = 2, b = 9, m = 4$, $s = 0.5$, $\| A = B \| = 0.987$.

is fully characterized by the spreads (a and b) and the modal value equal to n. Figure 3.7 provides some examples of A and B for selected values of the parameters and the resulting degrees of inclusion. They are intuitively appealing reflecting the nature of relationship (A is included in B).

3.3 ENERGY AND ENTROPY MEASURES OF FUZZINESS

We can offer a global view at the collection of membership grades conveyed by fuzzy sets by aggregating them in the form of so-called measures of fuzziness. Two main categories of such measures are known in the form of energy and entropy measures of fuzziness (De Luca and Termini, 1972, 1974).

3.3.1 Energy Measure of Fuzziness

Energy measure of fuzziness of a fuzzy set A in **X**, denoted by $E(A)$, is a functional of the membership degrees

$$E(A) = \sum_{i=1}^{n} e[A(x_i)] \qquad (3.13)$$

3.3 Energy and Entropy Measures of Fuzziness

if $\text{Card}(\mathbf{X}) = n$. In the case of the infinite space, the energy measure of fuzziness is the following integral:

$$E(A) = \int_X e[A(x)]dx \qquad (3.14)$$

The mapping $e: [0, 1] \to [0, 1]$ is a functional monotonically increasing over $[0,1]$ with the boundary conditions $e(0) = 0$ and $e(1) = 1$.

As the name of this measure stipulates, its role is to quantify a sort of energy associated with the given fuzzy set. The higher the membership degrees, the more essential are their contributions to the overall energy measure. In other words, by computing the energy measure of fuzziness we can compare fuzzy sets in terms of their overall count of membership degrees.

A particular form of the above functional comes with the identity mapping, that is, $e(u) = u$ for all u in $[0,1]$. We can see that in this case, (3.13) and (3.14) reduce to the cardinality of A,

$$E(A) = \sum_{i=1}^{n} A(x_i) = \text{Card}(A) \qquad (3.15)$$

The energy measure of fuzziness forms a convenient way of expressing a total mass of the fuzzy set. Since $\text{Card}(\varnothing) = 0$ and $\text{Card}(\mathbf{X}) = n$, the more a fuzzy set differ from the empty set, the larger its mass. Indeed, rewriting (3.15) we obtain

$$E(A) = \sum_{i=1}^{n} A(x_i) = \sum_{i=1}^{n} |A(x_i) - \varnothing(x_i)| = d(A, \varnothing) = \text{Card}(A) \qquad (3.16)$$

where $d(A, \varnothing)$ is the Hamming distance between fuzzy set A and the empty set.

Although the identity mapping (e) is the simplest alternative one could think of, in general, we can envision an infinite number of possible options. For instance, one could consider the functionals such as $e(u) = u^p$, $p > 0$, and $e(u) = \sin(\frac{\pi}{2}u)$. Note that by choosing a certain form of the functional, we accentuate a varying contribution of different membership grades. For instance, depending upon the form of e, the contribution of the membership grades close to 1 could be emphasized whereas those located close to 0 could be very much reduced. Figure 3.8 illustrates this effect by showing two different forms of the functional (e).

When each element x_i of $\mathbf{X}$ appears with some probability p_i, the energy measure of fuzziness of the fuzzy set A can include this probabilistic information in which case it assumes the following format:

$$E(A) = \sum_{i=1}^{n} p_i e[A(x_i)] \qquad (3.17)$$

A careful inspection of the above expression reveals that $E(A)$ is the expected value of the functional $e(A)$. For infinite $\mathbf{X}$, we use an integral format of the energy measure of fuzziness

$$E(A) = \int_X p(x)e[A(x)]dx \qquad (3.18)$$

54 Chapter 3 Characterization of Fuzzy Sets

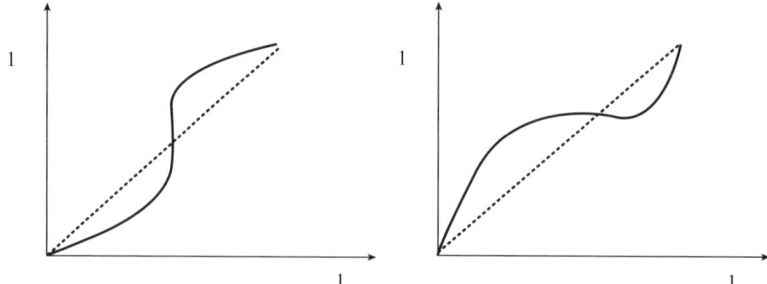

Figure 3.8 Two selected forms of the functional e. In the first case, high values of membership are emphasized (accentuated) (a), whereas the form of e shown in (b) puts emphasis on lower membership grades.

where $p(x)$ is the probability density function (pdf) defined over **X**. Again, $E(A)$ is the expected value of $e(A)$.

3.3.2 Entropy Measure of Fuzziness

The entropy measure of fuzziness of A, denoted by $H(A)$, is built upon the entropy functional h and comes in the form

$$H(A) = \sum_{i=1}^{n} h[A(x_i)] \qquad (3.19)$$

or in the continuous case of **X**

$$H(A) = \int_{\mathbf{X}} h(A(x)) dx \qquad (3.20)$$

where $h: [0, 1] \to [0, 1]$ is a functional such that (a) it is monotonically increasing in $[0, \frac{1}{2}]$ and monotonically decreasing in $[\frac{1}{2}, 1]$ and (b) it comes with the boundary conditions $h(0) = h(1) = 0$ and $h(\frac{1}{2}) = 1$. This functional emphasizes membership degrees around ½; in particular the value of ½ is stressed to be the most "unclear" (causing the highest level of hesitation with its quantification by means of the proposed functional).

3.4 SPECIFICITY OF FUZZY SETS

Quite often, we face the issue to quantify how much a single element of a universe could be regarded as a representative of a fuzzy set. If this fuzzy set is a singleton,

$$A(x) = \begin{cases} 1, & \text{if } x = x_0 \\ 0, & \text{if } x \neq x_0 \end{cases} \qquad (3.21)$$

3.4 Specificity of Fuzzy Sets

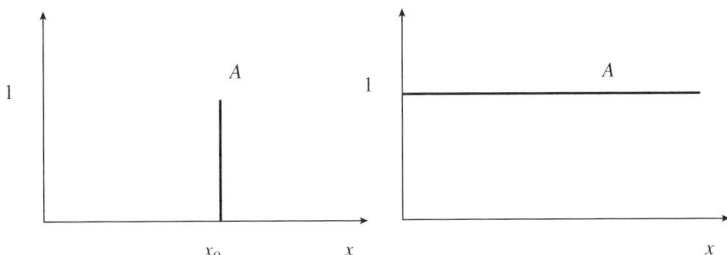

Figure 3.9 Examples of two extreme cases of sets exhibiting distinct levels of specificity.

then there is no hesitation in selecting x_0 as the sole representative of A. We say that A is very *specific* and its choice comes with no doubt. On the other extreme, if A covers the entire universe $\mathbf{X}$ and embraces all elements with the membership grade equal to 1, the choice of the only one representative of A comes with a great deal of hesitation that is triggered by a lack of specificity being faced in this problem. These two extreme situations are portrayed in Figure 3.9. Intuitively, we sense that the specificity is a concept that relates quite visibly to the cardinality of a set. The higher the cardinality of the set (viz. the more evident its abstraction), the lower its specificity. Having said that, we are interested in developing a measure that could be able to capture this effect of hesitation

One of the possible ways to quantify the notion of specificity of a fuzzy set was proposed by Yager (1983). The specificity of a fuzzy set A defined in $\mathbf{X}$, denoted by Spec(A), is a mapping from a family of normal fuzzy sets in $\mathbf{X}$ into nonnegative numbers such that the following conditions are satisfied:

1. Spec(A) $= 1$ if and only if there exists only one element x_0 of $\mathbf{X}$ for which $A(x_0) = 1$ and $A(x) = 0 \, \forall x \neq x_0$;
2. Spec(A) $= 0$ if and only if $A(x) = 0 \, \forall x \in \mathbf{X}$;
3. Spec(A_1) $\leq$ Spec(A_2) if $A_1 \supset A_2$.

Figure 3.10 illustrates the underlying concept of specificity.

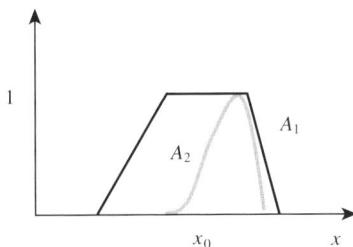

Figure 3.10 Expressing specificity of fuzzy sets: fuzzy set A_1 is less specific than A_2.

In particular, Yager (1983) introduced the specificity measure in the following form:

$$\text{Spec}(A) = \int_0^{\alpha_{\max}} \frac{1}{\text{Card}(A_\alpha)} d\alpha \qquad (3.22)$$

where $\alpha_{\max} = \text{hgt}(A)$. For finite universes, the integration is replaced by the sum

$$\text{Spec}(A) = \sum_{i=1}^{m} \frac{1}{\text{Card}(A_{\alpha_i})} \Delta\alpha_i \qquad (3.23)$$

where $\Delta\alpha_i = \alpha_i - \alpha_{i-1}$ with $\alpha_0 = 0$; m stands for the number of the membership grades of A.

3.5 GEOMETRIC INTERPRETATION OF SETS AND FUZZY SETS

In the case of finite universes of discourse **X**, we can arrive at an interesting and geometrically appealing interpretation of sets and fuzzy sets. Such an interpretation is also helpful in contrasting between sets and fuzzy sets, as it also visualizes interrelationships between them. For the n-element space **X**, any set can be represented as an n-dimensional vector **x** with the 0–1 values. The cardinality of the family of all sets defined in **X** is 2^n. The ith component of vector **x** is the value of the corresponding characteristic function of the ith element in the respective set. In the simplest case when $\mathbf{X} = \{x_1, x_2\}$, $n = 2$, the family of sets comprises the following elements, namely $\varnothing$, $\{x_1\}$, $\{x_2\}$, and $\{x_1, x_2\}$. The cardinality of **X** is $2^2 = 4$. Thus, each of the four elements of this family can be represented by a two-dimensional vector, say $\varnothing = [0,0]$, $\{x_1\} = [1,0]$, $\{x_2\} = [0,1]$, and $\{x_1, x_2\} = [1,1]$. Those sets are located at the corners of the unit square, as illustrated in Figure 3.11.

Owing to the values of the membership grades assuming any values in [0,1], fuzzy sets being two-dimensional vectors are distributed throughout the entire unit square. For instance, in Figure 3.11, fuzzy set A is represented as vector $\mathbf{x} = (0.25, 0.75)$. A family of fuzzy sets over $\mathbf{X} = \{x_1, x_2\}$ occupies the whole shaded area, including the borders and corners of the unit square. In general, proceeding with higher dimensionality of the space, we end up with a unit cube

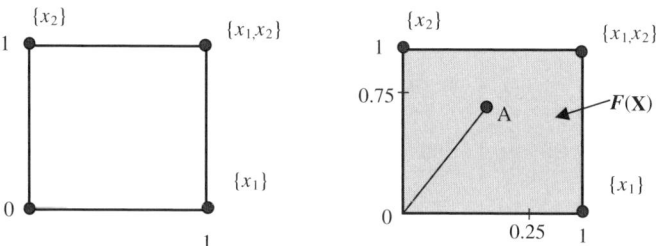

Figure 3.11 Sets and fuzzy sets represented as points in the unit square.

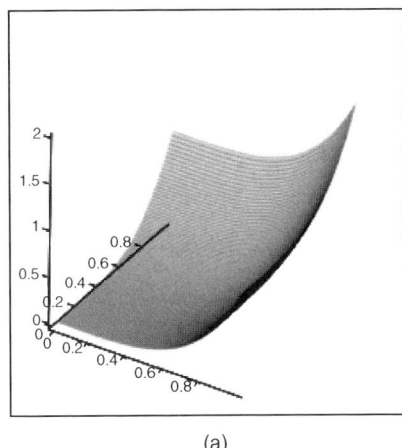 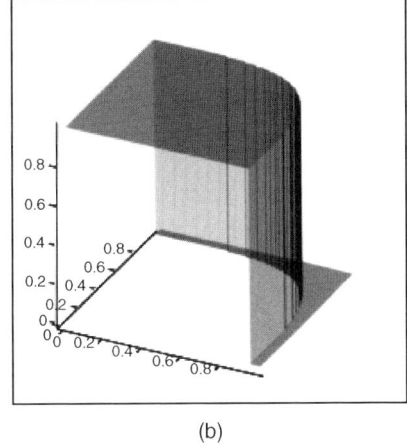

(a) (b)

Figure 3.12 Energy measure of fuzziness defined over the unit square of fuzzy sets (a) and a region of $[0,1]^2$—a collection of fuzzy sets satisfying the constraint $E(A) < 0.8$ (b).

($n = 3$) and unit hypercubes (for the dimensionality of the space of dimensionality higher than 3).

The geometric interpretation of fuzzy sets is interesting on its own by delivering a useful visualization vehicle. As we will show later, treating such points are the representatives of the corresponding fuzzy sets, we can easily show the distribution of results of operations on fuzzy sets (such as union, intersection, and complement). Associating with fuzzy sets, the measure of fuzziness, specificity, and alike, we can visualize how they behave depending upon the location of the point (fuzzy set) in the hypercube. For instance, the plot of the energy measure of fuzziness with the functional of the form $e(u) = u^4$ is shown in Figure 3.12. This plot could be used for different purposes. If we are required to identify all fuzzy sets whose energy measure of fuzziness does not exceed some threshold, say $\lambda, \lambda > 0$ the plot can help us locate all fuzzy sets satisfying this constraint; refer again to Figure 3.12.

3.6 GRANULATION OF INFORMATION

The notion of granulation emerges as a direct and immediate need to abstract and summarize information and data to support various processes of comprehension and decision-making. For instance, we often sample an environment for values of attributes of state variables, but we rarely process all details because of our physical and cognitive limitations. Quite often, just a reduced number of variables, attributes, and values are considered because those are the only features of interest given the task under consideration. To avoid all necessary and highly distractive details, we require an effective abstraction procedure. As discussed earlier, detailed numeric information is aggregated into a format of information granules where the granules themselves are regarded as collections of elements that are perceived as being indistinguishable, similar, close, or functionally equivalent.

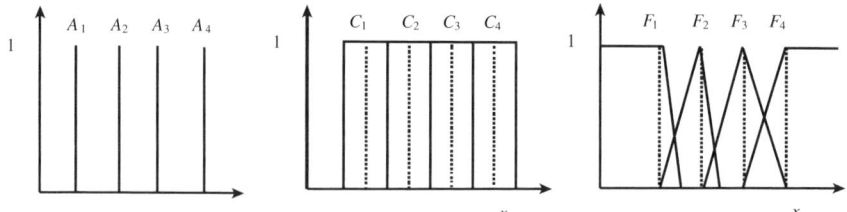

Figure 3.13 The concepts of discretization, quantization, and fuzzy granulation.

There are different formalisms and concepts of information granules. For instance, granules can be realized as sets (intervals), rough sets (Lin, 2004). Some typical examples of the granular constructs involve singletons and intervals. In these two special cases, we typically allude to the notion of discretization and quantization (Fig. 3.13). As the specificity of granules increases, intervals become singletons and in this limit case the quantization results in a discretization process.

Fuzzy sets are examples of information granules. When talking about a family of fuzzy sets, we are typically concerned with fuzzy partitions of **X**. Given the nature of fuzzy sets, fuzzy granulation generalizes the notion of quantization (Fig. 3.13) and emphasizes a gradual nature of transitions between neighboring information granules (Zadeh, 1999).

More generally, the mechanism of granulation can be formally characterized by a four-tuple of the form

$$\langle X, G, S, C \rangle \tag{3.24}$$

where **X** is a universe of discourse (space), G is a formal framework of granulation (resulting from the use of fuzzy sets, rough sets, etc.), S is a collection of information granules, and C is a transformation mechanism that realizes communication among granules of different nature and granularity levels (Pedrycz, 2005); see Figure 3.14. In Figure 3.14, notice the communication links that allow for communication between information granules expressed in the same formal framework but at different levels of granularity as well as communication links between information granules formed in different formal frameworks.

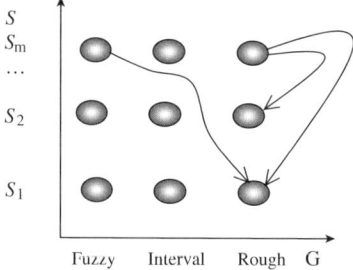

Figure 3.14 Granular Computing and communication mechanisms in the coordinates of formal frameworks (fuzzy sets, intervals, rough sets, etc.) and levels of granularity.

For instance, in the case of fuzzy granulation shown in Figure 3.14, if G is the formal framework of fuzzy sets, $S = \{S_1, S_2, S_3, S_m\}$, and C is a certain communication mechanism, then communicating the results of processing at the level of fuzzy sets to the framework of interval calculus, one could consider the use of some α-cuts. The pertinent computational details will be discussed later on.

3.7 CHARACTERIZATION OF THE FAMILIES OF FUZZY SETS

As we have already mentioned, when dealing with information granulation we often develop a family of fuzzy sets and move on with the processing that inherently uses all the elements of these families. Alluding to the existing terminology, we will be referring to such collections of information granules as frames of cognition. In what follows, we introduce the underlying concept and discuss its main properties.

3.7.1 Frame of Cognition

A frame of cognition is a result of information granulation in which we encounter a finite collection of fuzzy sets—information granules that "represent" the entire universe of discourse and satisfy a system of semantic constraints. The frame of cognition is a notion of particular interest in fuzzy modeling, fuzzy control, classification, and data analysis to name a few of the representative examples. In essence, the frame of cognition is crucial to all applications where local and globally meaningful granulation is required to capture the semantics of the conceptual and algorithmic settings in which problem solving has to be placed.

A frame of cognition consists of several labeled, normal fuzzy sets. Each of these fuzzy sets is treated as a reference for further processing. A frame of cognition can be viewed as a codebook of conceptual entities. Being more descriptive, we may view them as a family of linguistic landmarks, say *small*, *medium*, *high*, and so on. More formally, a frame of cognition Φ

$$\Phi = \{A_1, A_2, \ldots, A_m\} \tag{3.25}$$

is a collection of fuzzy sets defined in the same universe **X** that satisfies at least two requirements of coverage and semantic soundness.

3.7.2 Coverage

We say that Φ covers **X** if any element $x \in \mathbf{X}$ is compatible with at least one fuzzy set A_i in Φ, $i \in I = \{1, 2, \ldots, m\}$, meaning that it is compatible (coincides) with A_i to some nonzero degree, that is,

$$\forall_{x \in \mathbf{X}} \exists_{i \in I} A_i(x) > 0 \tag{3.26}$$

60 Chapter 3 Characterization of Fuzzy Sets

Being more strict, we may require a satisfaction of the so-called δ-level coverage, which means that for any element of **X**, fuzzy sets are activated to a degree not lower than δ

$$\forall_{x \in \mathbf{X}} \exists_{i \in I} A_i(x) > \delta \tag{3.27}$$

where $\delta \in [0, 1]$. Put it in a computational perspective: The coverage assures that each element of **X** is represented by at least one of the elements of Φ and guarantees any absence of gaps, namely, elements of **X** for which there is no fuzzy set being compatible with it.

3.7.3 Semantic Soundness

The concept of semantic soundness is more complicated and difficult to quantify. In principle, we are interested in information granules of Φ that are meaningful. Although there is far more flexibility in a way in which a suite of detailed requirements could be structured, we may agree upon a collection of several fundamental properties.

1. Each A_i, $i \in I$, is a unimodal and normal fuzzy set.
2. Fuzzy sets A_i, $i \in I$, are made disjoint enough to assure that they are sufficiently distinct to become linguistically meaningful. This imposes a maximum degree λ of overlap among any two elements of Φ. In other words, given any $x \in \mathbf{X}$, there is no more than one fuzzy set A_i such that $A_i(x) \geq \lambda$, $\lambda \in [0, 1]$.
3. The number of elements of Φ is kept low; following the psychological findings reported by Miller and others, we consider the number of fuzzy sets forming the frame of cognition to be maintained in the range of 7 ± 2 items.

Coverage and semantic soundness (Valente de Oliveira, 1993) are the two essential conditions that should be fulfilled by the membership functions of A_i to achieve interpretability. In particular, δ-coverage and λ-overlapping induce a minimal (δ) and maximal (λ) level of overlap between fuzzy sets (Fig. 3.15).

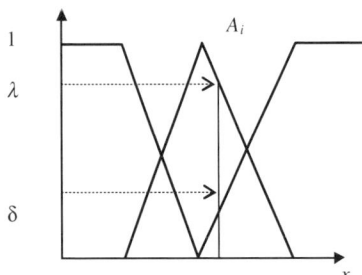

Figure 3.15 Coverage and semantic soundness of a cognitive frame.

3.7.4 Main Characteristics of the Frames of Cognition

Considering the families of linguistic labels and associated fuzzy sets embraced in a frame of cognition, several characteristics are worth emphasizing.

3.7.4.1 Specificity

We say that the frame of cognition Φ_1 is more specific than Φ_2 if all the elements of Φ_1 are more specific than the elements of Φ_2; for some illustration refer to Figure 3.16. Here the specificity of the fuzzy sets that compose the cognition frames can be evaluated using (3.22) or (3.23). The less specific cognition frames promotes granulation realized at the higher level of abstraction (generalization). Subsequently, we are provided with the description that captures less details.

3.7.4.2 Granularity

Granularity of a frame of cognition relates to the granularity of fuzzy sets used there. The higher the number of fuzzy sets in the frame, the finer the resulting granulation. Therefore, the frame of cognition Φ_1 is finer than Φ_2 if $|\Phi_1| > |\Phi_2|$. If the converse holds, Φ_1 is coarser than Φ_2 (Fig. 3.16).

3.7.4.3 Focus of Attention

A focus of attention (scope of perception) induced by a certain fuzzy set $A = A_i$ in Φ is defined as a certain α-cut of this fuzzy set. By moving A along $\mathbf{X}$ while keeping its membership function unchanged, we can focus attention on a certain selected region of $\mathbf{X}$, as portrayed in Figure 3.17.

3.7.4.4 Information Hiding

The idea of information hiding is closely related to the notion of focus of attention and manifests through a collection of elements that are hidden when viewed from the standpoint of membership functions. By modifying the membership function of $A = A_i$ in Φ, we can produce an equivalence of the elements positioned within

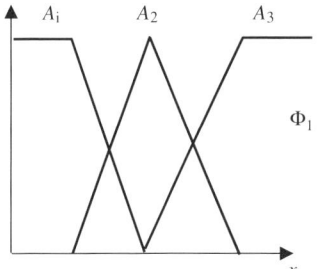 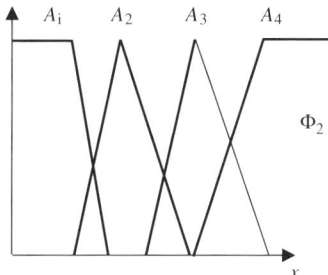

Figure 3.16 Examples of two frames of cognition; Φ_1 is coarser (more general) than Φ_2.

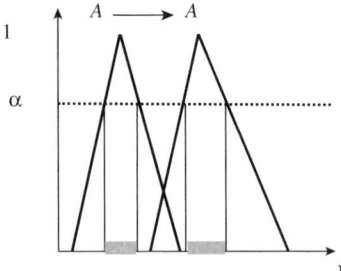

Figure 3.17 Focus of attention; two regions of focus of attention implied by the corresponding fuzzy sets are shown.

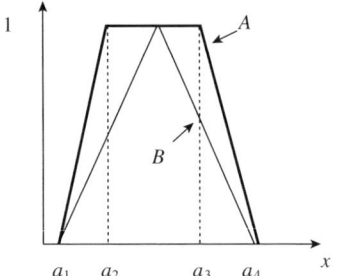

Figure 3.18 A concept of information hiding realized by the use of trapezoidal fuzzy set A: all elements in $[a_2, a_3]$ are made indistinguishable. The effect of information hiding is not present in the case of triangular fuzzy set B.

some region of **X**. For instance, consider a trapezoidal fuzzy set A on **R** and its 1-cut (viz. core), the closed interval $[a_2, a_3]$, as depicted in Figure 3.18.

All elements within the interval $[a_2, a_3]$ are made *indistinguishable*. Through the use of this specific fuzzy set they are made equivalent, namely, when expressed in terms of A. Hence, more detailed information, namely, a position of a certain point falling within this interval, is "hidden." In general, by increasing or decreasing the level of the α-cut we can accomplish a so-called α-information hiding through normalization.

3.8 FUZZY SETS, SETS AND THE REPRESENTATION THEOREM

Any fuzzy set can be viewed as a family of fuzzy sets. This is the essence of an identity principle known as the representation theorem. The representation theorem states that any fuzzy set A can be decomposed into a family of α-cuts,

$$A = \bigcup_{\alpha \in [0,1]} \alpha A_\alpha \qquad (3.28)$$

or, equivalently in terms of membership functions,

$$A(x) = \sup_{\alpha \in [0,1]} \alpha A_\alpha(x) \qquad (3.29)$$

Figure 3.19 illustrates the idea of the representation theorem.

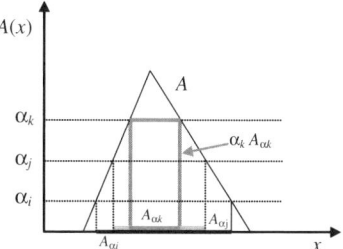

Figure 3.19 Illustration of the representation theorem.

Conversely, any fuzzy set can be reconstructed from a family of nested sets, assuming that they satisfy the consistency constraint, namely, if $\alpha_1 > \alpha_2$ then $A_{\alpha_1} \subset A_{\alpha_2}$.

The importance of the representation theorem lies in the its underscoring of the generalization nature introduced by fuzzy sets. Furthermore, the theorem implies that problems formulated in the framework of fuzzy sets (e.g. fuzzy optimization, decision making, information processing, data mining, etc.) can be solved by transforming these fuzzy sets into the corresponding families of nested α-cuts and determining solutions to each using standard, nonfuzzy techniques. Subsequently, all the partial results derived in this way can be merged, reconstructing a solution to the problem in its original formulation based on fuzzy sets. By increasing the number of α-levels of the membership values, that is, the α-cuts, the reconstruction can be made more detailed.

Example 3.3 Consider a the universe $\mathbf{X} = \{1, 2, 3, 4, 5\}$ and the fuzzy set A defined on $\mathbf{X}$, $A = \{0/1, 0.1/2, 0.3/3, 1/4, 0.3/5\}$ or, equivalently, $A = [0, 0.1, 0.3, 1, 0.3]$. Since the universe is finite, only the α-cuts for $\alpha_1 = 0.1, \alpha_2 = 0.1$, and $\alpha_3 = 1$ are of interest. They are as follows:

$A_{0.1} = \{0/1, 1/2, 1/3, 1/4, 1/5\} = [0, 1, 1, 1, 1] \rightarrow 0.1 A_{0.1} = [0, 0.1, 0.1, 0.1, 0.1]$
$A_{0.3} = \{0/1, 0/2, 1/3, 1/4, 1/5\} = [0, 0, 1, 1, 1] \rightarrow 0.3 A_{0.3} = [0, 0, 0.3, 0.3, 0.3]$
$A_1 = \{0/1, 0/2, 0/3, 1/4, 0/5\} = [0, 0, 0, 1, 0] \rightarrow 1.0 A_1 = [0, 0, 0, 1, 0]$

To recover the fuzzy set A from its α-cuts we first compute

$$0.1 A_{0.1} = [0, 0.1, 0.1, 0.1, 0.1]$$
$$0.3 A_{0.3} = [0, 0, 0.3, 0.3, 0.3]$$
$$1.0 A_1 = [0, 0, 0, 1, 0]$$

and find the membership function of A using (3.29) with sup replaced by max (because the universe is finite) and the max operation performed componentwise as follows

$A = \max\{0.1 A_{0.1}, 0.3 A_{0.3}, 1 A_1\}$
$= [\max(0, 0, 0), \max(0.1, 0, 0), \max(0.1, 0.3, 0), \max(0.1, 0.3, 1), \max(0.1, 0.3, 0)]$

Therefore $A = [0, 0.1, 0.3, 1, 0.3]$.

3.9 CONCLUSIONS

Fuzzy sets are fully described by their membership functions. Therefore, a way to view fuzzy sets and describing their key characteristics is to look at the attributes shown by their membership functions. We investigated the concepts of normality, convexity, cardinality, and specificity as the most generic descriptors of information granules. The descriptors such as energy and entropy measures of fuzziness are useful in discrete universes. In particular, fuzzy sets exhibit an interesting geometric interpretation when being treated as points of the unit hypercube.

For further processing and modeling activities, fuzzy sets rarely appear as single entities, but form collections of semantically meaningful entities usually referred to as frames of cognitions. We have investigated the fundamental properties of the frame of cognition and provided with their detailed quantification.

EXERCISES AND PROBLEMS

1. Consider the fuzzy set A with the following membership function:

$$A(x) = \begin{cases} x - 4/2, & \text{if } 4 \leq x \leq 5 \\ -x + 6/2, & \text{if } 5 < x \leq 6 \\ 0, & \text{otherwise} \end{cases}$$

   (a) Sketch the graph of the membership function
   (b) Is A normal? Does A have a core? What is the height of this fuzzy set?
   (c) Find the support of A. Is A a convex fuzzy set?

2. Assume a fuzzy set A whose membership functions are defined in the following form:

$$A(x) = \begin{cases} x - 4, & \text{if } 4 \leq x \leq 5 \\ 1, & \text{if } 5 < x \leq 6 \\ -x + 7, & \text{if } 6 < x \leq 7 \\ 0, & \text{otherwise} \end{cases}$$

   (a) Sketch the graph of the membership function
   (b) Find an analytic expression for its α-cuts
   (c) Is A a convex fuzzy set?

3. Demonstrate that if a fuzzy set is convex, then all its α-cuts are convex.

4. Consider the following fuzzy sets defined in the finite universe of discourse $\mathbf{X} = \{1, 2, 3, \ldots, 10\}$:

$$A = [0, 0, 0, 0, 0.4, 0.6, 0.8, 1, 0.8, 0.6]$$
$$B = [0, 0, 0, 0, 0.4, 0.5, 0.6, 1, 0.6, 0.4]$$
$$C = [0, 1, 0.2, 0.3, 0.4, 0.5, 0.6, 1, 0.5, 0]$$

(a) Is $A \subseteq B$? $B \subseteq A$?
(b) Is $C \subseteq A$? $C \subseteq B$?
(c) Quantify the findings obtained in (a) and (b).

5. Determine the cardinality of fuzzy sets A, B, and C discussed in problem 4.

6. Suppose that fuzzy sets A and B defined in $\mathbf{X} = \{x_1, x_2, x_3\}$ are represented as vectors whose components are the membership degrees of x_1, x_2, and x_3 in A and B. Plot A and B in the unit cube for each of the following cases:
   (a) $A = [1, 0, 0]$ and $B = [0, 1, 1]$
   (b) $A = [0, 1, 0]$ and $B = [1, 0, 1]$
   (c) $A = [0, 0, 1]$ and $B = [1, 1, 0]$
   (d) $A = [0.5, 0.5, 0.5]$ and $B = [0.5, 0.5, 0.5]$

7. Consider collections of information granules Γ_1 and Γ_2 of $\mathbf{X}$ as shown in the figures below. Do Γ_1 and Γ_2 qualify to be frames of cognition? Justify your statement. Which one is more specific? Comment on the granularity of the individual fuzzy sets of these families.

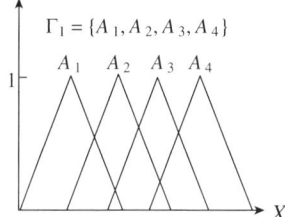
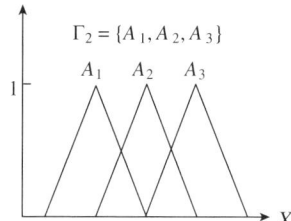

HISTORICAL NOTES

In the early 1970s, Zadeh (1971) suggested that the concept of a unit hypercube could be used as a model for all fuzzy sets. The idea was further developed by Kosko (1992) in which he referred to it as a set-as-points theory. Using the underlying geometry of the construct, Kosko argues that "*fuzziness is all about how much a thing and its opposite occur simultaneously; it is not how much the element belongs to the set. It is how much one set belongs to another set*" (McNeill and Freiberger, 1993).

The foundations of the theory of fuzzy information granulation were introduced by Zadeh (1973), whereas the concept of linguistic variables involving collections of semantically meaningful information granules in the form of fuzzy sets were studied in detail by Zadeh (1975, 1997).

The idea of granular computing (Lin, 2000; Bargiela and Pedrycz, 2003) encompasses various formal settings of information granules. We related to them by stressing that there is a genuine need to form suitable communication linkages that help complete processing in-between various platforms. This concerns the development of effective mechanisms of communication between fuzzy sets and interval analysis (in which case α-cuts play a predominant role) and fuzzy sets and rough sets (Pawlak, 1982).

REFERENCES

Bargiela, A., Pedrycz, W. *Granular Computing: An Introduction*, Kluwer Academic Publishers, Dordrecht, 2003.

De Luca, A., Termini, S. A definition of nonprobabilistic entropy in the setting of fuzzy sets, *Inf. Control* **20**, 1972, 301–312.

De Luca, A., Termini, S. Entropy of L-fuzzy sets, *Inf. Control* **24**, 1974, 55–73.

Klir, G., Yuan, B. *Fuzzy Sets and Fuzzy Logic: Theory and Applications*, Prentice-Hall, Upper Saddle River, NJ, 1995.

Kosko, B. *Neural Networks and Fuzzy Systems: A Dynamical Systems Approach to Machine Intelligence*, Prentice-Hall International, Englewood Cliffs, NJ, 1992.

Lin, T. Y. Data mining and machine oriented modeling: a granular computing approach, *J. Appl. Intell.*, **13**(2), 2000, 113–124.

Lin, T. Y. Granular computing: rough sets perspective, *IEEE Connections*, **2**(4), 2004, 10–13.

McNeill, D., Freiberger, P. *Fuzzy Logic: The Revolutionary Computer Technology that is Changing our World*, Touchstone Book, Simon and Shutter Publishers, New York, 1993.

Pawlak, Z. Rough sets, *Int. J. Inf. Comput. Sci.*, **11**(15), 1982, 341–356.

Pedrycz, W. From granular computing to computational intelligence and human-centric systems, *IEEE Connections* **3**(2), 2005, 6–11.

Valente de Oliveira, J. On optimal fuzzy systems with I/O interfaces, *Proceedings of the second IEEE International Conference on Fuzzy Systems*, San Francisco, USA, 1993, pp. 34–40.

Yager, R. Entropy and specificity in a mathematical theory of evidence, *Int. J. Gen. Syst.*, **9**, 1983, 249–260.

Zadeh, L. A. Toward a theory of fuzzy systems, in: R. E. Kalman and R. DeClaris (eds.), *Aspects of Networks and Systems*, Holt, Rinehart and Winston, New York, 1971.

Zadeh, L. A. Outline of a new approach to the analysis of complex system and decision process, *IEEE Trans. Syst. Man Cybern.*, **1**(1), 1973, 28–44.

Zadeh, L. A. The concept of linguistic variable and its application to approximate reasoning (part I), *Inf. Sci.*, **8**, 1975, 301–357.

Zadeh, L. A. Toward a theory of fuzzy information granulation and its centrality in human reasoning and fuzzy logic, *Fuzzy Set Syst.* **90**, 1997, 111–127.

Zadeh, L. A. Fuzzy logic = computing with words, in: L. A., Zadeh, J. Kacprzyk (eds.), *Computing with Words in Information and Intelligent Systems*, Physica-Verlag, Heidelberg, 1999, pp. 3–23.

Chapter 4

The Design of Fuzzy Sets

In this chapter, we focus on the development of fuzzy sets by presenting various ways of forming fuzzy sets and determining their membership functions. The subject of elicitation and interpretation of fuzzy sets (membership functions) is of paramount relevance from the conceptual, algorithmic, and application-oriented standpoints. There is a significant diversity of the methods that support the construction of membership functions. In general, one can clearly distinguish between user-driven and data-driven approaches with a number of techniques that share some features specific to both data- and user-driven techniques and hence are located somewhere in-between. The determination of membership functions has been a debatable issue for a long time almost since the very inception of fuzzy sets. In contrast to interval analysis and set theory where the estimation of bounds of the interval constructs has not attracted a great deal of attention and is seemed to be taken for granted, an estimation of membership degrees became essential and over time has led us to a suite of sound, well-justified, and algorithmically appealing estimation techniques.

4.1 SEMANTICS OF FUZZY SETS: SOME GENERAL OBSERVATIONS

Fuzzy sets are constructs that come with a well-defined meaning. They capture the semantics of the framework they intend to operate within. Fuzzy sets are the building conceptual blocks (generic constructs) that are used in problem description, modeling, control, and pattern classification tasks. Before discussing specific techniques of membership function estimation, it is worth casting the overall presentation in a certain context by emphasizing the aspect of the use of a finite number of fuzzy sets leading to some essential vocabulary reflective of the underlying domain knowledge. In particular, we are concerned with the related semantics, calibration capabilities of membership functions, and the locality of fuzzy sets.

Fuzzy Systems Engineering: Toward Human-Centric Computing, by Witold Pedrycz and Fernando Gomide
Copyright © 2007 John Wiley & Sons, Inc.

The limited capacity of a short term memory, as identified by Miller (1956) suggests that we could easily and comfortably handle and process 7 ± 2 items. This implies that the number of fuzzy sets to be considered as meaningful conceptual entities should be kept at the same level. The observation sounds reasonable—quite commonly in practice we witness situations in which this holds. For instance, when describing linguistically quantified variables, say error or change of error, we may use seven generic concepts (descriptors) labeling them as positive *large*, positive *medium*, positive *small*, *around* zero, and negative *small*, negative *medium*, and negative *large*. When characterizing speed, we may talk about its quite intuitive descriptors such as *low*, *medium*, and *high* speed. In the description of an approximation error, we may typically use the concept of a *small* error around a point of linearization (in all these examples, the terms are indicated in italics to emphasize the granular character of the constructs and the role being played there by fuzzy sets.) Although embracing very different tasks, all these descriptors exhibit a striking similarity. All of them are information granules, not numbers (whose descriptive power is very much limited). In modular software development when dealing with a collection of modules (procedures, functions, and alike), the list of their parameters is always limited to a few items, which is again a reflection of the limited capacity of the short term memory. The excessively long parameter list is strongly discouraged due to the possible programming errors and rapidly increasing difficulties of an effective comprehension of the software structure and ensuing flow of control.

In general, the use of an excessive number of terms does not offer any advantage. To the contrary, it remarkably clutters our description of the phenomenon and hampers further effective usage of such concepts we intend to establish to capture the essence of the domain knowledge. With the increase in the number of fuzzy sets, their semantics also becomes negatively impacted. Fuzzy sets may be built into a hierarchy of terms (descriptors), but at each level of this hierarchy (when moving down toward higher specificity that has an increasing level of detail), the number of fuzzy sets is kept relatively low.

Although fuzzy sets capture the semantics of the concepts, they may require some calibration depending upon the specification of the problem in hand. This flexibility of fuzzy sets should not be treated as any shortcoming, but rather viewed as a certain and fully exploited advantage. For instance, a term *low* temperature comes with a clear meaning yet it requires a certain calibration depending upon the environment and the context it was put into. The concept of *low* temperature is used in different climate zones and is of relevance to any communication between people, yet for each of the community the meaning of the term is different thereby requiring some calibration. This could be realized, for example, by shifting the membership function along the universe of discourse of temperature, affecting the universe of discourse by some translation, dilation, and alike. As a communication means, linguistic terms are fully legitimate and as such they appear in different settings. They require some refinement so that their meaning is fully understood and shared by the community of the users.

When discussing the methods aimed at the determination of membership functions or membership grades, (cf. Bortolan and Pedrycz, 2002; Civanlar and Trussell,

1986; Chen and Wang, 1999; Hong and Lee, 1996; Meson and Denoeux, 2006; Medaglia et al., 2002; Medasoui et al., 1998; Pedrycz, 1994;) it is worthwhile to underline the existence of the two main categories of approaches being reflective of the origin of the numeric values of membership. The first one captures the domain knowledge and opinions of experts. In the second one, we consider experimental data whose global characteristics become realized in the form and parameters of the membership functions. In the first group we can refer to the pairwise comparison (Saaty's approach, see Section 7) as one of the representative examples whereas fuzzy clustering is usually presented as a typical example of the data-driven method of membership function estimation. In what follows, we elaborate on several representative methods that will help us appreciate the level and flexibility of fuzzy sets.

4.2 FUZZY SET AS A DESCRIPTOR OF FEASIBLE SOLUTIONS

The aim of the method is to relate membership function to the level of feasibility of individual elements of a family of solutions associated with the problem in hand. Let us consider a certain function $f(x)$ defined in Ω, that is, $f: \Omega \to R$ where $\Omega \subset R$. Our intent is to determine its maximum, namely $x^{opt} = \arg\max_{x \in \Omega} f(x)$. On the basis of the values of $f(x)$, we can form a fuzzy set A describing a collection of feasible solutions that could be labeled as optimal. Being more specific, we use the fuzzy set to represent an extent (degree) to which some specific values of x could be sought as potential (optimal) solutions to the problem. Taking this into consideration, we relate the membership function of A with the corresponding value of $f(x)$ cast in the context of the boundary values assumed by f. For instance, the membership function of A could be expressed in the following form:

$$A(x) = \frac{f(x) - f_{min}}{f_{max} - f_{min}} \tag{4.1}$$

The boundary conditions are straightforward: $f_{min} = \min_{x \in \Omega} f(x)$ and $f_{max} = \max_{x \in \Omega} f(x)$, where the minimum and the maximum are computed over Ω. For other values of x where f attains its maximal value, $A(x)$ is equal to 1 and around this point, whereas the membership values are reduced when x is likely to be a solution to the problem $f(x) < f_{max}$. The form of the membership function depends upon the character of the function under consideration. The following examples illustrate the essence of the construction of membership functions.

EXAMPLE 4.1

Let us consider the problem of determining a maximum of the function $2\sin(0.5x)$ defined in $[\Omega = 0, 2\pi]$. The minimum and maximum of f in the range of the arguments between 0 and 2π is equal to 0 and 2, respectively; see also Figure 4.1. The maximal value of f is reached at $x* = \pi$. The membership function of the solution to the optimization problem is computed using (4.1) to be $A(x) = \sin(0.5x)$, Figure 4.1.

Linearization, its quality, and description of quality falls under the same banner as the optimization problem. When linearizing a function around some given point, a quality of such

70 Chapter 4 The Design of Fuzzy Sets

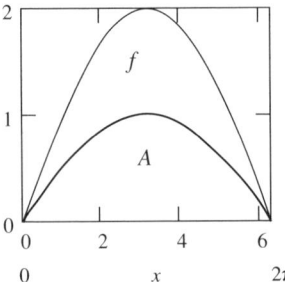

Figure 4.1 Function f and the induced membership function A.

linearization can be represented in a form of some fuzzy set. Its membership function attains one for all these points where the linearization error is equal to zero (in particular, this holds at the point around which the linearization is carried out). The following example illustrates this idea.

EXAMPLE 4.2

We are interested in the linearization of the function $y = g(x) = \exp(-x)$ around $x_0 = 1$ and assessing the quality of this linearization in the range $[-1, 7]$. The linearization formula reads as $y - y_0 = g'(x_0)(x - x_0)$ where $y_0 = g(x_0)$ and $g'(x_0)$ is the derivative of $g(x)$ at x_0. Given the form of the function, $g'(x) = -\exp(-x)$, the linearized version of the function reads as $\exp(-1)(2 - x)$. Next let us define the quality of this linearization by taking the absolute value of the difference between the original function and its linearization, $f(x) = |g(x) - \exp(-1)(2 - x)|$. As the fuzzy set A describes the quality of linearization, its membership function has to take into consideration the expression

$$A(x) = 1 - \frac{f(x) - f_{\min}}{f_{\max} - f_{\min}} \quad (4.2)$$

where $f_{\max} = f(7) = 1.84$ and $f_{\min} = 0.0$. When at some z, $f(z) = f_{\min}$, this means that $A(z) = 1$, which in the sequel indicates that the linearization at this point is perfect; no linearization error has been generated. The plot of the fuzzy set A is shown in Figure 4.2. We note that the higher quality of approximation is achieved for the arguments higher that the point at which the linearization has been completed.

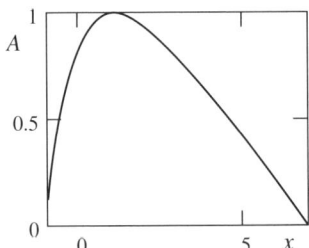

Figure 4.2 Fuzzy set A representing the quality of linearization of the function $\exp(-x)$ around the point $x_0 = 1$.

4.3 FUZZY SET AS A DESCRIPTOR OF THE NOTION OF TYPICALITY

Fuzzy sets address an issue of gradual *typicality* of elements to a given concept. They stress the fact that there are elements that fully satisfy the concept (are typical for it) and there are various elements that are allowed only with partial membership degrees. The form of the membership function is reflective of the semantics of the concept. Its details could be captured by adjusting the parameters of the membership function or choosing its form depending upon experimental data. For instance, consider a fuzzy set of squares. Formally, a rectangle includes a square shape as its special example when the sides are equal, $a = b$, Figure 4.3. What if $a = b + \varepsilon$, where ε is a very small positive number? Could this figure be sought as a square? It is very likely so. Perhaps the membership value of the corresponding membership function could be equal to 0.99. Our perception, which comes with some level of tolerance to imprecision, does not allow us to tell apart this figure from the ideal square, Figure 4.3.

Higher differences between a and b could result in lower values of the membership function. The definition of the fuzzy set square could be formed in a number of ways. Prior to the definition or even visualization of the membership function, it is important to formulate a space over which it will be defined. There are several intuitive alternatives worth considering:

(a) For each pair of values of the sides (a and b), collect an experimental assessment of membership of the rectangle to the category of squares. Here the membership function is defined over a Cartesian space of the spaces of lengths of sides of the rectangle. While selecting a form of the membership, we require that it assumes values at $a = b$ and is gradually reduced when the arguments start getting more different.

(b) We can define an absolute distance between a and b, $|a - b|$ and form a fuzzy set over this space $\mathbf{X}$; $\mathbf{X} = \{x | x = |a - b|\}$, $\mathbf{X} \subset \mathbf{R}^+$. The semantic constraints translate into the condition of $A(0) = 1$. For higher values of x we may consider monotonically decreasing values of A.

(c) We can envision ratios of a and b $x = a/b$ and construct a fuzzy set over the space of $\mathbf{R}^+$ such that $\mathbf{X} = \{x | x = a/b\}$. Here we require that $A(1) = 1$. We also anticipate lower values of membership grades when moving to the left

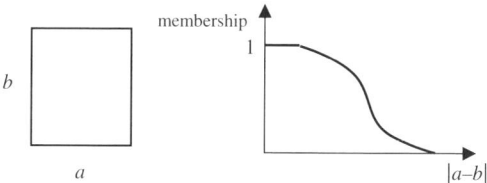

Figure 4.3 Perception of geometry of squares and its quantification in the form of membership function of the concept of fuzzy square.

and to the right from $x = 1$. Note that the membership function could be asymmetric, so we allow for different membership values for the same length of the sides, say $a = 6$ and $b = 5$, and $a = 6$ and $b = 5$ (the effect could be quite apparent due to the visual effects when perceiving geometric phenomena). The previous model of **X** as outlined in (b) cannot capture this effect.

Once the form of the membership function has been defined, it could be further adjusted by modifying the values of its parameters on the basis of some experimental findings. They come in the form of ordered triples or pairs, say (a, b, μ), $(a/b, \mu)$ or $(|a - b|, \mu)$ depending on the previously accepted definition of the universe of discourse. The membership values μ are those available from the expert offering an assessment of the likeness of the corresponding geometric figure.

4.4 MEMBERSHIP FUNCTIONS IN THE VISUALIZATION OF PREFERENCES OF SOLUTIONS

A simple electric circuit shown in Figure 4.4 helps us illustrate the underlying idea. Consider the problem of optimization of power maximization on the external resistance in the circuit.

The voltage source E is characterized by some internal resistance equal to r. The external resistance R is the one on which we want to maximize power dissipation. By straightforward calculations we compute the power dissipated on R to be given in the form

$$P = i^2 R = \left(\frac{E}{R + r}\right)^2 R \qquad (4.3)$$

The maximization of P with respect to R is determined by zeroing the derivative of P, $\frac{\partial P}{\partial R} = 0$, which leads to the optimal value of the resistance R^{opt}. Through simple derivations we obtain $R^{opt} = r$. It becomes evident that while the condition $R = r$ produces the maximum of P, this solution is not technically feasible as there is a substantial level of power dissipation on the internal resistance. If we plot the relationship of P versus R (Fig. 4.5), and treat it as a membership function of R (which requires a simple normalization of P by dividing it by the maximal value obtained for $R = r$), we note that the shape of this relationship is highly asymmetric:

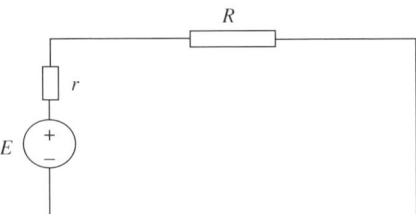

Figure 4.4 A simple electric circuit and the problem of maximization of power dissipation on the external resistance R.

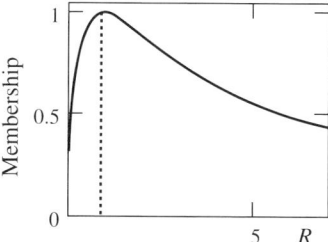

Figure 4.5 Membership function of the optimal power dissipation on external resistance R; the maximal value is achieved for $R = 1$ (the internal resistance r is equal to 1).

When increasing the value of resistance over the optimal value (R^{opt}), the membership function changes quite smoothly and the reduction of the membership grades is quite limited.

On the contrary, when moving toward lower values of R such that $R < R^{opt}$, the reduction in the membership grades is quite substantial. We can say that the membership function of the optimal resistance offers a highly visible and very much intuitive quantification of the notion of optimality. The asymmetric shape of the resulting fuzzy set delivers some guidance in the selection of possible suboptimal solution while the membership degree serves as an indicator of the suitability (degree of optimality) of the individual value of R.

4.5 NONLINEAR TRANSFORMATION OF FUZZY SETS

In many problems, we encounter a family of fuzzy sets defined in the same space. The family of fuzzy sets $\{A_1, A_2, \ldots, A_c\}$ is referred to as referential fuzzy sets. To form a family of semantically meaningful descriptors of the variable in hand, we usually require that these fuzzy sets satisfy the requirements of unimodality, limited overlap, and coverage. Technically, all of these features are reflective of our intention to provide this family of fuzzy sets with some semantics. These fuzzy sets could be sought as generic descriptors (say, *small*, *medium*, *high*, etc.) being described by some typical membership functions. For instance, those could be uniformly distributed triangular or Gaussian fuzzy sets with some standard level of overlap between the successive terms (descriptors).

As mentioned, fuzzy sets are usually subject to some calibration depending upon the character of the problem in hand. We may use the same terms of *small*, *medium*, and *large* in various contexts yet their detailed meaning (viz. membership degrees) has to be adjusted. For the given family of the referential fuzzy sets, their calibration could be accomplished by taking the space $\mathbf{X} = [a, b]$ over which they are originally defined and transforming it into itself, that is, $[a, b]$ through some nondecreasing monotonic and continuous function $\Phi(x, \mathbf{p})$ where $\mathbf{p}$ is a vector of some adjustable parameters bringing the required flexibility of the mapping. The nonlinearity of the mapping is such that some regions of $\mathbf{X}$ are contracted and some of them are stretched (expanded) and in this manner capture the required local context of the problem. This affects the membership functions of the referential fuzzy sets $\{A_1, A_2, \ldots, A_c\}$ whose membership functions are expressed now as $A_i\phi(x,p)$. The construction of the mapping Φ is optimized taking into account some experimental data concerning membership grades given at some points of $\mathbf{X}$. More specifically, the experimental data come in the form of the input–output pairs:

$$\begin{aligned}
x_1 &— \mu(1), \mu_2(1), \ldots, \mu_c(1) \\
x_2 &— \mu_1(2), \mu_2(2), \ldots, \mu_c(2) \\
&\ldots \\
x_N &— \mu_1(N), \mu_2(N), \ldots, \mu_c(N)
\end{aligned} \qquad (4.4)$$

where the kth input–output pair consists of x_k that denotes some point in $\mathbf{X}$ whereas $\mu_1(k), \mu_2(k), \ldots, \mu_c(k)$ are the numeric values of the corresponding membership degrees. The objective is to construct a nonlinear mapping, that is, optimizing it with respect to the available parameters $\mathbf{p}$. More formally, we could translate the problem into the minimization of the following sum of squared errors:

$$\sum_{i=1}^{c}(A_i(\Phi(x_1,\mathbf{p})) - \mu_i(1))^2 + \sum_{i=1}^{c}(A_i(\Phi(X_2,\mathbf{p})) - \mu_i(2))^2 + \cdots$$

$$+ \sum_{i=1}^{c}(A_i(\Phi(X_N,\mathbf{p})) - \mu_i(N))^2 \qquad (4.5)$$

One of the feasible mapping comes in the form of a piecewise linear function shown in Figure 4.6. Here the vector of the adjustable parameters $\mathbf{p}$ involves a collection of the split points $r_1, r_2, \ldots, r_L$ and the associated differences $D_1, D_2, \ldots, D_L$; hence $\mathbf{p} = [r_1, r_2, \ldots, r_L, D_1, D_2, \ldots, D_L]$. The regions of expansion or compression are used to affect the referential membership functions and adjust their values given the experimental data.

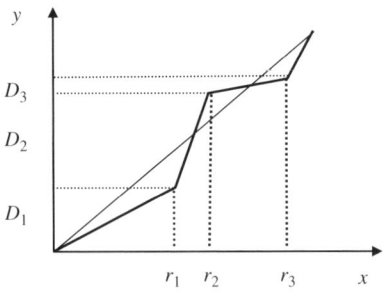

Figure 4.6 A piecewise linear transformation function Φ; a linear mapping not affecting the universe of discourse and not exhibiting any impact on the referential fuzzy sets is also shown. The proposed piecewise linear mapping is fully invertible.

4.5 Nonlinear Transformation of Fuzzy Sets

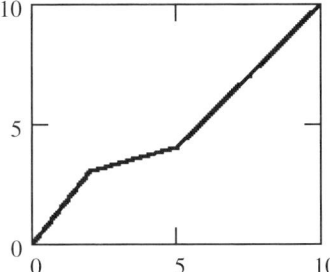

Figure 4.7 An example of the piecewise linear transformation.

EXAMPLE 4.3

We consider some examples of nonlinear transformations of Gaussian fuzzy sets through the piecewise linear transformations (here $L = 3$) shown in Figure 4.7.

Note a fact that some fuzzy sets become more specific while the others are made more general and expanded over some regions of the universe of discourse. This transformation leads to the membership functions illustrated in Figure 4.8.

Considering the same nonlinear mapping as before, two triangular fuzzy sets are converted into fuzzy sets described by piecewise membership functions as shown in Figure 4.9.

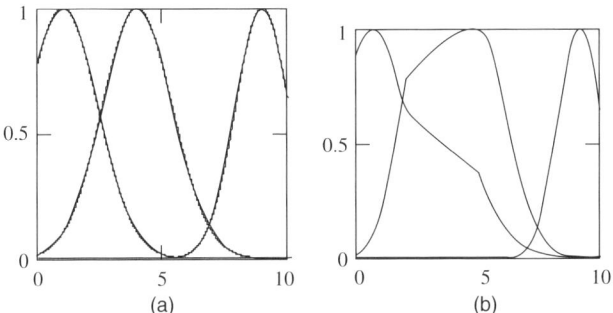

Figure 4.8 Examples of original membership functions (a) and the resulting fuzzy sets (b) after the piecewise linear transformation.

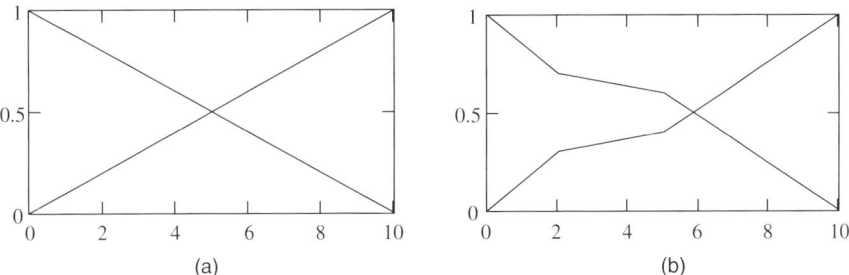

Figure 4.9 Two triangular fuzzy sets along with their piecewise linear transformation.

76 Chapter 4 The Design of Fuzzy Sets

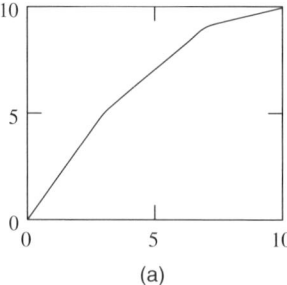 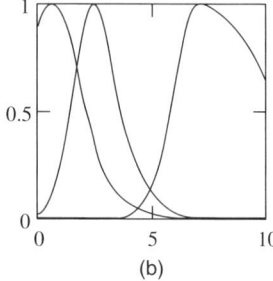

Figure 4.10 The piecewise linear mapping (a) and the transformed Gaussian fuzzy sets (b).

Some other examples of the transformation of fuzzy sets through the piecewise mapping are included in Figure 4.10.

4.6 VERTICAL AND HORIZONTAL SCHEMES OF MEMBERSHIP ESTIMATION

The vertical and horizontal modes of membership estimation are two standard approaches used in the determination of fuzzy sets. They reflect distinct ways of looking at fuzzy sets whose membership functions at some finite number of points are quantified by experts. In the horizontal approach we identify a collection of elements in the universe of discourse **X** and request that an expert answers the question

Does x belong to concept A?

The answers are expected to come in a binary (yes–no) format. The concept A defined in **X** could be any linguistic notion, say *high* speed, *low* temperature, and so on. Given n experts whose answers for a given point of **X** form a mix of yes–no replies, we count the number of "yes" answers and compute the ratio of the positive answers (p) versus the total number of replies (n), that is, p/n. This ratio (likelihood) is treated as a membership degree of the concept at the given point of the universe of discourse. When all experts accept that the element belongs to the concept, then its membership degree is equal to 1. Higher disagreement between the experts (quite divided opinions) results in lower membership degrees. The concept A defined in **X** requires collecting results for some other elements of **X** and determining the corresponding ratios as outlined in Figure 4.11.

If replies follow some, for example, binomial distribution, then we could determine a confidence interval of the individual membership grade. The standard deviation of the estimate of the positive answers associated with the point x, denoted here by σ, is given in the form

$$\sigma = \sqrt{\frac{p(1-p)}{n}} \qquad (4.6)$$

4.6 Vertical and Horizontal Schemes of Membership Estimation

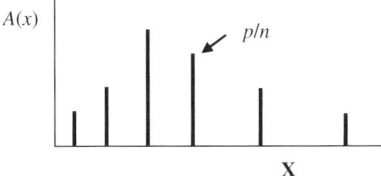

Figure 4.11 A horizontal method of the estimation of the membership function. Observe a series of estimates determined for selected elements of **X**. Note also that the elements of **X** need not be evenly distributed.

The associated confidence interval that describes a range of membership values is then determined as

$$[p - \sigma, p + \sigma] \tag{4.7}$$

In essence, when the confidence intervals are taken into consideration, the membership estimates become intervals of possible membership values, and this leads to the concept of so-called interval-valued fuzzy sets. By assessing the width of the estimates, we could control the execution of the experiment: when the ranges are too long, one could re-design the experiment and monitor closely the consistency of the responses collected in the experiment.

EXAMPLE 4.4

Let us consider responses of 10 experts who came up with the following assessment of the concept *high* interest rate (%) with the number of "yes" responses collected as follows:

x (%)	2	3	5	8	10
no. of "yes" replies	0	2	4	7	10

Following these responses, the membership function and its confidence values σ producing confidence intervals are given below.

x (%)	2	3	5	8	10
A(x) (*high* interest rate)	0.0	0.2	0.4	0.7	1.0
σ	0.0	0.126	0.155	0.144	0.0

The advantage of the method comes with its simplicity as the technique relies explicitly upon a direct counting of responses. The concept is also intuitively appealing. The probabilistic nature of the replies helps build confidence intervals that are essential to the assessment of the specificity of the membership quantification. A certain drawback is related with to the local character of the construct: As the estimates of the membership function are completed

78 Chapter 4 The Design of Fuzzy Sets

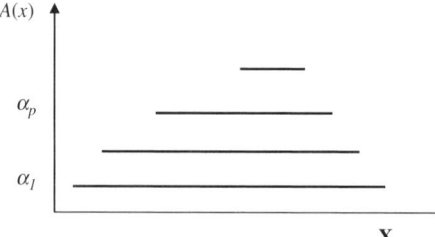

Figure 4.12 A vertical approach of membership estimation through the reconstruction of a fuzzy set through its estimated α-cuts.

separately for each element of the universe of discourse, they could exhibit a lack of continuity when moving from certain point to its neighbor. This concern is particularly valid in the case where **X** is a subset of real numbers.

The vertical mode of membership estimation is concerned with the estimation of the membership function by focusing on the determination of the successive α-cuts. The experiment focuses on the unit interval of membership grades. The experts involved in the experiment are asked the questions of the form

> what are the elements of **X** which belong to fuzzy set A at degree not lower than α?

where α is a certain level (threshold) of membership grades in [0,1]. The essence of the method is illustrated in Figure 4.12. Note that the satisfaction of the inclusion constraint is obvious: we envision that for higher values of α, the expert is going to provide more limited subsets of **X**; the vertical approach leads to the fuzzy set by combining the estimates of the corresponding α-cuts. Given the nature of this method, we are referring to the collection of random sets as these estimates appear in the successive stages of the estimation process.

The elements are identified by the expert as they form the corresponding α-cuts of A. By repeating the process for several selected values of α, we end up with the α-cuts and using them we reconstruct the fuzzy set. The simplicity of the method is its genuine advantage.

Like in the horizontal method of membership estimation, a possible lack of continuity is a certain disadvantage one has to be aware of. Here the selection of suitable levels of α needs to be carefully investigated. Similarly, an order at which different levels of α are used in the experiment could impact the estimate of the membership function.

4.7 SAATY'S PRIORITY METHOD OF PAIRWISE MEMBERSHIP FUNCTION ESTIMATION

The priority method introduced by Saaty (1980,1986) forms another interesting alternative used to estimate the membership function. There are several interesting extensions of this method (Buckley et al., 2001; Kulak and Kahraman, 2005; van Laarhoven and Pedrycz, 1983; Mikhailov and Tesvctinov, 2004; Pendharkar, 2003). To explain the essence of the method, let us consider a collection of elements $x_1, x_2, \ldots, x_n$ (those could be, for instance, some alternatives whose allocation to a certain fuzzy set is sought) for which membership grades $A(x_1), A(x_2), \ldots, A(x_n)$

4.7 Saaty's Priority Method of Pairwise Membership Function Estimation

are given. Let us organize them into a so-called reciprocal matrix of the following form:

$$R = [r_{ij}] = \begin{bmatrix} \frac{A(x_1)}{A(x_1)} & \frac{A(x_1)}{A(x_2)} & \cdots & \frac{A(x_1)}{A(x_n)} \\ \frac{A(x_2)}{A(x_1)} & \frac{A(x_2)}{A(x_2)} & \cdots & \frac{A(x_2)}{A(x_n)} \\ \cdots & \cdots & & \\ \frac{A(x_n)}{A(x_1)} & \frac{A(x_n)}{A(x_2)} & \cdots & \frac{A(x_n)}{A(x_n)} \end{bmatrix} = \begin{bmatrix} 1 & \frac{A(x_1)}{A(x_2)} & \cdots & \frac{A(x_1)}{A(x_n)} \\ \frac{A(x_2)}{A(x_1)} & 1 & \cdots & \frac{A(x_2)}{A(x_n)} \\ \cdots & & & \\ \frac{A(x_n)}{A(x_1)} & \frac{A(x_m)}{A(x_2)} & \cdots & 1 \end{bmatrix} \quad (4.8)$$

Noticeably, the diagonal values of R are equal to 1. The entries that are symmetrically positioned with respect to the diagonal satisfy the condition of reciprocity, that is, $r_{ij} = 1/r_{ji}$. Furthermore an important transitivity property holds, that is, $r_{ik} r_{kj} = r_{ij}$, for all indexes i, j, and k. This property holds because of the way in which the matrix has been constructed. By plugging in the ratios one gets $r_{ik} r_{kn} = \frac{A(x_i)}{A(x_k)} \frac{A(x_k)}{A(x_j)} = \frac{A(x_i)}{A(x_j)} = r_{ij}$. Let us now multiply the matrix by the vector of the membership grades $A = [A(x_1) A(x_2) \ldots A(x_n)]^T$. For the ith row of R (which is the ith entry of the resulting vector of results) we obtain

$$[RA]_i = \begin{bmatrix} \frac{A(x_i)}{A(x_1)} & \frac{A(x_i)}{A(x_2)} & \cdots & \frac{A(x_i)}{A(x_n)} \end{bmatrix} \begin{bmatrix} A(x_1) \\ A(x_2) \\ \cdots \\ A(x_n) \end{bmatrix} \quad (4.9)$$

where $i = 1, 2, \ldots, n$. Thus the ith element of the vector is equal to $nA(x_i)$. Overall once completing the calculations for all i, this leads us to the expression $RA = nA$. In other words, we conclude that A is the eigenvector of R associated with the largest eigenvalue of R that is equal to n. In the above scenario, we have assumed that the membership values $A(x_i)$ are given and then showed what form of results could they lead to. In practice the membership grades are not given and have to be looked for.

The starting point of the estimation process are entries of the reciprocal matrix that are obtained through collecting results of pairwise evaluations offered by an expert, designer, or user (depending on the character of the task in hand). Prior to making any assessment, the expert is provided with a finite scale with values spread in between 1 and 7. Some other alternatives of the scales such as those involving 5 or 9 levels could be sought as well. If x_i is strongly preferred to x_j when being considered in the context of the fuzzy set whose membership function we would like to estimate, then this judgment is expressed by assigning high values of the available scale, say 6 or 7. If we still sense that x_i is preferred to x_j yet the strength of this preference is lower in comparison with the previous case, then this is quantified using some intermediate values of the scale, say 3 or 4. If no difference is sensed, the values close to 1 are the preferred choice, say 2 or 1. The value of 1 indicates that x_i and x_j are equally preferred. On the contrary, if x_j is preferred to x_i, the corresponding entry assumes values below one. Given the reciprocal character of the assessment, once the preference of x_i to x_j has been quantified, the inverse of this number is plugged into the entry of the matrix that is located at the (j,i)th coordinate. As indicated earlier, the elements on the main diagonal are equal to 1. Next the maximal eigenvalue is computed along with its corresponding eigenvector. The

Chapter 4 The Design of Fuzzy Sets

normalized version of the eigenvector is then the membership function of the fuzzy set we considered when doing all pairwise assessments of the elements of its universe of discourse. The pairwise evaluations are far more convenient and manageable in comparison to any effort we make when assigning membership grades to all elements of the universe in a single step. Practically, the pairwise comparison helps the expert focus only on two elements once at a time thus reducing uncertainty and hesitation while leading to the higher level of consistency. The assessments are not free of bias and could exhibit some inconsistent evaluations. In particular, we cannot expect that the transitivity requirement could be fully satisfied. Fortunately, the lack of consistency could be quantified and monitored. The largest eigenvalue computed for R is always greater than the dimensionality of the reciprocal matrix (recall that in reciprocal matrices the elements positioned symmetrically along the main diagonal are inverse of each other), $\lambda_{max} > n$ where the equality $\lambda_{max} = n$ occurs only if the results are fully consistent. The ratio

$$\nu = (\lambda_{max} - n)/(n-1) \qquad (4.10)$$

can be regarded as an index of inconsistency of the data; the higher its value, the less consistent are the collected experimental results. This expression can be sought as the indicator of the quality of the pairwise assessments provided by the expert. If the value of ν is too high exceeding a certain superimposed threshold, the experiment may need to be repeated. Typically, if ν is less than 0.1, the assessment is sought to be consistent whereas higher values of ν call for a reexamination of the experimental data and a rerun of the experiment. To quantify how much the experimental data deviate from the transitivity requirement, we calculate the absolute differences between the corresponding experimentally obtained entries of the reciprocal matrix, namely r_{ik} and $r_{ij}r_{jk}$. The sum expressed in the form

$$V(i,k) = \sum_{j=1}^{n} |r_{ij}r_{jk} - r_{ik}| \qquad (4.11)$$

serves as a useful indicator of the lack of transitivity of the experimental data for the given pair of elements (i, k). If required, we may repeat the experiment if the above sum takes higher values. The overall sum $\sum_{i,k}^{n} V(i,k)$ becomes then a global evaluation of the lack of transitivity of the experimental assessment.

EXAMPLE 4.5

Let us estimate the membership function of the concept *hot* temperature for the space of temperatures consisting of 10, 20, 30, 30, 45°C. The scale in which the pairs of these elements are evaluated consists of 5 levels (say, 1, 2,..., 5). The experimental results of the pairwise comparison are collected in the reciprocal matrix R,

$$R = \begin{bmatrix} 1 & 1/2 & 1/4 & 1/5 \\ 2 & 1 & 1/3 & 1/4 \\ 4 & 3 & 1 & 1/3 \\ 5 & 4 & 3 & 1 \end{bmatrix}$$

Calculating the maximal eigenvalue, we obtain $\lambda_{max} = 4.114$, which is slightly higher than the dimension ($n = 4$) of the reciprocal matrix. The corresponding eigenvector is equal to [0.122, 0.195, 0.438, 0.869], which after normalization gives rise to the membership function of *hot* temperature to be equal to [0.14, 0.22, 0.50, 1.00]. The value of the inconsistency index n is equal to $(4.114 - 4)/3 = 0.038$ and is far lower than the threshold of 0.1.

EXAMPLE 4.6

Now let us consider some modified version of the previously discussed reciprocal matrix with the following entries:

$$\begin{bmatrix} 1 & 1/2 & 1/4 & 1/5 \\ 2 & 1 & 1/3 & 4 \\ 4 & 3 & 1 & 1/3 \\ 5 & 1/4 & 3 & 1 \end{bmatrix}$$

Now the maximal eigenvalue is far higher than the dimensionality of the problem, $\lambda_{max} = 5.426$. In this case given the high value of the inconsistency index, $v = (5.426 - 4)/3 = 0.475$, there is no point to compute the corresponding eigenvector. To fix the problem we could compute the lack of transitivity for the triples of indexes (i, j, k) and in this way highlight these assessments that tend to be highly inconsistent. These are the candidates whose evaluation has to be revised.

4.8 FUZZY SETS AS GRANULAR REPRESENTATIVES OF NUMERIC DATA

In general, a fuzzy set is reflective of numeric data that are put together in some context. Using its membership function we attempt to embrace them in a concise manner. The development of the fuzzy set is supported by the following experiment-driven and intuitively appealing rationale:

(a) First, we expect that fuzzy set (Bortokn and Pedrycz, 2002; Pedrycz and Vukovich, 2002) *A* reflects (or matches) the available experimental data to the highest extent, and

(b) Second, the fuzzy set is kept specific enough so that it comes with a well-defined semantics.

These two requirements point to the multiobjective nature of the construct: we want to maximize the coverage of experimental data (as articulated by (a)) and minimize the spread of the fuzzy set (as captured by (b)). These two requirements give rise to a certain optimization problem. Furthermore, which is quite legitimate, we assume that the fuzzy set to be constructed has a unimodal membership function or its maximal membership grades occupy a contiguous region in the universe of discourse in which this fuzzy set has been defined. This helps us build a membership function separately for its rising and declining sections. The core of the fuzzy set is determined first. Next, assuming the simplest scenario when using the linear type of membership functions, the essence of the optimization problem boils down to the rotation of the linear section of the membership function around the upper point of the core of *A*; for

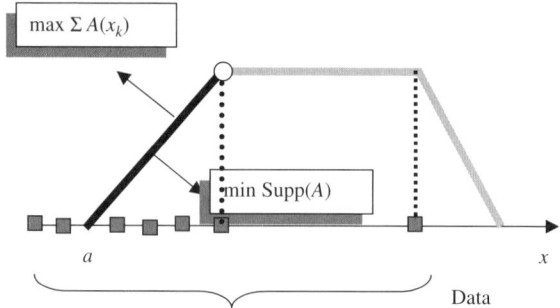

Figure 4.13 Optimization of the linear increasing section of the membership function of A; the positions of the membership function originating from the realization of the two conflicting criteria are highlighted.

illustration refer to Figure 4.13. The point of rotation of the linear segment of this membership function is marked by an empty circle. By rotating this segment, we intend to maximize (**a**) and minimize (**b**).

Before moving on with the determination of the membership function, we concentrate on the location of its numeric representative. Typically, one could view an average of the experimental data $x_1, x_2, \ldots, x_n$ to be its sound representative. Although its usage is quite common in practice, a better representative of the numeric data is the median value. There is a reason behind this choice. The median is a robust statistic meaning that it allows for a high level of tolerance to potential noise existing in the data. Its important ability is to ignore outliers. Given that the fuzzy set is sought to be a granular and "stable" representation of the numeric data, our interest is in the robust development not being affected by noise. Undoubtedly, the use of the median is a good starting point. Let us recall that the median is an order statistic and is formed on a basis of an ordered set of numeric values. In the case of the odd number of data in the data set, the point located in the middle of this ordered sequence is the median. When we encounter an even number of data in the granulation window, instead of picking up an average of the two points located in the middle, we consider these two points to form a core of the fuzzy set. Thus depending upon the number of data points, we either end up with triangular or trapezoidal membership function.

Having fixed the modal value of A (which could be a single numeric value, m or a certain interval $[m, n]$), the optimization of the spreads of the linear portions of the membership functions are carried out separately for their increasing and decreasing portions. We consider the increasing part of the membership function (the decreasing part is handled in an analogous manner). Referring to Figure 4.13, the two requirements (transformed into the corresponding multiobjective optimization problem) guiding the design of the fuzzy set are as outlined as follows:

(**a**) Maximize the experimental evidence of the fuzzy set; this implies that we tend to "cover" as many numeric data as possible, that is, the coverage has to be made as high as possible. Graphically, in the optimization of this requirement, we rotate the linear segment up (clockwise) as illustrated in Figure 4.13. Formally, the sum of the membership grades $A(x_k)$, $\sum_k A(x_k)$, where A is the linear membership function to be optimized and x_k is located to the left to the modal value has to be maximized

(b) Simultaneously, we would like to make the fuzzy set as specific as possible so that is comes with some well-defined semantics. This requirement is met by making the support of A as small as possible, that is, $\min_a |m - a|$

To accommodate the two conflicting requirements, we combine **(a)** and **(b)** in the form of the ratio that is maximized with respect to the unknown parameter of the linear section of the membership function

$$\max_a \frac{\sum_k A(x_k)}{|m-a|} \qquad (4.12)$$

The linearly decreasing portion of the membership function is optimized in the same manner. The overall optimization returns the parameters of the fuzzy number in the form of the lower and upper bound (*a* and *b*, respectively) and its support (*m* or [*m*,*n*]). We can write down such fuzzy numbers as $A(a, m, n, b)$. We exclude a trivial solution of $a = m$ in which case the fuzzy set collapses to a single numeric entity.

As an illustration, let us consider a scenario where experimental numeric data are governed by some uniform probability density function defined over the range [0, b], $b > 0$, that is, $p(x) = 1/b$ over the [0, b] and 0 otherwise. The linear membership function of A is the one of the form $A(x) = \max(0, 1 - x/a)$. The modal value of A is equal to zero. The optimization criterion (4.12) now reads as

$$V(a) = \frac{\int_0^a A(x)p(x)dx}{a} = \frac{1}{ab}\int_0^a \left(1 - \frac{x}{a}\right)dx = \frac{1}{ab}\left(b - \frac{b^2}{2a}\right) = \frac{2a - b}{2a^2} \qquad (4.13)$$

The plot of V regarded as a function of the optimized slope of A is shown in Figure 4.14; here the values of b were varied to visualize an effect of this parameter on the behavior of V.

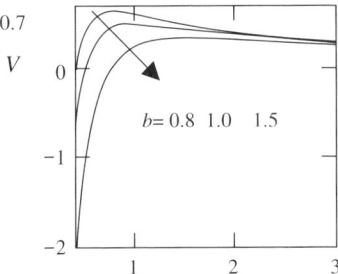

Figure 4.14 Plots of V versus *a* for selected values of *b*.

84 Chapter 4 The Design of Fuzzy Sets

The optimal value of a results from the relationship $\partial V/\partial a = 0$, and this leads to the equality $a = b$. The form of the relationship $V = V(a)$ is highly asymmetric; although the values of a higher than the optimal value (a^{opt}) leads to a very slow degradation of the performance (V changes slowly), the rapid changes in V are noted for the values of a that are lower than the optimal value.

EXAMPLE 4.7

We show the details of how the data-driven triangular fuzzy set is being formed. The dataset under discussion consists of the following numeric data:

$$\{-2.00 \; 0.80 \; 0.90 \; 1.00 \; 1.30 \; 1.70 \; 2.10 \; 2.60 \; 3.30\}$$

The values of the performance index obtained during the optimization of the left and right part slope of the triangular membership function and viewed as a function of the intercept are shown in Figure 4.15. The performance index shows a clear maximum for both the linear parts of the membership function. The final result coming in the form of the triangular fuzzy set (fuzzy number) is uniquely described by its bounds and the modal value; altogether described as the triangular fuzzy set $A(x, 0.51, 1.30, 2.96)$. This shows us how a sound compromise has been reached between the spread of the fuzzy set that helps us assure a solid coverage of the data while retaining its high specificity (limited spread). The result is quite appealing as the fuzzy set formed in this way nicely captures the core part of the numeric data.

EXAMPLE 4.8

Consider now another data set. It comes with a far higher dispersion (some points are sitting at the tails of the entire distribution):

$$\{1.1 \; 2.5 \; 2.6 \; 2.9 \; 4.3 \; 4.6 \; 5.1 \; 6.0 \; 6.2 \; 6.4 \; 8.1 \; 8.3 \; 8.5 \; 8.6 \; 9.9 \; 12.0\}$$

The plots of the optimized performance index V are included in Figure 4.16. The optimized fuzzy set comes in the form of the trapezoidal membership function $A(x, 0.61, 6.10, 6.20, 6.61)$. The location of several data points that are quite remote from the modal values makes

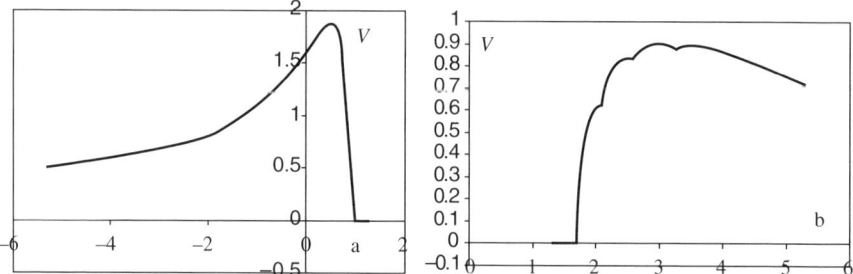

Figure 4.15 The values of the performance index V optimized for the linear sections of the membership function; in both cases we note a clearly visible maximum occurring at both sides of the modal value of the fuzzy set that determine the location of the bounds of the membership function ($a = 0.51$ and $b = 2.96$).

4.8 Fuzzy Sets as Granular Representatives of Numeric Data

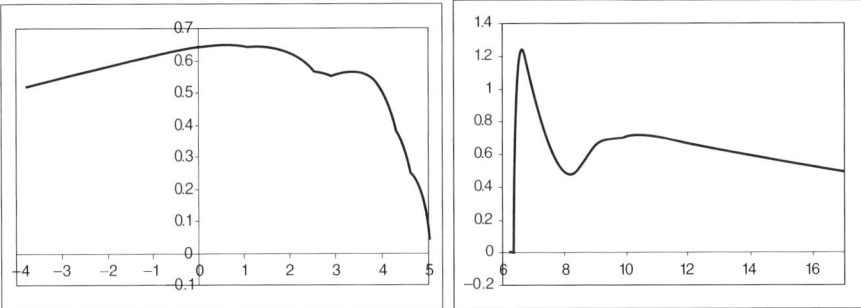

Figure 4.16 Performance index V computed separately for the linearly increasing and decreasing portions of the optimized fuzzy set.

substantial changes to the form of the membership function in which the left-hand side slope is pushed toward higher values of the arguments.

So far we have discussed the linear type of membership functions (viz. their linearly increasing or decreasing sections), however, any monotonically increasing or decreasing functions could be sought. In particular, a polynomial (monomial, to be more precise) type of relationships, say x^p with p being a positive integer could be of interest. The values of p impact the shape and more importantly, the spread of the resulting fuzzy set.

EXAMPLE 4.8

Let us consider a geometrical figure that resembles a fuzzy circle; see Figure 4.17. The coordinates of the central point are given as (x_0, y_0). Let us represent the figure as a fuzzy circle, that is, a circle whose radius is a fuzzy set (fuzzy number).

The membership of the fuzzy radius is determined on the basis of numeric values of the radii obtained for several successive discrete values of the angle ϕ_i, thus giving rise to the values $r_1, r_2, \ldots, r_n$. Next the determination of the fuzzy set of radius (fuzzy circle) is realized following the optimization scheme governed by (4.12).

In a similar way, we can define a fuzzy set of distance between a certain geometrical figure and a point. Although there are a number of definitions that attempt to capture the essence of the concept (say, a Hausdorff distance), they return a single numeric quantity. Consider a situation displayed in Figure 4.18.

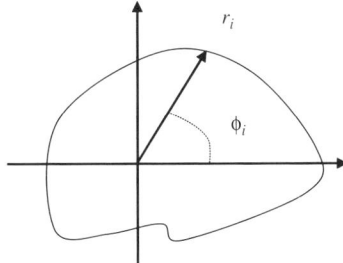

Figure 4.17 Example of figure to be represented as fuzzy circles.

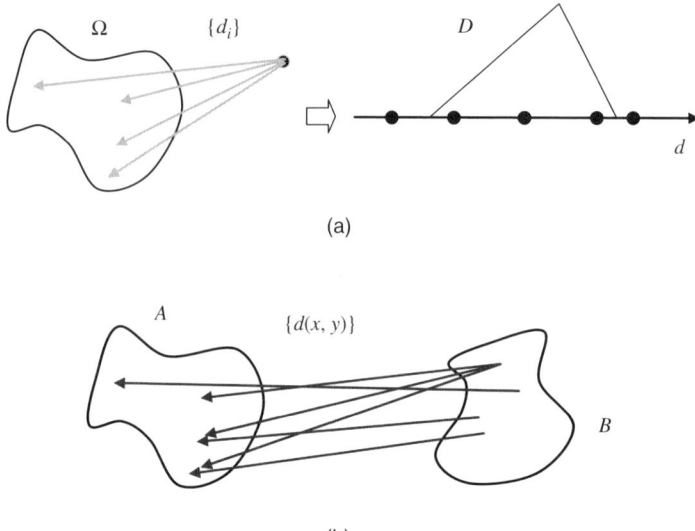

Figure 4.18 Computing a fuzzy set of distance between a point and some geometric figure Ω; note a sample of points located within the bounds of the figure and induced fuzzy set (a), and determining distance between two planar figures A and B (b).

In the computations of the distance we note that the concept of such distance is quite complex and describing it with the use of a single numeric quantity may not be fully reflective of the underlying concept. Given this, we proceed as follows. First we choose a sample of points that are located within the figure. Note that even though the object (geometrical figure) has clearly delineated boundaries (there is no uncertainty as to their position), the fuzzy set of distance is reflective of the complexity and nonuniqueness of the definition itself. Given the collection of numeric distances determined in this manner, say $\{d_1, d_2, \ldots, d_N\}$, we form a fuzzy set of distance by the maximization of (4.12). The character of the membership function, say piecewise linear, has to be specified in advance before starting the optimization procedure. Another generalization of this problem comes with the computing of distance between two figures; see Figure 4.18 (b). Although the Hausdorff distance $d(A, B)$ defined in the form

$$d_H(A,B) = \max\{\sup_{x \in A}[\min_{y \in B} d(x,y)], \sup_{y \in B}[\min_{x \in A} d(x,y)]\}$$

becomes available, its descriptive power could create some limitations. Instead, we may proceed as in the previous calculations of distance between a figure and a point: We sample the two figures producing a collection of points inside and then compute distances between pairs of $d(x_i, y_j)$, where x_i A and y_j B and then use them to form a fuzzy set of distance.

4.9 FROM NUMERIC DATA TO FUZZY SETS

Fuzzy sets can be formed on a basis of numeric data through their clustering (groupings). The groups of data give rise to membership functions that convey a more global abstract view of the available data. In this regard Fuzzy C-Means (FCM, in brief) is one of the commonly used mechanisms of fuzzy clustering (Bezdek, 1981; Pedrycz and Reformat, 2006).

Let us review its formulation, develop the algorithm and highlight the main properties of the fuzzy clusters. Given a collection of *n*-dimensional data set $\{\mathbf{x}_k\}$, $k = 1, 2, \ldots, N$, the task of determining its structure—a collection of *c* clusters, is expressed as a minimization of the following objective function (performance index) Q being regarded as a sum of the squared distances:

$$Q = \sum_{i=1}^{c} \sum_{k=1}^{N} u_{ik}^m \| \mathbf{x}_k - \mathbf{v}_i \|^2$$

Here $\mathbf{v}_i$'s are *n*-dimensional prototypes of the clusters, $i = 1, 2, .., c$ and $U = [u_{ik}]$ stands for a partition matrix expressing a way of allocation of the data to the corresponding clusters; u_{ik} is the membership degree of data $\mathbf{x}_k$ in the *i*th cluster. The distance between the data n_k and prototype $\mathbf{v}_i$ is denoted by $\|.\|$. The fuzzification coefficient m (>1.0) expresses the impact of the membership grades on the individual clusters.

A partition matrix satisfies two important properties:

$$0 < \sum_{k=1}^{N} u_{ik} < N, \quad i = 1, 2, \ldots, c \quad (4.14a)$$

$$\sum_{i=1}^{c} u_{ik} = 1, \quad k = 1, 2, \ldots, N \quad (4.14b)$$

Let us denote by **U** a family of matrices satisfying (a) and (b). The first requirement states that each cluster has to be nonempty and different from the entire set. The second requirement states that the sum of the membership grades should be confined to 1.

The minimization of Q completed with respect to $U \in \mathbf{U}$ and the prototypes $\mathbf{v}_i$ of $V = \{\mathbf{v}_1, \mathbf{v}_2, \ldots, \mathbf{v}_c\}$ of the clusters. More explicitly, we write it down as follows:

$$\min Q \text{ with respect to } U \in \mathbf{U}, \mathbf{v}_1, \mathbf{v}_2, \ldots, \mathbf{v}_c \in \mathbf{R}^n$$

From the optimization standpoint, there are two individual optimization tasks to be carried out separately for the partition matrix and the prototypes. The first one concerns the minimization with respect to the constraints given the requirement of the form (4.14b) that holds for each data point X_k. The use of Lagrange multipliers converts the problem into its constraint-free version. The augmented objective function formulated for each data point, $k = 1, 2, \ldots, N$, reads as

$$V = \sum_{i=1}^{c} u_{ik}^m d_{ik}^2 + \lambda \left(\sum_{i=1}^{c} u_{ik} - 1 \right) \quad (4.15)$$

88 Chapter 4 The Design of Fuzzy Sets

where

$d_{ik}^2 = \| \mathbf{x}_k - \mathbf{v}_i \|^2$. Proceeding with the necessary conditions for the minimum of V for $k = 1, 2, \ldots N$, one has

$$\frac{\partial V}{\partial u_{st}} = 0 \quad \frac{\partial V}{\partial \lambda} = 0 \tag{4.16}$$

$s = 1, 2, \ldots, c, t = 1, 2, \ldots, N$. Now we calculate the derivative of V with respect to the elements of the partition matrix in the following way:

$$\frac{\partial V}{\partial u_{st}} = m u_{st}^{m-1} d_{st}^2 + \lambda \tag{4.17}$$

From (4.17) we calculate u_{st} to be equal to

$$u_{st} = \left(-\frac{\lambda}{m}\right)^{\frac{1}{m-1}} \frac{2}{d_{st}^{m-1}} \tag{4.18}$$

Given the normalization condition $\sum_{j=1}^{c} u_{jt} = 1$ and plugging it into (4.18) one has

$$\left(-\frac{\lambda}{m}\right)^{\frac{1}{m-1}} \sum_{j=1}^{c} d_{jt}^{\frac{2}{m-1}} = 1 \tag{4.19}$$

we compute

$$\Rightarrow \left(-\frac{\lambda}{m}\right)^{\frac{1}{m-1}} = \frac{1}{\sum_{j=1}^{c} d_{jt}^{\frac{2}{m-1}}} \tag{4.20}$$

Inserting this expression into (4.18), we obtain the successive entries of the partition matrix

$$u_{st} = \frac{1}{\sum_{j=1}^{c} \left(\frac{d_{st}^2}{d_{jt}^2}\right)^{\frac{1}{m-1}}} \tag{4.21}$$

The optimization of the prototypes $\mathbf{v}_i$ is carried out assuming that the Euclidean distance between the data and the prototypes is $\| \mathbf{x}_k - \mathbf{v}_i \|^2 = \sum_{j=1}^{c} (x_{kj} - v_{ij})^2$. The objective function reads now as follows:

$$Q = \sum_{i=1}^{c} \sum_{k=1}^{N} u_{ik}^m \sum_{j=1}^{n} (x_{kj} - v_{ij})^2$$

and its gradient with respect to $\mathbf{v}_i$, $\nabla_{\mathbf{v}_i} Q$ made equal to zero yields the system of linear equations

$$\sum_{k=1}^{N} u_{ik}^m (x_{kt} - v_{st}) = 0 \tag{4.22}$$

$s = 1, 2, \ldots, c; \quad t = 1, 2, \ldots, n$

Thus

$$v_{st} = \frac{\sum_{k=1}^{N} u_{ik}^m x_{kt}}{\sum_{k=1}^{N} u_{ik}^m} \qquad (4.23)$$

Overall, the FCM clustering is completed through a sequence of iterations where we start from some random allocation of data (a certain randomly initialized partition matrix) and carry out the following updates by adjusting the values of the partition matrix and the prototypes. These steps are summarized next.

procedure FCM-CLUSTERING (**x**) **returns** prototypes and partition matrix
input: data $\mathbf{x} = \{\mathbf{x}_1, \mathbf{x}_2, \ldots, \mathbf{x}_k\}$
local: fuzzification parameter: m
              threshold: ε
              norm: $\|\cdot\|$
INITIALIZE-PARTITION-MATRIX
$t \leftarrow 0$
repeat
 for $i = 1 : c$ **do**

$$\mathbf{v}_i(t) \leftarrow \frac{\sum_{k=1}^{N} u_{ik}^m(t)\mathbf{x}_k}{\sum_{k=1}^{N} u_{ik}^m(t)} \quad \text{(compute prototypes)}$$

 for $i = 1 : c$ **do**
  for $k = 1 : N$ **do**
   update partition matrix

$$u_{ik}(t+1) = \frac{1}{\sum_{j=1}^{c} \left(\frac{\|\mathbf{x}_k - \mathbf{v}_i(t)\|}{\|\mathbf{x}_k - \mathbf{v}_j(t)\|}\right)^{2/(m-1)}} \quad \text{(update partition) matrix}$$

$t \leftarrow t + 1$
until $\| U(t+1) - U(t) \| \leq \varepsilon$
return U, V

Iteration is repeated until a certain termination criterion has been satisfied. Typically, the termination condition is quantified by looking at the changes in the membership values of the successive partition matrices. Denote by $U(t)$ and $U(t+1)$ the two partition matrices produced in two consecutive iterations of the algorithm. If the distance $\| U(t+1) - U(t) \|$ is les than a small predefined threshold ε, then we terminate the algorithm. Typically, one considers the Tchebyschev distance between the partition matrices meaning that the termination criterion reads as follows:

$$\max_{i,k} |u_{ik}(t+1) - u_{ik}(t)| \leq \varepsilon \qquad (4.24)$$

Table 4.1 The Main Features of the Fuzzy c-means (FCM) Clustering Algorithm.

Feature of the FCM algorithm	Representation and optimization aspects
Number of clusters (c)	Structure in the data set and the number of fuzzy sets estimated by the method; the increase in the number of clusters produces lower values of the objective function, however, given the semantics of fuzzy sets one should maintain this number quite low (5–9 information granules).
Objective function Q	Develops the structure aimed at the minimization of Q; iterative process supports the determination of the local minimum of Q.
Distance function $\|\cdot\|$	Reflects (or imposes) a geometry of the clusters one is looking for; essential design parameter affecting the shape of the membership functions.
Fuzzification coefficient (m)	Implies a certain shape of membership functions present in the partition matrix; essential design parameter. Low values of m (being close to 1.0) induce characteristic function. The values higher than 2.0 yield spiky membership functions.
Termination criterion	Distance between partition matrices in two successive iterations; the algorithm terminated once the distance below some assumed positive threshold (ε), that is, $\| U(t+1) - U(t) \| \varepsilon$.

The key components of the FCM and a quantification of their impact on the form of the produced results are summarized in Table 4.1.

The fuzzification coefficient exhibits a direct impact on the geometry of fuzzy sets generated by the algorithm. Typically, the value of m is assumed to be equal to 2.0. Lower values of m (which are closer to 1) yield membership functions that start resembling characteristic functions of sets; most of the membership values become localized around 1 or 0. The increase of the fuzzification coefficient ($m = 3, 4$, etc.) produces "spiky" membership functions with the membership grades equal to 1 at the prototypes and a fast decline of the values when moving away from the prototypes. Several illustrative examples of the membership functions are included in Figure 4.19. In addition to the varying shape of the membership functions, observe that the requirement put on the sum of membership grades imposed on the fuzzy sets yields some rippling effect: The membership functions are not unimodal, but may exhibit some ripples whose intensity depends upon the distribution of the prototypes and the values of the fuzzification coefficient.

The membership functions offer an interesting feature of evaluating the extent to which a certain data point is shared between different clusters and in this sense

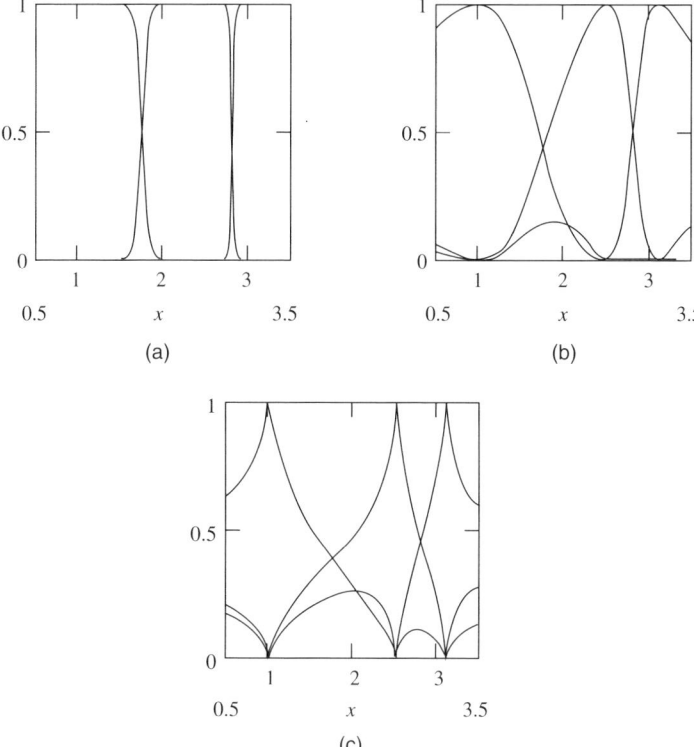

Figure 4.19 Examples of membership functions of fuzzy sets; the prototypes are equal to 1, 3.5, and 5 while the fuzzification coefficient assumes values of 1.2 (a), 2.0 (b), and 3.5 (c). The intensity of the rippling effect is affected by the values of m and increases with the higher values of m.

become difficult to allocate to a single cluster (fuzzy set). Let us introduce the following index that serves as a certain separation measure:

$$\varphi(u_1, u_2, \ldots, u_c) = 1 - c^c \prod_{i=1}^{c} u_i \qquad (4.25)$$

where $u_1, u_2, \ldots, u_c$ are the membership degrees for some data point. If only one of membership degrees, say $u_i = 1$, and the remaining are equal to zero, then the separation index attains its maximum equal to 1. On the other extreme, when the data point is shared by all clusters to the same degree equal to $1/c$, then the value of the index drops down to zero. This means that there is no separation between the clusters as reported for this specific point.

For instance, if $c = 2$, the above expression relates directly to the entropy of fuzzy sets (Chapter 3. We have $\varphi(u) = 1 - 4u(1-u)$, that is, $\varphi(u) = 1 - H(u)$.

Although the number of clusters is typically limited to a few information granules, we can easily proceed with successive refinements of fuzzy sets. This can be done by splitting fuzzy clusters of the highest heterogeneity (Pedrycz and Reformat, 2006). Let us assume that we have already constructed c fuzzy clusters. Each of them can be characterized by the performance index

$$V_i = \sum_{k=1}^{N} u_{ik}^m \parallel \mathbf{x}_k - \mathbf{v}_i \parallel^2 \qquad (4.26)$$

$i = 1, 2, \ldots, c$. The higher the value of V_i, the more heterogeneous the ith cluster. The one with the highest value of V_i, that is, the one for which we have $i_0 = \arg, \max_i V_i$ is refined by being split into two clusters. Denote the set of data associated with the i_0th cluster by $\mathbf{X}(i_0)$,

$$\mathbf{X}(i_0) = \{x_k \in \mathbf{X} \parallel u_{i_0 k} = \max_i u_{ik}\} \qquad (4.27)$$

We cluster the elements in $\mathbf{X}(i_0)$ by forming two clusters that leads to two more specific (detailed) fuzzy sets. This gives rise to a hierarchical structure of the family of fuzzy sets as illustrated in Figure 4.20. The relevance of this construct in the setting of fuzzy sets is that it emphasizes the essence of forming a hierarchy of fuzzy sets rather than working with a single level structure of a large number of components whose semantics could not be retained.

The process of further refinements is realized in the same manner by picking up the cluster of the highest heterogeneity and its split into two consecutive clusters.

It is worth emphasizing that the FCM algorithm is a highly representative method of membership estimation that profoundly dwells on the use of experimental data. In contrast to some other techniques presented so far that are also data-driven, FCM can easily cope with multivariable experimental data.

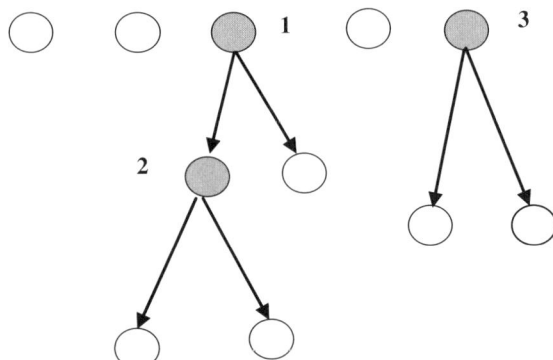

Figure 4.20 Successive refinements of fuzzy sets through fuzzy clustering applied to the clusters of the highest heterogeneity. The numbers indicate the order of the splits.

4.10 FUZZY EQUALIZATION

The underlying principle of this approach is based on an observation that any fuzzy set has to some extent be reflective of the existing numeric evidence. In other words, we anticipate that it has to be substantially supported by the existing numeric data. Although we have emphasized this aspect when dealing with the method outlined in Section 8, the concept to be discussed here is concerned with a family of fuzzy sets. (Pedrycz, 2001) The problem of fuzzy equalization can be outlined as follows:

> Given is a finite collection of numeric data $\{x_1, x_2, \ldots, x_N\}$, $x_i \in \mathbf{R}$. We consider that they are arranged in an nondecreasing order, that is, $x_1 \leq x_2 \leq \cdots \leq x_N$. Construct a family (partition) of triangular fuzzy sets $A_1, A_2, \ldots, A_c$ with an overlap of ½ between neighboring fuzzy sets; refer to Figure 4.21 such that each of them comes with the same experimental evidence. We require that the following system of equalities is satisfied:

$$\sum_{k=1}^{N} A_1(x_k) = \frac{N}{2(c-1)}$$

$$\sum_{k=1}^{N} A_2(x_k) = \frac{N}{(c-1)}$$

$$\ldots$$

$$\sum_{k=1}^{N} A_{c-1}(x_k) = \frac{N}{(c-1)}$$

$$\sum_{k=1}^{N} A_c(x_k) = \frac{N}{2(c-1)} \qquad (4.28)$$

In other words, we require that the σ-count computed for each fuzzy set is made the same. For the first and the last fuzzy set (A_1 and A_c) we require that that this σ-count is ½ of the one required for all remaining fuzzy sets. The essence of this construct is illustrated in Figure 4.21.

We can propose the following procedure to build the fuzzy sets $A_1, A_2, \ldots, A_{c-1}$ satisfying (4.28); the simplicity of the algorithm does not assure that the same numeric requirement holds for A_c. We elaborate on this in more detail

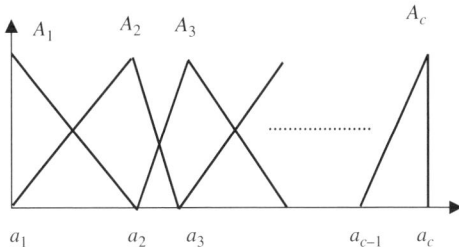

Figure 4.21 A collection of triangular fuzzy sets with equal experimental support provided by numeric data.

later on. The equalization is concerned with the determination of the modal values of the fuzzy sets. We start with A_1 and move to the right by choosing a suitable value of a_2 so that the sum of membership grades $\sum_{k=1}^{N} A_1(x_k)$ is equal to $N/2(c-1)$. The determination of the value of a_2 could be completed through a stepwise increment of its value. For instance, we start with $a_2 = a_1 + \varepsilon$ with $\varepsilon > 0$ being a small positive step, calculate the corresponding sum of the membership values, and if the condition of enough experimental evidence has not been met, we progress toward higher values of a_2 by moving with the assumed increment ε.

The modal value of A_1 is equal to the minimal value encountered in the data set, that is, $a_1 = x_1$. We assume here that the boundaries x_1 and x_N are not outliers; otherwise they have to be dropped and the construct should be based upon the other extreme points in the dataset. The experimental evidence of A_2 is made equal to $\frac{N}{(c-1)}$ by a proper choice of the upper bound of its membership function, namely a_3. Note that as the value of a_2 has been already selected, this implies the following level of the experimental evidence accumulated so far:

$$\sum_{x_k \in [a_1, a_2]} A_2(x_k) = \sum_{x_k \in [a_1, a_2]} (1 - A_2 1 \Rightarrow (x_k)) = \sum_{x_k \in [a_1, a_2]}$$
$$- \sum_{x_k \in [a_1, a_2]} A_2(x_k) = N_2 - \frac{N}{2(c-1)} = N_2' \quad (4.29)$$

where N_2 is the number of data points in $[a_1, a_2]$.

Given this, we require that the value of a_3 is chosen so that the following equality holds:

$$\sum_{x_k \in [a_2, a_3]} A_2(x_k) + N_2' = \frac{N}{c-1} \quad (4.30)$$

Note that depending upon the distribution of numeric data the resulting fuzzy set A_2 could be highly asymmetric.

To determine the parameters of the successive fuzzy sets, we repeat the same procedure moving toward higher values of x_k and determining the values of $a_3, a_4, \ldots$ One notes that the last fuzzy set A_c does not come with the required level of experimental evidence as we do not have any control over the sum of the corresponding membership grades. To alleviate this shortcoming, one may consider a replacement of the algorithm (whose advantage resides with its evident simplicity) by the minimization of the performance index V over the vector of the modal values $\mathbf{a} = [a_2, a_3 \ldots a_{c-1}]$

$$V = \left[\sum_{k=1}^{N} A_1(x_k) - \frac{N}{2(c-1)} \right]^2 + \left[\sum_{k=1}^{N} A_2(x_k) - \frac{N}{(c-1)} \right]^2 + \cdots$$
$$+ \left[\sum_{k=1}^{N} A_{c-1} - \frac{N}{(c-1)} \right]^2 + \left[\sum_{k=1}^{N} A_c(x_k) - \frac{N}{2(c-1)} \right]^2$$

that is,

$$\min_a V$$

4.11 LINGUISTIC APPROXIMATION

In many cases we are provided with a finite family of fuzzy sets $A_1, A_2, \ldots, A_c$ (whose membership functions could have been determined earlier) using which we would like to represent a certain fuzzy sets B. Quite often these fuzzy sets are referred to as a vocabulary of information granules. Furthermore we have at our disposal a finite collection of so-called linguistic modifiers (hedges) $\tau_1, \tau_2, \ldots, \tau_p$. Let us recall that we encounter two general categories of the modifiers realizing operations of concentration and dilution. Their semantics relates to the linguistic adjectives of the form *very* (concentration) and *more or less* (dilution). Given the semantics of A_i's and the available linguistic modifiers, the objective of the representation scheme is to capture the essence of B. Given the nature of the objects and the ensuing processing being used here we refer to this process as a linguistic approximation. There are several scenarios of the realization of the linguistic approximation. The scheme shown in Figure 4.22 comprises two phases: first we find the best match between B and A_i's (where the quality of matching is expressed in terms of some distance or similarity measure).

As the next step we refine the construct by applying one of the linguistic modifiers. The result of the linguistic approximation comes in the form $B \approx \tau_i(A_j)$ with the indexes i and j determined through the optimization of the matching mechanism.

4.12 DESIGN GUIDELINES FOR THE CONSTRUCTION OF FUZZY SETS

The considerations presented above give rise to a number of general guidelines supporting the development of fuzzy sets.

(a) Highly visible and well-defined semantics of information granules. No matter what the determination technique is, one has to become cognizant of the semantics of the resulting fuzzy sets. Fuzzy sets are interpretable information granules of a well-defined meaning and this aspect needs to be fully captured. Given this, the number of information granules has to be kept quite small with their number being restricted to 7 ± 2 fuzzy sets.

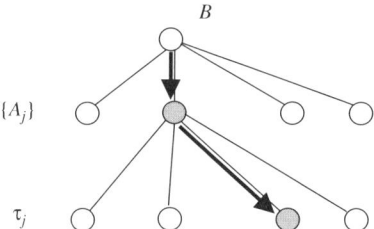

Figure 4.22 The process of linguistic approximation of B in terms of the elements of the vocabulary and the collection of the linguistic modifiers.

(b) There are several fundamental views of fuzzy sets and depending upon them, we could consider the use of various estimation techniques (e.g., by accepting the horizontal or the vertical view of fuzzy sets and adopting a pertinent technique).

(c) Fuzzy sets are context-sensitive constructs and as such require careful calibration. This feature of fuzzy sets should be treated as their genuine advantage. The semantics of fuzzy sets can be adjusted through shifting fuzzy sets or/and adjusting their membership functions. The nonlinear transformation we introduced here helps complete an effective adjustment of the membership functions making use of some "standard" membership functions. The calibration mechanisms being used in the design of the membership function are reflective of human-centricity of fuzzy sets.

(d) We have delineated between the two major categories of approaches supporting the design of membership functions, that is, data-driven and expert (user)-based. They are very different in the sense of the origin of the supporting evidence. Fuzzy clustering is a fundamental mechanism of the development of fuzzy sets. It is important in the sense that the method is equally suitable for one-dimensional and multivariable cases. The expert or simply user-based methods of membership estimation are important in the sense they offer some systematic and coherent mechanisms of elicitation of membership grades. With regard to consistency of the elicited membership grades, the pairwise estimation technique is of particular interest by providing well-quantifiable mechanisms of the assessment of the consistency of the produced membership grades. The estimation procedures underline some need of further development of higher type of constructs such as fuzzy sets of type-2 or higher and fuzzy sets of higher order that may be ultimately associated with constructs such as type-2 fuzzy sets or interval valued fuzzy sets (this particular construct is visible when dealing with the horizontal method of membership estimation that comes with the associated confidence intervals).

(e) The user-driven membership estimation uses the statistics of data yet it is done in an *implicit* manner. The granular term—fuzzy sets come into existence once there is some experimental evidence behind them (otherwise there is no point forming such fuzzy sets).

(f) The development of fuzzy sets can be carried out in an stepwise manner. For instance, a certain fuzzy set can be further refined, if required in the problem in hand. This could lead to several more specific fuzzy sets that are associated with the fuzzy set formed at the higher level. Being aware of the complexity of the granular descriptors, we should resist temptation of forming an excessive number of fuzzy sets at a single level as such fuzzy sets could be easily lacking any sound interpretation.

4.13 CONCLUSIONS

We have discussed various approaches and algorithmic aspects of the design of fuzzy sets. The estimation of membership functions is a multifaceted problem and the choice of a suitable method relies on the choice of the available experimental data and domain knowledge. For the user-driven approaches, it is essential to evaluate and flag the consistency of the results. Although some of the methods (the pairwise comparison) come with this essential feature, the results produced by the others have to be carefully inspected.

EXERCISES AND PROBLEMS

1. In the horizontal mode of construction of a fuzzy set of *safe* speed on a highway, the yes-no evaluations provided by the panel of 9 experts are the following:

x	20	50	70	80	90	100	110	120	130	140	150	160
No of yes responses	0	1	1	2	6	8	8	5	5	4	3	2

   Determine the membership function and assess its quality by computing the corresponding confidence intervals. Interpret the results and identify the points of the universe of discourse that may require more attention.

2. In the vertical mode of membership function estimation, we are provided with the following experimental data:

α	0.3	0.4	0.5	0.6	0.7	0.8	0.9	1.0
Range of **X**	[−2,13]	[−1,12]	[0, 11]	[1, 10]	[2, 9]	[3, 8]	[4, 7]	[5, 6]

   Plot the estimated membership function and suggest its analytical expression.

3. In the calculations of the distance between a point and a certain geometric figure (as discussed in Section 4.8), we assumed that the boundaries of the figure is well defined. How could you proceed with a more general case when the boundaries are not clearly defined, namely, the figure itself is defined by some membership function (Fig. 4.23). In other words, the figure is fully characterized by some membership function $R(\mathbf{x})$ where $\mathbf{x}$

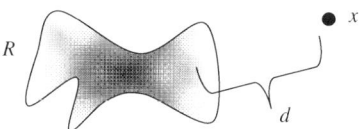

Figure 4.23 Forming a fuzzy set of distance between a geometric figure with fuzzy boundaries and a point.

98 Chapter 4 The Design of Fuzzy Sets

is a vector of coordinates of **x**. If $R(\mathbf{x}) = 1$, the point fully belongs to the figure while lower values of $R(\mathbf{x})$ indicate that **x** is closer to the boundary of R.

4. Construct a fuzzy set describing a distance between the point of (5, 5) from a circle $x^2 + y^2 = 4$.

5. We maximize a function $f(x) = (x - 6)^4$ in the range of [3, 10]. Suggest a membership function describing a degree of membership of the optimal solution that minimizes $f(x)$. What conclusion could you derive based on the obtained form of the membership function?

6. The results of pairwise comparisons of 4 objects being realized in the scale of 1–5 are given in the following matrix form:

$$\begin{bmatrix} 1 & 5 & 2 & 4 \\ 1/5 & 1 & 3 & 1/3 \\ 1/2 & 1/3 & 1 & 1/5 \\ 1/4 & 3 & 5 & 1 \end{bmatrix}$$

What is the consistency of the findings? Evaluate the effect of the lack of transitivity. Determine the membership function of the corresponding fuzzy set.

7. In the method of pairwise comparisons, we use different scales involving various levels of evaluation, typically ranging from 5 to 9. What impact could the number of these levels have on the produced consistency of the results? Could you offer any guidelines as how to achieve high consistency? What would be an associated tradeoff one should take into consideration here?

8. Construct a fuzzy set of *large* numbers for the universe of discourse of integer numbers ranging from 1 to 10. It has been found that the experimental results of the pairwise comparison could be described in the form

$$r(x, y) = \begin{cases} x - y & \text{if } x > y \\ 1 & \text{if } x = 1 \end{cases}$$

(for $x < y$ we consider the reciprocal of the above expression).

9. In the FCM algorithm, the shape of the resulting membership function depends upon the value of the fuzzification coefficient (m). How does the mean value of the membership function relate to the values of m. Run the FCM on the one-dimensional data set

$$\{1.3\ 1.9\ 2.0\ 5.5\ 4.9\ 5.3\ 4.5\ -1.3\ 0.0\ 0.3\ 0.8\ 5.1\ 2.5\ 2.4\ 2.1\ 1.7\}$$

considering $c = 3$ clusters. Next plot the relationship between the average of all membership grades and the associated fuzzification coefficient. For which values of m the average of membership grades differ from 0.33 for less than δ? Consider several values of δ, say 0.2, 0.1, and 0.05. What could you tell about the impact of m on the resulting average?

10. Consider a family of car makes, say $C_1, C_2, \ldots, C_n$. We are interested in forming fuzzy sets of economy, comfort, and safety, say $A_{economy}$, $A_{comfort}$, and A_{safety}. Use a method of a pairwise comparison to build the corresponding fuzzy sets. Next using the method of pairwise comparison, evaluate the car makes with respect to the overall quality (which involves economy, comfort, and safety). Given the already constructed fuzzy sets of the individual attributes and the overall quality $A_{overall}$, what relationship could you establish between them?

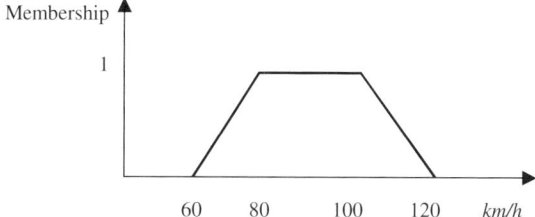

Figure 4.24 A fuzzy set of a *safe* speed on an average highway.

11. The method of membership estimation shown in Section 4.8 is concerned with one-dimensional data. How could you construct a fuzzy set over multidimensional data? Consider using one-dimensional constructs first.
12. Consider a fuzzy set of a *safe* speed on an average highway; refer to Figure 4.24.

How could this membership be affected when redefining this concept in the following settings of (a) autobahn (note that on these German highways there is no speed limit) and (b) a snowy country road. Elaborate on the impact of various weather conditions on the corresponding membership function. From the standpoint of the elicitation of the membership function, how could you transform the original membership function to address the needs of the specific context in which it is planned to be used?

HISTORICAL NOTES

The issue of membership elicitation has been an area of active research and numerous discussions almost since the inception of fuzzy sets. One may refer to very early studies reported by Dishkant (1981), Dombi (1990), Saaty (1986), and Turksen (1991). The horizontal and vertical views on the membership estimation appeared quite early yet their usage should be carefully planned to avoid potential inconsistencies in the estimates of the membership grades. The pairwise method of membership estimation (Saaty, 1980) has delivered an important feature of higher consistency of the results produced within the framework of this scheme. The properties of the families of fuzzy sets with triangular membership functions were investigated by Pedrycz (1994). The advantage of the data-driven approach relies on a direct calibration of the membership functions by the existing experimental data. Its visibility in the design of fuzzy sets started to grow in importance along with the more intensive use of fuzzy clustering in modeling and control. The idea of context-based adjustment of fuzzy sets was introduced and discussed by Pedrycz, Gudwin, and Gomide (1997).

REFERENCES

Bezdek, J. *Pattern Recognition with Fuzzy Objective Function Algorithms*, Plenum Press, New York, 1981.

Bortolan, G., Pedrycz, W. An interactive framework for an analysis of ECG signals, *Artif. Intell. Med.* **24**(2), 2002, 109–132.

Buckley, J. J., Feuring, T., Hayashi, Y. Fuzzy hierarchical analysis revisited, *European J. Op. Res.* **129**(1), 2001, 48–64.

Civanlar, M., Trussell, H. Constructing membership functions using statistical data, *Fuzzy Set. Syst.* **18**(1), 1986, 1–13.

Chen, M., Wang, S. Fuzzy clustering analysis for optimizing fuzzy membership functions, *Fuzzy Set. Syst.* **103**(2), 1999, 239–254.

Dishkant, H. About membership functions estimation, *Fuzzy Set. Syst.* **5**(2), 1981, 141–147.

Dombi, J. Membership function as an evaluation, *Fuzzy Set. Syst.* **35**(1), 1990, 1–21.

Hong, T., Lee, C. Induction of fuzzy rules and membership functions from training examples, *Fuzzy Set. Syst.* **84**(1), 1996, 389–404.

Kulak, O., Kahraman, C. Fuzzy multi-attribute selection among transportation companies using axiomatic design and analytic hierarchy process, *Inf. Sci.* **170**(2–4), 2005, 191–210.

van Laarhoven, P., Pedrycz, W. A fuzzy extension of Saaty's priority theory, *Fuzzy Set. Syst.* **11**(1–3), 1983, 199–227.

Masson, M., Denoeux, T. Inferring a possibility distribution from empirical data, *Fuzzy Set. Syst.* **157**(3), 2006, 319–340.

Medaglia, A., Fang, S., Nuttle, H., Wilson, J. An efficient and flexible mechanism for constructing membership functions, *Eur. J. Op. Res.* **139**(1), 2002, 84–95.

Medasani, S., Kim, J., Krishnapuram, R. An overview of membership function generation techniques for pattern recognition, *Int. J. Approxi. Reason.* **19**(3–4), 1998, 391–417.

Mikhailov, L., Tsvetinov, P. Evaluation of services using a fuzzy analytic hierarchy process, *Appl. Soft Comput.* **5**(1), 2004, 23–33.

Miller, G. A. The magical number seven plus or minus two: some limits of our capacity for processing information, *Psychol. Rev.* **63**, 1956, 81–97.

Pedrycz, W. Why triangular membership functions?, *Fuzzy Set. Syst.* **64**, 1994, 21–30.

Pedrycz, W. *Fuzzy Sets Engineering*, CRC Press, Boca Raton, FL, 1995.

Pedrycz, W. Fuzzy equalization in the construction of fuzzy sets, *Fuzzy Set. Syst.* **119**, 2001, 329–335.

Pedrycz, W., Gudwin, R., Gomide, F. Nonlinear context adaptation in the calibration of fuzzy sets, *Fuzzy Set. Syst.* **88**, 1997, 91–97.

Pedrycz, A., Reformat, M. Hierarchical FCM in a stepwise discovery of structure in data, *Soft Comput.* **10**, 2006, 244–256.

Pedrycz, W., Valente de Oliveira, J. An algorithmic framework for development and optimization of fuzzy models, *Fuzzy Set. Syst.* **80**, 1996, 37–55.

Pedrycz, W., Vukovich, G. On elicitation of membership functions, *IEEE Trans. Syst., Man, Cybern., Part A*, **32**(6), 2002, 761–767.

Pendharkar, P. Characterization of aggregate fuzzy membership functions using Saaty's eigenvalue approach, *Comput. Op. Res.* **30**(2), 2003, 199–212.

Saaty, T. *The Analytic Hierarchy Process,* McGraw Hill, New York, 1980.

Saaty, T. Scaling the membership functions, *Eur. J. Op. Res.* **25**(3), 1986, 320–329.

Simon, D. H_∞ estimation for fuzzy membership function optimization, *Int. J. Approx. Reason.* **40**(3), 2005, 224–242.

Turksen, I. Measurement of membership functions and their acquisition, *Fuzzy Set. Syst.* **40**(1), 1991, 5–138.

Yang, C., Bose, N. Generating fuzzy membership function with self-organizing feature map, *Pattern Recogn. Lett.*, **27**(5), 2006, 356–365.

Chapter 5

Operations and Aggregations of Fuzzy Sets

Similarly as in set theory, we operate with fuzzy sets to obtain new fuzzy sets. The operations must possess properties to match intuition, to comply with the semantics of the intended operation, and to be flexible to fit application requirements. This chapter covers set operations beginning with early fuzzy set operations and continuing with their generalization, interpretations, formal requirements, and realizations. We emphasize complements, triangular norms and conorms as unifying, general constructs of the complement, intersection, and union operations. Combinations of fuzzy sets to provide aggregations are also essential when operating with fuzzy sets. Analysis of fundamental properties and characteristics of operations with fuzzy sets are discussed thoroughly.

5.1 STANDARD OPERATIONS ON SETS AND FUZZY SETS

It is instructive to start with the familiar operations of intersection, union, and complement encountered in set theory. For instance, consider two sets $A = \{x \in \mathbf{R} | 1 \leq x \leq 3\}$ and $B = \{x \in \mathbf{R} | 2 \leq x \leq 4\}$, both being closed intervals in the real line. Their intersection is a set $A \cap B = \{x \in \mathbf{R} | 2 \leq x \leq 3\}$. Figure 5.1 illustrates the intersection operation represented in terms of the characteristic functions of A and B. Looking at the values of the characteristic function of $A \cap B$ that results when comparing the individual values of $A(x)$ and $B(x)$ at each $x \in \mathbf{R}$, we note that these are taken as the minimum between the values of $A(x)$ and $B(x)$.

In general, given the characteristic functions of A and B, the characteristic function of their intersection $A \cap B$ is computed in the following form:

$$(A \cap B)(x) = \min[A(x), B(x)] \ \forall x \in \mathbf{X} \qquad (5.1)$$

where $(A \cap B)(x)$ denotes the characteristic function of the intersection $A \cap B$.

Fuzzy Systems Engineering: Toward Human-Centric Computing, by Witold Pedrycz and Fernando Gomide
Copyright © 2007 John Wiley & Sons, Inc.

102 Chapter 5 Operations and Aggregations of Fuzzy Sets

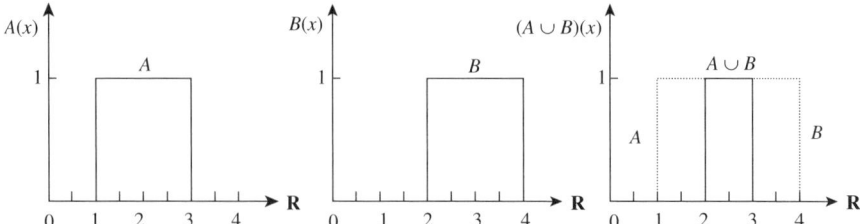

Figure 5.1 Intersection of sets represented in terms of their characteristic functions.

We consider the union of sets A and B and express its characteristic function in terms of the respective characteristic functions of A and B. For example, if A and B are the same intervals as presented before, then $A \cup B = \{x \in \mathbf{R} | 1 \leq x \leq 4\}$. We note that the value of the characteristic function of the union is taken as the maximum of corresponding values of the characteristic functions $A(x)$ and $B(x)$ at each point of the universe, see Figure 5.2.

Therefore, given the characteristic functions of A and B, we determine the characteristic function of the union as

$$(A \cup B)(x) = \max\,[A(x), B(x)] \;\; \forall x \in \mathbf{X} \tag{5.2}$$

where $(A \cup B)(x)$ denotes the characteristic function of the intersection $A \cup B$.

Likewise, as Figure 5.3 suggests, the complement $\overline{A}$ of set A, expressed in terms of its characteristic function, is the one-complement of the characteristic function of

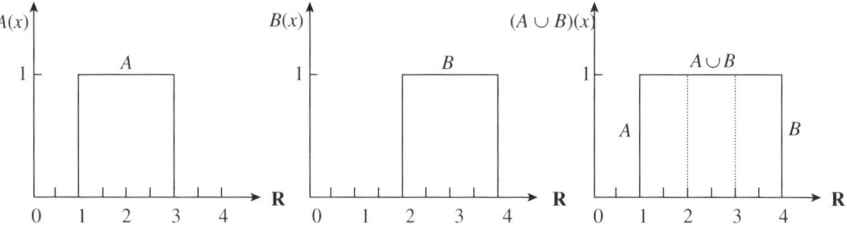

Figure 5.2 Union of two sets expressed in terms of their characteristic functions.

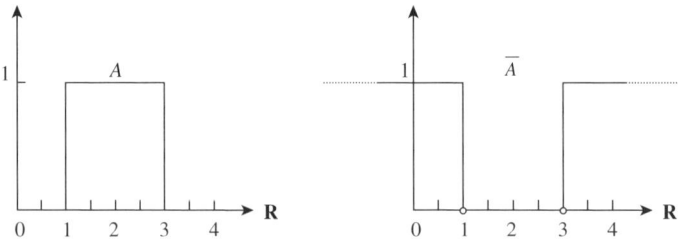

Figure 5.3 Complement of a set in terms of its characteristic function.

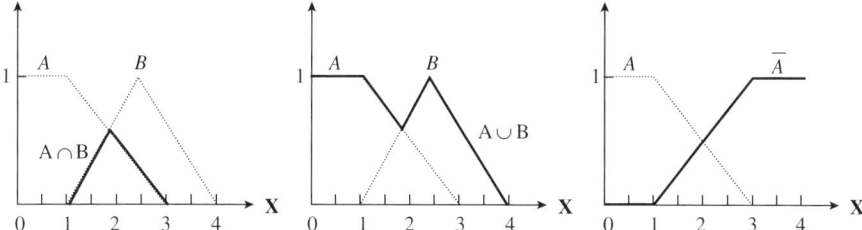

Figure 5.4 Operations on fuzzy sets realized with the use of min, max, and complement functions.

A. For instance, if $A = \{x \in \mathbf{R} | 1 \leq x \leq 3\}$, which is the same interval as discussed before, then $\overline{A} = \{x \in \mathbf{R} | 4 < x < 1\}$, Figure 5.3.

In general, the characteristic function of the complement of a set A is given in the form

$$\overline{A}(x) = 1 - A(x) \; \forall x \in \mathbf{X} \tag{5.3}$$

One may anticipate that since sets are particular instances of fuzzy sets, the operations of intersection, union, and complement as previously defined should equally extend to fuzzy sets. Indeed, when we use membership functions in expressions (5.1)–(5.3), these formulas serve as definitions of intersection, union, and complement of fuzzy sets. An illustration of these operations is included in Figure 5.4.

Standard set and fuzzy set operations fulfill the basic properties as summarized in Table 5.1.

Looking at Figures 5.3 and 5.4, however, we note that the laws of noncontradiction and excluded middle are satisfied by sets but not by fuzzy sets when using the operations shown in Table 5.2. Particularly worth noting is a violation of the noncontradiction law once it shows the issue of fuzziness from the point of view of the coexistence of a class and its complement, an issue we have already discussed in Chapter 2. This coexistence is impossible in set theory and a contradiction in conventional logic. Interestingly, if we consider a particular subnormal fuzzy set A

Table 5.1 Basic Properties of Set and Fuzzy Set Operations.

1. Commutativity	$A \cup B = B \cup A$
	$A \cap B = B \cap A$
2. Associativity	$A \cup (B \cup C) = (A \cup B) \cup C$
	$A \cap (B \cap C) = (A \cap B) \cap C$
3. Distributivity	$A \cup (B \cap C) = (A \cup B) \cap (A \cup C)$
	$A \cap (B \cup C) = (A \cap B) \cup (A \cap C)$
4. Idempotency	$A \cup A = A$
	$A \cap A = A$
5. Boundary conditions	$A \cup \phi = A$ and $A \cup \mathbf{X} = \mathbf{X}$
	$A \cap \phi = \phi$ and $A \cap \mathbf{X} = A$
6. Involution	$\overline{\overline{A}} = A$
7. Transitivity	if $A \subset B$ and $B \subset C$, then $A \subset C$

Table 5.2 Noncontradiction and Excluded Middle for Standard Operations.

	Sets	Fuzzy sets
8. Noncontradiction	$A \cap \overline{A} = \phi$	$A \cap \overline{A} \neq \phi$
9. Excluded middle	$A \cup \overline{A} = X$	$A \cup \overline{A} \neq X$

whose membership function is constant and equal to 0.5 for all elements of the universe, then using (5.1)–(5.3) we note that $A = A \cup \overline{A} = A \cap \overline{A} = \overline{A}$, a situation in which there is no way to distinguish the fuzzy set from its complement and any fuzzy set that results from standard operations with them. The value 0.5 is a crossover point representing a balance between membership and nonmembership at which we attain the highest level of fuzziness. We may visualize this fact geometrically as follows.

Let us consider a simple case in which $\mathbf{X} = \{x_1, x_2\}$. Thus, the power set $P(\mathbf{X})$ of $\mathbf{X}$ is $P(\mathbf{X}) = \{\phi, \{x_1\}, \{x_2\}, \{x_1, x_2\}\}, |P(\mathbf{X})| = 2^2 = 4$, and as shown in Figure 5.5, the elements of $P(\mathbf{X})$ lie at the corners of the unity square, the unit square is the set of all fuzzy sets of $\mathbf{X}$ denoted by $\mathbf{F}(\mathbf{X})$. In the figure we recognize the complement of a fuzzy set A as well as its intersection and union with A itself. A complement of any set in $F(\mathbf{X})$ is always symmetric to A along a diagonal. We note that $A = (0.5, 0.5)$ is located at the center of the square, a point at which

$$A = A \cup \overline{A} = A \cap \overline{A} = \overline{A} \tag{5.4}$$

This is a particular feature of the standard operations; they allow full coexistence of a sets and its complement.

The crossover point value of 0.5 between membership and nonmembership is where fuzziness is highest because a fuzzy set and its complement are indiscernible. In fact, as we shall see next, there exists a general class of operators that are qualified to act as intersection, union, and complement of fuzzy sets. Although for sets the general operators produce the same result as the standard operators, this is not the case with fuzzy sets. Therefore, (5.4) does not hold for any choice of intersection, union, and complement operators.

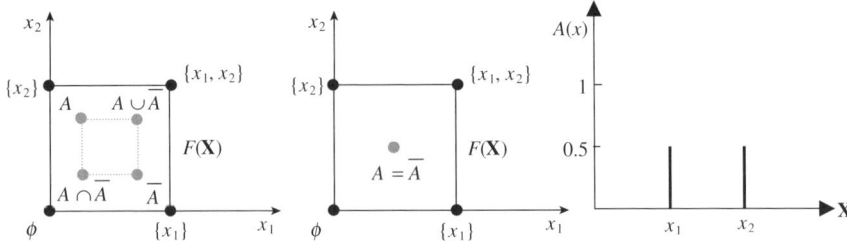

Figure 5.5 Geometric view of the standard operations with fuzzy sets.

5.2 GENERIC REQUIREMENTS FOR OPERATIONS ON FUZZY SETS

Operations on fuzzy sets concern manipulation of their membership functions. Therefore, they are domain dependent, and different contexts may require their different realizations. For instance, since operations provide ways to combine information, they can be performed differently in image processing, control, and diagnostic systems, for example. When contemplating the realization of operations of intersection and union of fuzzy sets, we should require a satisfaction of the following intuitively appealing properties:

(**a**) commutativity,

(**b**) associativity,

(**c**) monotonicity, and

(**d**) identity.

The last requirement of identity takes on a different form depending on the operation. More specifically, in the case of intersection, we anticipate that an intersection of any fuzzy set with the universe of discourse **X** should return this fuzzy set. For the union operations, the identity implies that the union of any fuzzy set and an empty fuzy set returns the fuzzy set.

Thus, any binary operator $[0, 1] \times [0, 1] \to [0, 1]$ that satisfies the collection of the requirements outlined above can be regarded as a potential candidate to realize the intersection or union of fuzzy sets. Note also that identity acts as boundary conditions meaning that when confining to sets, the above stated operations return the same results as encountered in set theory. In general, idempotency is not required; however, the realizations of union and intersection could be idempotent as happens for the operations of minimum and maximum where $(\min(a, a) = a$ and $\max(a, a) = a)$.

5.3 TRIANGULAR NORMS

In the theory of fuzzy sets, triangular norms offer a general class of operators of intersection and union. For instance, t-norms generalize intersection of fuzzy sets. Given a t-norm, a dual operator called a t-conorm (or s-norm) can be derived using the relationship $x s y = 1 - (1 - x) t (1 - y), \forall x, y \in [0, 1]$, the De Morgan law (Schweizer and Sklar, 1983), but t-conorm can also be described by an independent axiomatic definition (Klement et al., 2000). Triangular conorms provide generic models for the union of fuzzy sets.

5.3.1 Defining t-Norms

A triangular norm, t-norm for brief, is a binary operation $t : [0, 1] \times [0, 1] \to [0, 1]$ that satisfies the following properties:

1. Commutativity: $a\,t\,b = b\,t\,a$
2. Associativity: $a\,t\,(b\,t\,c) = (a\,t\,b)\,t\,c$
3. Monotonicity: if $b \leq c$, then $a\,t\,b \leq a\,t\,c$
4. Boundary conditions: $a\,t\,1 = a$
$a\,t\,0 = 0$

where $a, b, c \in [0, 1]$.

Let us elaborate on the meaning of these requirements vis-à-vis the use of t-norms as models of operators of intersection of fuzzy sets. There is a one-to-one correspondence between the general requirements outlined in the previous section and the properties of t-norms. The first three reflect the general character of set operations. Boundary conditions stress the fact all t-norms attain the same values at boundaries of the unit square $[0, 1] \times [0, 1]$. Thus, for sets, any t-norm produces the same result that coincides with the one one could have expected in set theory when dealing with intersection of sets, that is, $A \cap X = A, A \cap \varnothing = \varnothing$. Some commonly encountered examples of t-norms include the following operations:

1. Minimum: $a\,t_m\,b = \min(a, b) = a \wedge b$
2. Product: $a\,t_p\,b = ab$
3. Lukasiewicz: $a\,t_l\,b = \max(a + b - 1, 0)$
4. Drastic product: $a\,t_d\,b = \begin{cases} a & \text{if } b = 1 \\ b & \text{if } a = 1 \\ 0 & \text{otherwise} \end{cases}$

The plots of the operations of minimum(t_m), product (t_p), Lukasiewicz (t_l), and drastic product (t_d) operators are shown in Figure 5.6 together with a corresponding example of their application to triangular fuzzy sets on the closed interval $[0, 8], A = (x, 1, 3, 6)$ and $B = (x, 2.5, 5, 7)$. Triangular norms produce distinct results and to know how they behave helps to choose operators for specific applications. For instance, minimum, product, and Lukasiewicz are continuous whereas the drastic product is not. Minimum is idempotent whereas the remaining t-norms are not. In contrart, when we choose the drastic product, $A \cap \overline{A} = \phi$, and therefore this realization of the intersection of fuzzy sets recovers the noncontradiction property.

In general, t-norms cannot be linearly ordered. One can demonstrate that the min (t_m) t-norm is the largest t-norm, whereas the drastic product is the smallest one. They form the lower and upper bounds of the t-norms in the following sense:

$$a\,t_d\,b \leq a\,t\,b \leq a\,t_m\,b = \min(a, b) \tag{5.5}$$

In many applications, continuity of operations on fuzzy sets is a highly desirable feature. We do not have too much confidence in operations if they produce substantially different results if the valucs (membership grades) of arguments change slightly. On the contrary, one should emphasize the importance of continuity: we anticipate that small changes in the membership degrees of fuzzy sets A and B

5.3 Triangular Norms   107

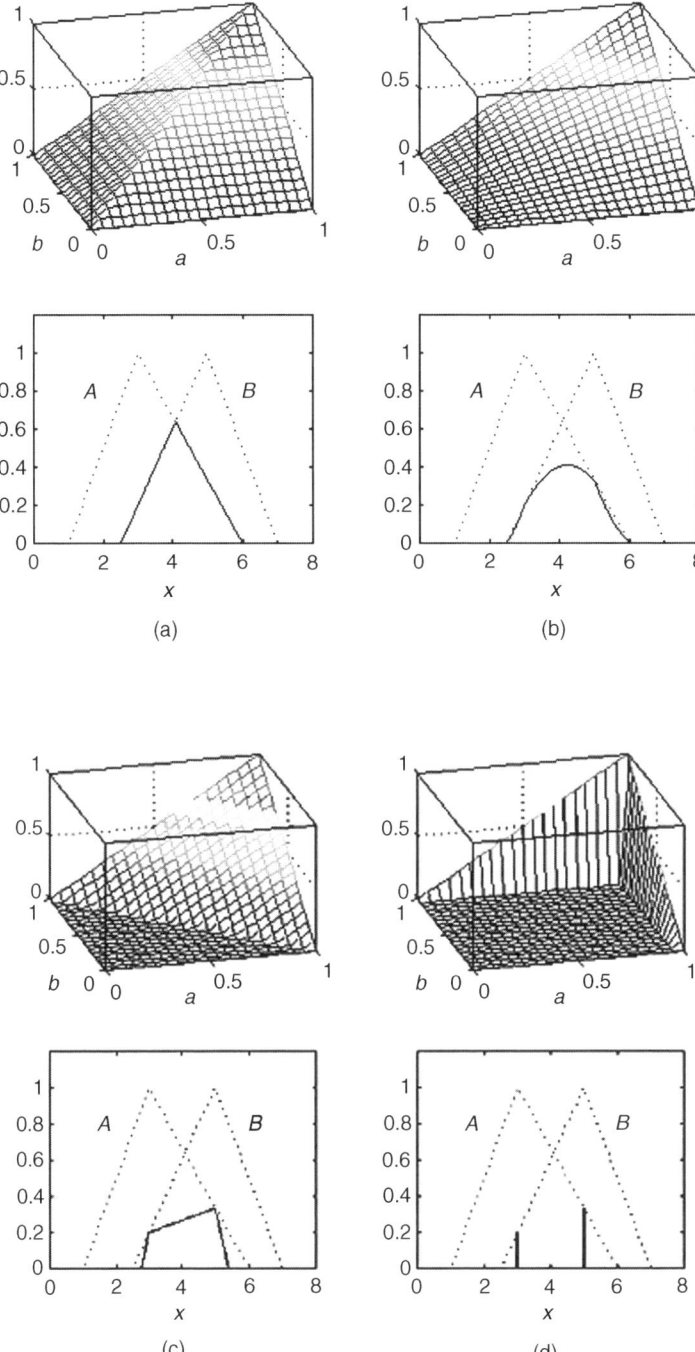

Figure 5.6 Examples of t-norms used in the realization of intersection of fuzzy sets A and B: (a) min, (b) product, (c) Lukasiewicz, and (d) drastic product.

produce small changes in the membership grades of the resulting intersection $A \cap B$. This is the case in many system modeling, optimization, control, decision-making, and data analysis applications. Under such circumstances, the use of a drastic product could be ruled out.

Since t-norms are monotonic and commutative, they are monotonic in both arguments:

$$a_1 \, t \, b_1 \leq a_2 \, t \, b_2 \quad \text{whenever} \quad a_1 \leq a_2 \quad \text{and} \quad b_1 \leq b_2$$

Hence, a t-norm is continuous if and only if it is continuous in each argument (Klement et al., 2000).

We say that a t-norm is Archimedean if it is continuous and $\forall a \in (0, 1), (a \, t \, a) < a$. Furthermore, we say that an Archimedean t-norm is strict if it is strictly monotonic on $(0, 1) \times (0, 1)$. For instance, the product and Lukasiewicz t-norms are Archimedean whereas the minimum and the drastic product are not.

Since t-norms are associative, without any misunderstanding we may use the notation $a^2 = a \, t \, a, a^2 \, t \, a = a^3, \ldots, a^{n-1} \, t \, a = a^n$. A t-norm is nilpotent if $a^n = 0$ and $a \neq 1$, where n is a positive integer, eventually depending on the value of a. For example, the Lukasiewicz t-norm is a nilpotent t-norm.

A general and detailed exposure to the subject of triangular norms and conorms, especially their characterization and properties, is given in Klement et al. (2000) and Jenei (2002, 2004).

5.3.2 Constructors of t-Norms

Often, various applications require different t-norms to attain domain context demands, and methods to construct t-norms are of utmost importance. There are three basic ways of forming new t-norms:

- construction of new t-norms on the basis of some given t-norms using monotonic transformations;
- the use of addition and multiplication of real numbers together with functions of one variable, called additive and multiplicative generators, respectively; and
- construction of new t-norms from a family of given t-norms based on a concept of a so-called ordinal sum.

Let us discuss these techniques in detail.

Monotonic Function Transformation

If t is a t-norm and $h : [0, 1] \to [0, 1]$ is strictly increasing bijection, then the operator

$$t_h(a, b) = h^{-1}(t(h(a), h(b))) \tag{5.6}$$

is a t-norm. Let us recall that a bijection (bijective function) is a function that is both injective (i.e., a one-to-one mapping) and surjective (a function that maps the domain of the function *onto* the range of the function).

Essentially, this method rescales the original values of membership degrees by regarding the function h as a scaling transformation. Constructor (5.6) preserves continuity, the Archimedean property, and strictness. Figure 5.7 illustrates the case when we choose $h(x) = x^2$ and apply this transformation to several t-norms, such as

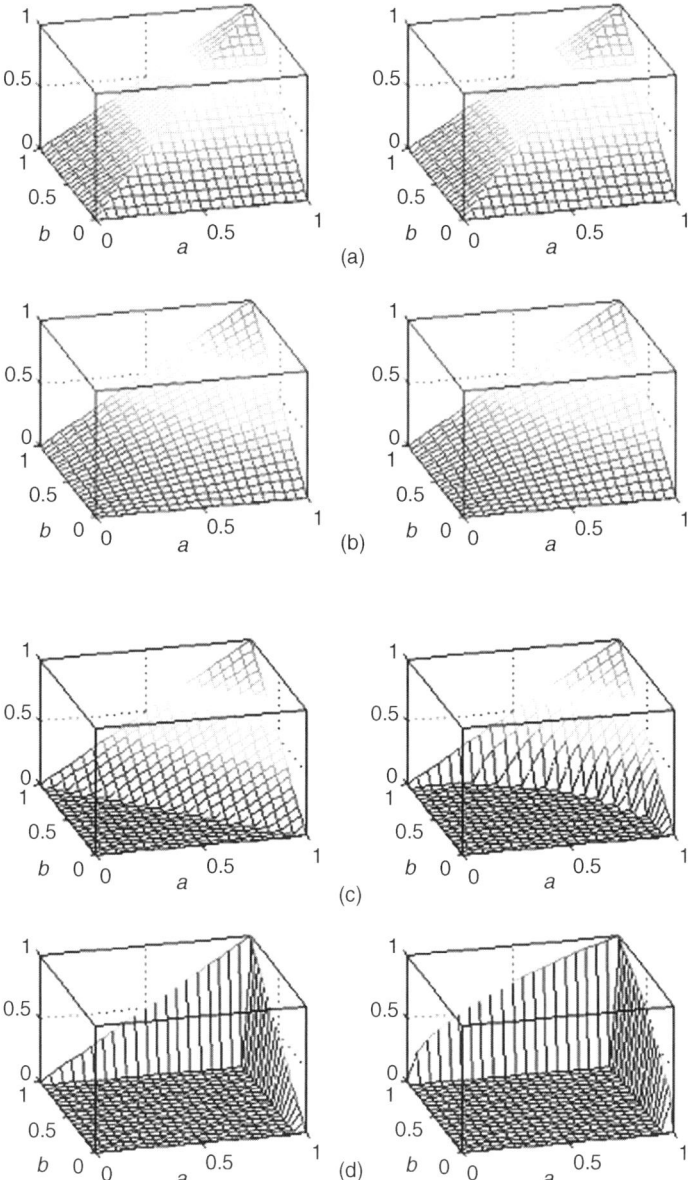

Figure 5.7 Monotonic transformations of selected t-norms: (a) min, (b) product, (c) Lukasiewicz, and (d) drastic product.

t_m, t_p, t_l, and t_d. As we see, t_m and t_p remain the same. Recall that t_m is continuous and t_p Archimedean. In the last two examples, we see an effect of rescaling over the original t-norms t_l and t_d.

Additive and Multiplicative Generators

An additive generator of a t-norm is a continuous and strictly decreasing function from [0,1] onto $\mathbf{R}^+ = [0,\infty)$ with $f(1) = 0$, $f : [0,1] \to [0,\infty)$. Given an additive generator, we can construct any Archimedean t-norm as follows:

$$a\, t_f\, b = f^{-1}(f(a) + f(b)) \tag{5.7}$$

And, conversely, any Archimedean t-norm comes with its additive generator. The t-norm is strict if and only if $f(0) = \infty$ and nilpotent if and only if $f(0) < \infty$. Figure 5.8 illustrates the idea of additive generators.

For example, the product $ab (= a\, t_p\, b)$ is an example of a strict Archimedean t-norm. Its additive generator is obtained by solving the following *functional equation* (recall that in functional equations, the unknown is just a function)

$$ab = f^{-1}(f(a) + f(b)).$$

We can verify that the function $f(x) = -\log(x)$ solves the functional equation. To demonstrate this note that

$$f(a) + f(b) = -\log(a) - \log(b) = -(\log(a) + \log(b)) = -\log ab$$

Therefore, $f^{-1}(u) = e^{-u}$ and hence $f^{-1}(f(a) + f(b)) = e^{\log ab} = ab$. Actually, in this case any function $f(x) = -c\log(cx)$, with constant $c > 0$, is also an additive generator. In general, we can note that additive generators are unique up to a *positive constant*.

A multiplicative generator of a t-norm is a continuous and strictly increasing function mapping [0,1] onto [0,1] with $g(1) = 1$, $g : [0,1] \to [0,1]$. Given a multiplicative generator we can construct an Archimedean t-norm as follows:

$$a\, t\, b = g^{-1}(g(a)g(b)) \tag{5.8}$$

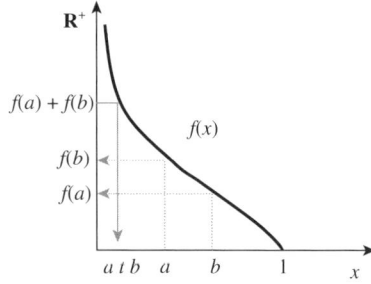

Figure 5.8 Additive generators of t-norms.

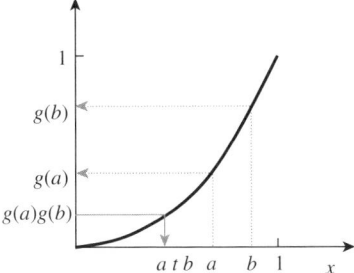

Figure 5.9. Multiplicative generators of t-norms.

and, conversely, any Archimedean t-norm has a multiplicative generator. Figure 5.9 illustrates the idea of multiplicative generators. For instance, one can show that the t-norm realized as product has a multiplicative generator in the form $g(x) = x^2$.

It is interesting to note that if we let $g(x) = e^{-f(x)}$, then additive and multiplicative generators (5.7) and (5.8) produce the same t-norms, as it can be verified, for example, with the generators of the product t-norm.

Ordinal Sums

Ordinal sum deals with a construction of new t-norms with the use of a family of given t-norms. Ordinal sum joins arbitrary, appropriately transformed t-norms along the diagonal of the unit square and takes the minimum otherwise.

Let $I = \{[\alpha_k, \beta_k], k \in K\}$ be a nonempty, countable family of pairwise disjoint open subintervals of $[0,1]$, and let $\tau = \{t_k, k \in K\}$ be a family of the corresponding t-norms. The ordinal sum of summands $\langle \alpha_k, \beta_k, t_k \rangle, k \in K$, denoted by $t_o = (\langle \alpha_k, \beta_k, t_k \rangle, k \in K)$, is the function $t_o : [0,1] \times [0,1] \to [0,1]$ defined by

$$t_o(a,b,I,\tau) = \begin{cases} \alpha_k + (\beta_k - \alpha_k) t_k \left(\dfrac{a - \alpha_k}{\beta_k - \alpha_k}, \dfrac{b - \alpha_k}{\beta_k - \alpha_k} \right), & \text{if } a, b \in [\alpha_k, \beta_k] \\ \min(a,b), & \text{otherwise} \end{cases} \quad (5.9)$$

It can be shown that t_o is a t-norm (Klement et al., 2000). Moreover, any continuous t-norm can be constructed as an ordinal sum of Archimedean t-norms (Jenei 2002). Several methods to construct left-continuous triangular norms are provided in Jenei, (2004).

Ordinal sum is a construct that accounts for the local properties of intersection operators. As such, ordinal sums offer a substantial level of modeling flexibility by capturing local properties of intersection of fuzzy sets depending upon the ranges of membership grades to be combined. For instance, if we choose t_p and t_l, the ordinal sum $t_o = (\langle 0.2, 0.4, t_p \rangle, \langle 0.5, 0.7, t_l \rangle)$ reads as follows (Fig. 5.10):

$$t_o(a,b) = \begin{cases} 0.2 + 5(a - 0.2)(b - 0.2), & \text{if } (a,b) \in [0.2, 0.4]^2 \\ 0.5 + \max(a + b - 1.2, 0), & \text{if } (a,b) \in [0.5, 0.7]^2 \\ \min(a,b), & \text{otherwise} \end{cases}$$

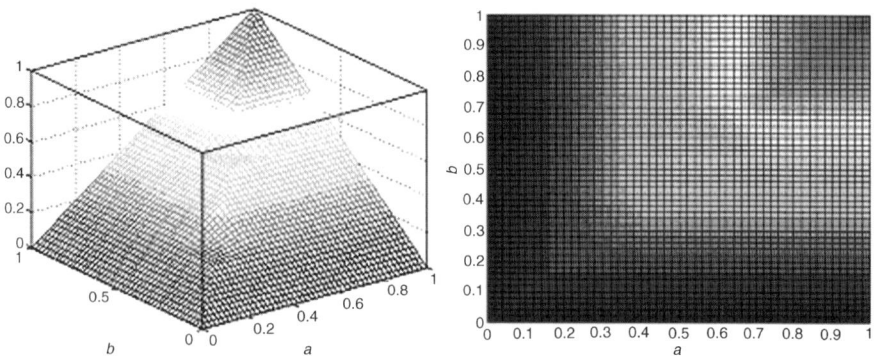

Figure 5.10 Ordinal sum constructed with the use of two t-norms (a) 3D plot, and (b) contour plot.

5.4 TRIANGULAR CONORMS

Triangular conorms are functions $s : [0, 1] \times [0, 1] \to [0, 1]$ that serve as generic realizations of the union operator on fuzzy sets. Similarly as triangular norms, conorms provide the highly desirable modeling flexibility needed to construct fuzzy models.

5.4.1 Defining t-Conorms

Triangular conorms can be viewed as dual operators to the t-norms and as such, explicitly defined with the use of De Morgan laws. We may characterize them in a fully independent manner by providing the following definition.

A triangular conorm (s-norm) is a binary operation $s : [0, 1] \times [0, 1] \to [0, 1]$ that satisfies the following requirements:

1. Commutativity: $\quad a\,s\,b = b\,s\,a$
2. Associativity: $\quad a\,s\,(b\,s\,c) = (a\,s\,b)\,s\,c$
3. Monotonicity: $\quad$ if $b \leq c$, then $a\,s\,b \leq a\,s\,c$
4. Boundary conditions: $\quad a\,s\,0 = a$
   $\quad a\,s\,1 = 1$
   where $\quad a, b, c \in [0, 1]$

One can show that $s : [0, 1] \times [0, 1] \to [0, 1]$ is a t-conorm if and only if there exists a t-norm (dual t-norm) such that $\forall a, b \in [0, 1]$ we have

$$a\,s\,b = 1 - (1 - a)\,t\,(1 - b) \qquad (5.10)$$

For the corresponding dual t-norm we have

$$a\,t\,b = 1 - (1 - a)\,s\,(1 - b) \qquad (5.11)$$

5.4 Triangular Conorms

The duality expressed by (5.10) and (5.11) can be viewed as an alternative definition of t-conorms. This duality allows us to deduce the properties of t-conorms on the basis of the analogous properties of t-norms. Notice that after rewriting (5.10) and (5.11), we obtain

$$(1-a)\,t\,(1-b) = 1 - a\,s\,b$$
$$(1-a)\,s\,(1-b) = 1 - a\,t\,b$$

These two relationships can be expressed symbolically as

$$\overline{A} \cap \overline{B} = \overline{A \cup B}$$
$$\overline{A} \cup \overline{B} = \overline{A \cap B}$$

that are nothing but the De Morgan laws.

The boundary conditions mean that all t-conorms behave similarly at the boundary of the unit square $[0, 1] \times [0, 1]$. Thus, for sets, any t-conorm returns the same result as encountered in set theory.

A list of commonly used t-conorms includes the following:

Maximum:	$a\,s_m\,b = \max(a,b) = a \vee b$
Probabilistic sum:	$a\,s_p\,b = a + b - ab$
Lukasiewicz:	$a\,s_l\,b = \min(a+b, 1)$
Drastic sum:	$a\,s_d\,b = \begin{cases} a, & \text{if } b = 0 \\ b, & \text{if } a = 0 \\ 1, & \text{otherwise} \end{cases}$

The characteristics of the maximum(s_m), probabilistic sum (s_p), Lukasiewicz (s_l), and drastic sum (s_d) operators are shown in Figure 5.11. We have included the union of two triangular fuzzy sets defined in $[0, 8]$, $A = (x, 1, 3, 6)$ and $B = (x, 2.5, 5, 7)$.

As visible in Figure 5.11, maximum, probabilistic sum, and Lukasiewicz t-conorm are continuous whereas the drastic sum is not. Maximum is idempotent whereas any other t-conorms are not. On the contrary, when we choose the drastic sum, the properties $A \cup \overline{A} = \mathbf{X}$ and the excluded middle are satisfied.

As in the case of t-norms, t-conorms cannot be linearly ordered, but as Figure 5.11 suggests, the max (s_m) t-conorm is the smallest, in the sense that it is a lower bound for all t-conorms, whereas the drastic sum is the upper bound t-conorms:

$$\max(a,b) = a\,s_m\,b \leq a\,s\,b \leq a\,s_d\,b \qquad (5.12)$$

As the drastic sum illustrates, t-conorms need not be continuous. Since t-conorms are monotonic and commutative, they are monotonic in both arguments, that is,

$$a_1\,s\,b_1 \leq a_2\,s\,b_2 \quad \text{whenever} \quad a_1 \leq a_2 \quad \text{and} \quad b_1 \leq b_2$$

Therefore, t-conorm is continuous if and only if it is continuous in each argument (Klement et al., 2000).

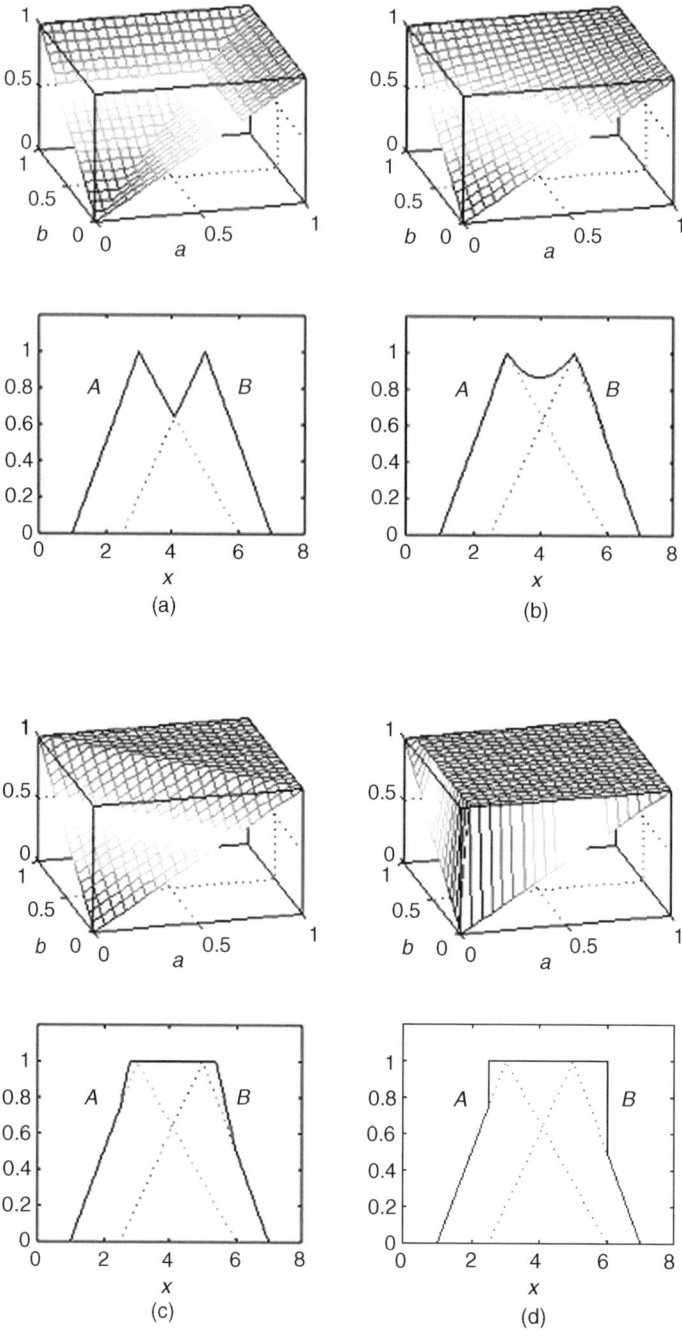

Figure 5.11 Examples of t-conorms as unions of fuzzy sets: (a) max, (b) probabilistic sum, (c) Lukasiewicz, and (d) drastic sum.

A t-conorm is Archimedean if it is continuous and $\forall a \in (0,1), (a\,s\,a) > a$. An Archimedean t-conorm is strict if it is strictly monotonic on $(0,1) \times (0,1)$. For instance, the probabilistic sum and Lukasiewicz t-conorms are Archimedean whereas the maximum and the drastic sum are not.

Since t-conorms are associative we may denote $a^2 = a\,s\,a, a^2 s\,a = a^3, \ldots,$ $a^{n-1} s\,a = a^n$. A t-conorm is nilpotent if $a^n = 1$ and $a \neq 1$, where n is a positive integer, eventually depending on a. A t-conorm is nilpotent if it is dual of a nilpotent t-norm. For example, the Lukasiewicz t-norm is a nilpotent t-conorm.

5.4.2 Constructors of t-Conorms

The construction of t-conorms proceeds in a similar way as discussed for t-norms. Again, three main categories of development methods are available. As before, given their importance, we focus on continuous t-conorms. For details, refer to Nguyen and Walker (1999), Klement et al. (2000), Jenei (2002, 2004).

Monotonic Function Transformation

If s is a t-conorm and $h : [0,1] \to [0,1]$ is strictly increasing bijection, then the operator

$$s_h(a,b) = h^{-1}(s(h(a), h(b))) \qquad (5.13)$$

is a t-conorm.

The method rescales the original membership values using h as a scaling transformation. Constructor (5.13) preserves continuity, the Archimedean property, and strictness. Figure 5.12 illustrates the case when $h(x) = x^2$ and we use s_m, s_p, s_l, and s_d. As we note, s_m remains the same. Also, s_m is continuous whereas s_p is Archimedean. The last three examples visualize the scaling effect over the original s-norms s_p, s_l, and s_d.

Additive and Multiplicative Generators

An additive generator of a given t-conorm is a continuous and strictly increasing mapping [0,1] onto $\mathbf{R}^+ = [0, \infty)$ with $f(0) = 0$, $f : [0,1] \to [0, \infty)$. Given an additive generator, we can construct any Archimedean t-conorm as follows:

$$a\,s_f\,b = f^{-1}(f(a) + f(b))) \qquad (5.14)$$

Conversely, any Archimedean t-conorm comes with its additive generator. The t-conorm is nilpotent if and only if $f(1) < \infty$. Figure 5.13 illustrates the idea of additive generators.

For example, the probabilistic sum $a + b - ab = a\,s\,b$ is an example of a strict Archimedean t-conorm. Its additive generator is obtained by solving the following functional equation:

$$a + b - ab = f^{-1}[f(1) \wedge f(a) + f(b)]$$

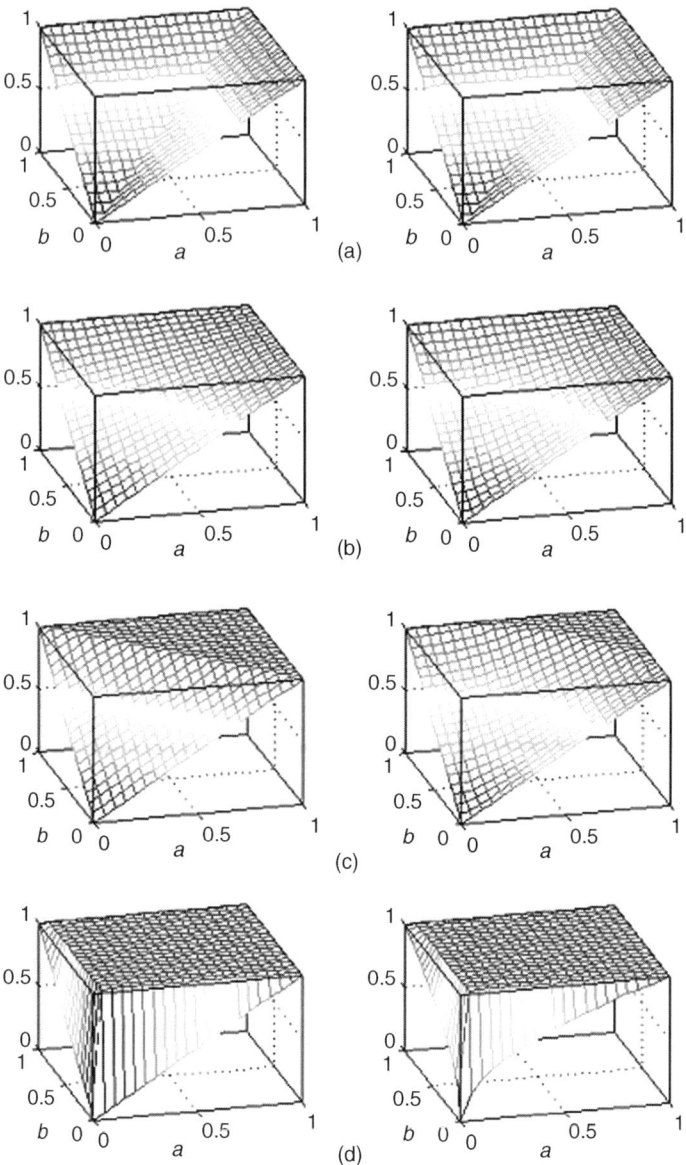

Figure 5.12 Examples of monotonic transformation of t-conorms; $h(x) = x^2$: (a) max, (b) probabilistic sum, (c) Lukasiewicz, and (d) drastic sum.

One can verify that the function $f(x) = -\log(1-x)$ is a solution to this functional equation. In fact, we have

$$f(a) + f(b) = -\log(1-a) - \log(1-b) = -(\log(1-a) + \log(1-b))$$
$$= -\log(1-a)(1-b)$$

5.4 Triangular Conorms

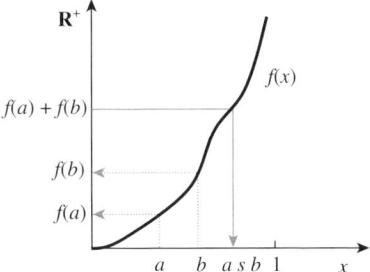

Figure 5.13 Additive generator of t-conorms.

Therefore, $f^{-1}(u) = 1 - e^{-u}$ and hence $f^{-1}(f(a) + f(b)) = 1 - e^{\log(1-a)(1-b)}$ $= 1 - (1-a)(1-b) = a + b - ab$. As before, the additive generators are unique up to some positive constant.

A multiplicative generator of a t-conorm is a continuous and strictly decreasing mapping $[0,1]$ onto $[0,1]$ with $g(0) = 1, g : [0, 1] \to [0, 1]$. Given a multiplicative generator, we can construct an Archimedean t-norm in the following way:

$$a \, t \, b = g^{-1}(g(a)g(b)) \quad (5.15)$$

And, conversely, any Archimedean t-conorm has a multiplicative generator. The t-conorm is nilpotent if and only if $g(1) > 0$ and is strict if $g(1) = 0$ (Nguyen and Walker, 1999). Figure 5.14 illustrates the idea of multiplicative generators. For instance, tone can show that the probabilistic sum has $g(x) = 1 - x$ as a multiplicative generator.

Also, if we let $g(x) = e^{-f(x)}$, then additive and multiplicative generators (5.14) and (5.15) produce the same t-conorms, as it can be verified, for example, with the generators of the probabilistic sum.

Ordinal Sums

An ordinal sum involves a construction of new t-conorms on the basis of a family of some given t-conorms. The construct is analogous to the one presented before for t-norms. The only difference is that the regions outside the main diagonal along

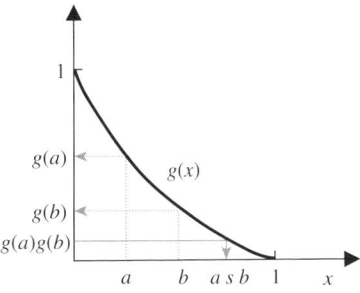

Figure 5.14 An example of multiplicative generators of some t-conorm.

118   Chapter 5   Operations and Aggregations of Fuzzy Sets

which different t-conorms are located, we use the maximum operation instead of the minimum operator used in the ordinal sum of t-norms.

Let $I = \{[\alpha_k, \beta_k], k \in K\}$ be a nonempty, countable family of pairwise disjoint open subintervals of [0,1], and let $\sigma = \{s_k, k \in K\}$ be a family of the corresponding t-conorms. Then the ordinal sum of summands $\langle \alpha_k, \beta_k, s_k \rangle, k \in K$, denoted $s_o = (\langle \alpha_k, \beta_k, s_k \rangle, k \in K)$, is the function $s_o : [0, 1] \times [0, 1] \to [0, 1]$ defined by

$$s_o(a, b, I, \sigma) = \begin{cases} \alpha_k + (\beta_k - \alpha_k) s_k \left(\dfrac{a - \alpha_k}{\beta_k - \alpha_k}, \dfrac{b - \alpha_k}{\beta_k - \alpha_k} \right), & \text{if } a, b \in [\alpha_k, \beta_k] \\ \max(a, b), & \text{otherwise} \end{cases}$$

(5.16)

It can be shown that s_o is a t-conorm (Klement et al., 2000). Moreover, any continuous t-conorm can be constructed as an ordinal sum of Archimedean t-conorms.

For instance, if we choose s_p and s_l, the ordinal sum $s_o = (\langle 0.2, 0.4, s_p \rangle, \langle 0.5, 0.7, s_l \rangle)$ is as follows:

$$s_o(a,b) = \begin{cases} 0.2 + (a - 0.2) + (b - 0.2) - 5(a - 0.2)(b - 0.2), & \text{if } (a, b) \in [0.2, 0.4]^2 \\ 0.5 + 0.2 \min\{5(a - 0.2) + 5(b - 0.2), 1\}, & \text{if } (a, b) \in [0.5, 0.7]^2 \\ \max(a, b), & \text{otherwise} \end{cases}$$

The plots of the ordinal sum are shown in Figure 5.15.

5.5 TRIANGULAR NORMS AS A GENERAL CATEGORY OF LOGICAL OPERATORS

Fuzzy propositions involve combination of given linguistic statements (or their symbolic representations) such as in

1. temperature is *low* and humidity is *mild* and
2. velocity is *high* or noise level is *low*.

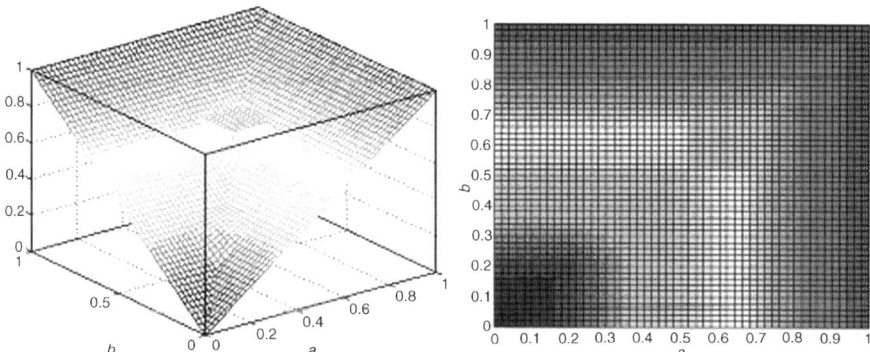

Figure 5.15   Ordinal sum using t-conorms: (a) 3D plot and (b) contour plot.

5.5 Triangular Norms as a General Category of Logical Operators

The distinctive feature of these propositions is the use of logical operations $\wedge$ (and), $\vee$ (or) to combine linguistic statements into propositions. For instance, in the first example we have a conjunction (and, $\wedge$) of linguistic statements whereas in the second there is a disjunction (or, $\vee$) of the statements. Given the truth values of each statement, the question is how to determine the truth value of the composite statement or, equivalently, the truth value of the proposition.

Let us denote by truth$(P) = p \in [0, 1]$, the truth value of proposition P. The value of p equal to 0 means that the proposition is false. The value of p equal to 1 states that P is true. Intermediate values $p \in (0, 1)$ indicate partial truth of the proposition. To compute the truth value of composite propositions coming in the form of $P \wedge Q, P \vee Q$ given the truth values p and q of its components, we have to come up with operations that transforms truth values p and q into the corresponding truth values $p \wedge q$ and $p \vee q$. To make these operations meaningful, we require that they satisfy some basic requirements. For instance, we require that $p \wedge q$ and $q \wedge p$ (similarly, $p \vee q$ and $q \vee p$) produce the same truth values. Likewise, we require that the truth value of $(p \wedge q) \wedge r$ is the same as the following combination $p \wedge (q \wedge r)$. In other words, the conjunction and disjunction operations are to be commutative and associative. Also, it seems natural that when the truth values of an individual statement increase, the truth values of their combinations also increase. This brings the requirement of monotonicity. Moreover, if P is absolutely false, $p = 0$, then $P \wedge Q$ should also be false no matter what the truth value of Q is. Furthermore, the truth value of $P \vee Q$ should coincide with the truth value of Q. On the contrary, if P is absolutely true, $p = 1$, then the truth value of $P \wedge Q$ should coincide with the truth value of Q, while $P \vee Q$ should also be true. As we have already discussed, triangular norm and conorm are the general families of logic connectives that comply with these requirements. Triangular norms provide a general category of logical connectives in the sense that t-norms are used to model conjunction operators while t-conorms serve as models of disjunctions.

Let $L = \{P, Q, \ldots\}$ be a set of single (atomic) statements $P, Q, \ldots$ and truth: $L \to [0, 1]$ a function that assigns truth values $p, q, \ldots \in [0, 1]$ to each element of L. Then we have

$$\text{truth}(P \text{ and } Q) \equiv \text{truth }(P \wedge Q) \to p \wedge q = p\, t\, q$$
$$\text{truth }(P \text{ or } Q) \equiv \text{truth }(P \vee Q) \to p \vee q = p\, s\, q$$

Table 5.3 includes examples of truth values for $P, Q, P \wedge Q$, and $P \vee Q$, when we selected the min and product t-norms, and the max and probabilistic sum t-conorms, respectively. For $p, q \in \{0, 1\}$, the results coincide with the classic interpretation of conjunction and disjunction for any choice of the triangular norm and conorm. The differences are present when $p, q \in (0, 1)$.

A point worth noting here concerns the interpretation of set operations in terms of logical connectives. By being supported by the isomorphism between set theory and propositional two-valued logic, the intersection and union can be identified with conjunction and disjunction, respectively. This can also be realized with triangular norms viewed as general conjunctive and disjunctive connectives within the

Table 5.3 Triangular Norms as Generalized Logical Connectives.

p	q	min(p, q)	max(p, q)	pq	p + q − pq
1	1	1	1	1	1
1	0	0	1	0	1
0	1	0	1	0	1
0	0	0	0	0	0
0.2	0.5	0.2	0.5	0.1	0.6
0.5	0.8	0.5	0.8	0.4	0.9
0.8	0.7	0.7	0.8	0.56	0.94

Table 5.4 φ Operator in Case of Boolean Values of Its Arguments.

a	b	$a \Rightarrow b$	$a \varphi b$
0	0	1	1
0	1	1	1
1	0	0	0
1	1	1	1

framework of multivalued logic (Klir and Yuan, 1995). Triangular norms also play a key role in different fuzzy logics (Klement and Navara, 1999).

Given a continuous t-norm t, let us define the following φ operator:

$$a \varphi b = \sup \{c \in [0,1] | a\, t\, c \leq b\}, \quad \text{for all } a, b \in [0, 1]$$

This operation can be interpreted as an implication induced by some t-norm,

$$a \varphi b = a \Rightarrow b$$

and therefore it is, like implication, an inclusion relation. The operator φ generalizes the classic implication. As Table 5.4 shows, the two-valued implication arises as a special case of the φ operator in case when $a, b \in \{0, 1\}$.

Note that $a \varphi b (a \Rightarrow b)$, returns 1 whenever $a \leq b$. If we interpret these two truth values as membership degrees, we conclude that $a \varphi b$ models a multivalued inclusion relationship (viz. values of a higher than b produce lower values of the result).

5.6 AGGREGATION OPERATIONS

Several fuzzy sets can be combined together (aggregated) thus leading to a single fuzzy set forming the result of such an aggregation operation. For instance, when we compute intersection and union of fuzzy sets, the result is a fuzzy set whose membership function captures information conveyed by the original fuzzy sets. This fact suggests a general view of aggregation of fuzzy sets as certain transformations performed on their membership functions. In general, we encounter a wealth of aggregation operations (Dubois and Prade, 1985; Bouchon-Meunier, 1998; Calvo et al., 2002; Dubois and Prade, 2004).

5.6 Aggregation Operations

Formally, an aggregation operation is a n-ary function $g : [0, 1]^n \to [0, 1]$ satisfying the following requirements:

1. Monotonicity $\quad g(x_1, x_2, \ldots, x_n) \geq g(y_1, y_2, \ldots, y_n) \quad \text{if} \quad x_i > y_i$
2. Boundary conditions $\quad g(0, 0, \ldots, 0) = 0$
$\quad g(1, 1, \ldots, 1) = 1$

An element $e \in [0, 1]$ is called a neutral element of the aggregation operation g, and an element $l \in [0, 1]$ is called an annihilator (absorbing element) of the aggregation operation g if for each $i = 1, 2, \ldots, n, n \geq 2$ and for all $x_1, x_2, \ldots, x_{i-1}, x_{i+1}, \ldots, x_n \in [0, 1]$ we have

1. $g(x_1, x_2, \ldots, x_{i-1}, e, x_{i+1}, \ldots, x_n) = g(x_1, x_2, \ldots, x_{i-1}, x_{i+1}, \ldots, x_n)$
2. $g(x_1, x_2, \ldots, x_{i-1}, l, x_{i+1}, \ldots, x_n) = l$

Since triangular norms and conorms are monotonic, associative, and satisfy the boundary conditions, they provide a wide class of associative aggregation operations whose neutral elements are equal to 1 and 0, respectively. We are, however, not restricted to those as the only available alternatives.

5.6.1 Averaging Operations

In addition to monotonicity and the satisfaction of the boundary conditions, averaging operations are idempotent and commutative. They can be described in terms of the generalized mean (Dyckhoff and Pedrycz, 1984)

$$g(x_1, x_2, \ldots, x_n) = \sqrt[p]{\frac{1}{n} \sum_{i=1}^{n} (x_i)^p}, p \in \mathbf{R}, p \neq 0.$$

Interestingly, generalized mean subsumes some well-known cases of averages such as

$p = 1 \quad g(x_1, x_2, \ldots, x_n) = \dfrac{1}{n} \sum_{i=1}^{n} x_i \quad$ arithmetic mean

$p \to 0 \quad g(x_1, x_2, \ldots, x_n) = \sqrt[n]{\prod_{i=1}^{n} x_i} \quad$ geometric mean

$p = -1 \quad g(x_1, x_2, \ldots, x_n) = \dfrac{n}{\sum_{i=1}^{n} 1/x_i} \quad$ harmonic mean

$p \to -\infty \quad g(x_1, x_2, \ldots, x_n) = \min(x_1, x_2, \ldots, x_n) \quad$ minimum

$p \to \infty \quad g(x_1, x_2, \ldots, x_n) = \max(x_1, x_2, \ldots, x_n) \quad$ maximum

The following containment relationship holds:

$$\min(x_1, x_2, \ldots, x_n) \leq g(x_1, x_2, \ldots, x_n) \leq \max(x_1, x_2, \ldots, x_n) \qquad (5.17)$$

Therefore, generalized means range over the values not being covered by triangular norms and conorms. An illustration of the arithmetic, geometric, and harmonic means is presented in Figure 5.16.

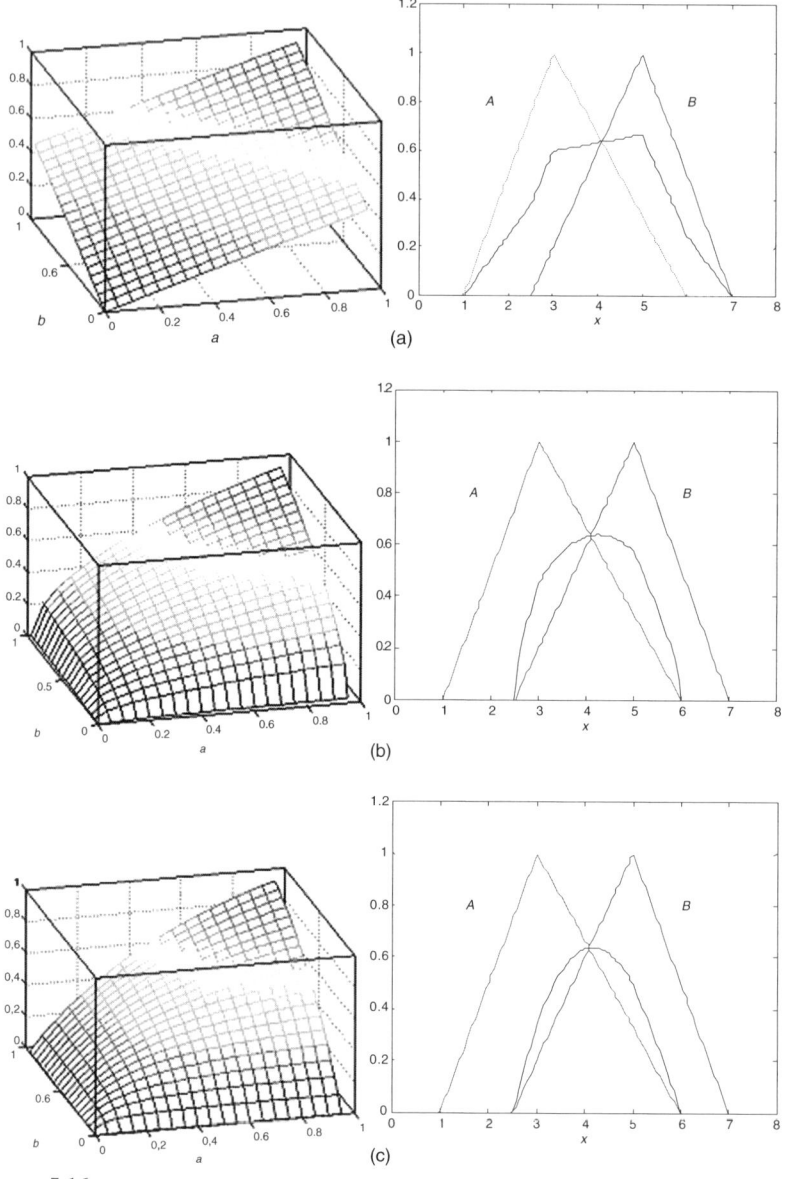

Figure 5.16 Examples of (a) arithmetic, (b) geometric, and (c) harmonic mean for two arguments.

5.6.2 Ordered Weighted Averaging Operations

Ordered weighted averaging (OWA) is a weighted sum whose arguments are ordered (Yager, 1988). Let $\mathbf{w} = [w_1, w_2, \ldots w_n], w_i \in [0,1]$, be weights such that

$$\sum_{i=1}^{n} w_i = 1$$

Let a sequence of membership values $\{A(x_i)\}$ be ordered as follows:

$$A(x_1) \leq A(x_2) \leq \ldots \leq A(x_n)$$

Then, a family of OWA(A, $\mathbf{w}$) is defined as

$$\text{OWA}(A,\mathbf{w}) = \sum_{i=1}^{n} w_i A(x_i) \qquad (5.18)$$

By choosing certain forms of $\mathbf{w}$, we can show that OWA includes several special cases of aggregation operators studied before:

1. if $\mathbf{w} = [1, 0, \ldots, 0]$, then OWA($A$, $\mathbf{w}$) = $\min(A(x_1), A(x_2), \ldots, A(x_n))$
2. if $\mathbf{w} = [0, 0, \ldots, 1]$, then OWA($A$, $\mathbf{w}$) = $\max(A(x_1), A(x_2), \ldots, A(x_n))$
3. if $\mathbf{w} = [1/n, , 1/n]$, then OWA($A$,$\mathbf{w}$) = $\frac{1}{n}\sum_{i=1}^{n} A(x_i)$, which is the arithmetic mean

It is easy to show that OWA is a continuous, symmetric, and idempotent operator. Varying the values of the weights w_i results in aggregation values located in-between between min and max,

$$\min(A(x_1), A(x_2), \ldots, A(x_n)) \leq \text{OWA}(A, \mathbf{w}) \leq \max(A(x_1), A(x_2), \ldots, A(x_n))$$

OWA behaves as a compensatory operator, similar to the generalized mean. Figure 5.17 illustrates the OWA operator. An illustration of the characteristics of the OWA is included in Figure 5.17.

5.6.3 Uninorms and Nullnorms

Triangular norms provide one of the possible ways to aggregate membership grades. By definition, the identity elements are 1 (t-norms) and 0 (t-conorm). When used in the aggregation operations, these elements do not affect the result of aggregation (i.e., at$1 = a$ and at $aso = a$). It can be shown that triangular norms are monotonic when dealing with the number of its arguments (Yager, 1993; Fodor et al., 1997), that is,

$$a_1\, t\, a_2\, t \ldots t\, a_n \geq a_1\, t\, a_2\, t \ldots t\, a_n\, t\, a_{n+1}$$
$$a_1\, s\, a_2\, s \ldots s\, a_n \leq a_1\, s\, a_2 s \ldots s\, a_n\, s\, a_{n+1}$$

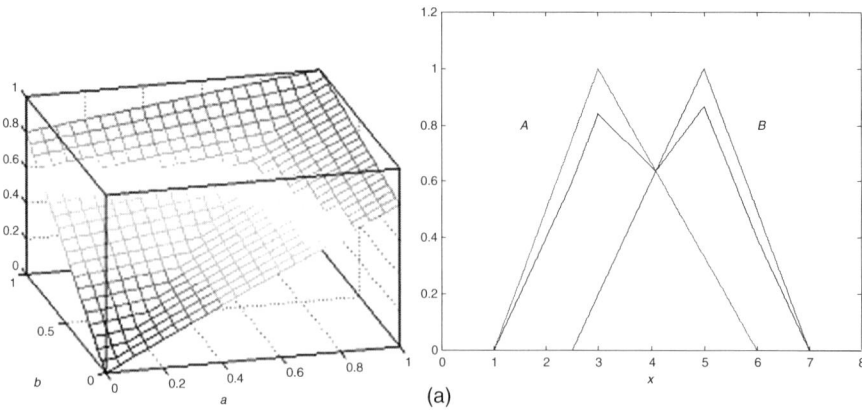

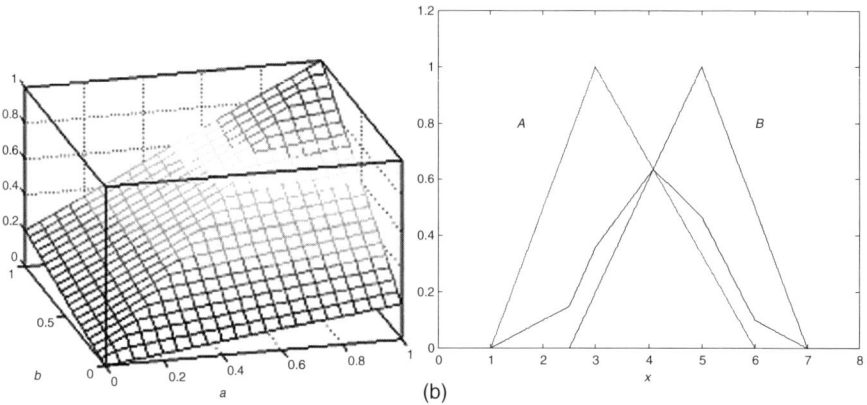

Figure 5.17 Characteristics of OWA operator for selected combinations of the values of the weight vector **w**: (a) $w_1 = 0.8, w_2 = 0.2$, and (b) $w_1 = 0.2, w_2 = 0.8$.

that means that increasing the number of elements in the t-norm aggregation does not increase the result of the aggregation. Expanding the number of arguments in the t-conorm aggregation does not decrease the result of this aggregation.

Uninorms unify and generalize triangular norms by allowing the identity element to be any number in the unit interval, that is, $e \in (0, 1)$. In this sense, uninorms become t-norms when $e = 1$ and t-conorms when $e = 0$. They exhibit some intermediate characteristics for all remaining values of e. Therefore, uninorms share the same properties as triangular norms with the exception of the identity (Yager and Rybalov, 1996).

5.6 Aggregation Operations

More formally, a uninorm is a binary operation $u : [0, 1] \times [0, 1] \to [0, 1]$ that satisfies the following requirements:

1. Commutativity: $\quad a\,u\,b = b\,u\,a$
2. Associativity: $\quad a\,u\,(b\,u\,c) = (a\,u\,b)\,u\,c$
3. Monotonicity: $\quad$ if $b \leq c$, then $a\,u\,b \leq a\,u\,c$
4. Identity: $\quad a\,u\,e = a, \forall a \in [0, 1]$

where $a, b, c \in [0, 1]$.

In Fodor et al. (1997) and Klement et al. (2000), presented were a number of interesting results on uninorms.

1. Let u be a uninorm with neutral element $e \in (0, 1)$ and $t_u, s_u : [0, 1] \times [0, 1] \to [0, 1]$ be functions such that $\forall a, b \in [0, 1]$,

$$a\,t_u\,b = \frac{(ea)u(eb)}{e} \quad (5.19)$$

$$a\,s_u\,b = \frac{(e + (1-e)a)u(e + (1-e)b) - e}{1 - e} \quad (5.20)$$

Thus, t_u and s_u are t-norm and t-conorm, respectively.

2. If $a \leq e \leq b$ or $a \geq e \geq b$, then $\min(a, b) \leq a\,u\,b \leq \max(a, b)$.

This result shows that on the squares $[0, e] \times [e, 1]$ and $[e, 1] \times [0, e]$, uninorms act as a compensatory aggregators similar to generalized means, see (5.17). Uninorms behave as t-norms in the square $[0, e] \times [0, e]$ and as t-conorms in $[e, 1] \times [e, 1]$.

3. For any uninorm u with $e \in (0, 1)$, we have

$$a\,u_w\,b \leq a\,u\,b \leq a\,u_s\,b$$

where u_w and u_s are the weakest and strongest uninorm, respectively.

$$a\,u_w\,b = \begin{cases} 0, & \text{if } 0 \leq a, b \leq e \\ \max(a, b), & \text{if } e \leq a, b \leq 1 \\ \min(a, b), & \text{otherwise} \end{cases}$$

$$a\,u_s\,b = \begin{cases} \min(a, b), & \text{if } 0 \leq a, b \leq e \\ 1, & \text{if } e \leq a, b \leq 1 \\ \max(a, b), & \text{otherwise} \end{cases}$$

4. Conjunctive u_c and disjuntive u_d forms of uninorms can be obtained in terms of triangular norms as follows:

   (a) If $(0\ u\ 1) = 0$, then

   $$a\,u_c\,b = \begin{cases} e\left(\dfrac{a}{e}\right)t\left(\dfrac{b}{e}\right), & \text{if } 0 \le a,b \le e \\ e + (1-e)\left(\dfrac{a-e}{1-e}\right)s\left(\dfrac{b-e}{1-e}\right), & \text{if } e \le a,b \le 1 \\ \min(a,b), & \text{otherwise} \end{cases}$$

   (b) If $(0\ u\ 1) = 1$, then

   $$a\,u_d\,b = \begin{cases} e\left(\dfrac{a}{e}\right)t\left(\dfrac{b}{e}\right), & \text{if } 0 \le a,b \le e \\ e + (1-e)\left(\dfrac{a-e}{1-e}\right)s\left(\dfrac{b-e}{1-e}\right), & \text{if } e \le a,b \le 1 \\ \max(a,b), & \text{otherwise} \end{cases}$$

where t is a t-norm and s is a t-conorm. Figure 5.18 depicts several examples of conjunctive and disjunctive uninorms with $e = 0.5$ when using the product and probabilistic sum.

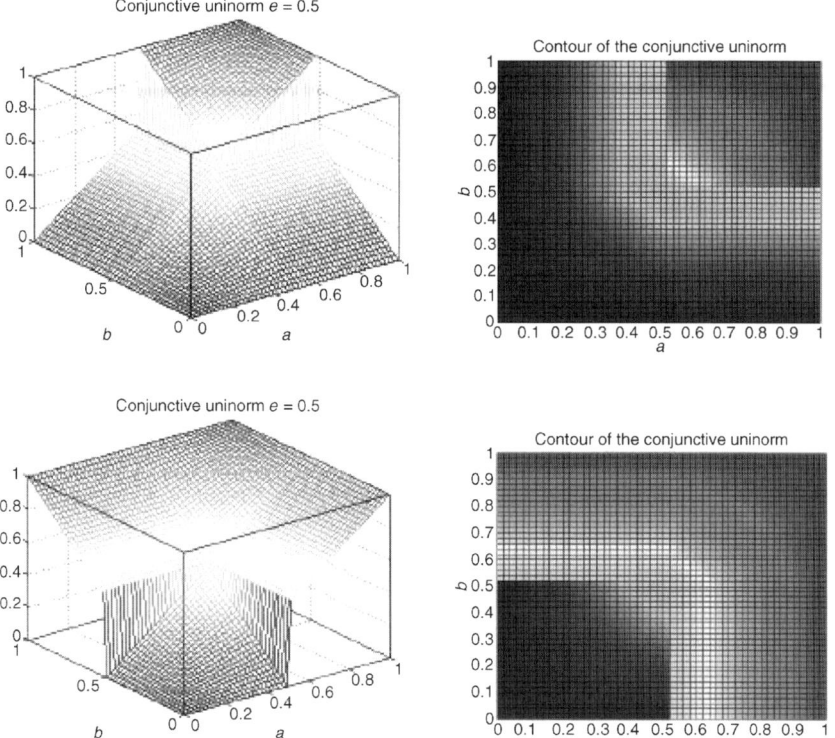

Figure 5.18 Examples of conjunctive and disjunctive uninorms; see detailed description in the text.

5. A uninorm is Archimedian if it is almost continuous (i.e., continuous on $[0,1] \times [0,1]$ except for $\{(0,1), (1,0)\}$) and satisfies

   (a) $a\, u\, a < a$ for $0 < a < e$ and
   (b) $a\, u\, a > a$ for $e < a < 1$.

   It can be shown (Fodor et al., 1997) that a uninorm is Archimedian if and only if the underlying triangular norms in (5.19) and (5.20) are Archimedian.

6. Let f be a strictly increasing continuous function $f : [0,1] \to (-\infty, +\infty)$ with $f(e) = 0$. Then there exists an almost continuous uninorm u_f such that

$$a\, u_f\, b = f^{-1}(f(a) + f(b))$$

Moreover, if $g(x) = e^{f(x)}$, then there exists an almost continuous uninorm u_g such that

$$a\, u_g\, b = g^{-1}(g(a) g(b))$$

Functions f and g are called the additive and multiplicative generators of uninorm u_f and u_g, respectively.

7. Finally, pseudo-continuous uninorms on $[0,1]$, the ones which are continuous in $[0,1]$ except in the set $\{(a,b) | a = e \text{ or } b = e\}$, can be expressed in terms of conjunctive and disjunctive ordinal sums (Fodor et al. 1997) using Archimedian triangular norms as follows.

Let $\iota = \{[\alpha_k, \beta_k], k \in K\}$ and be a nonempty, countable family of pairwise disjoint open subintervals of $[0,1]$, $\iota_1 = \{[\alpha_k, \beta_k[\in \iota | b_k \leq e\}$, $\iota_2 = \{[\alpha_k, \beta_k] \in \iota | a_k \geq e\}$, $\tau = \{t_k, k \in K\}$, and $\sigma = \{s_k, k \in K\}$ a family of corresponding t-norms and t-conorms.

The conjunctive ordinal sum of summands $\langle \alpha_k, \beta_k, t_k, s_k \rangle$, $k \in K$, denoted $u_{co} = (\langle \alpha_k, \beta_k, t_k, s_k \rangle, k \in K)$, is the function $u_{co} : [0,1] \times [0,1] \to [0,1]$ defined by

$$u_{co}(a, b, \iota, \tau, \sigma) = \begin{cases} \alpha_k + (\beta_k - \alpha_k) t_k \left(\dfrac{a - \alpha_k}{\beta_k - \alpha_k}, \dfrac{b - \alpha_k}{\beta_k - \alpha_k} \right), & \text{if } a, b \in [\alpha_k, \beta_k] \in \iota_1 \\ \alpha_k + (\beta_k - \alpha_k) s_k \left(\dfrac{a - \alpha_k}{\beta_k - \alpha_k}, \dfrac{b - \alpha_k}{\beta_k - \alpha_k} \right), & \text{if } a, b \in [\alpha_k, \beta_k] \in \iota_2 \\ \max(a, b), & \text{if } a, b \notin [\alpha_k, \beta_k] \text{ and } a, b \geq e \\ \min(a, b), & \text{otherwise} \end{cases}$$

Similarly, the conjunctive ordinal sum of summands $\langle \alpha_k, \beta_k, t_k, s_k \rangle, k \in K$, denoted $u_{do} = (\langle \alpha_k, \beta_k, t_k, s_k \rangle, k \in K)$, is the function $u_{do} : [0,1] \times [0,1] \to [0,1]$ defined in the following manner:

$$u_{do}(a, b, \iota, \tau, \sigma) = \begin{cases} \alpha_k + (\beta_k - \alpha_k) t_k \left(\dfrac{a - \alpha_k}{\beta_k - \alpha_k}, \dfrac{b - \alpha_k}{\beta_k - \alpha_k} \right), & \text{if } a, b \in [\alpha_k, \beta_k) \in \iota_1 \\ \alpha_k + (\beta_k - \alpha_k) s_k \left(\dfrac{a - \alpha_k}{\beta_k - \alpha_k}, \dfrac{b - \alpha_k}{\beta_k - \alpha_k} \right), & \text{if } a, b \in [\alpha_k, \beta_k) \in \iota_2 \\ \min(a, b), & \text{if } a, b \notin [\alpha_k, \beta_k] \text{ and } a, b \leq e \\ \max(a, b), & \text{otherwise} \end{cases}$$

As Figure 5.18 suggests, there is no uninorm that is continuous (Klement et al., 2000) on the whole unit square.

Nullnorms (Calvo et al., 2001) are another way of generalization of triangular norms. Similarly as it was in the case of uninorms, in nullnorms we relax an assumption about the values of the identity element $e \in (0, 1)$. They become t-norms when $e = 0$, and t-conorms when $e = 1$.

Formally, a nullnorm is a binary operation $v : [0, 1] \times [0, 1] \to [0, 1]$ that satisfies the following requirements:

1. Commutativity: $\quad a\,v\,b = b\,v\,a$
2. Associativity: $\quad a\,v(b\,v\,c) = (a\,v\,b)\,v\,c$
3. Monotonicity: $\quad$ if $b \leq c$, then $a\,v\,b \leq a\,v\,c$
4. Absorbing element $\quad (a\,v\,e) = e, \forall a \in [0, 1]$
5. Boundary conditions: $\quad a\,v\,0 = a, \forall a \in [0, e]$ and $a\,v\,1 = a, \forall a \in [e, 1]$

where $a, b, c \in [0, 1]$.

We can express nullnorms with its absorbing element (annihilator) e in terms of some t-norm and t-conorm

$$a\,t_v\,b = \frac{(e + (1 - e)a)v(e + (1 - e)b) - e}{1 - e}$$

$$a\,s_v\,b = \frac{(ea)v(eb)}{e}$$

Thus, as Figure 5.19 illustrates, nullnorms behave as t-conorms in $[0, e] \times [0, e]$, exhibit t-norms characteristics in $[e, 1] \times [e, 1]$, and assume a constant value e in the rest of the unit square.

5.6.4 Symmetric Sums

Symmetric sums provide yet another alternative to aggregate membership values.

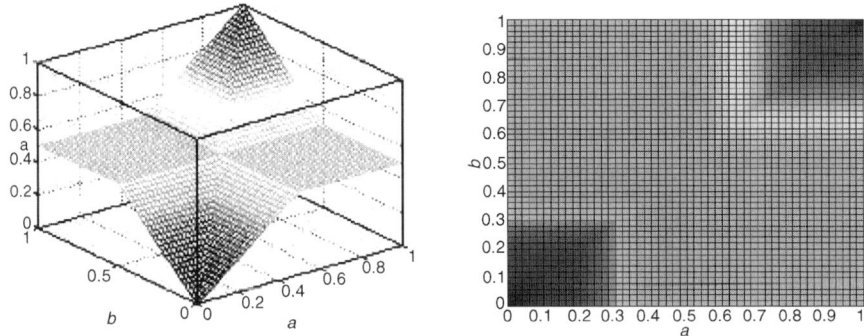

Figure 5.19 3D and contour plots of nullnorm ($e = 0.5$, t-conorm: maximum., t-norm: minimum).

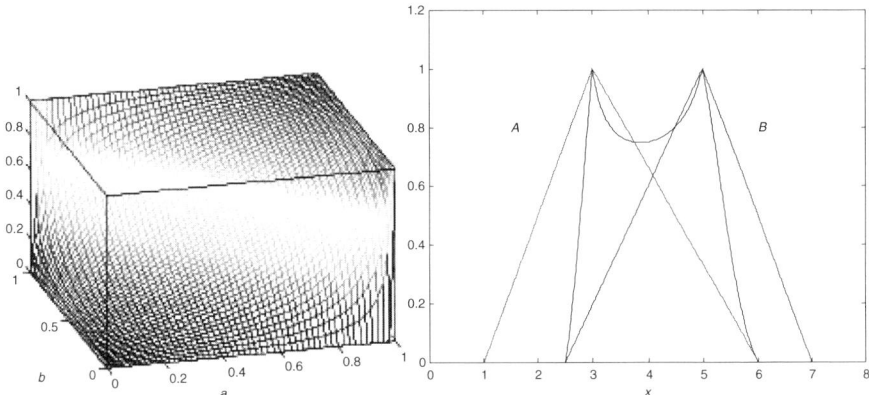

Figure 5.20 Plots of the symmetric sum.

They are *n*-ary functions such that, in addition to fulfilling boundary conditions and monotonicitiy, they are continuous, commutative, and auto-dual. We say that a symmetric sum σ_s is auto-dual if, for all $a_i \in [0, 1]$, we have

$$\sigma_s(a_1, a_2, \ldots, a_n) = 1 - \sigma_s(1 - a_1, 1 - a_2, \ldots, 1 - a_n)$$

It can be shown (Dubois and Prade, 1980) that any symmetric sum can be represented in the form

$$\sigma_s(a_1, a_2, \ldots, a_n) = \left[1 + \frac{f(1 - a_1, 1 - a_2, \ldots, 1 - a_n)}{f(a_1, a_2, \ldots a_n)}\right]^{-1}$$

where f is the generator, namely any increasing continuous function with $f(0, 0, \ldots, 0) = 0$. Figure 5.20 shows an example of the symmetric sum $\sigma_s(a, b)$ when choosing $f(a, b) = a^2 + b^2$ and triangular fuzzy sets, respectively.

5.6.5 Compensatory Operations

Set-theoretic operations or their corresponding logic operations may not fit well experimental data (Greco and Rocha, 1984). For instance, minimum has been shown not to model well intersection (*and*) and conjunction (*or*), whereas product appeared to be very conservative (Zimmermann and Zysno, 1980). Quite often, experimental results may suggest considering a sort of intermediate style of aggregation of fuzzy sets in which we consider combination of *and* and *or* type of combination of membership grades. For instance, introducing a compensation factor $\gamma \in [0, 1]$ indicating a "position" of the actual operator somewhere in between some t-norm and t-conorm, we may consider the following compensatory operations:

$$a \odot b = (a\,t\,b)^{1-\gamma}(a\,s\,b)^{\gamma} \qquad \text{compensatory product}$$
$$a \boxplus b = (1 - \gamma)(a\,t\,b) + \gamma(a\,s\,b) \qquad \text{compensatory sum}$$

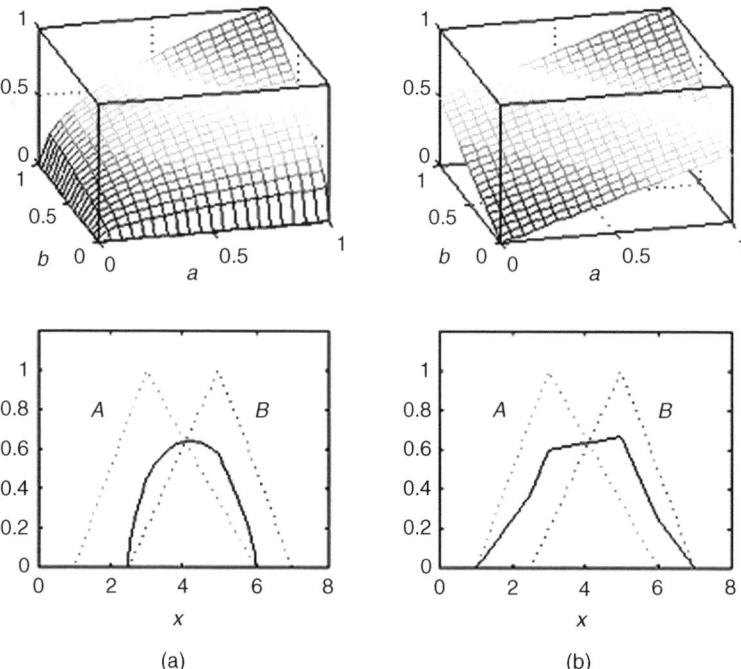

Figure 5.21 Plots of characteristics of compensatory operations: (a) compensatory product and (b) compensatory sum.

where $a, b \in [0, 1]$. For instance, Figure 5.21 shows the compensation effect of such compensatory operators when choosing $\gamma = 0.5$ and implementing the t-norm and t-conorm as the minimum and maximum operator, respectively.

5.7 FUZZY MEASURE AND INTEGRAL

Fuzzy measure and the related concept of a fuzzy integral give rise to another class of aggregation operators. Fuzzy measure provides a way to handle a type of uncertainty that results from information insufficiency and is substantially different from fuzziness being viewed as lack of sharp boundaries (see the historical notes of Chapter 2). To stress the essence of the concept, in our investigations we focus on a finite case. A; detailed treatment of the subject can be found in Wang and Klir (1992).

Given a finite universe **X** and a nonempty family Ω of subsets of **X**, a fuzzy measure is a set function $g : \Omega \to [0, 1]$ that satisfies

1. $g(\phi) = 0$ and $g(\mathbf{X}) = 1$,   boundary conditions.
2. If $A \subset B$, then $g(A) \leq g(B)$,   monotonicity.

Fuzzy measures come with an interesting interpretation in the context of sensor fusion and system diagnostics. For instance, consider an autonomous vehicle operating in an unknown environment. To safely navigate in such an environment, the vehicle needs a suite of sensors to detect obstacles, identify surface of terrain, and collect other information for navigational purposes. When viewed collectively, sensors provide evidence that help make suitable decisions as to path planning and current navigation. Sensor information is rarely precise, and some uncertainty always remains. A way to handle uncertainty is to assign a value (confidence measure) to each possible set of sensor readings. Within this setting, the fundamental conditions of fuzzy measures come with a straightforward interpretation. If there are no readings of sensors available, our confidence to make decision is zero, that is, $g(\phi) = 0$. If the readings of all sensors are available, then $g(\mathbf{X}) = 1$. The monotonicity property is also quite appealing: the more sensors become available to us, the higher the confidence about the state of the environment.

Although the definition of the fuzzy measure presented above is conceptually sound, it is not fully operational. For instance, it does not tell us how to compute fuzzy measures for two disjoint sets whose fuzzy measures $g(A)$ and $g(B)$ have been provided. A way to alleviate this shortcoming is a λ-fuzzy measure (Sugeno, 1974) g_λ. We will be also using a shorthand notation g if this does not produce any misunderstanding. Given two disjoint sets A and B, $A \cap B = \phi$, the fuzzy measure of their union, $g(A \cup B)$ is computed as follows:

$$g(A \cup B) = g(A) + g(B) + \lambda g(A)g(B), \quad \lambda > -1 \qquad (5.21)$$

The parameter λ present in the above expression is used to quantify interaction between the sets that are combined. Some particular cases are worth distinguishing.

1. If $\lambda = 0$, then $g(A \cup B) = g(A) + g(B)$. Here the fuzzy measure reduces to an *additive* measure.
2. If $\lambda > 0$, then $g(A \cup B) \geq g(A) + g(B)$. The fuzzy measure is *super-additive* and quantifies a synergy between A and B. In more descriptive way, we say that an evidence associated with the union of two sources of information (A and B) is greater than the sum of the evidences of the individual sources when being treated separately.
3. If $\lambda < 0$, then $g(A \cup B) \leq g(A) + g(B)$. The fuzzy measure becomes *sub-additive* We say that the two sources of evidence are in competition or are redundant, resulting in the union of evidences being less than the sum of the individual evidences.

In general, the determination of λ is obtained by considering the boundary condition of the fuzzy measure, that is, $g(\mathbf{X}) = 1$; see Pedrycz and Gomide (1998) for a detailed discussion.

Now, let $(\mathbf{X}, \Omega)$ be a measurable space, that is, a pair consisting of $\mathbf{X}$ and Ω, a σ-algebra of $\mathbf{X}$. Recall that σ-algebra is concerned with a family of subsets of $\mathbf{X}$ that is closed under countable union and complement, that is, if A and B are members of the family, then their union and complement are also members of the family.

Let $h : \mathbf{X} \rightarrow [0, 1]$ be a Ω-measurable function. The fuzzy integral of h computed with respect to the fuzzy measure g over A is expressed as

$$\int_A h(x) \circ g() = \sup_{\alpha \in [0,1]} \{\min[\alpha, g(A \cap H_\alpha)]\} \tag{5.22}$$

where H_α denotes a subset of $\mathbf{X}$ for which $h(x)$ assumes values no lower than α, $H_\alpha = \{x | h(x) \geq \alpha\}$. In this sense, regarding h as some fuzzy set (membership function), H_α is it's α-cut.

In the context of the autonomous navigation, if g describes the relevance of the individual sources of information (sensors) collected so far and h denotes the results the sensors have reported, then the fuzzy integral can be interpreted as a nonlinear aggregation of the readings of the sensors. In particular, if $A = \mathbf{X}$ then we are concerned with integration of h with respect to the fuzzy measure.

The computing of fuzzy integral is significantly simplified when we consider finite universe $\mathbf{X} = \{x_1, x_2, \ldots, x_n\}$, the one that is of interest here. Let us also assume that the following sequence of inequalities hold:

$$h(x_1) \geq h(x_2) \geq \ldots h(x_n)$$

Notice that this assumption does not limit the generality of the construct since its satiafaction requires a straight rearrangement of the elements of x_i so that the above inequalities are satisfied.

Let us now define the following sequence of nested sets.

$$A_1 = \{x_1\}, A_2 = \{x_1, x_2\} \ldots A_p = \{x_1, x_2, \ldots, x_p\}, \ldots, A_n = \{x_1, x_2, \ldots, x_n\} = \mathbf{X}.$$

In virtue of the monotonicity property of the fuzzy measure, we have

$$g(A_1) \leq \ldots \leq g(A_p) \leq \ldots \leq g(A_n) = 1$$

The calculations of the fuzzy integral described by (5.22) is realized through the standard max-min composition applied to the two sequences of membership values, that is, $\{h(x_i)\}$ and $\{g(A_i)\}$:

$$\int h(x) \circ g() = \max_{i=1,\ldots,n} \{\min[h(x_i), g(A_i)]\} \tag{5.23}$$

As a continuation of the example, let us consider five sensors of the autonomous vehicle, that is, $\mathbf{X} = \{x_1, x_2, \ldots, x_5\}$. The quality of information generated by each sensor is expressed by the values of the fuzzy measure specified for each x_i, $g(\{x_i\})$. The current readings of the sensors are $h(x_1) = 0.1, h(x_2) = 0.4, h(x_3) = 0.3, h(x_4) = 0.7$, and $h(x_5) = 0.05$. After the rearrangement of these values, we get $\{h(x_i)\} = \{0.7, 0.4, 0.3, 0.1, 0.05\}$. Given the values of $g(\{x_i\})$, the fuzzy measures computed over the corresponding nested sets A_i produce the values $g(A_1) = 0.210, g(A_2) = 0.492, g(A_3) = 0.520, g(A_4) = 0.833$, and $g(A_5) = 1$. For

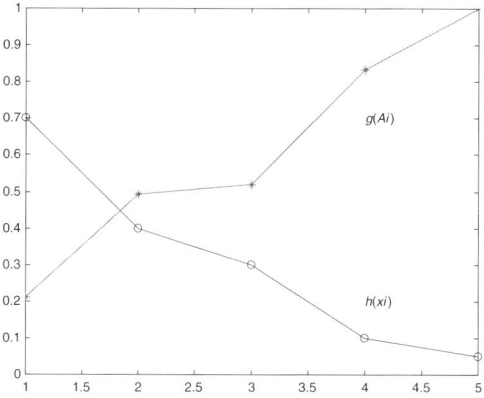

Figure 5.22 Computing the fuzzy integral.

the details of this computing, refer to Exercise 9 at the end of this chapter. Inserting the corresponding values of $g(A_i)$ and $h(x_i)$ into (5.23), we obtain

$\min(h(x_1), g(A_1)) = 0.21,$
$\min(h(x_2), g(A_2)) = 0.40,$
$\min(h(x_3), g(A_3)) = 0.30,$
$\min(h(x_4), g(A_4)) = 0.10,$
$\min(h(x_5), g(A_5)) = 0.05.$

Hence, the maximum taken over the partial results is equal to
$\max\{0.21, 0.4, 0.3, 0.1, 0.05\} = 0.40$, see also Figure 5.22.

We can envision some possible generalizations of the generic definition of the fuzzy integral (5.23). The one that is quite straightforward concerns a replacement of the min operation by any t-norm:

$$\int h(x) \circ g() = \max_{i=1,\dots,n} \{h(x_i) t g(A_i)\} \quad (5.24)$$

The Choquet integral (1953) is another concept that closely relates to the fuzzy measure. It is defined as follows:

$$(\text{Ch}) \int f \circ g = \sum_{i=1}^{n} \{h(x_i) - h(x_{i+1})\} g(A_i), h(x_{n+1}) = 0 \quad (5.25)$$

For instance, using the same values of the fuzzy measure and the same function h to be integrated as used in the previous example, we obtain

$$(\text{Ch}) \int f \circ g = (0.7 - 0.4)0.21 + (0.4 - 0.3)0.492 + (0.3 - 0.1)0.520$$
$$+ (0.1 - 0.005)0.833 + (0.05 - 0)1.0 = 0.3079$$

In contrast with the Sugeno fuzzy integral (5.23), the Choquet integral (5.25) is more intuitive and coincides with our standard understanding of usual notion of integral.

Primary applications of fuzzy measures and integrals include computer vision, prediction, assessment of human reliability, and multi-attribute decision-making. An overview of aggregation operators emphasizing application-related issues are given in Torra (2005).

5.8 NEGATIONS

Contrary to the set-theoretic (logic) operations discussed so far, negations are unary (single-argument) operations whose purpose is to generalize the standard notion of complement, that is, in the sense of "one-complement." In general, negations are functions $N : [0, 1] \to [0, 1]$ that satisfy the following conditions:

1. Monotonicity: N is nonincreasing
2. Boundary conditions: $N(0) = 1$ and $N(1) = 0$.

Further conditions may be required; the ones deemed essential are continuity and involution.

3. Continuity: N is a continuous function.
4. Involution: $N(N(x)) = x, \quad \forall x \in [0, 1]$.

An example of a negation operation is the threshold function with $a \in [0, 1]$

$$N(x) = \begin{cases} 1, & \text{if } x < a \\ 0, & \text{if } x \geq a \end{cases}$$

It is monotonic, satisfies the boundary conditions but is not involutive. Further examples include

$$N(x) = \begin{cases} 1 & \text{if } x = 0 \\ 0 & \text{if } x > 0 \end{cases}$$

$$N(x) = \frac{1-x}{1+\lambda x}, \quad \lambda \in (-1, \infty), \qquad \text{Sugeno}$$

$$N(x) = \sqrt[w]{1 - x^w}, \quad w \in (0, \infty), \qquad \text{Yager}$$

The last two examples are involutive. For $\lambda = 0$ and $w = 1$, they become the standard complement function, $N(x) = 1 - x$, Figure 5.22.

A formal system of logic operations formed by triangular norms and negations (t, s, N) involves a t-norm and a t-conorm dual with respect to N, that is, the triangular norms are such that $\forall x, y \in [0, 1]$,

$$x \, s \, y = N(N(x) \, t \, N(y))$$
$$x \, t \, y = N(N(x) \, s \, N(y)).$$

Examples of (t, s, N) systems include

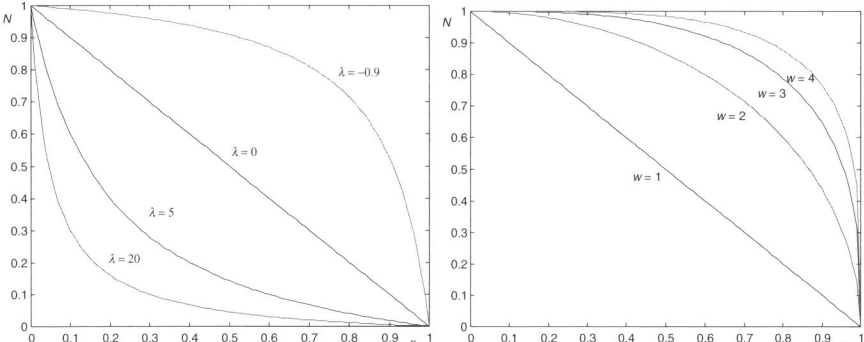

Figure 5.22 Examples of negations: Sugeno and Yager.

1. $x t y = \min(x, y)$

   $x s y = \max(x, y)$

   $N(x) = 1 - x.$

2. $x t y = \max\left(0, \dfrac{x + y - 1 + \lambda xy}{1 + \lambda}\right)$

   $x s y = \min(1, x + y - 1 + \lambda xy)$

   $N(x) = \dfrac{1 - x}{1 + \lambda x}.$

5.9 CONCLUSIONS

Operations involving membership values are behind operations involving fuzzy sets. These include standard intersection, union, and complement and their generalization via triangular norms and negations. The concepts of ordinal sum, uninorm, and nullnorm put triangular norms within a fairly general framework to operate with fuzzy sets. Besides set operations, triangular norms also provide a mechanism for information fusion when interpreted as aggregation operators. In this sense, generalized means and ordered weight average extend aggregation operations once they cover values between the lower and upper bounds identified by triangular norms. Fuzzy measures and integrals give a distinct treatment for aggregation. These operations have been introduced and discussed to provide a sense of their potential and flexibility in applications.

HISTORICAL NOTES

Triangular norms, as introduced by Menger (1942), are constructs originally developed in the setting of probabilistic metric spaces with the purpose to generalize the triangle inequality (if lengths of a triangle sides are x, y, and z, then $x \leq y + z$). Their role in fuzzy sets is profound as they help address a semantic diversity of logic operators in fuzzy sets. While the lattice (minimum and maximum) operators were introduced at the time of inception of fuzzy sets, it became clear that those are only one possibility of realization of union and intersection of fuzzy sets. There have been several experimental studies reported by Zimmermann and Zysno (1980) that confirmed a need for various aggregation operations.

Uninorms appeared as a generalization of a class of aggregation operators called monotonic identity commutative aggregation (MICA) operators introduced by Yager (1994a) where it was shown that triangular norms are particular instances of MICA. Yager (1994b) also suggested an approach for weighted min and max aggregation, median aggregation and developed a procedure to carry out aggregation on ordinal scales(Kelman and Yoga, 1995).

EXERCISES AND PROBLEMS

1. Consider two fuzzy sets with triangular membership functions $A(x; 1, 2, 3)$ and $B(x; 2, 3, 4)$.
   (a) Find their intersection and union, and express analytically the resulting membership functions.
   (b) Find the complements of A and B and compute their intersection and union with the original fuzzy sets using the Lukasiewicz triangular norms.
2. Show that the drastic product and drastic sum satisfy the law of excluded middle and the law of contradiction. Are they related via De Morgan law?
3. Show that a function $s : [0, 1] \times [0, 1] \to [0, 1]$ is a t-conorm if and only if there exists a t-norm t such that for all $a, b \in [0, 1]$, $a\,s\,b = 1 - (1 - a)t\,t(1 - b)$.
4. Assume that $f : [0, 1] \to [0, 1]$ is a one-to-one and onto increasing function. Define $a\,t\,b = f^{-1}(f(a)f(b))$. Show that t is an Archimedean t-norm.
5. Consider a uninorm u_p with identity element $e_p \in [0, 1]$ and define $a\,u_d\,b = 1 - (1 - a)u_p(1 - b)$. Show that u_d is a uninorm with identity $e_d = 1 - e_p$.
6. Let u be a uninorm with identity element e. Show that
   (a) if $a_{n+1} < e$, then $a_1\,u\,a_2\,u..u a_n \geq a_1\,u a_2\,u..u\,a_n\,u a_{n+1}$.
   (b) if $a_{n+1} > e$, then $a_1\,u\,a_2\,u..u\,a_n \leq a_1\,u a_2\,u..u\,a_n\,u\,a_{n+1}$.
7. Show that De Morgan laws are satisfied for the standard intersection and union when using the following negation operators.

   (a) $N(x) = \dfrac{1-x}{1+\lambda x}$, $\lambda \in (-1, \infty)$.

   (b) $N(x) = \sqrt[w]{1 - x^w}$, $w \in (0, \infty)$.
8. Show whether the function $N(x) = \frac{1}{2}\{1 + \sin[(2x + 1)\pi/2]\}$ qualifies as a negation. Justify your answer.
9. The fuzzy measure, g, can be expressed in the following form

$$g(\bigcup_{i \in I} \{x_i\}) = \frac{1}{\lambda}\left[\prod_{i \in I}(1 + \lambda g_i) - 1\right]$$

for given values of the fuzzy measure for the individual elements of the universe of discourse, that is, $g_i = g(\{x_i\})$.

(a) Discuss how to determine the value of the parameter of the measure (λ).
(b) For the values $g_1 = 0.7$, $g_2 = 0.05$, $g_3 = 0.6$, and $g_4 = 0.12$, determine the values of the fuzzy measure over $\{x_1, x_4\}$, $\{x_2, x_3, x_4\}$.
(c) Compute the fuzzy integral over **X** considering that $h(x_1) = 0.85$, $h(x_2) = 0.72$, $h(x_3) = 0.30$, and $h(x_4) = 0.07$.

REFERENCES

Bouchon-Meunier, B. *Aggregation and Fusion of Imperfect Information*, Physica-Verlag, Heidelberg, 1998.

Calvo, T., De Baets, B., Fodor, J. The functional equations of Frank and Alsina for uninorms and nullnorms, *Fuzzy Set Syst.* **120**, 2001, 385–394.

Calvo, T., Kolesárová, A., Komorníková, M., Mesiar, R. Aggregation operators: properties, classes and construction methods, *Aggregation Operators: New Trends and Applications*, Physica-Verlag, Heildelberg, 2002, 1–104.

Choquet, G. theory of capacities. Amm. Inst. Fourier 5, 131–295, 1953.

Dubois, D., Prade, H. *Fuzzy Sets and Systems*: *Theory and Applications*, Academic Press, New York, NY, 1980.

Dubois, D., Prade, H. A review of fuzzy set aggregation connectives, *Information Sciences.* **36**, 1985, 85–121.

Dubois, D., Prade, H. On the use of aggregation operations in information fusion processes, *Fuzzy Set and system.* **142**, 2004, 143–161.

Dyckhoff, H., Pedrycz, W. Generalized means as a model of compensative connectives, *Fuzzy Set Syst.* **14**, 1984, 143–154.

Fodor, J., Yager, R., Rybalov, A. Structure of uninorms, *Int. J. Uncert., Fuzziness Knowl.-Based Syst.* **5**(4), 1997, 113–126.

Greco, D., Rocha, A. The fuzzy logic of text understanding, *Fuzzy Set Syst.* **23**, 1984, 143–154.

Jenei, S. A note on the ordinal sum theorem and its consequence for the construction of triangular norms, *Fuzzy Set Syst.* **126**, 2002, 199–205.

Jenei, S. How to construct left-continuous triangular norms—state of the art, *Fuzzy Set Syst.* **143**, 2004, 27–45.

Kelman, A., Yager, R. On the application of a class of MICA operators, *Int. J. Uncert., Fuzziness Knowl.-Based Syst.* **3**(2), 1995, 113–126.

Klement, E., Navara, M. A survey on different triangular norm-based fuzzy logics, *Fuzzy Set Syst.* **101**, 1999, 241–251.

Klement, E., Mesiar, R., Pap, E. *Triangular Norms*, Kluwer Academic Publishers, Dordrecht, 2000.

Klir, G., Yuan, B. *Fuzzy Sets and Fuzzy Logic: Theory and Applications*, Prentice-Hall, Upper Saddle River, NJ, 1995.

Menger, K. Statistical metrics, *Proc. Nat. Acad. Sci. USA* **8**, 1942, 535–537.

Nguyen, H., Walker, E. *A First Course in Fuzzy Logic*, Chapman Hall, CRC Press, Boca Raton, FL, 1999.

Pedrycz, W., Gomide, F. *An Introduction to Fuzzy Sets: Analysis and Design*, MIT Press, Cambridge, MA, 1998.

Schweizer, B., Sklar, A. *Probabilistic Metric Spaces*, North-Holland, New York, NY, 1983.

Sugeno, M. *Theory of fuzzy integrals and its applications*, Ph.D. dissertation, Tokyo Institute of Technology, Tokyo, Japan, 1974.

Torra, V. Aggregation operators and models, *Fuzzy Set Syst.* **156**, 2005, 407–410.

Wang, Z., Klir, G. *Fuzzy Measure Theory*, Plenum Press, New York, NY, 1992.

Yager, R. On ordered weighted averaging aggregation operations in multicriteria decision making, *IEEE Trans. Syst., Man Cybern.* **18**, 1988, 183–190.

Yager, R. MAM and MOM operators for aggregation, *Inf. Syst.* **69**(1), 1993, 259–273.

Yager, R. Aggregation operators and fuzzy systems modeling, *Fuzzy Set Syst.* **67**, 1994a, 129–146.

Yager, R. On weighted median aggregation, *Int. J. Uncert. Fuzziness Knowl.-Based Syst.* **2**(1), 1994b, 101–114.

Yager, R, Rybalov, A. Uninorm aggregation operators, *Fuzzy Set Syst.* **80**, 1996, 111–120.

Zimmermann, H., Zysno, P. Latent connectives in human decision making, *Fuzzy Set Syst.* **4**, 1980, 37–51.

Chapter 6

Fuzzy Relations

Relations represent and quantify associations between objects. They provide a vehicle to describe interactions and dependencies between variables, components, modules, and so on. Fuzzy relations generalize the concept of relations in the same manner as fuzzy sets generalize the fundamental idea of sets. Fuzzy relations are instrumental in problems of information retrieval, pattern classification, control, and decision-making. Here, we introduce the idea of fuzzy relations, present some illustrative examples, discuss the main properties of fuzzy relations, and provide with some interpretation. Subsequently we look at some ideas of relational calculus and its algorithms.

6.1 THE CONCEPT OF RELATIONS

Before proceeding with fuzzy relations, we provide a few introductory lines about relations. Relations capture the associations between objects. For instance, consider the space of documents **X** and a space of keywords **Y** that these documents contain. Now form a Cartesian product of **X** and **Y**, that is, $\mathbf{X} \times \mathbf{Y}$. Recall that the Cartesian product of **X** and **Y**, denoted as $\mathbf{X} \times \mathbf{Y}$, is the set of all pairs (x,y) such that $x \in \mathbf{X}$ and $y \in \mathbf{Y}$. We define a relation R as the set of pairs of documents and keywords, $R = \{(d_i, w_j) | d_i \in \mathbf{X} \text{ and } w_j \in \mathbf{Y}\}$. In terms of the characteristic function we express this as follows: $R(d_i, w_j) = 1$ if keyword w_j is in document d_i, and $R(d_i, w_j) = 0$ otherwise.

More generally, a relation R defined over the Cartesian product of **X** and **Y** is a collection of selected pairs (x, y) where $x \in \mathbf{X}$ and $y \in \mathbf{Y}$. Equivalently, it is a mapping:

$$R : \mathbf{X} \times \mathbf{Y} \to \{0, 1\}$$

The characteristic function of R is such that if $R(x, y) = 1$, then we say that the two elements x and y are related. If $R(x, y) = 0$, we say that these two elements (x and y) are unrelated. For example, suppose that $\mathbf{X} = \mathbf{Y} = \{2, 4, 6, 8\}$. The relation "equal

---

Fuzzy Systems Engineering: Toward Human-Centric Computing, by Witold Pedrycz and Fernando Gomide
Copyright © 2007 John Wiley & Sons, Inc.

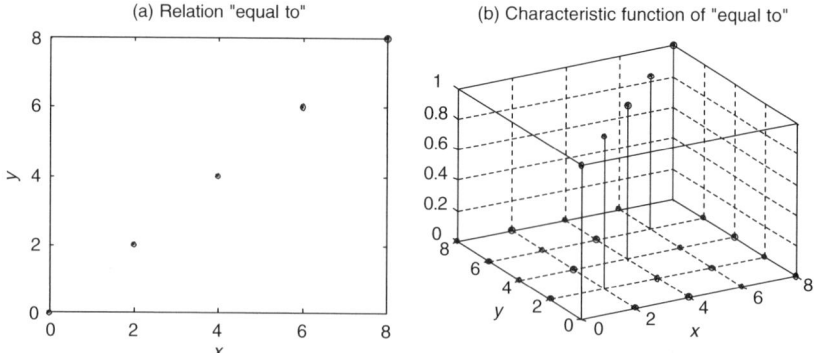

Figure 6.1 Relation "equal to" and its characteristic function.

to" formed over $\mathbf{X} \times \mathbf{X}$ is the set of pairs $R = \{(x, y) \in \mathbf{X} \times \mathbf{X} | x = y\} = \{(2,2), (4,4), (6,6), (8,8)\}$; refer to Figure 6.1(a). Its characteristic function is equal to

$$R(x, y) = \begin{cases} 1, & \text{if } x = y \\ 0, & \text{otherwise} \end{cases}$$

The plot of this characteristic function is included in Figure 6.1(b).

Depending on the nature of the universe, being either finite or infinite, relations can be represented in a tabular or matrix form, or analytically. For instance, the set $\mathbf{X} = \{2, 4, 6, 8\}$ is finite and the relation "equal to" in $\mathbf{X} \times \mathbf{X}$ has a representation in the (4×4) matrix:

$$R = \begin{bmatrix} 1 & 0 & 0 & 0 \\ 0 & 1 & 0 & 0 \\ 0 & 0 & 1 & 0 \\ 0 & 0 & 0 & 1 \end{bmatrix}$$

In general, if $\mathbf{X}$ and $\mathbf{Y}$ are finite, say $\mathrm{Card}(\mathbf{X}) = n$ and $\mathrm{Card}(\mathbf{Y}) = m$, then R is an $(n \times m)$ matrix $R = [r_{ij}]$ with the entries r_{ij} being equal to 1 if and only if $(x_i, y_j) \in R$. Elementary geometry provides examples of relations on infinite universes such as $\mathbf{R} \times \mathbf{R} = \mathbf{R}^2$. In these cases, characteristic functions can, in general, be expressed analytically:

$$R(x, y) = \begin{cases} 1, & \text{if } |x| \leq 1 \text{ and } |y| \leq 1 \\ 0, & \text{otherwise} \end{cases} \quad \text{square}$$

$$R(x, y) = \begin{cases} 1, & \text{if } x^2 + y^2 = r^2 \\ 0, & \text{otherwise} \end{cases} \quad \text{circle}$$

Relations subsume functions but not vice versa; all functions are relations, but not all relations are functions. For instance, the relation "equal to" shown above is a function, but the relations "square" and "circle" are not. A relation is a function if and only if for every x in $\mathbf{X}$ there is only a single element $y \in \mathbf{Y}$ such that $R(x, y) = 1$. Therefore, functions are directional constructs, clearly implying some certain direction, for example, from $\mathbf{X}$ to $\mathbf{Y}$, say

$$f : \mathbf{X} \to \mathbf{Y}$$

If the mapping f is a function, there is no guarantee that the mapping $f^{-1} : \mathbf{Y} \to \mathbf{X}$ is also a function, except in some case when f^{-1} exists. In contrast, relations are direction free as there is no specific direction identified. Being more descriptive, they can be accessed from any direction. This makes a significant conceptual and computational difference.

When a space under discussion involves n universes as its coordinate, an n-ary relation is any subset of the Cartesian product of these universes:

$$R : \mathbf{X}_1 \times \mathbf{X}_2 \times \cdots \times \mathbf{X}_n \to \{0, 1\}$$

If $\mathbf{X}_1, \mathbf{X}_2, \ldots, \mathbf{X}_n$ are finite and $\text{Card}(\mathbf{X}_1) = n_1 \ldots \text{Card}(\mathbf{X}_n) = n_p$, then R can be written as a $(n_1 \times \ldots \times n_p)$ matrix $R = [r_{ij..k}]$ with $r_{ij..k} = 1$ if and only if $(x_i, x_j, \ldots, x_k) \in R$.

6.2 FUZZY RELATIONS

Fuzzy relations generalize the concept of relations by admitting the notion of partial association between elements of universes. Given two universes $\mathbf{X}$ and $\mathbf{Y}$, a fuzzy relation R is any fuzzy subset of the Cartesian product of $\mathbf{X}$ and $\mathbf{Y}$ (Zadeh, 1971). Equivalently, a fuzzy relation on $\mathbf{X} \times \mathbf{Y}$ is a mapping:

$$R : \mathbf{X} \times \mathbf{Y} \to [0, 1]$$

The membership function of R for some pair (x, y), $R(x, y) = 1$, denotes that the two elements x and y are fully related. On the other hand, $R(x, y) = 0$ means that these elements are unrelated while the values in-between, $0 < R(x, y) < 1$, underline a partial association. For instance, if d_{fs}, d_{nf}, d_{ns}, d_{gf} are documents whose subjects concern mainly fuzzy systems, neural fuzzy systems, neural systems, and genetic fuzzy systems, with keywords w_f, w_n, and w_g, respectively, then a relation R on $\mathbf{D} \times \mathbf{W}$, $\mathbf{D} = \{d_{fs}, d_{nf}, d_{ns}, d_{gf}\}$, and $\mathbf{W} = \{w_f, w_n, w_g\}$, can assume the matrix form with the following entries:

$$R = \begin{bmatrix} 1 & 0 & 0.6 \\ 0.8 & 1 & 0 \\ 0 & 1 & 0 \\ 0.8 & 0 & 1 \end{bmatrix}$$

Since the universes are discrete, R can be represented as a (4×3) matrix (four documents and three keywords) and entries, for example, $R(d_{fs}, w_f) = 1$ means that the document content d_{fs} is fully compatible with the keyword w_f, whereas $R(d_{fs}, w_n) = 0$ and $R(d_{fs}, w_g) = 0.6$ indicates that d_{fs} does not mention neural systems, but does have genetic systems as part of its content, refer to

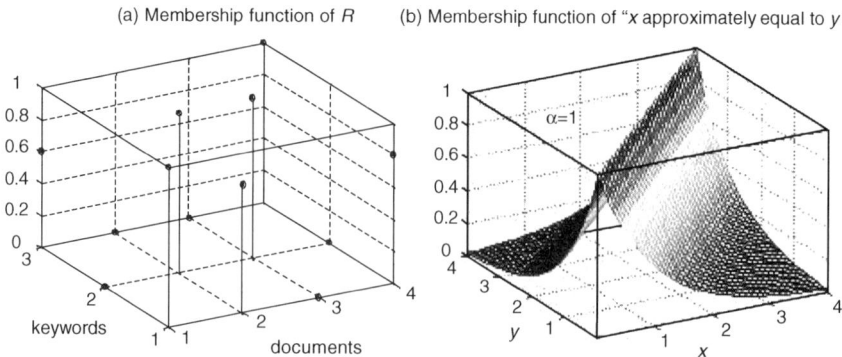

Figure 6.2 Membership functions of the relation R (a) and "x approximately equal to y" (b).

Figure 6.2(a). As with relations, when $\mathbf{X}$ and $\mathbf{Y}$ are finite with $\text{Card}(\mathbf{X}) = n$ and $\text{Card}(\mathbf{Y}) = m$, then R can be arranged into a certain $(n \times m)$ matrix $R = [r_{ij}]$, with $r_{ij} \in [0,1]$ being the corresponding degrees of association between x_i and y_j.

Fuzzy relations defined on some continuous spaces such as $\mathbf{R}^2$, say "much smaller than," "approximately equal," and "similar" could, for instance, be characterized by the following membership functions:

$$R_m(x,y) = \begin{cases} 1 - \exp(-|y-x|), & \text{if } x \leq y \\ 0, & \text{otherwise} \end{cases} \qquad x \text{ much smaller than } y$$

$$R_e(x,y) = \exp\left\{\frac{-|x-y|}{\alpha}\right\}, \alpha > 0 \qquad x \text{ approximately equal to } y$$

$$R_s(x,y) = \begin{cases} \exp[-(x-y)/\beta], & \text{if } |x-y| \leq 5 \\ 0, & \text{if } |x-y| \geq 5 \end{cases}, \beta > 0 \qquad x \text{ and } y \text{ similar}$$

Figure 6.2(b) displays the membership function of the relation "x approximately equal to y" on $\mathbf{X} = \mathbf{Y} = [0,4]$ assuming that $\alpha = 1$.

6.3 PROPERTIES OF THE FUZZY RELATIONS

6.3.1 Domain and Codomain of Fuzzy Relations

The domain, dom$R$, of a fuzzy relation R defined in $\mathbf{X} \times \mathbf{Y}$ is a fuzzy set whose membership function is equal to

$$\text{dom}R(x) = \sup_{y \in \mathbf{Y}} R(x,y)$$

while its codomain, cod$R$, is a fuzzy set whose membership function is given as

$$\text{cod}R(y) = \sup_{x \in \mathbf{X}} R(x,y)$$

Considering finite universes of discourse, domain and codomain can be viewed as the height of the rows and columns of the fuzzy relation matrix (Zadeh, 1971).

6.3.2 Representation of Fuzzy Relations

Similar to the case of fuzzy sets, fuzzy relations can be represented by their α-cuts, that is,

$$R = \bigcup_{\alpha \in [0,1]} \alpha R_\alpha$$

or, in terms of the membership function $R(x, y)$ of R

$$R(x, y) = \sup_{\alpha \in [0,1]} \{\min[\alpha, R(x, y)]\}$$

6.3.3 Equality of Fuzzy Relations

We say that two fuzzy relations P and Q defined in the same Cartesian product of spaces $\mathbf{X} \times \mathbf{Y}$ are equal if and only if their membership functions are identical, that is,

$$P(x, y) = Q(x, y) \quad \forall \ (x, y) \in \mathbf{X} \times \mathbf{Y}$$

6.3.4 Inclusion of Fuzzy Relations

A fuzzy relation P is included in Q, denoted by $P \subseteq Q$, if and only if

$$P(x, y) \leq Q(x, y) \quad \forall \ (x, y) \in \mathbf{X} \times \mathbf{Y}$$

Similarly, as it was presented in the case of relations, given n-fold Cartesian product of these universes, we define the fuzzy relation in the form

$$R : \mathbf{X}_1 \times \mathbf{X}_2 \times \cdots \times \mathbf{X}_n \to [0, 1]$$

If the spaces $\mathbf{X}_1, \mathbf{X}_2, \ldots, \mathbf{X}_n$ are finite with $\text{Card}(\mathbf{X}_1) = n_1 \ldots \text{Card}(\mathbf{X}_n) = n_n$, then R can be expressed as an n-fold ($n_1 \times \cdots \times n_p$) matrix $R = [r_{ij..k}]$ with $r_{ij..k} \in [0, 1]$ being the degree of association assigned to the n-tuple $(x_i, x_j, \ldots, x_k) \in \mathbf{X}_1 \times \mathbf{X}_2 \times \cdots \times \mathbf{X}_n$. If $\mathbf{X}_1, \mathbf{X}_2, \ldots, \mathbf{X}_n$ are infinite, then the membership function of R is a certain function of many variables. The concepts of equality and inclusion of fuzzy relations could be easily extended for relations defined in multidimensional spaces.

6.4 OPERATIONS ON FUZZY RELATIONS

The basic operations on fuzzy relations, say union, intersection, and complement, conceptually follow the corresponding operations on fuzzy sets once fuzzy relations are fuzzy sets formed on multidimensional spaces. For illustrative purposes the definitions of union, intersection, and complement below involve two-argument fuzzy relations. Without any loss of generality, we can focus on binary fuzzy relations P, Q, R defined in $\mathbf{X} \times \mathbf{Y}$. As in the case of fuzzy sets, all definitions are defined pointwise.

6.4.1 Union of Fuzzy Relations

The union R of two fuzzy relations P and Q defined in $\mathbf{X} \times \mathbf{Y}$, $R = P \cup Q$, is defined with the use of the following membership function:

$$R(x, y) = P(x, y) \, s \, Q(x, y) \quad \forall \, (x, y) \in \mathbf{X} \times \mathbf{Y}$$

recall that s stands for some t-conorm.

6.4.2 Intersection of Fuzzy Relations

The intersection R of fuzzy relations P and Q defined in $\mathbf{X} \times \mathbf{Y}$, $R = P \cap Q$, is defined in the following form:

$$R(x, y) = P(x, y) t Q(x, y) \quad \forall \, (x, y) \in \mathbf{X} \times \mathbf{Y}$$

where t is a t-norm.

6.4.3 Complement of Fuzzy Relations

The complement $\overline{R}$ of the fuzzy relation R is defined by the membership function

$$\overline{R}(x, y) = 1 - R(x, y) \quad \forall \, (x, y) \in \mathbf{X} \times \mathbf{Y}$$

6.4.4 Transpose of Fuzzy Relations

Given a fuzzy relation R, its transpose, denoted by R^{T}, is a fuzzy relation on $\mathbf{Y} \times \mathbf{X}$ such that the following relationship holds:

$$R^{\mathrm{T}}(y, x) = P(x, y) \quad \forall \, (x, y) \in \mathbf{X} \times \mathbf{Y}$$

If R is a relation defined in some finite space, then R^{T} is the transpose of the corresponding $(n \times m)$ matrix representation of R. Therefore, the form of R^{T} is an $(m \times n)$ matrix whose columns are now the rows of R.

The following properties are direct consequences of the definitions provided above:

$$(R^{\mathrm{T}})^{\mathrm{T}} = R$$
$$(\overline{R})^{\mathrm{T}} = \overline{(R^{\mathrm{T}})}$$

6.5 CARTESIAN PRODUCT, PROJECTIONS, AND CYLINDRICAL EXTENSION OF FUZZY SETS

A mechanism to construct fuzzy relations is through the use of the concept of Cartesian product extended to fuzzy sets. The concept closely follows the one adopted for sets once they involve the pairs of points of the underlying universes, added with a membership degree.

6.5.1 Cartesian Product

Given fuzzy sets $A_1, A_2, \ldots, A_n$ defined in universes $\mathbf{X}_1, \mathbf{X}_2, \ldots, \mathbf{X}_n$, respectively, their Cartesian product $A_1 \times A_2 \times \cdots \times A_n$ is a fuzzy relation R on $\mathbf{X}_1 \times \mathbf{X}_2 \times \cdots \times \mathbf{X}_n$ with the following membership function:

$$R(x_1, x_2, \ldots, x_n) = \min\{A_1(x_1), A_2(x_2), \ldots, A_n(x_n)\} \quad \forall x_1 \in \mathbf{X}_1, \; \forall x_2 \in \mathbf{X}_2, \ldots, \forall x_n \in \mathbf{X}_n$$

In general, we can generalize the concept of this Cartesian product by using some t-norms.

$$R(x_1, x_2, \ldots, x_n) = A_1(x_1) \, t \, A_2(x_2) \, t \ldots t \, A_n(x_n) \quad \forall x_1 \in \mathbf{X}_1, \forall x_2 \in \mathbf{X}_2, \ldots, \forall x_n \in \mathbf{X}_n$$

6.5.2 Projection of Fuzzy Relations

In contrast to the concept of the Cartesian product, the idea of projection is to construct fuzzy relations on some subspaces of the original relation. Projection reduces the dimensionality of the original space over which the original fuzzy relation is defined.

Given R being a fuzzy relation defined in $\mathbf{X}_1 \times \mathbf{X}_2 \times \cdots \times \mathbf{X}_n$, its projection on $\mathbf{X} = \mathbf{X}_i \times \mathbf{X}_j \times \cdots \times \mathbf{X}_k$, where $I = \{i, j, \ldots, k\}$ is a subsequence of the set of indexes $N = \{1, 2, \ldots, n\}$, is a fuzzy relation R_X with the membership function (Zadeh, 1975a,b).

$$R_X(x_i, x_j, \ldots, x_k) = \text{Proj}_X R(x_1, x_2, \ldots, x_n) = \sup_{x_t, x_u, \ldots, x_v} R(x_1, x_2, \ldots, x_n)$$

where $J = \{t, u, \ldots, v\}$ is a subsequence of N such that $I \cup J = N$ and $I \cap J = \emptyset$. In other words, J is the complement of I with respect to N. Notice that the above expression is computed for all values of $(x_1, x_2, \ldots, x_n) \in \mathbf{X}_i \times \mathbf{X}_j \times \cdots \times \mathbf{X}_k$.

For instance, Figure 6.3 shows the projections R_X and R_Y of a Gaussian, binary fuzzy relation R defined in $\mathbf{X} \times \mathbf{Y}$ with $\mathbf{X} = [0, 8]$ and $\mathbf{Y} = [0, 10]$, whose membership function is equal to $R(x, y) = \exp\{-\alpha[(x-4)^2 + (y-5)^2]\}$. In this case the projections are formed as

$$R_X(x) = \text{Proj}_X R(x, y) = \sup_y R(x, y)$$

$$R_Y(y) = \text{Proj}_Y R(x, y) = \sup_x R(x, y)$$

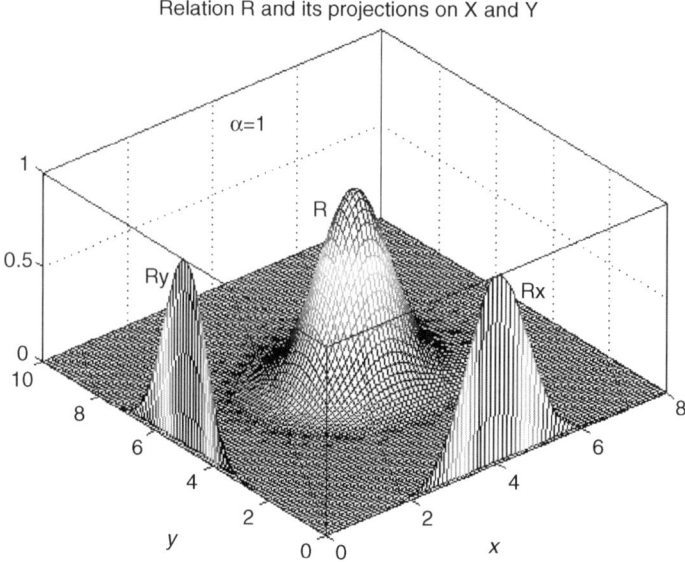

Figure 6.3 Fuzzy relation R along with its projections on **X** and **Y**.

To find projections of the fuzzy relations defined in some finite spaces, the maximum operation replaces the sup operation occurring in the definition provided above. For example, for the fuzzy relation $R : \mathbf{X} \times \mathbf{Y} \to [0, 1]$ with $\mathbf{X} = \{1, 2, 3\}$ and $\mathbf{Y} = \{1, 2, 3, 4, 5\}$,

$$R(x, y) = \begin{bmatrix} 1.0 & 0.6 & 0.8 & 0.5 & 0.2 \\ 0.6 & 0.8 & 1.0 & 0.2 & 0.9 \\ 0.8 & 0.6 & 0.8 & 0.3 & 0.9 \end{bmatrix}$$

The three elements of the projection R_X are taken as the maximum computed for each of the three rows of R.

$R_X = [\max(1,0,0.6,0.8,0.5,0.2), \max(0.6,0.8,1.0,0.2,0.9), \max(0.8,0.6,0.8,0.3,0.9)]$
$= [1.0, 1.0, 0.9]$

Similarly, the five elements of R_Y are taken as the maximum among the entries of the five columns of R. Figure 6.4 shows R and its projections R_X and R_Y.

$$R_Y = [1.0, \quad 0.8, \quad 1.0, \quad 0.5, \quad 0.9]$$

Note that domain and codomain of the fuzzy relation are examples of its projections.

6.5.3 Cylindrical Extension

The cylindrical extension increases the number of coordinates of the Cartesian product over which the fuzzy relation is formed and of the set on which it

6.5 Cartesian Product, Projections, and Cylindrical Extension of Fuzzy Sets

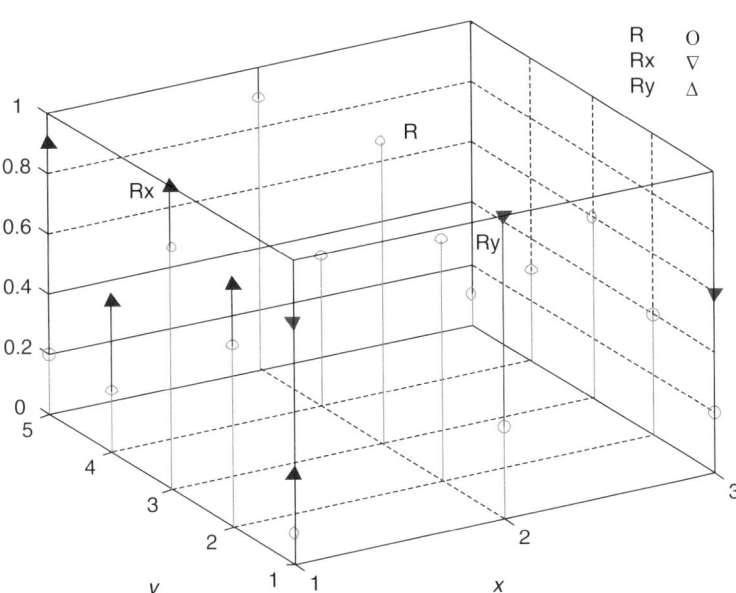

Figure 6.4 Fuzzy relation R and its projections on **X** and **Y**.

operates by expanding a fuzzy set into a binary relation, a two-dimensional relation into three-dimensional counterpart, and so forth. In this sense, cylindrical extension is an operation that is complementary to the projection operation (Zadeh, 1975a,b).

The cylindrical extension on $\mathbf{X} \times \mathbf{Y}$ of a fuzzy set A of **X** is a fuzzy relation cyl$A$ whose membership function is equal to

$$\text{cyl}A(x, y) = A(x), \quad \forall\, x \in \mathbf{X}, \quad \forall\, y \in \mathbf{Y}$$

If the fuzzy relation is viewed as a two-dimensional matrix, the operation of cylindrical extension forms identical columns indexed by the successive values of $y \in \mathbf{Y}$. The main intent of cylindrical extensions is to achieve compatibility of spaces over which fuzzy sets and fuzzy relations are formed. For instance, let A be a fuzzy set of **X** and R a fuzzy relation on $\mathbf{X} \times \mathbf{Y}$, Figure 6.5(a) and (b), respectively. Suppose we attempt to compute union and intersection of A and R. Because the universes over which A and R are defined are different, we cannot carry out any set-based operations on A and R. The cylindrical extension of A, denoted by cyl$A$, Figure 6.5(c) provides the compatibility required. Then the operations such as (cyl$A$) $\cup$ R and (cyl$A$) $\cap$ R make sense, see Figure 6.5(d) and (e).

The concept of cylindrical extension can be easily generalized to multidimensional cases.

148 Chapter 6 Fuzzy Relations

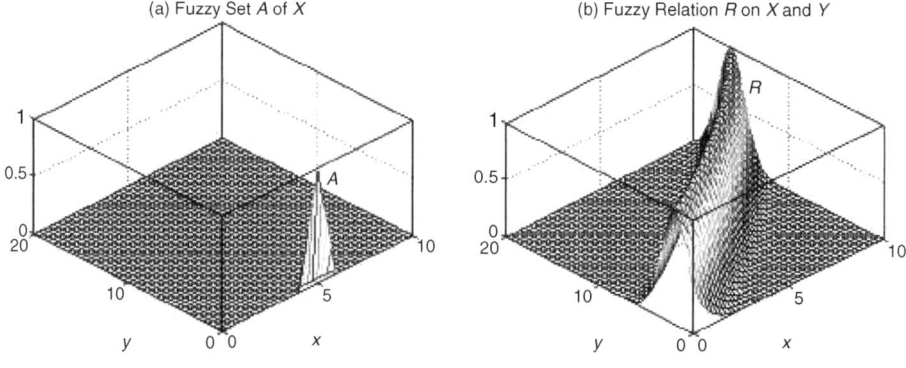

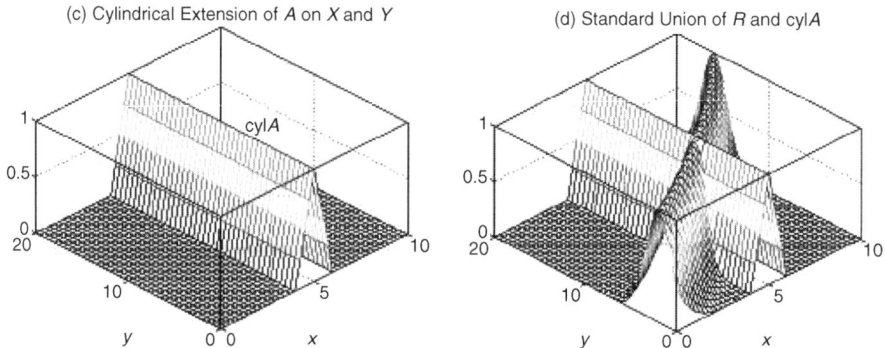

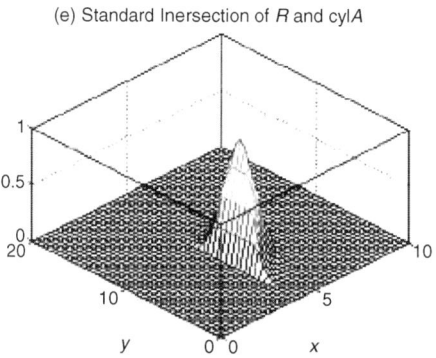

Figure 6.5 Cylindrical extension of a fuzzy set.

6.6 RECONSTRUCTION OF FUZZY RELATIONS

Projections do not retain complete information conveyed by the original fuzzy relation. This means that in general one can not faithfully reconstruct a relation from its projections. In other words, projections $\text{Proj}_X R$ and $\text{Proj}_Y R$ of some fuzzy relation R do not necessarily lead to the original fuzzy relation R. In general, the reconstruction of a relation via the Cartesian product of its projections is a relation that includes the original relation, that is,

$$\text{Proj}_X R \times \text{Proj}_Y R \supseteq R$$

If, however, in the above relationship the equality holds, then we call the relation R to be noninteractive. Figure 6.6 shows an example of a noninteractive fuzzy relation R_m defined in $\mathbf{X} \times \mathbf{Y}$. Clearly, the Cartesian product of its projections $\text{Proj}_X R_m$ and $\text{Proj}_Y R_m$ recover the original relation, see Figure 6.6(b) and (e).

Figure 6.7 shows an example of an interactive fuzzy relation R_p. In this case, the Cartesian product of the projections $\text{Proj}_X R_p$ and $\text{Proj}_Y R_p$ does not recover the original relation as Figure 6.7(b) and (e) demonstrates; we have $\text{Proj}_X R_p \times \text{Proj}_Y R_p \supset R_p$.

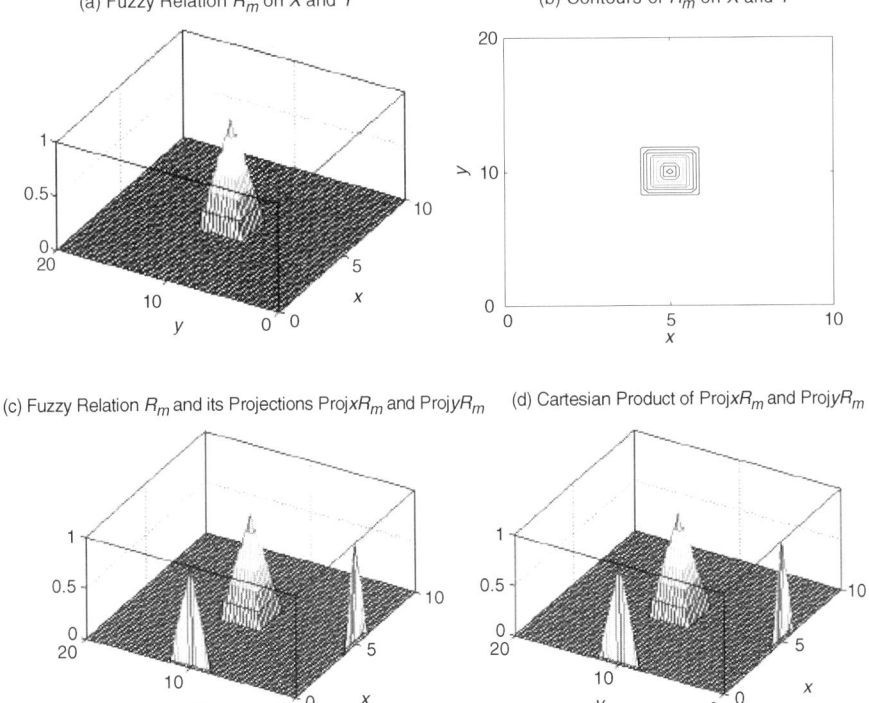

Figure 6.6 Reconstruction of noninteractive fuzzy relation R_m.

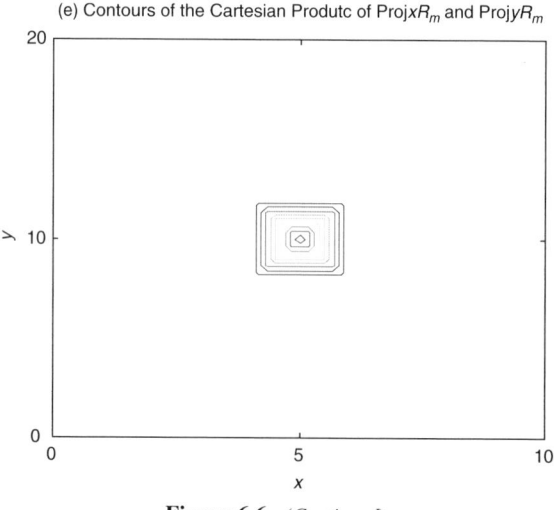

Figure 6.6 (*Continued*)

Fuzzy relations certainly are an efficient mechanism that is used to define, construct, and handle multidimensional fuzzy sets. Among the *n*-dimensional relations, binary relations are of particular interest especially when they involve a Cartesian product of **X**. Notice that, when talking about fuzzy relations, the notion of binary relations concerns the universes of discourse in which the fuzzy relations are defined, rather than the membership values they assume.

6.7 BINARY FUZZY RELATIONS

A binary fuzzy relation R on $\mathbf{X} \times \mathbf{X}$ is defined as follows:
$$R : \mathbf{X} \times \mathbf{X} \to [0, 1]$$
There are several important features of binary fuzzy relations.

  (a) Reflexivity: $R(x, x) = 1 \quad \forall \, x \in \mathbf{X}$, refer to Figure 6.8(a). When **X** is finite, $R \supseteq I$ where I is an identity matrix, $I(x, y) = 1$ if $x = y$ and $I(x, y) = 0$ otherwise. Reflexivity can be relaxed by admitting a concept of the so-called ε-reflexivity, $\varepsilon \in [0, 1]$. This means $R(x, x) \geq \varepsilon$. When $R(x, x) = 0$ the fuzzy relation is irreflexive. A fuzzy relation is locally reflexive if, for any $(x, y) \in \mathbf{X}$, $\max\{R(x, y), R(y, x)\} \leq R(x, x)$.

  (b) Symmetry: $R(x, y) - R(y, x) \quad \forall \, (x, y) \subset X \times X$, refer to Figure 6.8(b). For finite **X**, the matrix representing R has entries distributed symmetrically along the main diagonal. Clearly, if R is symmetric, then $R^T = R$.

6.7 Binary Fuzzy Relations 151

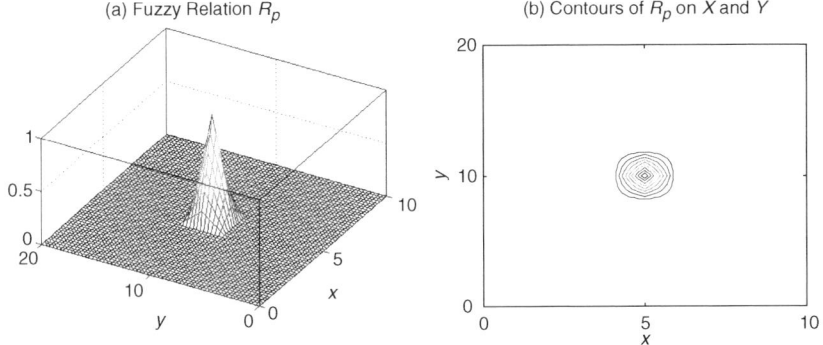

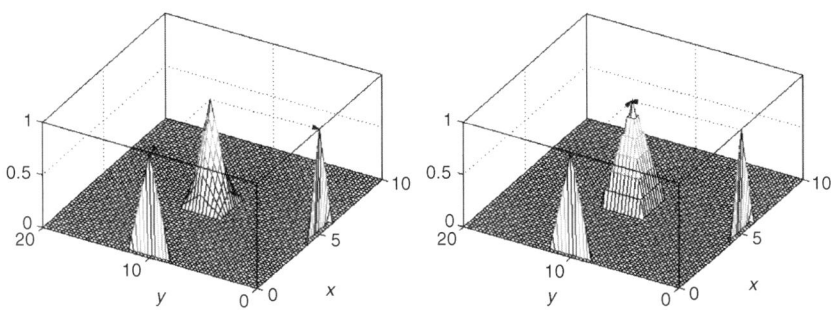

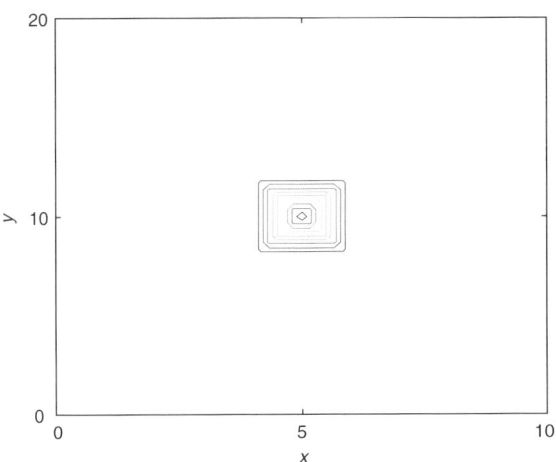

Figure 6.7 Reconstruction of interactive fuzzy relation R_p.

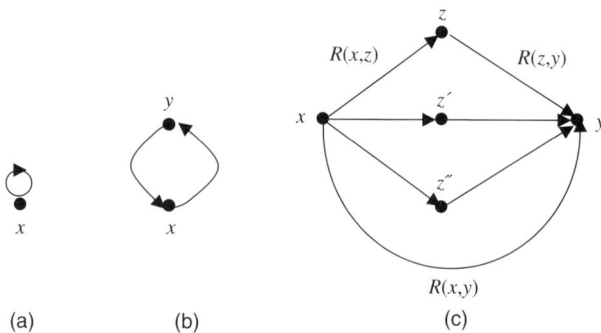

Figure 6.8 Main characteristics of binary fuzzy relations; see the details in the text.

(c) Transitivity: $\sup_{z \in X}\{R(x,z) t R(z,y)\} \leq R(x,y) \forall x, y, z \in X$. In particular, if this relationship holds for $t = \min$, then the relation is called sup-min transitive. Looking at the levels of associations $R(x,z)$ and $R(z,y)$ occurring between x, and z, and z and y, the property of transitivity reflects the maximal strength among all possible links arranged in series (such as ($R(x,z)$ and $R(z,y)$)) that does not exceed the strength of the direct link $R(x,z)$, refer to Figure 6.8(c).

6.7.1 Transitive Closure

Given a binary fuzzy relation in a finite universe **X**, there exists a unique fuzzy relation $\overleftrightarrow{R}$ on **X**, called transitive closure of R, which contains R and itself is included in any transitive fuzzy relation on **X** that contains R (De Baets and Meyer, 2003). Therefore, if R is defined on a finite universe of cardinality n, the transitive closure is given by

$$\mathrm{trans}(R) = \overleftrightarrow{R} = R \cup R^2 \cup \cdots \cup R^n$$

where, by definition,

$$R^2 = R \circ R \cdots R^p = R \circ R^{p-1}$$
$$R \circ R(x,y) = \max_z \{R(x,z) t R(z,y)\}$$

Notice that $R \circ R$ can be computed similarly as encountered in matrix algebra by replacing the ordinary multiplication by some t-norm and the sum by the max operations. In other words, if $r_{ij}^2 = [R^2]_{ij} = [R \circ R]_{ij}$, then

$$r_{ij}^2 = \max_k (r_{ik} t r_{kj})$$

If R is reflexive, then

$$I \subseteq R \subseteq R^2 \subseteq \cdots \subseteq R^{n-1} = R^n$$

The transitive closure of the fuzzy relation R can be found by computing the successive k max-t products of R until $R^k = R^{k-1}$. A procedure whose complexity is $O(n^3 \log_2 n)$ in time and $O(n^2)$ in space (Naessens et al., 2002) is as follows:

procedure TRANSITIVE-CLOSURE-Z (R) **returns** transitive fuzzy relation
static: fuzzy relation $R = [r_{ij}]$

for $i = 1 \ldots$ **do**
    $\overrightarrow{R} \leftarrow R \cup (R \circ R)$
    if $\overrightarrow{R} = R$ **then return** $\overrightarrow{R}$
    $R \leftarrow \overrightarrow{R}$

An algorithm that computes the transitive closure of a fuzzy relation in $O(n^3)$ in time and $O(n^2)$ in space was suggested in (Kandel and Yelowitz, 1974). This algorithm is a modification of the Floyd–Warshall algorithm originally developed to solve all-to-all shortest path problem (Rardin, 1998). The Floyd–Warshall procedure to compute the transitive closure of an $(n \times n)$ fuzzy relation $R = [r_{ij}]$ is as follows:

procedure TRANSITIVE-CLOSURE-W (R) **returns** transitive fuzzy relation
static: fuzzy relation $R = [r_{ij}]$

for $i = 1 : n$ **do**
    for $j = 1 : n$ **do**
        for $k = 1 : n$ **do**
            $\overrightarrow{r}_{jk} \leftarrow \max(r_{jk}, r_{ji} t\, r_{ik})$
return $\overrightarrow{R}$

Alternative algorithms with similar complexity are given in (Naessens et al., 2002). Binary fuzzy relations on universes with large cardinality are often represented by sparse matrices. In these circumstances more effective algorithms to compute transitive closures can be obtained exploring appropriate representations for sparse matrices. Procedures with time complexity averaging $n \log_4 n$ have been developed for sparse relations (Wallace et al., 2006).

6.7.2 Equivalence and Similarity Relations

Equivalence relations are relations that are reflexive, symmetric, and transitive. Suppose that one of the arguments of $R(x, y)$, x for example, has been fixed. Thus, all elements related to x constitute a set called an equivalence class of R with respect to x, denoted by

$$A_x = \{y \in \mathbf{Y} | R(x, y) = 1\}$$

The family of all equivalence classes of R, denoted by $\mathbf{X}/R$, is a partition of $\mathbf{X}$. In other words, $\mathbf{X}/R$ is a family of pairwise disjoint nonempty subsets of $\mathbf{X}$ whose union is $\mathbf{X}$. Equivalence relations can be viewed as a generalization of the equality relations in the sense that members of an equivalence class can be considered equivalent to each other under the relation R.

Similarity relations are fuzzy relations that are reflexive, symmetric, and transitive. Like any fuzzy relation, a similarity relation can be represented by a nested family of its α-cuts, R_α. Each α-cut constitutes an equivalence relation and forms a partition of $\mathbf{X}$. Therefore, each similarity relation is associated with a set $P(R)$ of partitions of $\mathbf{X}$:

$$P(R) = \{X/R_\alpha | \alpha \in [0,1]\}$$

Partitions are nested in the sense that, if $\alpha > \beta$, then the partition $\mathbf{X}/R_\alpha$ is finer than the partition $\mathbf{X}/R_\beta$. For example, consider the relation defined on $\mathbf{X} = \{a, b, c, d, e\}$ in the following way:

$$R = \begin{bmatrix} 1.0 & 0.8 & 0 & 0 & 0 \\ 0.8 & 1.0 & 0 & 0 & 0 \\ 0 & 0 & 1.0 & 0.9 & 0.5 \\ 0 & 0 & 0.9 & 1.0 & 0.5 \\ 0 & 0 & 0.5 & 0.5 & 1.0 \end{bmatrix}$$

One can verify that R is a symmetric matrix, has a values of 1 at its main diagonal, and is max-min transitive. Therefore, R is a similarity relation. The levels of refinement of the similarity relation R can be represented in the form of partition tree in which each node corresponds to a fuzzy relation on $\mathbf{X}$ whose degrees of association between the elements are greater than or equal to the threshold value α. For instance, we have the following fuzzy relations for $\alpha = 0.5$, 0.8, and 0.9, respectively:

$$R_{0.5} = \begin{bmatrix} 1 & 1 & 0 & 0 & 0 \\ 1 & 1 & 0 & 0 & 0 \\ 0 & 0 & 1 & 1 & 1 \\ 0 & 0 & 1 & 1 & 1 \\ 0 & 0 & 1 & 1 & 1 \end{bmatrix}, R_{0.8} = \begin{bmatrix} 1 & 1 & 0 & 0 & 0 \\ 1 & 1 & 0 & 0 & 0 \\ 0 & 0 & 1 & 1 & 0 \\ 0 & 0 & 1 & 1 & 0 \\ 0 & 0 & 0 & 0 & 1 \end{bmatrix}, R_{0.9} = \begin{bmatrix} 1 & 0 & 0 & 0 & 0 \\ 0 & 1 & 0 & 0 & 0 \\ 0 & 0 & 1 & 1 & 0 \\ 0 & 0 & 1 & 1 & 0 \\ 0 & 0 & 0 & 0 & 1 \end{bmatrix}$$

Notice that $R = \cup_{\alpha \in \Lambda} \alpha R_\alpha$ where $\cup = \max$ and $\Lambda = \{0.5, 0.8, 0.9, 1.0\}$ is the level set of R. Also, notice that the greater the value of α, the finer the classes are, as Figure 6.9 shows.

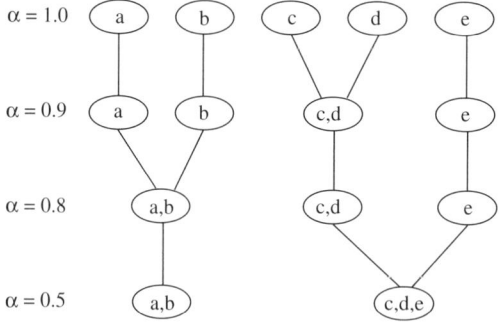

Figure 6.9 Partition tree induced by binary fuzzy relation R.

6.7.3 Compatibility and Proximity Relations

Compatibility relations are reflexive and symmetric relations. Associated with any compatibility relation are sets called compatibility classes. A compatibility class is a subset A of a universe $\mathbf{X}$ such that $R(x,y) = 1$ for all $x, y \in A$.

Proximity relations are reflexive and symmetric fuzzy relations. Let A be a subset of a universe $\mathbf{X}$. Thus, A is a ε-proximity class of R if $R(x,y) \geq \varepsilon$ for all $(x,y) \in A$. For instance, the relation R on $\mathbf{X} = \{1,2,3,4,5\}$

$$R = \begin{bmatrix} 1.0 & 0.7 & 0 & 0 & 0.6 \\ 0.7 & 1.0 & 0.6 & 0 & 0 \\ 0 & 0.6 & 1.0 & 0.7 & 0.4 \\ 0 & 0 & 0.7 & 1.0 & 0.5 \\ 0.6 & 0 & 0.4 & 0.5 & 1.0 \end{bmatrix}$$

has the unity in its main diagonal and is symmetric. Therefore, R is a proximity relation. Compatibility classes and α-compatibility classes do not necessarily induce partitions of X (Klir and Yuan, 1995).

Proximity is an important concept in pattern recognition being used in contexts such as visual images as under these circumstances human subjectivity leads to some useful information that could be represented in the form of proximity relations (Yang and Shih, 2001).

6.8 CONCLUSIONS

Fuzzy relations generalize the concept of fuzzy sets to multidimensional universes and introduce the notion of association degrees between the elements of some universe of discourse. Fuzzy relations are subject to the same type of set operations as we studied for fuzzy sets. However, additional operations can be performed to either lower dimensions via projections or increase dimensions via cylindrical extension and Cartesian products. Operations with fuzzy relations are important to process fuzzy models constructed via fuzzy relations.

Relations are associations and remain at the very basis of most methodological approaches of science and engineering. As fuzzy relations are more general constructs than functions, they allow dependencies between several variables to be captured without necessarily committing to any particular directional association of the variables being involved.

EXERCISES AND PROBLEMS

1. Let $R_\alpha = \{(x,y) \in \mathbf{X} \times \mathbf{Y} | R(x,y) \geq \alpha\}$ be the α-cut of the fuzzy relation R. Show that any fuzzy relation $R: \mathbf{X} \times \mathbf{Y} \to [0,1]$ can be represented in the following form:

$$R = \cup_{\alpha \in (0,1]} \alpha R_\alpha$$

where $\cup$ denotes standard union operation, and αR_α is a subnormal fuzzy relation whose membership function is α if $(x, y) \in R_\alpha$ and zero otherwise.

2. How can the algorithm to compute the transitive closure of a fuzzy relation be used to verify if a fuzzy relation is transitive or not? Use your answer to verify that the fuzzy relation R in Section 6.6 is actually max-min transitive.
3. Show that if R is a similarity relation, then each of its α-cut R_α is an equivalence relation.
4. Verify that the transitive closure of a fuzzy proximity relation is a similarity relation.
5. A tolerance relation R in $\mathbf{X} \times \mathbf{Y}$ is a reflexive and symmetric ordinary relation. Show that if R is a proximity relation, then for any $0 < \alpha \leq 1$, R_α is a tolerance relation.

HISTORICAL NOTES

The notions of fuzzy relation, similarity relation, ordering relation and transitivity, and the basic theory of fuzzy relation were introduced in Zadeh (1971). The ideas of projection and cylindrical extension appeared in Zadeh (1975a,b).

Abstract, *natural* interpretations of fuzzy sets, fuzzy relations, and fuzzy mappings have been discussed by Shinoda (2002, 2003). By *natural* the author meant an interpretation in a Heyting value model for intuitionistic set theory.

The characterization of the main classes of fuzzy relations using fuzzy modal operations has been presented by Radzikowska and Kerre (2005). Fuzzy modal operators are binary fuzzy relations that transform a fuzzy set to another one.

REFERENCES

De Baets, B., Meyer, H. On the existence and construction of T-transitive closures, *Inf. Sci.* **152**, 2003, 167–179.

Kandel, A., Yelowitz, L. Fuzzy chains, *IEEE Trans. Syst. Man Cybern.* SMC-4, 1974, 472–475.

Klir, G., Yuan, B. *Fuzzy Sets and Fuzzy Logic: Theory and Applications*, Prentice-Hall, Upper Saddle River, NJ, 1995.

Naessens, H., Meyer, H., De Baets, B. Algorithms for the computation of T-transitive closures, *IEEE Trans. Fuzzy Syst.* **10**(4), 2002, 541–551.

Radzikowska, A., E. Kerre, Characterization of main classes of fuzzy relations using fuzzy modal operators, *Fuzzy Set Syst.* **152**, 2005, 223–247.

Rardin, R. *Optimization in Operations Research*, Prentice-Hall, Upper Saddle River, NJ, 1998.

Shinoda, M. *A natural interpretation of fuzzy sets and fuzzy relations*, *Fuzzy Set Syst.* **128**, 2002, 135–147.

Shinoda, M. *A natural interpretation of fuzzy mappings*, *Fuzzy Set Syst.* **138**, 2003, 67–82.

Wallace, M., Acrithis, Y., Kollias, S. Computationally efficient sup-t transitive closure for sparse fuzzy binary relations, *Fuzzy Set Syst.* **157**, 2006, 341–372.

Yang, M., Shih, H. Cluster analysis based on fuzzy relations, *Fuzzy Set Syst.* **120**, 2001, 197–212.

Zadeh, L. A. Similarity relations and fuzzy orderings, *Inf. Sci.*, **3**, 1971, 177–200.

Zadeh, L. A. Fuzzy logic and approximate reasoning, *Synthese*, **30**(1), 1975a, 407–428.

Zadeh, L. A. The concept of linguistic variables and its application to approximate reasoning I, II, III, *Inf. Sci.* **8**(9), 1975b, 199–251, 301–357, 43–80.

Chapter 7

Transformations of Fuzzy Sets

Transformations of elements (points) through functions are omnipresent. An immediate generalization of such point transformations involves set transformations between spaces. Mappings of fuzzy sets between universes constitute another generalization of mapping sets between spaces. Thus, point transformations can be expanded to cover transformations involving fuzzy sets. Transformations of this nature can be realized using either functions or relations. In both cases these transformations constitute an essential component of various pursuits including system modeling and control applications, pattern recognition and information retrieval, just to name a few representative areas. This chapter introduces two important mechanisms to transform fuzzy sets, namely, the extension principle and the calculus of fuzzy relations. We elaborate on their essential properties, present algorithmic aspects, and discuss various interpretations of the resulting constructs.

7.1 THE EXTENSION PRINCIPLE

The extension principle is a fundamental construct that enables extensions of point operations to operations involving sets and fuzzy sets. Intuitively, the idea is as follows: Given a function (mapping) from some domain $\mathbf{X}$ to codomain (range) $\mathbf{Y}$, the extension principle offers a mechanism to transform a fuzzy set defined in $\mathbf{X}$ to some fuzzy set defined in $\mathbf{Y}$.

Let $f: \mathbf{X} \rightarrow \mathbf{Y}$ be a function. Given any $x \in \mathbf{X}$, $y = f(x)$ denotes the image of x under f, namely, the point transformation of x under f, refer to Figure 7.1. This is the straightforward idea that the customary notion of any function conveys. Pointwise transformations can be extended to handle transformations of sets.

Let $P(\mathbf{X})$ and $P(\mathbf{Y})$ be the power sets of $\mathbf{X}$ and $\mathbf{Y}$ and $A \in P(\mathbf{X})$ a set. The image of A under f can be determined by realizing point transformations $y = f(x)$ for all $x \in A$. In this sense, the image of A under f is the set B that arises in the following form:

$$B = f(A) = \{y \in \mathbf{Y} | y = f(x), \quad \forall \, x \in A\}$$

Fuzzy Systems Engineering: Toward Human-Centric Computing, by Witold Pedrycz and Fernando Gomide
Copyright © 2007 John Wiley & Sons, Inc.

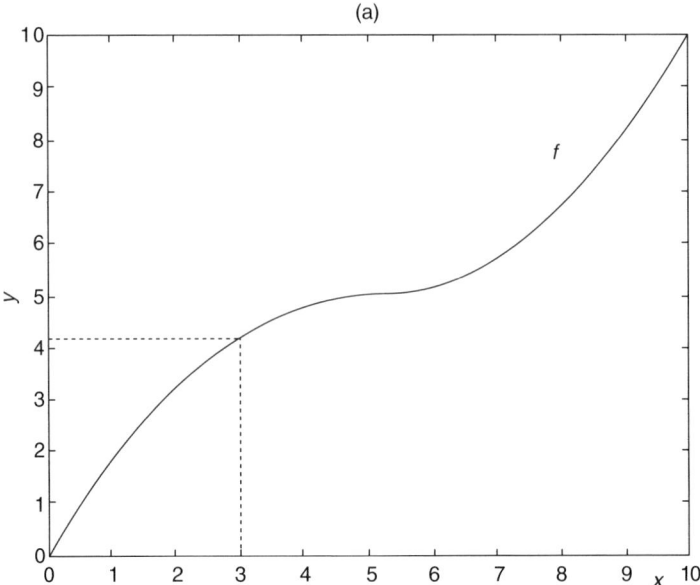

Figure 7.1 An example of function f along with its point transformation.

Since A and B are sets, they can be expressed in terms of their characteristic functions as follows:

$$B(y) = \sup_{x|y=f(x)} A(x)$$

as displayed in Figure 7.2. Notice that this mechanism provides a way to extend the notion of functions regarded as point transformations to the notion of set functions. Once viewed in terms of characteristic functions, it becomes natural to extend this notion to fuzzy sets as follows:

Let $F(\mathbf{X})$ and $F(\mathbf{Y})$ denote the families of all fuzzy sets defined in $\mathbf{X}$ and $\mathbf{Y}$, respectively and $f\colon \mathbf{X} \to \mathbf{Y}$ be a function. Function f induces a mapping $f\colon F(\mathbf{X}) \to F(\mathbf{Y})$ such that if A is a fuzzy set in $\mathbf{X}$, then its image under f is a fuzzy set $B = f(A)$ whose membership function is expressed as (Zadeh, 1975)

$$B(y) = \sup_{x/y=f(x)} A(x) \qquad (7.1)$$

Figure 7.3 illustrates the extension principle in the case where A is a triangular fuzzy set with membership function $A = A(x, 3, 5, 8)$ and the function f of the form

$$f(x) = \begin{cases} -0.2(x-5)^2 + 5, & \text{if } 0 \leq x \leq 5 \\ 0.2(x-5)^2 + 5, & \text{if } 5 < x \leq 10 \end{cases}$$

If $A = \{A(x_1)/x_1, A(x_2)/x_2, \ldots, A(x_n)/x_n\}$ is a fuzzy set in $\mathbf{X} = \{x_1, x_2, \ldots, x_n\}$ and $y = f(x)$ with $y \in \mathbf{Y} = \{y_1, y_2, \ldots, y_m\}$, that is, the universes $\mathbf{X}$ and $\mathbf{Y}$ are finite, then

7.1 The Extension Principle 159

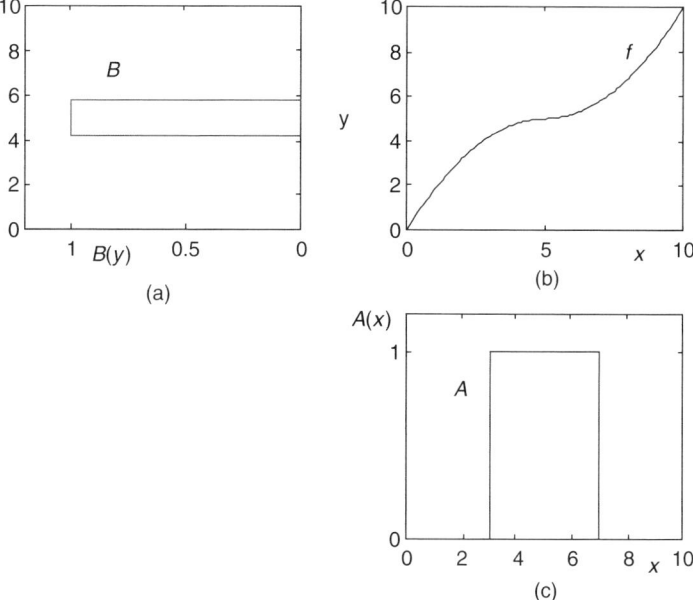

Figure 7.2 Set transformation.

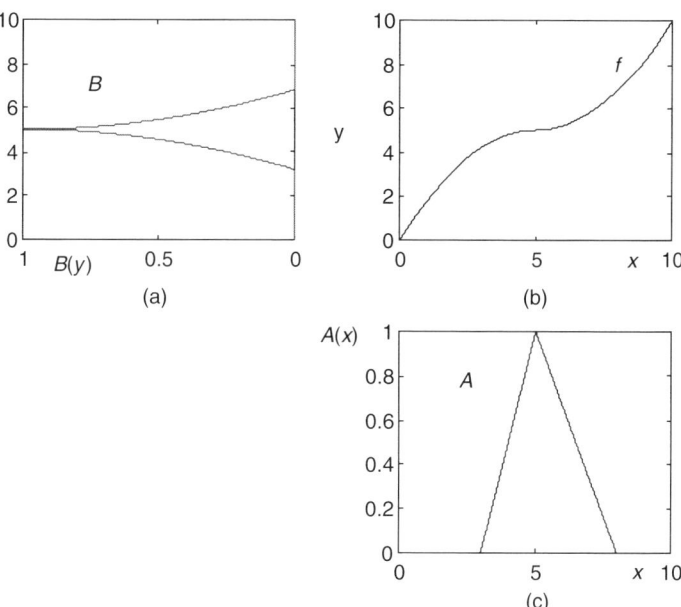

Figure 7.3 An illustration of the extension principle; a nonlinear transformation of the triangular fuzzy number is shown.

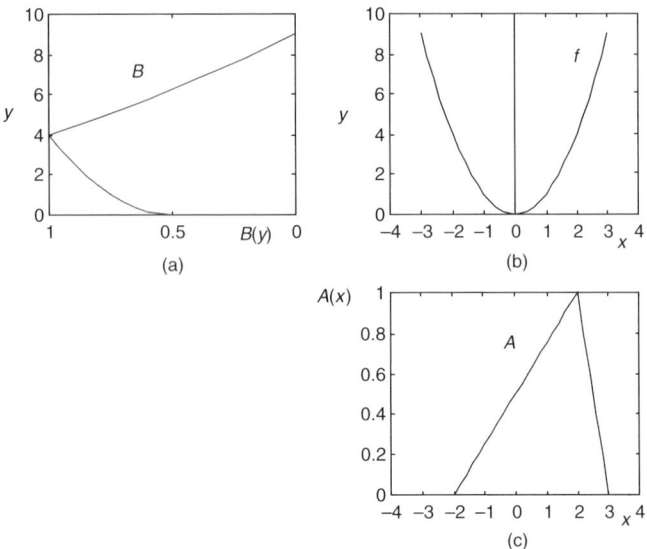

Figure 7.4 Extension principle applied to the case of many-to-one mapping.

the extension principle becomes

$$B(y_i) = \max_{x_j / y_i = f(x_j)} A(x_j) \qquad (7.2)$$

The sup and max operations in (7.1) and (7.2) are needed when function f is a many-to-one mapping. In this case, there exist several points x (say, x_i and x_j) for which $y = f(x_i) = f(x_j)$ and the membership grade assigned to y is the greatest among the corresponding membership values $A(x_i)$ and $A(x_j)$.

Figure 7.4 illustrates the case where $y = f(x) = x^2$, a two-to-one mapping, and A is a triangular fuzzy set with membership function $A = A(x, -2, 2, 3)$.

For finite universes, consider $\mathbf{X} = \{-3, -2, -1, 0, 1, 2, 3\}$ and $y = f(x) = x^2$. Given the fuzzy set $A = \{0/-3, 0.1/-2, 0.3/-1, 1/0, 0.2/1, 0/2, 0/3\}$ defined in $\mathbf{X}$, the image $B = f(A)$ is a fuzzy set in $\mathbf{Y} = \{y|y = x^2\} = \{0, 1, 4, 9\}$ whose membership function is $B = \{1/0, \max(0.2, 0.3)/1, \max(0, 0.1)/4, 0/9\} = \{1/0, 0.3/1, 0.1/4, 0/9\}$ as shown in Figure 7.5.

The extension principle generalizes to functions of many variables as follows: Let $\mathbf{X}_i$, $i = 1, \ldots, n$ and $\mathbf{Y}$ be universes and $\mathbf{X} = \mathbf{X}_1 \times \mathbf{X}_2 \times \cdots \times \mathbf{X}_n$. Consider fuzzy sets A_i on $\mathbf{X}_i$, $i = 1, \cdots n$, and a function $y = f(\mathbf{x})$ with $\mathbf{x} = [x_1, x_2, \ldots, x_n]$ a point of $\mathbf{X}$. Fuzzy sets $A_1, A_2, \ldots, A_n$ can be transformed through f to give a fuzzy set $B = f(A_1, A_2, \ldots, A_n)$ in $\mathbf{Y}$ with the membership function

$$B(y) = \sup_{\mathbf{x}|y = f(\mathbf{x})} \{\min[A_1(x_1), A_2(x_2), \ldots, A_n(x_n)]\} \qquad (7.3)$$

In (7.3), the min operation is a choice within the family of triangular norms and any t-norm can be adopted because each component x_i occurs concurrently in $\mathbf{x}$. As discussed in Chapter 5, t-norms capture the idea of conjunction or coincidence, of elements.

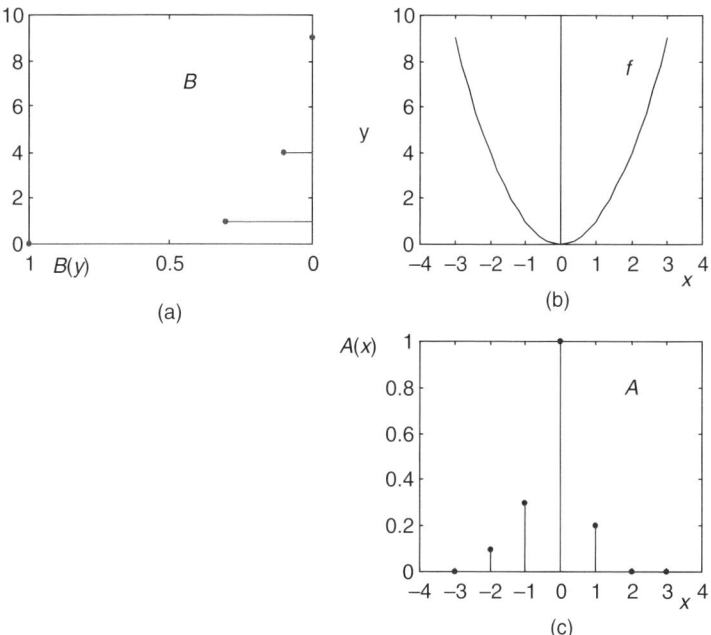

Figure 7.5 Extension principle applied in the case of a certain many-to-one mapping and finite universes.

Transformations of fuzzy sets through the extension principle produce fuzzy sets that satisfy certain properties. The most important are as follows (Klir and Yuan, 1995; Nguyen and Walker, 1999): Let $f : \mathbf{X} \to \mathbf{Y}$ be an arbitrary function and consider fuzzy sets $A_i, A_i \in F(\mathbf{X})$ and $B_i, B_i \in F(\mathbf{Y}), B_i = f(A_i), i = 1, 2, \ldots, n$.

1. $B_i = \emptyset$ if and only if $A_i = \emptyset$
2. $A_1 \subseteq A_2 \Rightarrow B_1 \subseteq B_2$
3. $f(\bigcup_{i=1}^{n} A_i) = \bigcup_{i=1}^{n} B_i$
4. $f(\bigcap_{i=1}^{n} A_i) \subseteq \bigcap_{i=1}^{n} B_i$

If $A \in F(\mathbf{X})$ and $B \in F(\mathbf{Y})$, with $B = f(A)$, and if $A_\alpha, A_\alpha^+, B_\alpha, B_\alpha^+$ are their corresponding α-cuts and strong α-cuts, respectively, then we obtain

5. $B_\alpha \supseteq f(A_\alpha)$
6. $B_\alpha^+ = f(A_\alpha^+)$

7.2 COMPOSITIONS OF FUZZY RELATIONS

The extension principle is a fundamental concept of fuzzy set theory as it offers a general vehicle to extend theoretical and applied notions as well as procedures to the cases involving fuzzy sets. In many cases transformations used in practice may not be

162 Chapter 7 Transformations of Fuzzy Sets

functions. Instead we may envision that the relationships between elements are captured in terms of relations or fuzzy relations. In these circumstances transformations are completed through their compositions.

Fuzzy relations can be composed with the use of different set theoretic operations, and triangular norms, in particular. Different families of composition operators arise that depend upon the choice of some specific t-norms and t-conorms. Two most important compositions come in the form of a sup-t composition and an inf-s composition, respectively.

7.2.1 Sup-t Composition

The sup-t composition of fuzzy relations $G : \mathbf{X} \times \mathbf{Z} \to [0, 1]$ and W: $\mathbf{Z} \times \mathbf{Y} \to [0, 1]$ is a fuzzy relation $R : \mathbf{X} \times \mathbf{Y} \to [0, 1]$ whose membership function $R(x, y)$ is given as

$$R(x, y) = \sup_{z \in Z} \{G(x, z) t W(z, y)\}, \quad \forall\, x \in \mathbf{X} \text{ and } \forall\, y \in \mathbf{Y} \quad (7.4)$$

The sup-t composition of G and W is denoted symbolically by $R = G \circ W$.

For instance, let the membership function of relation G describing the concept "*close to*" be expressed as

$$G(x, z) = \exp[-(x - z)^2]$$

Likewise, the fuzzy relation W of the same semantics (*close to*) as G is defined as

$$W(z, y) = \exp[-(z - y)^2]$$

Let us now compute the composition of G and W, $R = G \circ W$ with the use of the algebraic product (t-norm). Using (7.4) we obtain

$$R(x, y) = \sup_{z \in Z}\{e^{-(x-z)^2} e^{-(z-y)^2}\} = \max_{z \in Z}\{e^{-(x-z)^2} e^{-(z-y)^2}\}, \quad \forall\, x \in \mathbf{X} \text{ and } \forall\, y \in \mathbf{Y}$$

Since the Gaussian function is continuous and unimodal, we replace the supremum by the max operation and the membership function of R is found computing the value of z that maximizes the product of the membership functions of G and W, namely, $z = (x + y)/2$. Therefore,

$$R(x, y) = \exp[-(x - y)^2/2]$$

Interestingly, R may also be interpreted as a relation with semantics "*x close to y.*" This intuitively agrees with the idea that composition of values close to each other should result in the values that are also close. Figure 7.6 illustrates the fuzzy relations G and R. Notice that the shape of the membership function of R is

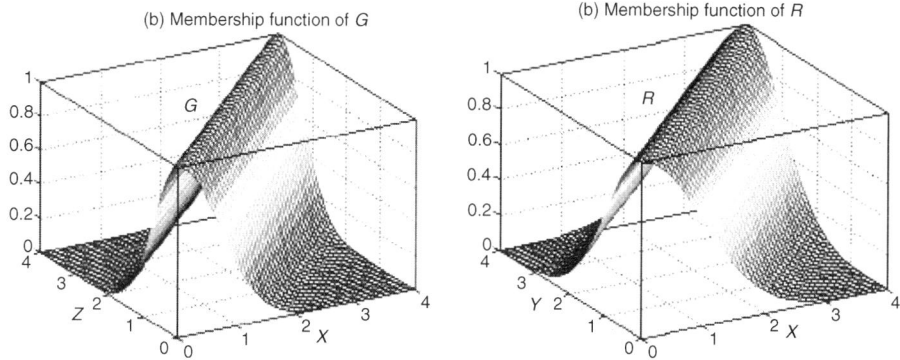

Figure 7.6 Sup-product composition of fuzzy relations *"close to"*.

identical with the one of G. They are different with respect to the values of the spread. In essence, the composition of W and G results in the fuzzy relation of increased spread value.

The sup-min composition is another particular example of the family of sup-t compositions (Zadeh, 1965) being widely used in practice

$$R(x,y) = \sup_{z \in Z}\{G(x,z) \wedge W(z,y)\}, \quad \forall\, x \in \mathbf{X} \text{ and } \forall\, y \in \mathbf{Y} \tag{7.5}$$

Contrary to the sup-product composition, sup-min composition is not very amenable for analytical developments. Often detailed mathematical analysis of the resulting expressions involving sup-min composition becomes less transparent.

The composition of fuzzy relations defined in finite universes is easily completed by considering their matrix representations. In this case, the sup-*t* composition becomes the max-*t* composition, and its computation follows the rules as the usual matrix calculation except that the algebraic product is replaced by the t-norm and the algebraic sum by the max operation. Let G and W be represented by $(n \times p)$ and $(p \times m)$ matrices, respectively. The steps below form the procedure to compute the $(n \times m)$ relational matrix $R = G \circ W$.

procedure SUP-T-COMPOSITION (G,W) **returns** composition of fuzzy relations
static: fuzzy relations: $G = [g_{ik}]$, $W = [w_{kj}]$
    0_{nm} : $n \times m$ matrix with all entries equal to zero
    t: a *t*-norm
$R = 0_{nm}$
for $i = 1 : n$ **do**
  for $j = 1 : m$ **do**
    for $k = 1 : p$ **do**
      tope $\leftarrow g_{ik}\ t\ w_{kj}$
      $r_{ij} \leftarrow \max(r_{ij},\ \text{tope})$
return R

EXAMPLE 7.1

Let us consider fuzzy relations G and W given as the following (3×4) and (4×2) matrices:

$$G = \begin{bmatrix} 1.0 & 0.6 & 0.5 & 0.5 \\ 0.6 & 0.8 & 1.0 & 0.2 \\ 0.8 & 0.3 & 0.4 & 0.3 \end{bmatrix} \quad W = \begin{bmatrix} 0.6 & 0.1 \\ 0.5 & 0.7 \\ 0.7 & 0.8 \\ 0.3 & 0.6 \end{bmatrix}$$

Using (7.5), the max–min composition $R = G \circ W$, $R = [r_{ij}]$ with min ($\wedge$) t-norm gives rise to the expression

$r_{11} = \max(1.0 \wedge 0.6, 0.6 \wedge 0.5, 0.5 \wedge 0.7, 0.5 \wedge 0.3) = \max(0.6, 0.5, 0.5, 0.3) = 0.6$
$r_{21} = \max(0.6 \wedge 0.6, 0.8 \wedge 0.5, 1.0 \wedge 0.7, 0.2 \wedge 0.3) = \max(0.6, 0.5, 0.7, 0.2) = 0.7$
...
$r_{32} = \max(0.1 \wedge 0.8, 0.3 \wedge 0.7, 0.4 \wedge 0.8, 0.3 \wedge 0.6) = \max(0.1, 0.3, 0.4, 0.3) = 0.4$

$$R = \begin{bmatrix} 0.6 & 0.6 \\ 0.7 & 0.8 \\ 0.6 & 0.4 \end{bmatrix}$$

Likewise, the max-product composition of G and W is computed using the algebraic product (t-norm) produces

$r_{11} = \max(1.0 \cdot 0.6, 0.6 \cdot 0.5, 0.5 \cdot 0.7, 0.5 \cdot 0.3) = \max(0.6, 0.3, 0.35, 0.15) = 0.60$
$r_{21} = \max(0.6 \cdot 0.6, 0.8 \cdot 0.5, 1.0 \cdot 0.7, 0.2 \cdot 0.3) = \max(0.36, 0.4, 0.7, 0.06) = 0.70$
...
$r_{32} = \max(0.1 \cdot 0.8, 0.3 \cdot 0.7, 0.4 \cdot 0.8, 0.3 \cdot 0.6) = \max(0.08, 0.21, 0.32, 0.18) = 0.32$

$$R = \begin{bmatrix} 0.60 & 0.42 \\ 0.70 & 0.80 \\ 0.48 & 0.32 \end{bmatrix}$$

The sup-t composition exhibits a number of interesting properties. Let P, Q and S, and R be fuzzy relations defined in $\mathbf{X} \times \mathbf{Y}$, $\mathbf{Y} \times \mathbf{Z}$, and $\mathbf{Z} \times \mathbf{W}$, respectively. Here the $\cup$ and $\cap$ are the standard operations of union and intersection:

1. associativity $\qquad P \circ (Q \circ R) = (P \circ Q) \circ R$
2. distributivity over union $\qquad P \circ (Q \cup R) = (P \circ Q) \cup (P \circ R)$
3. weak distributivity over intersection $\qquad P \circ (Q \cap R) \subseteq (P \circ Q) \cap (P \circ R)$
4. monotonicity $\qquad$ if $Q \subseteq S$ then $P \circ Q \subseteq P \circ S$

It is worth analyzing several particular instances of the sup-t composition to substantiate its semantics and reveal links between fundamental constructs of fuzzy computing and composition of fuzzy relations. In what follows, $R_y(x)$ denotes the unary fuzzy relation indexed by y, namely, a slice of the fuzzy relation $R(x, y)$ in $\mathbf{X} \times \mathbf{Y}$ whose location is specified by the value assumed by the second variable (y).

1. Let A be a fuzzy set in $\mathbf{X}$. The sup-t composition of A and R_y is a unary fuzzy set B in $\mathbf{Y}$ whose membership function value at y is equal to

$$B(y) = \sup_{x \in \mathbf{X}}[A(x) t R_y(x)]$$

becomes the possibility measure of fuzzy set A determined with respect to fuzzy set R_y, a measure that quantifies the extent to which A and R_y overlap.

2. The supremum operation implements the existential quantifier $\exists$ (there exists), while the t-norm is viewed as some *and* connective. Then, in this case $B(y)$ is a truth value of the statement $A(x)$ and $R_y(x)$:

$$B(y) = \text{truth}(\exists x | \ A(x) \text{ and } Ry(x))$$

3. The sup-t composition can be regarded as a special version of the projection operation focused or directed by the fuzzy set A. In particular, if A is the entire universe $\mathbf{X}$, then the sup-t composition reduces to the projection operator, namely

$$B(y) = \sup_{x \in \mathbf{X}}[\mathbf{X}(x) t R(x,y)] = \sup_{x \in \mathbf{X}}[1 t R(x,y)] = \sup_{x \in \mathbf{X}} R(x,y)])$$

7.2.2 Inf-s Composition

The inf-s composition of fuzzy relations $G : \mathbf{X} \times \mathbf{Z} \to [0, 1]$ and $W : \mathbf{Z} \times \mathbf{Y} \to [0, 1]$ is a fuzzy relation $R : \mathbf{X} \times \mathbf{Y} \to [0, 1]$, whose membership function $R(x, y)$ is computed as

$$R(x, y) = \inf_{z \in Z} \{G(x,z) s W(z,y)\}, \quad \forall \ x \in \mathbf{X} \text{ and } \forall \ y \in \mathbf{Y} \quad (7.6)$$

The inf-s composition of G and W is denoted symbolically by $R = G \bullet W$. Opposing the sup-t composition, inf-s composition is not common in practice and is analytically more complex. For fuzzy relations on finite universes, they can be computed using their matrix representations. In this case inf-s becomes the min-s composition, and its computation follows the rules as the usual matrix calculation except that the algebraic product is replaced by the s-norm and the algebraic sum by the min operation. If G and W are represented by $(n \times p)$ and $(p \times m)$ matrices, the computational steps to compute the $(n \times m)$ relational matrix $R = G \bullet W$ are

procedure INF-S-COMPOSITION (G, W) **returns** composition of fuzzy relations
static: fuzzy relations: $G = [g_{ik}], W = [w_{kj}]$
    $1_{nm} : n \times m$ matrix with all entries equal to unity
    s: a s-norm
$R = 1_{nm}$
for $i = 1 : n$ **do**
  for $j = 1 : m$ **do**
    for $k = 1 : p$ **do**
      sope $\leftarrow g_{ik} \ s \ w_{kj}$
      $r_{ij} \leftarrow \min(r_{ij}, \text{sope})$
return R

EXAMPLE 7.2

Let G and W be as shown below.

$$G = \begin{bmatrix} 1.0 & 0.6 & 0.5 & 0.5 \\ 0.6 & 0.8 & 1.0 & 0.2 \\ 0.8 & 0.3 & 0.4 & 0.3 \end{bmatrix} \quad W = \begin{bmatrix} 0.6 & 0.1 \\ 0.5 & 0.7 \\ 0.7 & 0.8 \\ 0.3 & 0.6 \end{bmatrix}$$

Performing the inf-s composition of G and W with the aid of the probabilistic sum (t-conorm), we obtain:

$$r_{11} = \min(1.0 + 0.6 - 0.6, 0.6 + 0.5 - 0.6 \cdot 0.5, 0.5 + 0.7 - 0.5 \cdot 0.7, 0.5 + 0.3 - 0.5 \cdot 0.3)$$
$$= \min(1.0, 0.8, 0.85, 0.65) = 0.65$$
$$r_{21} = \min(0.6 + 0.6 - 0.6 \cdot 0.6, 0.8 + 0.5 - 0.8 \cdot 0.5, 1.0 + 0.7 - 0.7, 0.2 + 0.3 - 0.2 \cdot 0.3)$$
$$= \min(0.84, 0.9, 1.0, 0.44) = 0.44$$

...

$$r_{32} = \min(0.8 + 0.1 - 0.8 \cdot 0.1, 0.3 + 0.7 - 0.3 \cdot 0.7, 0.4 + 0.8 - 0.4 \cdot 0.8, 0.3 + 0.6 - 0.3 \cdot 0.6)$$
$$= \min(0.82, 0.79, 0.88, 0.72) = 0.72$$

$$R = \begin{bmatrix} 0.65 & 0.80 \\ 0.44 & 0.64 \\ 0.51 & 0.72 \end{bmatrix}$$

EXAMPLE 7.3

Considering the discrete version of the fuzzy relations $G(x,z) = \exp[-(x-z)^2]$ and $W(z,y) = \exp[-(z-y)^2]$ computed for $x, y, z \in [0, 4]$ in steps of size $\delta = 0.1$, G, and W have the same matrix (41×41) representation. The membership function of G and the resulting inf-probabilistic sum composition $R(x, y)$ of G and W are shown in Figure 7.7(a) and (b), respectively.

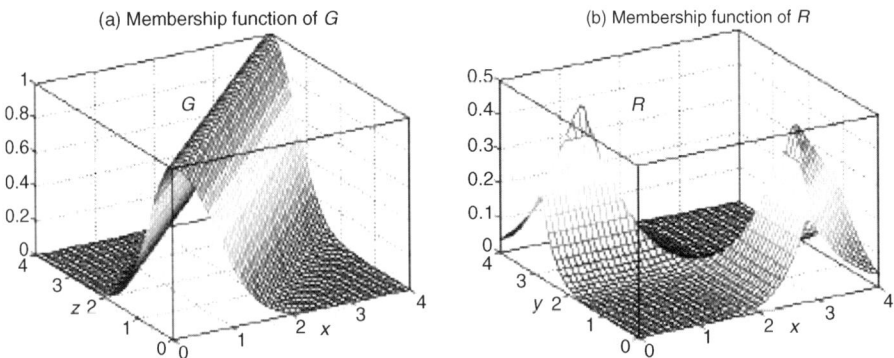

Figure 7.7 Inf-probabilistic sum composition of fuzzy relations "*close to*."

The following are the basic properties of the inf-s composition. Let P, Q and S, and R be fuzzy relations on $\mathbf{X} \times \mathbf{Y}$, $\mathbf{Y} \times \mathbf{Z}$, and $\mathbf{Z} \times \mathbf{W}$, and consider $\cup$ and $\cap$ as standard union and intersection, respectively.

1. associativity $P \bullet (Q \bullet R) = (P \bullet Q) \bullet R$
2. weak distributivity over union $P \bullet (Q \cup R) \supseteq (P \bullet Q) \cup (P \bullet R)$
3. distributivity over intersection $P \bullet (Q \cap R) = (P \bullet Q) \cap (P \bullet R)$
4. monotonicity   if $Q \subseteq S$ then $P \bullet Q \supseteq P \bullet S$

Particular instances of the inf-s composition reveal links between its semantics and constructs of fuzzy computing. As before, $R_y(x)$ denotes the unary fuzzy relation indexed by y, namely, the fuzzy relation $R(x, y)$ on $\mathbf{X} \times \mathbf{Y}$, when the second variable is kept fixed.

1. Let A be a fuzzy set on $\mathbf{X}$. The inf-s composition of A and R_y is a unary fuzzy set B on Y whose membership function value at y is

$$B(y) = \inf_{x \in \mathbf{X}}[A(x) s \, R_y(x)] = \inf_{x \in \mathbf{X}}[R_y(x) s A(x)] = \inf_{x \in \mathbf{X}}[R_y(x) s \overline{\overline{A}}(x)]$$

the necessity measure of fuzzy set A with respect to fuzzy set R_y, a measure that quantifies the degree with which the complement of A, $\overline{A}$, is included in R_y.

2. The infimum implements the universal quantifier $\forall$ (for all) while the s-norm is viewed as the *or* connective. Thus, in this case $B(y)$ is a truth value of the statement $A(x)$ or $R_y(x)$:

$$B(y) = \text{truth}(\forall \, x | A(x) \text{ or } Ry(x))$$

7.2.3 Inf-φ Composition

Given a continuous t-norm t, let $a \varphi b = \{c \in [0, 1] | \, a \, t \, c \leq b\}$ for all $a, b \in [0, 1]$. As discussed in Chapter 5, this operation may be interpreted as an implication induced by some t-norm. In this sense, this φ operator models an operation of inclusion.

The inf-φ composition of fuzzy relations $G : \mathbf{X} \times \mathbf{Z} \rightarrow [0, 1]$ and $W : \mathbf{Z} \times \mathbf{Y} \rightarrow [0, 1]$ is a fuzzy relation $R : \mathbf{X} \times \mathbf{Y} \rightarrow [0, 1]$ whose membership function $R(x, y)$ is

$$R(x, y) = \inf_{z \in \mathbf{Z}}\{G(x, z) \varphi W(z, y)\}, \quad \forall \, x \in \mathbf{X} \text{ and } \forall \, y \in \mathbf{Y} \tag{7.7}$$

The inf-φ composition of G and W is denoted symbolically by $R = G \varphi W$. Similarly as in the previous cases, composition of fuzzy relations on finite universes can be computed using their matrix representations, and in this way the inf-φ composition becomes the min-φ one. The computational steps follow the usual rules for matrix product calculation except that the algebraic product is replaced by the φ operation and the algebraic sum by the min operation.

EXAMPLE 7.4

If G and W are (3×4) and (4×2) matrices

$$G = \begin{bmatrix} 1.0 & 0.6 & 0.5 & 0.5 \\ 0.6 & 0.8 & 1.0 & 0.2 \\ 0.8 & 0.3 & 0.4 & 0.3 \end{bmatrix} \quad W = \begin{bmatrix} 0.6 & 0.1 \\ 0.5 & 0.7 \\ 0.7 & 0.8 \\ 0.3 & 0.6 \end{bmatrix}$$

then performing the inf-φ composition of G and W and using the bounded difference (t-norm), $a \, t \, b = \max(0, a + b - 1)$ we get $a \, \varphi \, b = \min(1, 1 - a + b)$, which is the Lukasiewicz implication. Therefore,

$r_{11} = \min(1.0 \, \varphi \, 0.6, 0.6 \, \varphi \, 0.5, 0.5 \, \varphi \, 0.7, 0.5 \, \varphi \, 0.3) = \min(0.6, 0.9, 1.0, 0.8) = 0.6$

$r_{21} = \min(0.6 \, \varphi \, 0.6, 0.8 \, \varphi \, 0.5, 1.0 \, \varphi \, 0.7, 0.2 \, \varphi \, 0.3) = \min(1.0, 0.7, 0.7, 1.0) = 0.7$

...

$r_{32} = \min(0.8 \, \varphi \, 0.1, 0.3 \, \varphi \, 0.7, 0.4 \, \varphi \, 0.8, 0.3 \, \varphi \, 0.6) = \min(0.3, 1.0, 1.0, 1.0) = 0.3$

$$R = \begin{bmatrix} 0.6 & 0.10 \\ 0.7 & 0.50 \\ 0.8 & 0.3 \end{bmatrix}$$

If P, Q and S, and R are fuzzy relations on $\mathbf{X} \times \mathbf{Y}$, $\mathbf{Y} \times \mathbf{Z}$, and $\mathbf{Z} \times \mathbf{W}$, $\cup$ is the standard union, and $\cap$ is the standard intersection, then the basic properties of the inf-φ composition are

1. associativity $\quad P \varphi (Q \varphi R) = (P \circ Q) \varphi R$
2. weak distributivity over union $\quad P \varphi (Q \cup R) \supseteq (P \varphi Q) \cup (P \varphi R)$
3. distributivity over intersection $\quad P \varphi (Q \cap R) = (P \varphi Q) \cap (P \varphi R)$
4. monotonicity $\quad$ if $Q \subseteq S$ then $P \varphi Q \subseteq P \varphi S$

Again, if $R_y(x)$ denotes the unary fuzzy relation indexed by y, then the inf-φ composition, because of the implication models inclusion, becomes

$$B(y) = \inf_{x \in \mathbf{X}} [A(x) \varphi R_y(x)] = \inf_{x \in \mathbf{X}} [A(x) \Rightarrow R_y(x)] = \inf_{x \in \mathbf{X}} [A(x) \subset R_y(x)]$$

which is a minimal (pessimistic) degree of inclusion of $A(x)$ in the respective slice of fuzzy relation R_y. In this case the inf-φ composition denotes a truth value of the statement A is included in R_y and $B(y) = \forall \, x(A(x) \Rightarrow R_y(x))$.

7.3 FUZZY RELATIONAL EQUATIONS

Fuzzy relational equations are closely associated with the notion of composition of fuzzy relations. One can view a fuzzy relation R in $\mathbf{X} \times \mathbf{Y}$ as a model of a fuzzy system whose input is a fuzzy set U on $\mathbf{X}$ and output is a fuzzy set V on $\mathbf{Y}$, as shown in Figure 7.8. Fuzzy sets are unary relations. The fuzzy relation R describes the dependencies (relationships) between system input and output.

Consider the simplest fuzzy relational model of the form

$$V = U \circ R$$

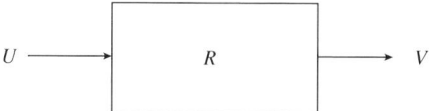

Figure 7.8 Single-input–single-output fuzzy system.

where the operator denote by ∘ is either a sup-t or inf-s composition discussed in Section 2. Clearly, given an input U and a fuzzy relation R, the output V is found using the definitions of the operator. Two fundamental problems arise:

1. Given U and V, determine R
2. Given V and R, determine U

The first is referred to as an *estimation* problem once it is concerned with determination of the parameters of the relational model. The second is an *inverse* problem once it aims at computing the input, given an output and the relation between input and output.

Consider, for instance, an information retrieval system. In information retrieval systems a set of index terms $\mathbf{X} = \{x_1, x_2, \ldots, x_n\}$ and a set of relevant documents $\mathbf{Y} = \{y_1, y_2, \ldots, y_m\}$ can be associated through a $(n \times m)$ fuzzy relational matrix $R = [r_{ij}]$ whose membership value r_{ij} specifies the relevance degree of index term x_i of document y_j. If R is known, then given an input query U on $\mathbf{X}$ its corresponding output V contains the membership values attached to the documents in $\mathbf{Y}$ that reflects the query U through its composition with the relational matrix R. In this direct case, the membership values of V rank the documents consistent with query U. Given samples of queries and outputs, the estimation problem concerns the construction of the relational matrix that represents their association. Therefore, the estimation problem is fundamental to the design tools and implementation mechanisms of fuzzy information retrieval systems (Pereira et al., 2006). The inverse problem in an information retrieval system assumes an interesting interpretation in this context: We are to determine a query that produces a given set of ranked documents.

In image compression and reconstruction, binary fuzzy relation matrices encode images by normalizing the intensity range of gray-scale pixels. Image compression requires composition of the image with reference fuzzy sets to produce a lower dimensional representation of the image. Reconstruction of the original image requires solution of the inverse problem involving the compressed image and the respective reference fuzzy sets (Nobuhara et al., 2002; Hirota and Pedrycz, 2002).

In what follows, the focus is on composition of fuzzy relations on finite universes because this is the most important instance of composition of fuzzy relations in practice. The emphasis is on the sup-t form because this is the one that has been most extensively studied.

7.3.1 Solutions to the Estimation Problem

7.3.1.1 Single-Input–Single-Output Fuzzy Relational Equations

Initially, let $U : \mathbf{X} \to [0, 1]$ and $V : \mathbf{Y} \to [0, 1]$ be unary fuzzy relations, or equivalently, fuzzy sets on finite universes $\mathbf{X} = \{x_1, x_2, \ldots, x_n\}$ and $\mathbf{Y}\{= y_1, y_2, \ldots, y_m\}$ represented by a $(1 \times n)$ vector $[u_i]$ and by a $(1 \times m)$ vector $[v_j]$, respectively. Let $R : \mathbf{X} \times \mathbf{Y} \to [0, 1]$ be a fuzzy relation represented by the $(n \times m)$ fuzzy relational matrix $R = [r_{ij}]$.

To address the solution of the estimation problem, denote by S_e the family of fuzzy relations R that satisfies

$$V = U \circ R \tag{7.8}$$

namely, $S_e = \{R \in F(\mathbf{X}) \times F(\mathbf{Y}) | V = U \circ R\}$. Consider the operator $\varphi : [0, 1] \to [0, 1]$ such that

$$a \varphi b = \sup\{c \in [0, 1] |\ a\ t\ c \leq b\} \tag{7.9}$$

PROPOSITION

If $S_e \neq \phi$, then the unique maximal solution $\hat{R}$ of the sup-t relational equation $V = U \circ R$ is

$$\hat{R} = U^T \varphi V \tag{7.10}$$

$\hat{R}$ is maximal in the sense that if R is an element of S_e, then $R \subseteq \hat{R}$.

PROOF

Notice that $U^T \varphi (U \circ R) = U^T \varphi\ V = \hat{R} \supset R$ and, since the sup-t composition is monotonic, $U \circ \hat{R} \supset U \circ R = V$. Moreover, $U \circ (U^T \varphi\ V) \subset V$ means that $U \circ \hat{R} \subset V$. Thus,

1. $U \circ \hat{R} \supset V$
2. $U \circ \hat{R} \subset V$

and from (1) and (2), $V = U \circ \hat{R}$. Also, if $\hat{R}$ is not unique, then there exists a maximal solution $\hat{R}'$ such that $V = U \circ \hat{R}'$ and $\hat{R} \subseteq \hat{R}'$. But, since $\hat{R}$ is a maximal solution, $\hat{R}' \subseteq \hat{R}$. Therefore $\hat{R}' = \hat{R}$; in other words, the maximal solution is unique.

The procedure to solve the estimation problem is summarized next.

procedure ESTIMATE-SOLUTION (U, V) **returns** fuzzy relation
static: fuzzy unary relations $U = [u_i], V = [v_j]$
      t: a t-norm
define φ operator
for $i = 1 : n$ **do**
    for $j = 1 : m$ **do**
        $r_{ij} \leftarrow u_i \varphi v_j$
return R

EXAMPLE 7.5

Consider two fuzzy sets

$$U = [0.8, 0.5, 0.3] \text{ and } V = [0.4, 0.2, 0.0, 0.7]$$

and specify the t-norm t = min. Then, using (7.9) we have

$$a\varphi b = \begin{cases} 1 & \text{if } a \leq b \\ b & \text{if } a > b \end{cases}$$

and from (7.10)

$$\hat{R} = \begin{bmatrix} 0.8 \\ 0.5 \\ 0.3 \end{bmatrix} \varphi [0.4 \ 0.2 \ 0.0 \ 0.7] = \begin{bmatrix} 0.8\varphi 0.4 & 0.8\varphi 0.2 & 0.8\varphi 0.0 & 0.8\varphi 0.7 \\ 0.5\varphi 0.4 & 0.5\varphi 0.2 & 0.5\varphi 0.0 & 0.5\varphi 0.7 \\ 0.3\varphi 0.4 & 0.3\varphi 0.2 & 0.3\varphi 0.0 & 0.3\varphi 0.7 \end{bmatrix}$$

Therefore,

$$\hat{R} = \begin{bmatrix} 0.4 & 0.2 & 0.0 & 0.7 \\ 0.4 & 0.2 & 0.0 & 1.0 \\ 1.0 & 0.2 & 0.0 & 1.0 \end{bmatrix} \quad (7.11)$$

One can verify that the fuzzy relations

$$\hat{R}_1 = \begin{bmatrix} 0.4 & 0.2 & 0.0 & 0.7 \\ 0.0 & 0.0 & 0.0 & 0.5 \\ 0.3 & 0.0 & 0.0 & 0.5 \end{bmatrix} \text{ and } \hat{R}_2 = \begin{bmatrix} 0.0 & 0.2 & 0.0 & 0.7 \\ 0.4 & 0.0 & 0.0 & 0.2 \\ 0.6 & 0.2 & 0.0 & 1.0 \end{bmatrix}$$

are also elements of S_e, and clearly $\hat{R}_1 \subseteq \hat{R}$ and $\hat{R}_2 \subseteq \hat{R}$. However, $\hat{R}_1$ and $\hat{R}_2$ are not comparable and cannot be ordered linearly by the inclusion relation. In general, the family of solutions can be viewed as shown in Figure 7.9: It clearly involves a unique maximal solution and several incomparable minimal solutions.

Existence of and procedures to compute minimal solutions for general sup-t compositions still remain a challenge. Most of the current work concentrates on max–min relational equations for which the existence and procedures of some solution methods have already

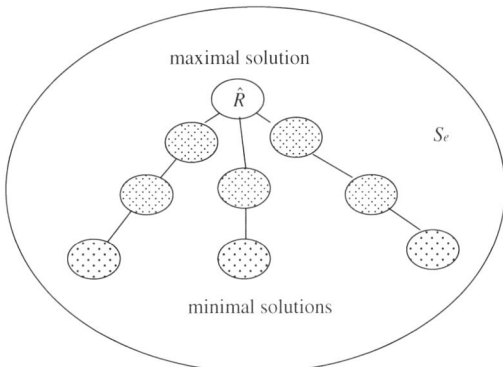

Figure 7.9 A structure of solutions to the estimation problem with sup-t composition.

been established (Higashi and Klir, 1984). Computational procedures to compute the maximal and all minimal solutions of max–min (Chen and Wang, 2002), and max–min and max-product fuzzy relational equations have been developed (Luoh et al., 2003). In particular, Chen and Wang (2002) state that the problem of constructing all minimal solutions of max–min fuzzy relational equations is NP-hard. Thus, no polynomial time algorithm is likely to exist.

7.3.2 Fuzzy Relational System

The estimation problem can be extended to cover a set of relational equations that are collectively termed a fuzzy relational system. Let U_k and V_k denote fuzzy sets in finite universes $\mathbf{X}$ and $\mathbf{Y}$, respectively

$$V_k = U_k \circ R, \; k = 1, 2, \ldots, N \tag{7.12}$$

The problem is to estimate R such that the system of fuzzy relational equation (7.12) is satisfied. The solution can be inferred assuming that there exists a maximal solution $\hat{R}_k$ for each of the kth relational equation and that the fuzzy relational system has a solution, namely

$$S_e^k = \{R \in F(\mathbf{X}) \times F(\mathbf{Y}) | V_k = U_k \circ R\} \neq \phi$$

$$S_e^N = \bigcap_{k=1}^{N} S_e^k \neq \phi$$

The maximal solution $\hat{R}$ of the system of equations is then computed intersecting the N maximal solutions $\hat{R}_k, \; k = 1, 2, \ldots, N$

$$\hat{R} = \bigcap_{k=1}^{N} \hat{R}_k$$

$$\hat{R}_k = U_k^T \varphi V_k \tag{7.13}$$

7.3.3 Relation-Relation Fuzzy Equations

Next, consider the instance in which $U : \mathbf{Z} \times \mathbf{X} \to [0, 1]$ and $V : \mathbf{Z} \times \mathbf{Y} \to [0, 1]$ are fuzzy relations on finite universes $\mathbf{X} = \{x_1, x_2, \ldots, x_n\}$, $\mathbf{Z} = \{z_1, z_2, \ldots, z_p\}$, $\mathbf{Y} = \{y_1, y_2, \ldots, y_m\}$, represented by $(p \times n)$ and $(p \times m)$ fuzzy relational matrices $[u_{ki}]$ and $[v_{kj}]$, respectively. Let $R - [r_{ij}]$ be the $(n \times m)$ fuzzy relational matrix associated with a fuzzy relation $R : \mathbf{X} \times \mathbf{Y} \to [0, 1]$. The fuzzy relational matrix becomes

$$V = U \circ R \tag{7.8}$$

Denoting by U^k the kth row of U and likewise by V^k the kth row of $V, k = 1, 2, \ldots, p$, and R^j the jth column of $R, j = 1, \ldots, m$ (7.8) can be rewritten as follows:

$$\begin{bmatrix} V^1 \\ V^2 \\ \vdots \\ V^p \end{bmatrix} = \begin{bmatrix} U^1 \\ U^2 \\ \vdots \\ U^p \end{bmatrix} \circ \begin{bmatrix} R^1 & R^2 & \cdots & R^m \end{bmatrix} = \begin{bmatrix} U^1 \circ R^1 & U^1 \circ R^2 & \cdots & U^1 \circ R^m \\ U^2 \circ R^1 & U^2 \circ R^2 & \cdots & U^2 \circ R^m \\ \cdots & \cdots & \cdots & \cdots \\ U^p \circ R^1 & U^p \circ R^2 & \cdots & U^p \circ R^m \end{bmatrix}$$

Therefore,

$$V^1 = [U^1 \circ R^1 \quad U^1 \circ R^2 \quad \cdots \quad U^1 \circ R^m] = U^1 \circ R$$
$$V^2 = [U^2 \circ R^1 \quad U^2 \circ R^2 \quad \cdots \quad U^2 R^m] = U^2 \circ R$$
$$\vdots$$
$$V^p = [U^p \circ R^1 \quad U^p \circ R^2 \quad \cdots \quad U^p R^m] = U^p \circ R \quad (7.14)$$

This means that (7.8) can be partitioned into a collection of equations of the form (7.12), and each of these equations can be solved independently. Thus the problem of solving (7.8) reduces to the fuzzy relational system (7.14) and, using (7.13)

$$\hat{R} = \bigcap_{k=1}^{p} \hat{R}_k$$
$$\hat{R}_k = U^k \circ R$$

7.3.4 Multi-Input, Single-Output Fuzzy Relational Equations

A natural extension of single-input–single-output systems of Figure 7.8 involves multi-input–single-output systems, and multivariable fuzzy relational equations as suggested in Figure 7.10 (Pedrycz and Gomide, 1998).

In this case, the inputs and outputs are fuzzy sets $U_i \in F(\mathbf{X}_i), i = 1, \ldots, p, V \in F(\mathbf{Y})$, and relation $R \in F(\mathbf{X}_1 \times \mathbf{X}_2 \times ; \cdots \times \mathbf{X}_p \times \mathbf{Y})$, with all universes being finite. Then

$$V = U_1 \circ U_2 \circ \cdots \circ U_p \circ R \quad (7.15)$$

Expressing (7.15) in terms of the membership functions of the fuzzy sets, we derive,

$$V = \sup_{\mathbf{x} \in \mathbf{X}}[U_1(x_1) t U_2(x_2) t \cdots t U_p(x_p) t R(x_1, x_2, \cdots, x_p, y)]$$

where $\mathbf{x} = [x_1, x_2, \ldots, x_p]$ and $\mathbf{X} = \mathbf{X}_1 \times \mathbf{X}_2 \times \cdots \times \mathbf{X}_p$. The estimation problem is as follows: given $U_1, U_2, \ldots, U_p$ and V, estimate R. Defining U as $U = U_1 t U_2 t \ldots t U_p$, (7.15) reduces to the form

$$V = U \circ R$$

In the sequel, its solution is computed using (7.10).

Figure 7.10 Multi-input fuzzy system.

7.3.5 Solution of the Estimation Problem for Equations with Inf-s Composition

To summarize the solution procedure for the estimation problem involving inf-s composition in finite universes, let us denote by S_e^s a family of fuzzy relations R that satisfy the relationship

$$V = U \bullet R \qquad (7.16)$$

namely, $S_e^s = \{R \in F(\mathbf{X}) \times F(\mathbf{Y}) | V = U \bullet R\}$
Consider the following operator from [0,1] to [0,1], β: [0,1] → [0,1]

$$a \, \beta \, b = \inf\{c \in [0,1] | a \, s \, c \geq b\} \qquad (7.17)$$

PROPOSITION

If $S_e^s \neq \phi$, then the unique minimal solution $\hat{R}$ of the inf-s relational equation $V = U \bullet R$ is

$$\hat{R} = U^T \beta V \qquad (7.18)$$

The proof of the proposition is analogous to the one produced in the case of the equations with the sup-t composition.

Similarly as in the case of sup-t composition, $\hat{R}$ is minimal in the sense that, if R is an element of S_e^s, then $\hat{R} \subseteq R$. The proof is completed in a similar way as presented before.

When solving inf-s fuzzy relational equations, we consider dual operations to those encountered when dealing with the sup-t case. For instance, in the case of a system of fuzzy relational equations involving U_k and V_k, fuzzy sets on finite universes $\mathbf{X}$ and $\mathbf{Y}$, respectively, such that

$$V_k = U_k \bullet R, \qquad k = 1, 2, \ldots, N$$

the standard intersection of the analogous solution of the sup-t composition is replaced by standard union in the solution as follows:

$$\hat{R} = \bigcup_{k=1}^{N} \hat{R}_k$$

where $\hat{R}_k = U_k^T \beta V_k$

EXAMPLE 7.6

Let us consider two fuzzy sets

$$U = [0.7, 0.5, 0.3] \text{ and } V = [0.4, 0.5, 1.0, 0.6]$$

and specify the t-conorm as the maximum. Thus (7.17) gives rise to the expression

$$a \beta b = \begin{cases} b & \text{if } a \leq b \\ 0 & \text{if } a > b \end{cases}$$

Making use of (7.18) we obtain

$$\hat{R} = \begin{bmatrix} 0.7 \\ 0.5 \\ 0.3 \end{bmatrix} \beta [0.4 \ \ 0.5 \ \ 1.0 \ \ 0.6] = \begin{bmatrix} 0.7\beta 0.4 & 0.7\beta 0.5 & 0.7\beta 1.0 & 0.7\beta 0.6 \\ 0.5\beta 0.4 & 0.5\beta 0.5 & 0.5\beta 1.0 & 0.5\beta 0.6 \\ 0.3\beta 0.4 & 0.3\beta 0.5 & 0.3\beta 1.0 & 0.3\beta 0.6 \end{bmatrix}$$

Therefore,

$$\hat{R} = \begin{bmatrix} 0.0 & 0.0 & 1.0 & 0.0 \\ 0.0 & 0.5 & 1.0 & 0.6 \\ 0.4 & 0.5 & 1.0 & 0.6 \end{bmatrix}$$

7.3.6 Solution of the Inverse Problem

7.3.6.1 Single-Input–Single-Output Fuzzy Relational Equations

To proceed the solution of the inverse problem, denote by S_i the family of fuzzy relations U that, given V and R, satisfies

$$V = U \circ R$$

namely,

$$S_i = \{U \in F(\mathbf{X}) | V = U \circ R\}.$$

Consider the operator θ such that each element u_i of $U = V\theta S$, assuming S is an $(m \times n)$ fuzzy relational matrix, that is

$$u_i = v_j \, \theta \, sj_i = \min{(v_j \varphi s_{ji}, j = 1, \ldots, m)}, i = 1, \ldots, n \quad (7.19)$$

PROPOSITION
If $S_i \neq \phi$, then the unique maximal solution $\hat{U}$ of the sup-t relational equation $V = U \circ R$ is given in the form

$$\hat{U} = V\theta R^T \quad (7.20)$$

PROOF
First, note that $(V \theta R^T) \circ R = \hat{U} \circ R \subset V$. Moreover, $U \subset (U \circ R)\theta R^T$ and since the sup-t composition is monotonic $V = U \circ R \subset \hat{U} \circ R$, which means that $\hat{U} \circ R \supset V$. Thus

1. $\hat{U} \circ R \subset V$
2. $\hat{U} \circ R \supset V$

and from (1) and (2), $V = \hat{U} \circ R$. Similarly as in the estimation problem, $\hat{U}$ is maximal in the sense that, if U is any element of S_i, then $U \subseteq \hat{U}$. Thus if $\hat{U}$ is not unique, there exists a $\hat{U}'$ such that $V = \hat{U}' \circ R$ and $\hat{U} \subseteq \hat{U}'$. But, since $\hat{U}$ also is maximal $\hat{U}' \subseteq \hat{U}$, which means that the maximal solution is unique. The steps to solve the inverse problem are summarized below, where M is a large number.

procedure INVERSE-SOLUTION (R,V) **returns** fuzzy unary relation
static: fuzzy relations: $R = [r_{ij}]$, $V = [v_j]$
   M: large number
   t: a t-norm
define: φ operator
for $i = 1 : n$ **do**
   $u \leftarrow M$
   for $j = 1 : m$ **do**
      $u \leftarrow \min(u, v_j \, \varphi \, r_{ji})$
$u_i \leftarrow u$
return U

EXAMPLE 7.7

Consider the fuzzy relation in (7.11) and fuzzy set $V = [0.4, 0.2, 0.0, 0.7]$, the same output of the previous example. Assume the t-norm t = min. Then, from (7.9) we get

$$a \, \varphi \, b = \begin{cases} 1 & \text{if } a \leq b \\ b & \text{if } a > b \end{cases}$$

and from (7.19) and (7.20) we obtain

$$\hat{U} = (0.4 \; 0.2 \; 0.0 \; 0.7) \theta \begin{bmatrix} 0.4 & 0.4 & 1.0 \\ 0.2 & 0.2 & 0.2 \\ 0.0 & 0.0 & 0.0 \\ 0.7 & 1.0 & 1.0 \end{bmatrix} = \min \left\{ (0.4 \; 0.2 \; 0.0 \; 0.7) \varphi \begin{bmatrix} 0.4 & 0.4 & 1.0 \\ 0.2 & 0.2 & 0.2 \\ 0.0 & 0.0 & 0.0 \\ 0.7 & 1.0 & 1.0 \end{bmatrix} \right\}$$

$$= \begin{pmatrix} \min(0.4 \, \varphi \, 0.4, 0.2 \, \varphi \, 0.2, 0.0 \, \varphi \, 0.0, 0.7 \, \varphi \, 0.7) \\ \min(0.4 \, \varphi \, 0.4, 0.2 \, \varphi \, 0.2, 0.0 \, \varphi \, 0.0, 1.0 \, \varphi \, 0.7) \\ \min(1.0 \, \varphi \, 0.4, 0.2 \, \varphi \, 0.2, 0.0 \, \varphi \, 0.0, 1.0 \, \varphi \, 0.7) \end{pmatrix}^T$$

therefore,

$$\hat{U} = [1.0, 0.7, 0.4]$$

Many other fuzzy sets also are solutions of this inverse problem. For instance, as it can be easily checked, $\hat{U}_1 = [1, 0, 0]$ and $\hat{U}_2 = [0, 0.7, 0]$ are solutions, but both $\hat{U}_1$ and $\hat{U}_2$ are included in $\hat{U}$.

7.3.7 Relation—Relation Fuzzy Equations

If $U : \mathbf{Z} \times \mathbf{X} \to [0, 1]$ and $V : \mathbf{Z} \times \mathbf{Y} \to [0, 1]$ are fuzzy relations on finite universes $\mathbf{X} = \{x_1, x_2, \ldots, x_n\}$, $\mathbf{Z} = \{z_1, z_2, \ldots, z_p\}$, and $\mathbf{Y} = \{y_1, y_2, \ldots, y_m\}$, represented by $(p \times n)$ and $(p \times m)$ fuzzy relational matrices $[u_{ki}]$ and $[v_{kj}]$, respectively, and $R = [r_{ij}]$ is the $(n \times m)$ fuzzy relational matrix associated with a fuzzy relation R: $\mathbf{X} \times \mathbf{Y} \to [0, 1]$, then the fuzzy relation becomes

$$V = U \circ R$$

Similarly as before, denote by U^k the kth row of U and by V^k the kth row of V, $k = 1, 2, \ldots, p$. Let R^j be the jth column of R, $j = 1, \ldots, m$. Equation $V = U \circ R$ can be rewritten as follows:

$$\begin{bmatrix} V^1 \\ V^2 \\ \vdots \\ V^p \end{bmatrix} = \begin{bmatrix} U^1 \\ U^2 \\ \vdots \\ U^p \end{bmatrix} \circ [R^1 \; R^2 \; \cdots \; R^m] = \begin{bmatrix} U^1 \circ R^1 & U^1 \circ R^2 & \cdots & U^1 \circ R^m \\ U^2 \circ R^1 & U^2 \circ R^2 & \cdots & U^2 \circ R^m \\ \cdots & \cdots & \cdots & \cdots \\ U^p \circ R^1 & U^p \circ R^2 & \cdots & U^p \circ R^m \end{bmatrix}$$

Therefore,

$$V^1 = [U^1 \circ R^1 \quad U^1 \circ R^2 \quad \cdots \quad U^1 \circ R^m] = U^1 \circ R$$
$$V^2 = [U^2 \circ R^1 \quad U^2 \circ R^2 \quad \cdots \quad U^2 R^m] = U^2 \circ R$$
$$\vdots$$
$$V^p = [U^p \circ R^1 \quad U^p \circ R^2 \quad \cdots \quad U^p R^m] = U^p \circ R \quad (7.21)$$

Again, this means that $V = U \circ R$ can be partitioned into a set of independent equations. Thus solving $V = U \circ R$ reduces to the fuzzy relational system (7.21) and, using (7.20)

$$\hat{U}^i = V^i \theta R^T, \quad i = 1, 2, \ldots, p$$

7.3.8 Multi-Input, Single-Output Fuzzy Relational Equations

The case of multi-input–single-output fuzzy systems and the corresponding multivariable fuzzy relational equations involve inputs and outputs, the fuzzy sets $U_i \in F(\mathbf{X}_i), i = 1, \ldots, p$ and $V \in F(\mathbf{Y})$, respectively, and the relation $R \in F(\mathbf{X}_1 \times \mathbf{X}_2 \times \cdots \times \mathbf{X}_p \times \mathbf{Y})$. As before, here all universes are finite. Thus, the fuzzy multivariable system is governed by

$$V = U_1 \circ U_2 \circ \cdots \circ U_p \circ R \quad (7.22)$$

which can be expressed in terms of its membership function as follows:

$$V(\mathbf{y}) = \sup_{\mathbf{x} \in \mathbf{X}} [U_1(x_1) t \; U_2(x_2) t \cdots t \; U_p(x_p) t \; R(x_1, x_2, \cdots, x_p, \mathbf{y})]$$

where
$\mathbf{x} = (x_1, x_2, \ldots, x_p)$, $\mathbf{X} = \mathbf{X}_1 \times \mathbf{X}_2 \times \ldots \times \mathbf{X}_p$, and $\mathbf{y} \in \mathbf{Y}$

For multivariable fuzzy systems the inverse problem can be formulated in many distinct forms. For instance, the problem can be solved with respect to a single variable. This implies different individual versions of the inverse problem. Therefore, given $U_1, U_2, \ldots, U_{i-1}, U_{i+1}, \ldots, U_p$, V, and R, the question is to determine U_i. The solution of this problem comes in the form of the maximal fuzzy set

$$\hat{U}_i = V \theta R_i^T$$
$$R_i = U_1 \circ U_2 \circ \cdots \circ U_{i-1} \circ U_{i+1} \circ \cdots \circ U_p \circ R$$

Another formulation of the inverse problem involves an ensemble of fuzzy sets. Let U_I, $I = \{i, j, \ldots, k\}$ denote a fuzzy relation composed by the Cartesian product of some fuzzy sets computed with the same t-norm as in the original fuzzy relational equation

$$U_I = U_i \times U_j \times \cdots \times U_k$$

Let $J = \{t, u, \ldots, v\}$ be the collection of the remaining indices of $\{1, 2, \ldots, p\}$, that is, J is such that $I \cup J = \{1, 2, \ldots, p\}$ and $I \cap J = \emptyset$. Rewriting (7.22) to distinguish the ensemble of fuzzy sets U_I

$$V = U_I \circ U_J \circ R = U_I \circ R_I$$

where $R_I = U_J \circ R$. Then, from (7.20)

$$\hat{U}_I = V \theta R_I^T$$

Explicit computations of the fuzzy relation U_I may require its decomposability if $\hat{U}_I$ is decomposable, meaning that the following holds

$$\hat{U}_I = \hat{U}_i \times \hat{U}_j \times \cdots \times \hat{U}_k$$

In other words, we have

$$\hat{U}_I(x_i, x_j, \cdots, x_k) = \hat{U}_i(x_i) t \hat{U}_j(x_j) t \cdots t \hat{U}_k(x_k) \quad (7.23)$$

If (7.23) is not satisfied, the only result comes in the form of the fuzzy relation $\hat{U}_I$.

7.3.9 Solvability Conditions for Maximal Solutions

The methodology developed so far is valid assuming that S_e or S_i are nonempty, so that solutions for the equations do exist. Although this may not always be the case, it is worth to look at conditions that guarantee the solvability of the fuzzy relational equations. In general these conditions are not obvious or easy to find, but in some circumstances nonempty solution sets exist under quite mild conditions.

For the estimation problem the condition is simple,

$$hgt(U) \geq hgt(V)$$

In particular, if U is normal, then one is guaranteed to find a fuzzy relation satisfying $V = U \circ R$ (Pedrycz and Gomide, 1998). The general solvability conditions for the system of equations are more demanding and cannot be stated explicitly, but often it requires normal and pairwise disjoint input fuzzy sets (Chung and Lee, 1998).

For the particular case of $V = U \circ R$, $V \in F(\mathbf{Y})$ and $U \in F(\mathbf{X})$, $\mathbf{X}$ and $\mathbf{Y}$ finite, viewed as max–min composition, if

$$\max_i r_{ij} < v_j$$

for some $j = 1, 2, \ldots, m$, then the inverse problem has no solution, otherwise it is only a necessary condition for the existence of a solution (Klir and Yuan, 1995). In general. Concise and practically relevant solvability conditions for the inverse problem are difficult to obtain (Gottwald, 1984).

Properties of the solution set of single-input, single-output of min–max relation equations from the viewpoint of attainability have been studied for finite and infinite universes (Imai et al., 2002). Attainability concerns the existence of a subset of the output universe for which, given a solution for the inverse problem, the relational equation remains satisfied.

If the system is not solvable, then the only approach is to look for approximate solutions (Chung and Lee, 1998; Pedrycz and Gomide, 1998; Chen and Wang, 2006).

7.4 ASSOCIATIVE MEMORIES

Associative systems are entities whose input–output behavior is governed by a memory constructed by the association of sample patterns. Association may take one of the two fundamental forms, that is, autoassociation or heteroassociation. Autoassociation requires that a set of input patterns, represented by vectors, be encoded and stored by the system. Subsequently, the system is presented by a description of an original pattern stored in it, and the task is to retrieve that particular pattern using a certain recall process. Therefore, in autoassociation, a pattern vector is associated with itself. Accordingly, the input and output spaces have the same dimensionality. Contrary to autoassociation, in heteroassociation a set of input patterns are paired with another set of output patterns and stored as such pairs of items. When the system receives an input pattern, the task is to produce the corresponding output pattern using the association encoded in memory. The input and output spaces may have different dimensions. Often, in both cases a stored vector may be recalled from memory by applying a partial or noisy version of the input originally associated with the desired output vector.

7.4.1 Sup-t Fuzzy Associative Memories

Consider a pair of fuzzy patterns symbolized by a $(1 \times n)$ vector $\mathbf{u}_i = [u_i]$ and by a $(1 \times m)$ vector $\mathbf{v}_i = [v_j]$, representing fuzzy sets U and V in universes $\mathbf{X} = \{x_1, x_2, \ldots, x_n\}$ and $\mathbf{Y} = \{y_1, y_2, \ldots, y_m\}$, respectively. The solution of the estimation problem (7.10), repeated below, suggests that there exists a fuzzy relation $R = [r_{ij}]$ in $\mathbf{X} \times \mathbf{Y}$

$$\hat{R} = U^T \varphi\ V$$
$$\varphi : [0, 1] \to [0, 1], a\ \varphi\ b = \sup \{c \in [0, 1] | \ a\ t\ c \leq b\}$$

that satisfies the fuzzy relational (7.8), namely

$$V = U \circ R$$

Therefore, given a t-norm t, (7.10) define a φ-encoding mechanism to store an association between fuzzy patterns U and V in a fuzzy associative memory (FAM) represented by the binary fuzzy relation R. Subsequently, when the fuzzy system is presented by a description of the original pattern U, it retrieves the particular pattern V using a recall process viewed as the sup-t composition (7.8) of U and R.

Similarly, if U_k and $V_k, k = 1, 2, \ldots, N$ denote fuzzy sets on finite universes $\mathbf{X}$ and $\mathbf{Y}$, respectively, they can be encoded in the fuzzy relation $\hat{R}$

$$\hat{R} = \bigcap_{k=1}^{N} \hat{R}_k$$

$$\hat{R}_k = U_k^T \varphi V_k$$

Given some pattern U_k, the recall process assumes the form of the composition

$$V_k = U_k \circ R$$

Despite the aforementioned features, the FAM storage and recall depend on how the fuzzy sets $U_1, U_2, \ldots, U_N$ and $V_1, V_2, \ldots, V_N$ overlap in their respective universes. If the input fuzzy sets U_k form a fuzzy partition and overlap at the level of ½ (semioverlap), that is, they satisfy the conditions

$$hgt(U_k \cap U_{k-1}) = 0.5 \quad \text{and} \quad \sum_{k=1}^{N} U_k(\mathbf{x}) = 1, \forall\, \mathbf{x} \in \mathbf{X}$$

where U_k and U_{k-1} are adjacent normal fuzzy sets, then the following proposition holds.

PROPOSITION

If the fuzzy patterns U_k are semioverlapped, then the pairwise encoding of U_k and $V_k, k = 1, 2, \ldots, N$, using

$$\hat{R} = \bigcap_{k=1}^{N} \hat{R}_k, \quad \hat{R}_k = U_k^T \varphi V_k$$

produces perfect recall realized as

$$V_k = U_k \circ R$$

The proof of the proposition above is a retranslation of the one developed by Chung and Lee (1997; 1998) for the systems of fuzzy relational equations.

When semioverlapping does not hold, perfect recall can be achieved when each pair of patterns U_k and V_k are replaced by their respective sharpened versions U'_k and V'_k of the following form

$$U'_k(\mathbf{x}) = \begin{cases} U_k(\mathbf{x}), & \text{if } U_k(\mathbf{x}) \geq \gamma \\ 0, & \text{otherwise} \end{cases}$$

where

$$\gamma = \max_{\forall k} \{\max U_i(\mathbf{x})\}$$
$$i \neq k$$
$$x \in \{x | U_i(\mathbf{x}) = 1\}$$

Similarly, we introduce a sharpened version of V_k, that is V'_k. Encoding and recall processes are carried out as follows

$$\hat{R} = \bigcap_{k=1}^{N} \hat{R}_k, \qquad \hat{R}_k = U'^T_k \varphi V'_k$$

and

$$V_k = U_k \circ R$$

Thus, the output of the sharpened approach can be exactly the same as the original ones (Chung and Lee, 1998).

It is worth noting that associative memories are bidirectional in the sense that, given an output pattern V, the corresponding input pattern U can be found from the solution of the inverse problem (7.20), namely

$$U_k = V_k \, \theta \, R^T$$

7.4.2 Inf-s Fuzzy Associative Memories

Similarly as sup-t associative memories, their duals, inf-s associative memories, are developed under the framework of inf-s relational equations. Thus, if U_k and $V_k, k = 1, 2, \ldots, N$, are fuzzy patterns on finite universes $\mathbf{X}$ and $\mathbf{Y}$, respectively, encoding and recall are completed as follows:

encoding $\qquad \hat{R} = \bigcup_{k=1}^{N} \hat{R}_k \qquad$ where $\qquad \hat{R}_k = U_k^T \beta V_k.$

recall $\qquad V_k = U_k \bullet R, \; U_k = V_k \beta R^T, \; k = 1, 2, \ldots, N$

Clearly, when U_k and $V_k, k = 1, 2, \ldots, N$ are distinct fuzzy patterns $\hat{R}$ is a hetero-association, and when $U_k = V_k, k = 1, 2, \ldots, N$, $\hat{R}$ is an autoassociation (Valle et al., 2004) (Sussner et al., 2006).

7.5 FUZZY NUMBERS AND FUZZY ARITHMETIC

7.5.1 Algebraic Operations on Fuzzy Numbers

In Chapter 3, it was noted that a membership function $A: \mathbf{X} \to [0, 1]$ is upper semicontinuous if the set $\{x \in \mathbf{X} | A(x) \geq \alpha\}$ is closed, that is, the α-cuts are closed intervals and, therefore, convex sets. If the universe $\mathbf{X}$ is the set $\mathbf{R}$ of real numbers and membership function is normal, $A(x) = 1 \; \forall \; x \in [b, c]$, then $A(x)$ is a model of a fuzzy interval, with monotone increasing function $f_A : [a, b) \to [0, 1]$, monotone

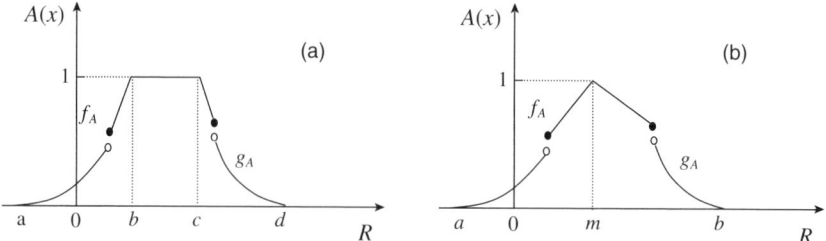

Figure 7.11 Canonical form of a fuzzy interval (a) and fuzzy number (b).

decreasing function $g_A : (c, d] \to [0, 1]$, and null otherwise. Fuzzy intervals $A(x)$ have the following canonical form:

$$A(x) = \begin{cases} f_A(x), & \text{if } x \in [a, b) \\ 1, & \text{if } x \in [b, c] \\ g_A(x), & \text{if } x \in (c, d] \\ 0, & \text{otherwise} \end{cases}$$

where $a \leq b \leq c \leq d$; see Figure 7.11(a).

When $b = c$ and $A(x) = 1$ for exactly one element of **X**, the fuzzy quantity is called a fuzzy number; see Figure 7.11(b).

In general, the functions f_A and g_A are semicontinuous from the right and left, respectively. From the pragmatic point of view, fuzzy intervals and numbers are mappings from the real line **R** to the unit interval that satisfy a series of properties such as normality, unimodality, continuity, and boundness of support. As Figure 7.12 suggests, fuzzy intervals and numbers model our intuitive notion of approximate intervals and approximate numbers.

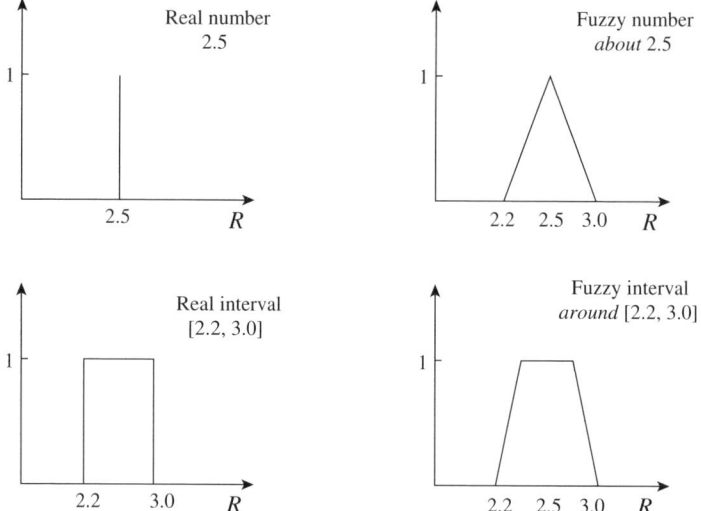

Figure 7.12 Real and fuzzy numbers and intervals.

7.5.2 Computing with Fuzzy Numbers

Before we move on to operations on fuzzy numbers, let us introduce a few examples that motivate their use.

> Consider that you traveled for 2 h at a speed of *about* 110 km/h. What was the distance you traveled? The speed is described in the form of some fuzzy set S whose membership function is given.

The next example is a more general version of the above problem.

> You traveled at a speed of *about* 110 km/h for *about* 3 h. What was the distance traveled? We assume that both the speed and time of travel are described by fuzzy sets.
>
> In a certain manufacturing process, there are five operations completed in series. Given the nature of the manufacturing activities, the duration of each of them can be characterized by fuzzy sets $T_1, T_2, \ldots, T_5$. What is the time of realization of this process?

Basically, there exist two fundamental methods to carry out algebraic operations on fuzzy numbers. The first method is based on interval arithmetic and α-cuts, while the second one employs the extension principle. The fundamentals of these two methods are discussed next.

7.5.3 Interval Arithmetic and α-Cuts

The first approach to compute with fuzzy numbers exhibits its roots in the framework of interval analysis (Moore, 1979), a branch of mathematics developed to deal with the calculus of tolerances. In this framework, the interest is in intervals of real numbers, $[a, b], a, b \in \mathbf{R}$, such as $[4, 6]$, $[-1.5, 3.2]$ and so forth. The formulas developed to perform the basic arithmetic operations, namely, addition, subtraction, multiplication, and division are as follows (assuming that $c, d \neq 0$ for the division operation):

1. addition: $[a, b] + [c, d] = [a + c, b + d]$
2. subtraction: $[a, b] - [c, d] = [a - d, b - c]$
3. multiplication: $[a, b].[c, d] = [\min(ac, ad, bc, bd), \max(ac, ad, bc, bd)]$
4. division: $[a, b]/[c, d] = \left[\min\left(\frac{a}{c}, \frac{a}{d}, \frac{b}{c}, \frac{b}{d}\right), \max\left(\frac{a}{c}, \frac{a}{d}, \frac{b}{c}, \frac{b}{d}\right)\right]$

Now, let A and B be two fuzzy numbers and let $*$ be any of the four basic arithmetic operations. Thus, for any $\alpha \in (0, 1]$ the fuzzy set $A * B$ is computed using the α-cuts A_α and B_α of A and B, respectively,

$$(A * B)_\alpha = A_\alpha * B_\alpha$$

Recall that, by definition, the α-cuts A_α and B_α are closed intervals, and therefore the formulas of interval operations can be applied for each value of α. When $*$ is / (division operation), we must require that $0 \notin B_\alpha$, and $\forall\, \alpha \in (0,1]$.

After the interval operation is performed for α-cuts, the representation theorem (Chapter 9) leads to the well-known relationship

$$A * B = \bigcup_{\alpha \in [0,1]} (A * B)_\alpha$$

In terms of the membership functions, we obtain

$$(A * B)(x) = \sup_{\alpha \in [0,1]} [\alpha(A * B)_\alpha(x)] = \sup_{\alpha \in [0,1]} [(A * B)^f_\alpha(x)]$$

where $(A * B)^f_\alpha(x) = \alpha(A * B)_\alpha(x)$.

Therefore, the interval arithmetic-α-cut method to perform fuzzy arithmetic is a generalization of interval arithmetic.

EXAMPLE 7.8

If A and B are two triangular fuzzy numbers, denoted as $A(x, a, m, b)$ and $B(x, c, n, d)$, then their α-cuts are determined as

$$A_\alpha = [(m-a)\alpha + a, (m-b)\alpha + b]$$
$$B_\alpha = [(n-c)\alpha + c, (n-d)\alpha + d]$$

Now let $A = A(x, 1, 2, 3)$ and $B = B(x, 2, 3, 5)$. Then, the corresponding α-cuts are equal to

$$A_\alpha = [\alpha + 1, -\alpha + 3]$$
$$B_\alpha = [\alpha + 2, -2\alpha + 5]$$

Therefore,

$$(A+B)_\alpha = [2\alpha + 3, -3\alpha + 8]$$
$$(A-B)_\alpha = [3\alpha - 4, -2\alpha + 1]$$
$$(AB)_\alpha = [(\alpha + 1)(\alpha + 2), (-\alpha + 3)(-2\alpha + 5)]$$
$$(A/B)_\alpha = [(\alpha + 1)/(-2\alpha + 5), (-\alpha + 3)(\alpha + 2)]$$
$$\left(\frac{B}{A}\right)_\alpha = [(\alpha + 2)/(-\alpha + 3), (-2\alpha + 5)/(\alpha + 1)]$$

Figure 7.13 shows the resulting fuzzy numbers $A + B$, $A - B$, AB and B/A, respectively.

The extension of the interval arithmetic and the use of α-cuts and the representation of fuzzy sets means that each fuzzy number can be regarded as a family of nested α-cuts. Subsequently, these α-cuts are used to reconstruct the resulting fuzzy number. In essence, the use of α-cuts is a sort of a brute-force method of computing with fuzzy numbers. However, α-cuts are becoming important to develop parametric representation of fuzzy numbers to control their shapes and associated approximation error (Stefanini et al., 2006).

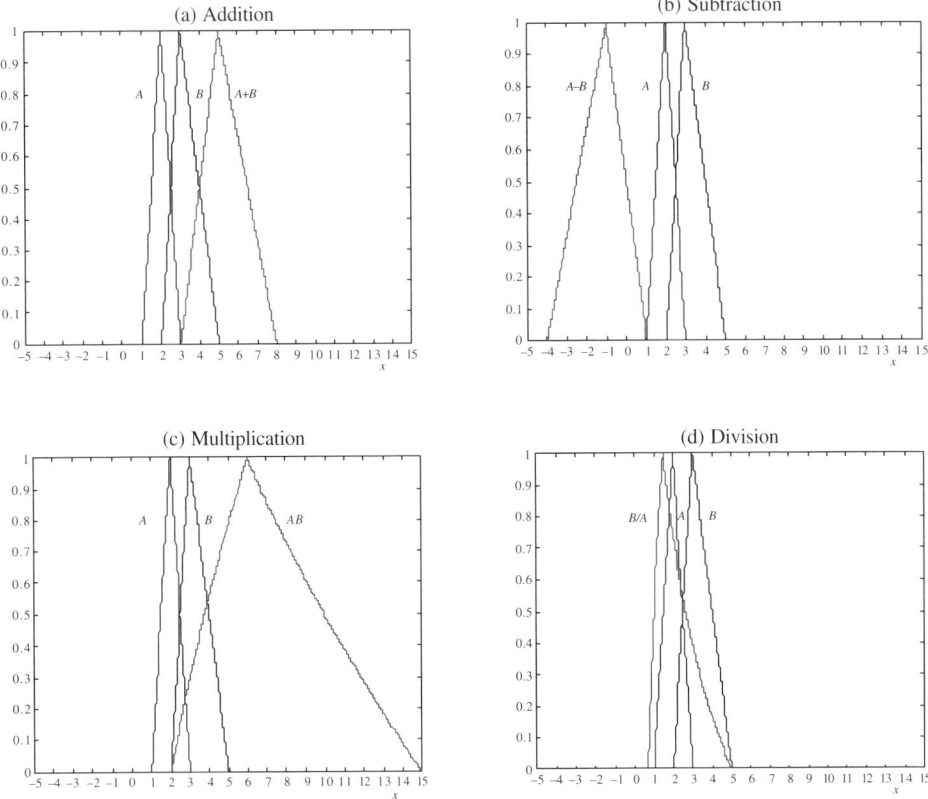

Figure 7.13 Algebraic operations on triangular fuzzy numbers.

7.5.4 Fuzzy Arithmetic and the Extension Principle

The second method of computing with fuzzy numbers dwells on the extension principle to extend standard operations on real numbers to fuzzy numbers. Here the fuzzy set $A * B$ on R is defined using (7.7)

$$(A * B)(z) = \sup_{z=x*y} \min[A(x), B(y)], \quad \forall\, z \in \mathbf{R} \tag{7.25}$$

In general, if t is a t-norm and $*\colon \mathbf{R}^2 \to \mathbf{R}$ is an operation on the real line, then operations on fuzzy numbers become

$$(A * B)(z) = \sup_{z=x*y} [A(x) t B(y)], \quad \forall\, z \in \mathbf{R}$$

Figure 7.14 illustrates the addition $(A + B)$ of triangular fuzzy numbers A and B using the minimum t-norm $t_{m\ (A+B)}$ and the drastic product $t_{d\ (A+B)}$ t-norm, respectively.

186   Chapter 7   Transformations of Fuzzy Sets

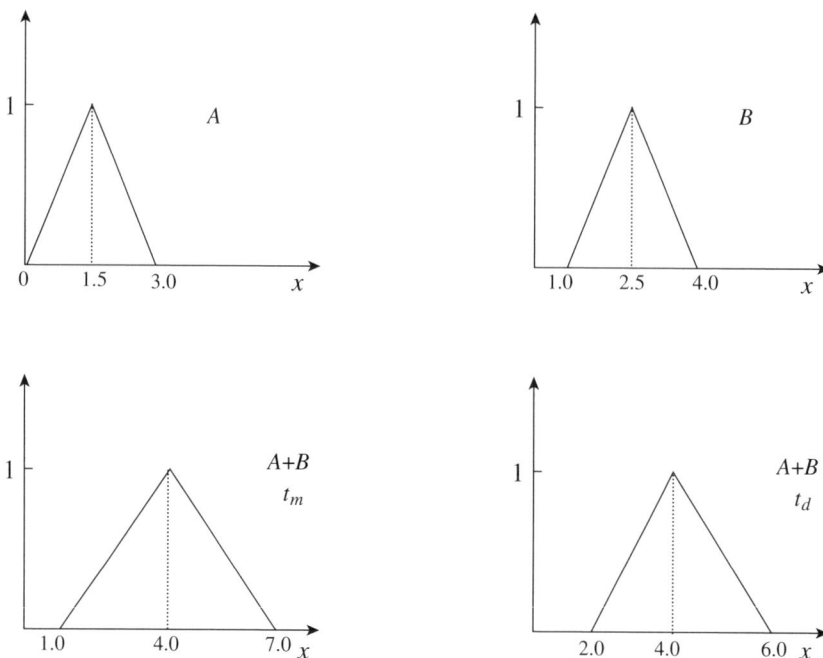

Figure 7.14 Algebraic operations: the use of the extension principle with different triangular norms.

Clearly, different choices of t-norms produce different results. In general, if $t_1 \leq t_2$ in the sense that $a\ t_1\ b \leq a\ t_2\ b,\ \forall a, b \in [0, 1]$, then

$$\sup_{z=x*y}[A(x)t_d B(y)] \leq \sup_{z=x*y}[A(x)tB(y)] \leq \sup_{z=x*y}[A(x)t_m B(y)] \quad \forall z \in \mathbf{R}$$

therefore,

$$^{t_d}(A*B), \leq {}^{t}(A*B)(z) \leq {}^{t_m}(A*B), \quad \forall z \in \mathbf{R}$$

In the special case of the largest t-norm which is minimum, t_m, the one we will concentrate on the remaining of part this section, property 6 of Section 7.1 suggests a fundamental result as a basis to compute with fuzzy numbers under the framework of the extension principle.

PROPOSITION

For any fuzzy numbers A and B and a continuous monotone binary operation $*$ on R, the following equality holds for all α-cuts with $\alpha \in [0, 1]$:

$$(A*B)_\alpha = A_\alpha * B_\alpha$$

The proof of this proposition is given in (Nguyen and Walker, 1999). There are important consequences of the proposition.

   1. Since A_α and B_α are closed and bounded for all α, $(A*B)\alpha$ also is closed and bounded;

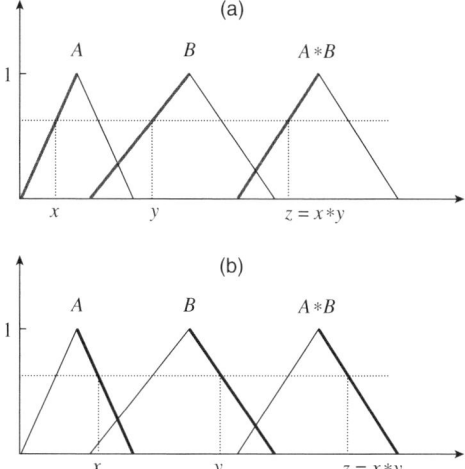

Figure 7.15 Combining increasing and decreasing parts of the membership functions of the fuzzy numbers A and B.

2. Because A and B are fuzzy numbers, they are normal and therefore $A * B$ is also normal.

These two observations clearly demonstrate that the extension principle produces a transformation that is a fuzzy number and, therefore, is a sound mechanism to perform algebraic operations with fuzzy numbers.

3. Computation of $A * B$ can be done by combining the increasing and decreasing parts of the membership functions of A and B.

Figure 7.15 provides a graphical visualization of the above statement.

The results above can be generalized to broader classes and choices of t-norms and operations with fuzzy quantities (Di Nola et al., 1985; Mares, 1997; De Baets, 2000; Klement et al., 2000). Moreover, approximation schemes were developed in the framework of interpolation of a fuzzy function (Perfilieva, 2004). In what follows we detail the basic operations with triangular fuzzy numbers because they are, by far, the most commonly used ones in practice. Moreover, the analysis focusing this class of fuzzy numbers reveals the most visible properties of fuzzy arithmetic.

7.5.5 Computing with Triangular Fuzzy Numbers

Consider two triangular fuzzy numbers $A(x, a, m, b)$ and $B(x, c, n, d)$. More specifically, A and B are described by the following piecewise membership functions:

$$A(x) = \begin{cases} \dfrac{x-a}{m-a} & \text{if } x \in [a, m) \\ \dfrac{b-x}{b-m} & \text{if } x \in [m, b] \\ 0 & \text{otherwise} \end{cases} \qquad B(x) = \begin{cases} \dfrac{x-c}{n-c} & \text{if } x \in [c, n) \\ \dfrac{d-x}{d-n} & \text{if } x \in [n, d] \\ 0 & \text{otherwise} \end{cases}$$

188 Chapter 7 Transformations of Fuzzy Sets

Let us recall that the modal values m and n identify a dominant, typical value, while the lower and upper bounds, a or c and b or d, reflect the spread of the number. To simplify computing, for the time being we consider fuzzy numbers with positive lower bounds $a, c > 0$.

7.5.5.1 Addition

The extension principle (7.25) applied to A and B to compute $C = A + B$ yields

$$C(z) = \sup_{z=x+y} \min[A(x), B(y)], \quad \forall\, z \in \mathbf{R}$$

The resulting fuzzy number is normal, that is, $C(z) = 1$ for $z = m + n$. Computations of the spreads of C can be done, according to statement 3 mentioned in Section 7.5.4, by treating the increasing and decreasing parts of the membership functions of A and B separately.

Consider first that $z < m + n$. In this situation, the calculation involves the increasing parts of the membership function of A and B. According to statement 3 and Figure 7.15(a), there exist values x and y such that $x < m$ and $y < n$ for which we have

$$A(x) = B(y) = \alpha, \alpha \in [0, 1]$$

Based on this we derive

$$\frac{x - a}{m - a} = \alpha$$

along with

$$\frac{y - c}{n - c} = \alpha$$

for $x \in [a, m]$ and $y \in [c, n]$. Expressing x and y as functions of α we get

$$x = (m - a)\alpha + a$$
$$y = (n - c)\alpha + c$$

which are the same as the lower bounds of the intervals we get using interval analysis, as they should be. Replacing the values of x and y in $z = x + y$ we have

$$z = x + y = (m - a)\alpha + a + (n - c)\alpha + c$$

that is,

$$\alpha = \frac{z - (a + c)}{(m + n) - (a + c)} \tag{7.26}$$

Notice that z has, as expected, the same lower bound of the corresponding interval associated with the α-cut we use with interval analysis.

Proceeding similarly for the decreasing portions of the membership functions we obtain

$$\frac{b - x}{b - m} = \alpha$$

along with

$$\frac{d-y}{d-n} = \alpha$$

for $x \in [m,b]$ and $y \in [n,d]$. Again, expressing x and y as functions of α we get

$$x = (m-b)\alpha + b$$
$$y = (n-d)\alpha + d$$

Furthermore, replacing the values of x and y in $z = x + y$ we have

$$z = x + y = (m-b)\alpha + b + (n-d)\alpha + d$$

that is

$$\alpha = \frac{(b+d) - z}{(b+d) - (m+n)} \tag{7.27}$$

As expected, z has the same upper bound as the corresponding interval associated with the α-cut we use with interval analysis.

Finally, from (7.26) and (7.27) we obtain the membership function of $C = A + B$:

$$C(x) = \begin{cases} \dfrac{z - (a+c)}{(m+n) - (a+c)}, & \text{if } z < m+n \\ 1, & \text{if } z = m+n \\ \dfrac{(b+d) - z}{(b+d) - (m+n)}, & \text{if } z > m+n \end{cases}$$

Interestingly, C is also a triangular fuzzy number, as shown in Figure 7.13(a). To emphasize this fact, we use a concise notation

$$C = C(x, a+c, m+n, b+d)$$

Whenever several triangular fuzzy numbers are added, the result also is a triangular fuzzy number. In general, however, shape preserving does not hold for any shape of the fuzzy number and t-norm adopted in the extension principle.

7.5.5.2 Multiplication

As with the addition, we look first at the increasing parts of the membership functions from which we get

$$x = (m-a)\alpha + a$$
$$y = (n-c)\alpha + c$$

The product z of x and y becomes

$$z = xy = [(m-a)\alpha + a][(n-c)\alpha + c]$$
$$z = (m-a)(n-c)\alpha^2 + (m-a)\alpha c + a(n-c)\alpha + ac = f_1(\alpha)$$

If $ac \leq z \leq mn$, then the membership function of the fuzzy number $D = AB$ is an inverse of the function $f_1(\alpha)$, namely

$$D(z) = f_1^{-1}(z)$$

Similarly, consider the decreasing parts of the fuzzy numbers A and B

$$x = (m-b)\alpha + b$$
$$y = (n-d)\alpha + d$$
$$z = xy = [(m-b)\alpha + b][(n-d)\alpha + d]$$
$$z = (m-b)(n-d)\alpha_2 + (m-b)\alpha d + b(n-d)\alpha + bd = f_2(\alpha)$$

As before, for any $mn \leq z \leq bd$ we have

$$D(z) = f_2^{-1}(z)$$

Notice that in this case the fuzzy number D is does not have a triangular membership function, which means that multiplication of triangular fuzzy numbers does not preserve the original shape. Instead, multiplication of piecewise linear membership functions produces a quadratic form of the resulting fuzzy number. Figure 7.13(c) shows an example of the multiplication.

7.5.5.3 Division

Like multiplication, for the increasing parts of the membership functions

$$x = (m-a)\alpha + a$$
$$y = (n-c)\alpha + c$$

we compute the division $z = x/y$ which after replacing x and y

$$z = \frac{x}{y} = \frac{(m-a)\alpha + a}{(n-c)\alpha + c} = g_1(\alpha)$$

so that, for $a/c \leq z \leq m/n$, the fuzzy number $E = A/B$ has the following membership function:

$$E(z) = g_1^{-1}(\alpha)$$

Analogously, for the decreasing parts of the membership functions

$$x = (m-b)\alpha + b$$
$$y = (n-d)\alpha + d$$

and we obtain

$$z = \frac{x}{y} = \frac{(m-b)\alpha + b}{(n-d)\alpha + d} = g_2(\alpha)$$

Thus, for $m/n \leq z \leq b/d$, the membership function of $E = A/B$ is

$$E(z) = g_2^{-1}(\alpha)$$

Clearly, the membership function of E is a rational function. Hence, division, like multiplication, does not preserve shape of the triangular membership functions.

7.6 CONCLUSIONS

Transformation of fuzzy sets in the form of the extension principle and composition generalizes similar transformations that can be performed with sets. They play an important role to provide further transformations through fuzzy relational equations, associative memories, and algebraic operations with fuzzy numbers. Fuzzy relations and associative memories are important to model, design, and develop applications in many areas, *inter alia*, information search and image processing systems, fuzzy modeling and control, to name a few.

Fuzzy numbers are convex and normal fuzzy sets on the set of real numbers. Operations with fuzzy numbers can be developed with the help of extension principle. In particular, standard fuzzy arithmetic can be approached choosing the min t-norm. Several other choices are possible, but practice has shown that standard fuzzy arithmetic is still one of the highest applicability.

EXERCISES AND PROBLEMS

1. Consider $\mathbf{X} = \{1,2,3,4\}$ and the fuzzy set $A = \{0.1/1, 0.2/2, 0.7/3, 1.0/4\}$ defined in this universe. Also, let $\mathbf{Y} = \{1,2,3,4,5,6\}$. Given is a function f: $\mathbf{X} \to \mathbf{Y}$ such that $y = f(x) = x + 2$. Show that $B = f(A) = \{0.1/3, 0.2/4, 0.7/5, 1.0/6\}$.

2. Determine the α-cuts of the fuzzy set A whose membership function is equal to

$$A(x) = \begin{cases} 2x - x^2 & \text{if } 0 \leq x \leq 1 \\ 0 & \text{otherwise} \end{cases}$$

Let $f(x) = 2x - x^2$. Compute the image of the α-cuts of the fuzzy set A under f. Sketch the transformations of the α-cuts graphically.

3. If A and B are fuzzy sets of $\mathbf{X}$ and $\mathbf{Y}$, respectively, show that $\overline{A \circ B} = \overline{A} \bullet \overline{B}$ where the t-norm and t-conorm of $\circ$ and $\bullet$ are dual, that is, $x \, s \, y = 1 - (1 - x) \, t \, (1 - y)$, $x, y \in [0, 1]$.

4. Perform the max-t and min-s compositions of the fuzzy relations G and W shown below using different t-norms and t-conorms. Compare the obtained results.

$$G = \begin{bmatrix} 0.5 & 1.0 & 0.7 & 0.9 \\ 0.4 & 1.0 & 0.2 & 0.1 \\ 0.6 & 0.9 & 1.0 & 0.4 \end{bmatrix} \quad W = \begin{bmatrix} 0.9 & 0.3 & 0.1 & 0.7 & 0.6 & 1.0 \\ 0.1 & 0.1 & 0.9 & 1.0 & 1.0 & 0.4 \\ 0.0 & 0.3 & 0.6 & 0.9 & 1.0 & 0.0 \\ 1.0 & 0.0 & 0.0 & 0.0 & 1.0 & 1.0 \end{bmatrix}$$

5. Let U_k and $V_k, k = 1, 2$ denote fuzzy sets defined in finite universes **X** and **Y**

$$U_1 = [1.0, 0.4, 0.5, 0.8, 0.0] \quad V_1 = [0.5, 0.7, 0.3, 0.1]$$
$$U_2 = [0.1, 0.9, 1.0, 0.2, 0.0] \quad V_2 = [1.0, 0.3, 0.1, 0.0]$$

Solve the system of fuzzy relational equations with respect to R. Once solved, use this relation to verify if, for each U_k, V_k is a solution for the respective inverse problem.

6. Consider the $(m \times n)$ fuzzy relation $W = [w_{ij}]$ such that $w_{ij} = \min_{k=1,...,N}(x_j^k \Rightarrow x_i^k)$, where $(x \Rightarrow y) = \sup\{c \in [0, 1], x\ t\ c \leq y\}$. Show that W is sup-t idempotent, namely $W = W \circ W$.

7. Develop, analytically, the membership function of the fuzzy number F that is the subtraction of fuzzy numbers A and B, namely, $F = A - B$.

8. Consider fuzzy numbers A and B whose membership functions are given in the form

$$A(x) = \begin{cases} e^{-(x-m)^2/k}, & a \leq x \leq b \\ 0, & \text{otherwise} \end{cases}$$

$$B(x) = \begin{cases} 0, & \text{if } x \leq a \\ \dfrac{x-a}{m-a}, & \text{if } x \in (a, m] \\ \dfrac{b-x}{b-m}, & \text{if } x \in [m, b) \\ 0, & \text{if } x \geq b \end{cases}$$

Show that their α-cuts are given in the form

$$A_\alpha = \begin{cases} \left[m - \sqrt{\ln\left(\dfrac{1}{\alpha^k}\right)}, m - \sqrt{\ln\left(\dfrac{1}{\alpha^k}\right)}\right] & \text{if } \alpha \geq e^{-\left(\dfrac{-(a-m)^2}{k}\right)} \\ [a, b] & \text{if } f\alpha < e^{-\left(\dfrac{-(a-m)^2}{k}\right)} \end{cases}$$

and

$$B_\alpha = [(m-a)\alpha + a, (m-b)\alpha + b], \ \forall\ \alpha \in [0, 1]$$

Sketch the membership functions of fuzzy sets of the addition, subtraction, multiplication, and division of A and B.

9. Are the parabolic fuzzy numbers $A, B, C...$ whose membership functions come in the form

$$P(x, m, a) = \begin{cases} 1 - \left(\dfrac{x-m}{a}\right)^2, & \text{if } x \in [m-a, m+a] \\ 0, & \text{otherwise} \end{cases}$$

closed under addition operation? Justify your answer.

HISTORICAL NOTES

The principle of extension has its roots in (Zadeh, 1965). According to Zadeh (1975), the extension principle is implicit in 1965. The notions of fuzzy relation and composition of fuzzy relations were introduced by Zadeh (1965).

Fuzzy relational equations emerged in the 1970s as a result of the work of Sanchez (1976). Sanchez also addressed the solvability issue of the equations with sup-min and inf-max composition operators. Interestingly, fuzzy relational equations can be regarded as a generalization of well-known Boolean equations (Rudeanu, 1974). Zadeh and Desoer (1963) first stressed the correspondence between relations and general system theory.

The composition of intuitionistic fuzzy relations is a concept introduced by Deschrijver and Kerre (2003). Here intuitionistic fuzzy relations are intuitionistic fuzzy sets formed in a Cartesian product of some universes.

The problem of solvability and unique solvability of fuzzy equations in the framework of max–min fuzzy algebra is addressed in Gavalec (2001). Max–min fuzzy equations form an alternative approach to address fuzzy equations whose nature differs from the fuzzy relational equations derived from compositions of fuzzy relations.

Research on fuzzy associative memory models originated in early 1990s with the work of Kosko (1992, 1997). The approach uses correlation-min and correlation-product encoding and max–min and max-product composition recall. Despite successful applications, these models of associative memories suffer from a relatively low storage capacity.

Different forms of fuzzy arithmetic, than the one addressed in this chapter, have been proposed using parametric form of fuzzy numbers through location index numbers and two nondecreasing left continuous functions (Ma et al., 1999).

REFERENCES

De Baets, B. Analytical solution methods for fuzzy relational equations, in: D. Dubois and H. Prade (eds.), *Foundations of Fuzzy Sets, The Handbook of Fuzzy Sets Series*, Kluwer Academic Publishers, Dordrecht, 2000, pp. 291–340.

Chen, L., Wang, P. Fuzzy relation equations (I): The general and specialized solving algorithms, *Soft Comput.*, **6**, 2002, 428–435.

Chen, L., Wang, P. Fuzzy relation equations (II): The branch-point solutions and the categorized minimal solutions, *Soft Comput.*, DOI 10.1007/s00500-006-0050-1, 2006.

Chung, F., Lee, T. A new look at solving a system of fuzzy relational equations, analytical resolution and numerical identification of fuzzy relational systems, *Fuzzy Set Syst.* **88**, 1997, 343–353.

Chung, F., Lee, T. Analytical resolution and numerical identification of fuzzy relational systems, *IEEE Trans. Syst., Man Cybern.* B: Cybern, **28**(6), 1998, 919–994.

Gavalec, M. Solvability and unique solvability of max–min fuzzy equations, *Fuzzy Set Syst.* **124**, 2001, 385–393.

Deschrijver, G., Kerre, E. On the composition of intuitionistic fuzzy relations, *Fuzzy Set Syst.* **136**, 2003, 333–361.

Di Nola, A., Pedrycz, W., Sessa, S., Processing of fuzzy numbers, *Kybanetics*, **15**, 1985, 43–47.

Gottwald, S. On the existence of solutions of systems of fuzzy equations, *Fuzzy Set Syst.* **12**, 1984, 301–302.

Hirota, K., Pedrycz, W. Data compression with fuzzy relational equations, *Fuzzy Set Syst.* **126**, 2002, 325–335.

Higashi, M., Klir, G. Resolution of finite fuzzy relation equations, *Fuzzy Set Syst.* **13**, 1984, 65–82.

Imai, H., Miyakoshi, M., Sato, Y. Properties of the solution set of a fuzzy relation equation from the point of view of attainability, *Soft Comput.*, **6**, 2002, 87–91.

Klement, P., Mesiar, R., Pap, E. *Triangular Norms*, Kluwer Academic Publishers, Dordrecht, 2000.

Klir, G., Yuan, B. *Fuzzy Sets and Fuzzy Logic*: *Theory and Applications*, Prentice-Hall, Upper Saddle River, NJ, 1995.

Kosko, B. *Neural Networks and Fuzzy Systems*: *A Dynamical Systems Approach to Machine Intelligence*, Prentice-Hall, Upper Saddle River, NJ, 1992.

Kosko, B. *Fuzzy Engineering*. Prentice-Hall, Upper Saddle River, NJ, 1997.

Luoh, L., Wang, W., Liaw, Y. Matrix-pattern-based computer algorithm for solving fuzzy relation equations, *IEEE Trans. on Fuzzy Syst.* **11**(1), 2003, 100–108.

Ma, M., Friedman, M., Kandel, A. A new fuzzy arithmetic, *Fuzzy Set Syst.*, **108**, 1999, 83–90.

Mares, M. Weak arithmetics of fuzzy numbers, *Fuzzy Set Syst.* **91**, 1997, 143–153.

Moore, R. *Methods and Application of Interval Analysis*, SIAM, Philadelphia, PA, 1979.

Nguyen, H., Walker, E. *A First Course in Fuzzy Logic*, Chapman Hall, CRC Press, Boca Raton, FL, 1999.

Nobuhara, H., Pedrycz, W., Hirota, K. Fast solving method of fuzzy relational equation and its application to lossy image compression/reconstruction, *IEEE Trans. on Fuzzy Syst.* **18** (3), 2002, 325–134.

Pedrycz, W., Gomide, F. *An Introduction to Fuzzy Sets: Analysis and Design*, MIT Press, Cambridge, MA, 1998.

Pereira, R., Ricarte, I., Gomide, F. Fuzzy relational ontological model in information search and retrieval systems, in: E. Sanchez (ed.), *Fuzzy Logic and the Semantic Web*, Elsevier, Amsterdam, 2006, 395–412.

Perfilieva, I. Fuzzy function as an approximate solution to a system of fuzzy relation equations, *Fuzzy Set Syst.*, **147**, 2004, 363–383.

Rudeanu, S. *Boolean Functions and Equations*, North Holland, Amsterdam, 1974.

Sanchez, E. Resolution of composite fuzzy relation equations, *Inf. Control*, **34**, 1976, 38–48.

Stefanini, L., Sorini, L., Guerra, M. Parametric representations of fuzzy numbers and application to fuzzy calculus, *Fuzzy Set Syst.* **157**, 2006, 2423–2455.

Sussner, P., Valle, M. Implicative fuzzy associative memories, *IEEE Trans. Fuzzy Systems*, **14** (6), 2006, 793–807.

Valle, M., Sussner, P., Gomide, F. Introduction to implicative fuzzy associative memories, *Proceedings of the IEEE International Joint Conference on Neural Networks*, Budapest, Hungary, 2004, pp. 925–931.

Zadeh, L. A., Desoer, C. *Linear System Theory*, McGraw Hill, New York, 1963.

Zadeh, L. A. Fuzzy sets, *Inf. Control*, **8**, 1965, 338–353.

Zadeh, L. A. The concept of linguistic variables and its application to approximate reasoning I, II, III, *Inf. Sci.* **8**, **9**, 1975, 199–251, 301–357, 43–80.

Chapter 8

Generalizations and Extensions of Fuzzy Sets

In this chapter, we expand and generalize the notion of fuzzy sets. We also put their development in the framework of granular computing and show linkages of fuzzy sets with rough sets. The essential developments of fuzzy sets occur along the line of their generalizations into more abstract constructs usually referred to as higher order fuzzy sets, in particular fuzzy sets of second order. Various implementation and conceptual issues arising around numeric values of membership functions give rise to a collection of concepts of granular membership grades. In the simplest scenario they give rise to an idea of interval-valued fuzzy sets. More refined versions of the construct produce type-2 fuzzy sets and fuzzy sets of higher type. Fuzzy sets and probabilistic constructs are orthogonal concepts. We clarify the origin of this orthogonality and discuss various formal approaches that lead to hybrid constructs of fuzzy probabilities. We also cover the ideas of rough sets and show how the formalism of these information granules can be combined with the ideas of fuzzy sets leading to the hybrids in the form of rough fuzzy and fuzzy rough sets.

8.1 FUZZY SETS OF HIGHER ORDER

In Chapter 1 (Section 1.3.2), we have shown that there is a distinction between *explicit* and *implicit* manifestations of fuzzy sets. This observation triggers further conceptual investigations and leads to the concept of fuzzy sets of higher order. Let us recall that a fuzzy set is defined in a certain universe of discourse $\mathbf{X}$ so that for each element of it we come up with the corresponding membership degree, which is interpreted as a degree of compatibility of this specific element with the concept conveyed by the fuzzy set under discussion. The essence of a fuzzy of second order is that it is defined over a collection of some other generic fuzzy sets. As an illustration, let us consider a concept of a *comfortable* temperature that we define over a finite

---

Fuzzy Systems Engineering: Toward Human-Centric Computing, by Witold Pedrycz and Fernando Gomide
Copyright © 2007 John Wiley & Sons, Inc.

196 Chapter 8 Generalizations and Extensions of Fuzzy Sets

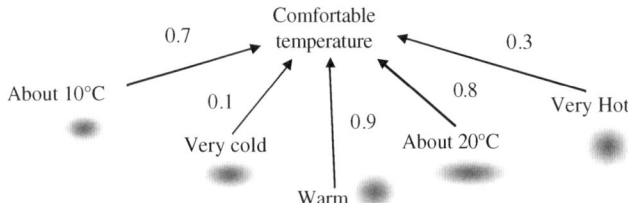

Figure 8.1 An example of second order fuzzy set of *comfortable* temperature defined over a collection of generic fuzzy sets (graphically displayed as small clouds); shown are also corresponding membership degrees.

collection of some generic fuzzy sets, say *about* 10°C, *warm, hot, cold, about* 20°C,... and so on. We could easily come to a quick conclusion that the term *comfortable* sounds more "descriptive" and hence semantically more advanced in comparison to the generic terms using which we describe it. An illustration of this second order fuzzy set is shown in Figure 8.1. To make a clear distinction, fuzzy sets studied so far could be referred to as fuzzy sets of the first order.

Using the membership degrees as shown in Figure 8.1, we can write down the membership of *comfortable* temperature in the vector form as [0.7, 0.1, 0.9, 0.8, 0.3]. It is understood that the corresponding entries of this vector pertain to the generic fuzzy sets we started with when forming the fuzzy set. Figure 8.2 graphically emphasizes the difference between fuzzy sets (which in this context could be referred to as fuzzy sets of the first order) and fuzzy sets of order 2. For the order 2 fuzzy set, we can use the notation $A = [\lambda_1, \lambda_2, \lambda_3]$ given that the reference fuzzy sets are B_1, B_2, and B_3.

Fuzzy sets of order 2 could be also formed on a Cartesian product of some families of generic fuzzy sets. Consider, for instance, a concept of a *preferred* car. To everybody this term could mean something else, yet all of us agree that the concept itself is quite complex and definitely multifaceted. We easily include several aspects such as economy, reliability, depreciation, acceleration, and others. For each of these aspects, we might have a finite family of fuzzy sets, say when talking about economy, we may use descriptors such as *about* 10 $\frac{1}{100}$ km (or expressed in mpg), *high* fuel

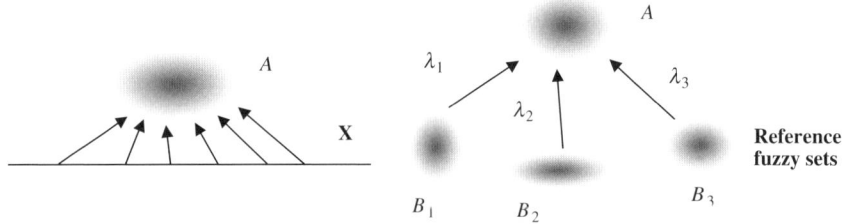

Figure 8.2 Contrasting fuzzy sets of (a) order 1 and (b) order 2. Note a role of reference fuzzy sets played in the development of order 2 fuzzy sets.

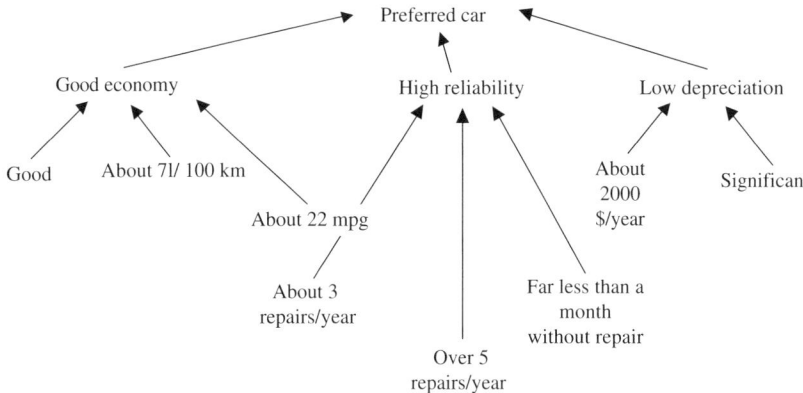

Figure 8.3 Fuzzy set of order 2 of *preferred* cars; note a number of descriptors quantified in terms of fuzzy sets and contributing directly to its formation.

consumption, *about* 30 mpg, and so on. For the given families of generic fuzzy sets in the vocabulary of generic descriptors, we combine them in a hierarchical manner as illustrated in Figure 8.3.

In a similar way, we can propose fuzzy sets of higher order, say third order or higher. They are formed in a recursive manner. Although conceptually appealing and straightforward, its applicability could become an open issue. One may not venture in allocating more effort into their design unless there is a legitimate reason behind the further usage of fuzzy sets of higher order.

Nothing prevents us from building fuzzy sets of second order on a family of generic terms that are not only fuzzy sets. One might consider a family of information granules such as sets over which a certain fuzzy set is being formed.

8.2 ROUGH FUZZY SETS AND FUZZY ROUGH SETS

Rough sets are information granules that arise in description of concepts using a finite vocabulary of existing concepts (information granules). As highlighted in Chapter 1, rough sets arise when describing information granules (and sets, in particular) using elements of some limited vocabulary. This description is usually imprecise. Intuitively, such description may lead to some approximations, called lower and upper bounds. These approximations lead us to the concept of rough sets introduced by Pawlak (1982, 1991); refer also to Skowron (1989) and Polkowski and Skowron (1998). Interesting generalizations, conceptual insights, and algorithmic investigations are offered in a series of papers authored by Pawlak and Skowron (2007a, b, c).

To explain the concept of rough sets and show what they are to offer in terms of representing information granules, we use an illustrative example. Consider a description of environmental conditions expressed in terms of temperature and pressure. For each of these factors, we fix several ranges of possible values where each of such ranges comes with some interpretation such as "values below," "values in-between," "values above," and so on. By admitting such selected ranges in both variables, we construct a grid of concepts formed in the Cartesian product of the spaces of temperature and pressure, refer to Figure 8.4. Being more descriptive, this grid forms a vocabulary of generic terms (information granules) using which we would like to describe all new information granules. As illustrated in Figure 8.4, there is a finite family of those, say $A_1, A_2, \ldots, A_{12}$.

Now let us consider that the environmental conditions monitored over some time have resulted in some values of temperature and pressure ranging in-between some lower and upper bound as illustrated in Figure 8.4. Denote this result as X. It becomes obvious that when describing it in the terms of the information granules of the vocabulary, we end up with a collection of elements that are fully included in X. They form a lower bound of description of X when being completed in presence of the given vocabulary. Likewise, we may identify elements of the vocabulary that have a nonempty overlap with X and in this sense constitute an upper bound of the description of the given environmental conditions. Along with the vocabulary, the description forms a certain rough set. More formally, we describe an upper bound by enumerating elements of A_is that have a nonzero overlap with X, that is,

$$X_+ = \{A_i | A_i \cap X \neq \varnothing\} \tag{8.1}$$

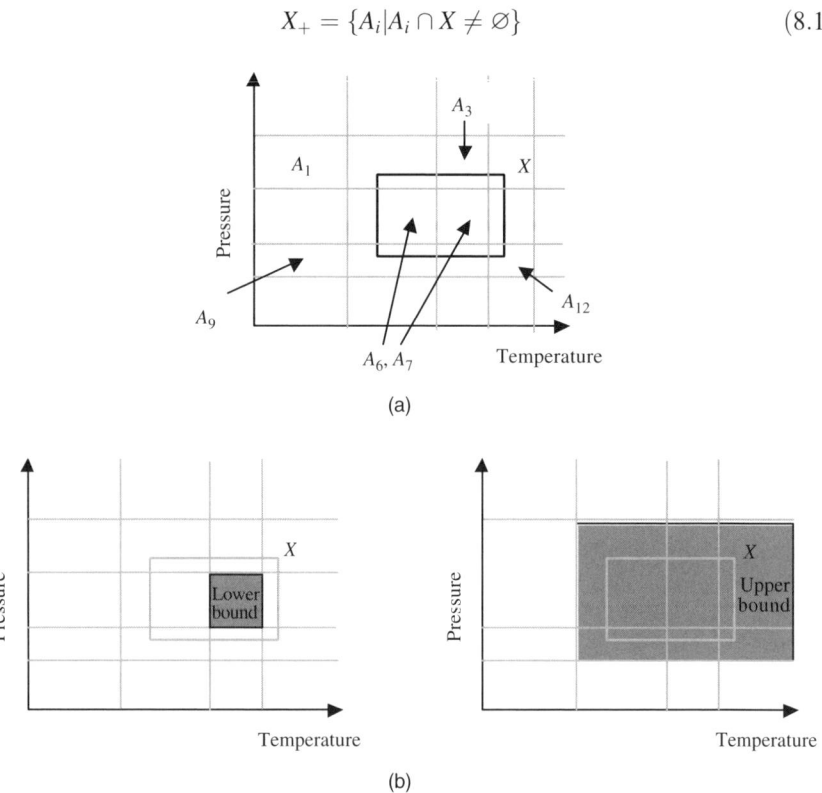

Figure 8.4 A collection of generic information granules forming the vocabulary and their use in the problem description. Environmental conditions X result in some interval of possible values (a). In the sequel, this gives rise to the concept of a rough set with the roughness of the description being captured by the lower and upper bounds (approximations) as illustrated in (b).

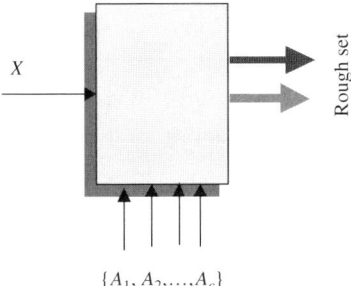

Figure 8.5 Rough set as a result of describing X in terms of some fixed vocabulary $\{A_1, A_2, \ldots, A_c\}$; the lower and upper bounds are results of the description.

More specifically, in Figure 8.4 we $X_+ = \{A_2, A_3, A_4, A_6, A_7, A_8, A_{10}, A_{11}, A_{12}\}$.

The lower bound of X involves all A_i such that they are fully included within X), namely

$$X_- = \{A_i | A_i \subset X\} \quad (8.2)$$

Here $X_- = \{A_7\}$. As succinctly visualized in Figure 8.4, we are concerned with a description of a given concept X realized in the language of a certain collection (vocabulary) of rather generic and simple terms $A_1, A_2, \ldots, A_c$. The lower and upper boundaries (approximation) are reflective of the resulting imprecision caused by the conceptual incompatibilities between the concept itself and the existing vocabulary, see Figure 8.5.

It is interesting to note that the vocabulary used in the above construct could comprise information granules being expressed in terms of any other formalism, say fuzzy sets. Quite often, we can encounter constructs like rough fuzzy sets and fuzzy rough sets in which both fuzzy sets and rough sets are put together.

These constructs rely on the interaction between fuzzy sets and sets being used in the construct. Let us consider a finite collection of sets $\{A_i\}$ and use them to describe some fuzzy set X. In this scheme, we arrive at the concept of a certain fuzzy rough set, refer to Figure 8.6. The upper bound of this fuzzy rough set is computed as in the previous case (8.1), yet given the membership function of X, the detailed calculations return membership degrees rather than 0–1 values. Given the binary character of A_is, the above expression for the upper bound reads in the form

$$X_+(A_i) = \sup_x[\min(A_i(x), X(x))] = \sup_{x \in \text{Supp}(A_i)} X(x) \quad (8.3)$$

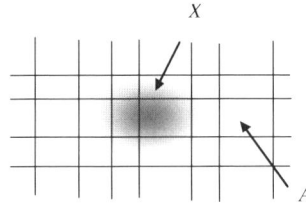

Figure 8.6 The development of the fuzzy rough set.

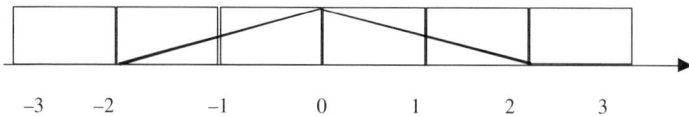

Figure 8.7 Family of generic descriptors, fuzzy set, and its representation in the form of some rough set.

The lower bound of the resulting fuzzy rough set is taken in the form

$$X_-(A_i) = \inf_x[\max(1 - X(x), A_i(x))] \tag{8.4}$$

EXAMPLE 8.1

Let us consider a universe of discourse $\mathbf{X} = [-3, 3]$ and a collection of intervals regarded as basic descriptors, see Figure 8.7.

The fuzzy set A with a triangular membership function distributed between -2 and 2 gives rise to some rough set with the lower and upper approximation of the form $X_+ = [0, 0.5, 1, 1, 0.5, 0], X_- = [0, 0, 0, 0, 0, 0]$.

We can also consider another combination of information granules in which $\{A_i\}$ is a family of fuzzy sets and X is a set, see Figure 8.8. This leads us to the concept of rough fuzzy sets.

Alternatively, we can envision a situation in which both $\{A_i\}$ and X are fuzzy sets. The result comes with the lower and upper bound whose computing follows the formulas presented above.

8.3 INTERVAL-VALUED FUZZY SETS

When defining or estimating membership functions or membership degrees, one may argue that characterizing membership degrees as single numeric values could be counterproductive and even somewhat counterintuitive given the nature of fuzzy sets themselves. Some remedy could be sought along the line of capturing the semantics of fuzzy sets through intervals of possible membership grades rather than single numeric entities (for instance, Sambuc, 1975; Wagenknecht and Hartmann, 1988; Pedrycz,

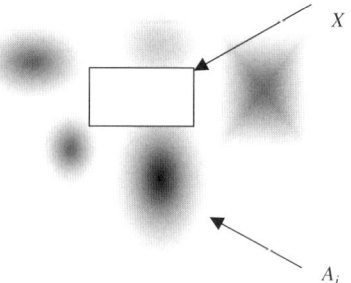

Figure 8.8 The concept of a rough fuzzy set.

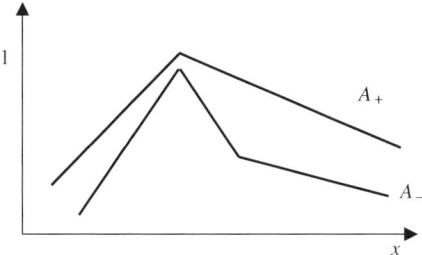

Figure 8.9 An illustration of an interval-valued fuzzy set; note that the lower and upper bound of possible membership grades could differ quite sunbstantially across the universe of discourse.

1990; Gau and Buehrer, 1993, Gehrke et al. 1996). This gives rise to the concept of so-called interval-valued fuzzy sets. Formally, an interval-valued fuzzy set A is defined by two mappings from $\mathbf{X}$ to the unit interval $A = (A_-, A_+)$, where A_- and A_+ are the lower and upper bound of membership grades, $A_-(x)\ A_+(x)$ for all $x \in \mathbf{X}$, where $A_-(x) \leq A_+(x)$. The bounds are used to capture an effect of a lack of uniqueness of numeric membership—not knowing the detailed numeric values we admit bounds of possible membership grades. Hence, the name of the interval-valued fuzzy sets is very much descriptive of the essence of the construct. The broader the range of the membership values, the less specific we are about membership degree of the element to the information granule. An illustration of the interval-valued fuzzy set is included in Figure 8.9. We already built interval-valued fuzzy sets when constructing membership functions using a horizontal approach and accepting confidence intervals produced by the statistical assessment of the results produced by a group of experts.

In particular, when $A_-(x) = A_+(x)$, we end up with a "standard" fuzzy set. The operations on interval-valued fuzzy sets are defined by considering separately the lower and upper bounds describing ranges of membership degrees. Given two interval-valued fuzzy sets $A = (A_-, A_+)$ and $B = (B_-, B_+)$, their union, intersection, and complement are introduced as follows:

$$
\begin{aligned}
&\text{union} && (A \cup B)(x) = ((\min(A_+(x), B_+(x)), \max(A_-(x), B_-(x))) \\
&\text{intersection} && (A \cap B)x = ((\max(A_+(x), B_+(x)), \min(A_-(x), B_-(x))) \\
&\text{complement} && \overline{A}(x) = (A_-(x), A_+(x))
\end{aligned}
\tag{8.5}
$$

8.4 TYPE-2 FUZZY SETS

Type-2 fuzzy sets form an intuitively appealing generalization of interval-valued fuzzy sets. Instead of the intervals of numeric values of membership degrees, we allow for the characterization of membership by fuzzy sets themselves. Consider a certain element of the universe of discourse, say x. The membership of x to A is captured by a certain fuzzy set formed over the unit interval. This construct generalizes the fundamental idea of a fuzzy set and helps us relieve from the restriction

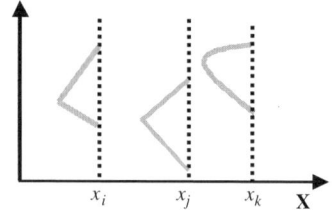

Figure 8.10 An illustration of type-2 fuzzy set; for each element of **X** there is a corresponding fuzzy set of membership grades.

of having single numeric values describing a given fuzzy set (Mendel, 2001; Mendel and John, 2002). An example of type-2 fuzzy set is illustrated in Figure 8.10.

With regard to these forms of generalizations of fuzzy sets, there are two important facets that should be taken into consideration. First, there should be a clear motivation and a straightforward need to develop and use them. Second, it is imperative that there is sound membership determination procedure in place using which we can construct the pertinent fuzzy set.

To elaborate on these two issues, let us discuss a situation in which we deal with several databases populated by data coming from different regions of the same country. Using them we build a fuzzy set describing a concept of *high* income where the descriptor *high* is modeled as a certain fuzzy set. Given the experimental evidence, the estimation method described in Pedrycz (1990) could be a viable alternative to pursue. By being induced by some locally available data, the concept could exhibit some level of variability, yet we may anticipate that all membership functions might be quite similar as being reflective of some general commonalities. Given that the individual estimated membership functions are trapezoidal (or triangular), we can consider two alternatives to come up with some aggregation of the individual fuzzy sets, see Figure 8.11.

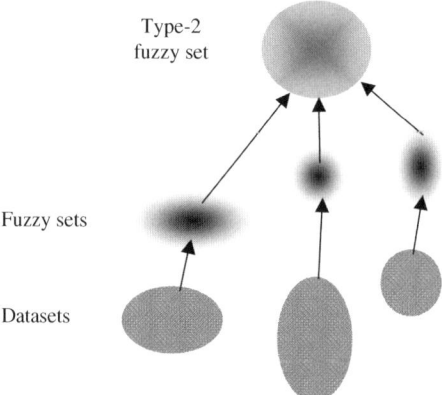

Figure 8.11 A scheme of aggregation of fuzzy sets induced by *P* datasets.

To fix the notation, let for P databases, the corresponding estimated trapezoidal membership functions are denoted by $A_1(x, a_1, m_1, n_1, b_1), A_2(x, a_2, m_2, n_2, b_2),$ $\ldots, A_P(x, a_P, m_P, n_P, b_P)$, respectively. The first aggregation alternative leads to the emergence of an interval-valued fuzzy set $A(x)$. Its membership function assumes interval values where for each x the interval of possible values of the membership grades is given in the form $[\min_i A_i(x, a_i, m_i, n_i, b_i), \max_i A_i(x, a_i, m_i, n_i, b_i)]$.

It is worth noting that both the upper and lower bounds associated with the intervals formed in this way do not form any longer triangular or trapezoidal membership functions.

EXAMPLE 8.2

Let us consider four triangular fuzzy sets $A_1(x, 2, 5, 9)$, $A_2(x, 1, 7, 11)$, $A_3(x, 1.5, 8, 10)$, and $A_4(x, 0, 7, 10)$. The resulting membership function of the interval-valued fuzzy set A is illustrated in Figure 8.12. One should note that A is not necessarily either a triangular or a trapezoidal fuzzy set as we may encounter a substantial level of diversity among fuzzy sets. The form of the interval-valued fuzzy set may be advantageous in further computing, yet the estimation process could be very conservative leading to very broad ranges of membership grades (which is particularly visible when dealing with different data and fuzzy sets induced on their basis).

(b) Being aware of the drawbacks of the conservative way in which the membership function of the interval-valued fuzzy set has been estimated, we may refine the estimation process, and instead relying on the minimal and maximal membership grades, take advantage of the statistical characteristics of the collection of the membership grades. This implies that for each x we collect the membership grades and apply to them the estimation procedure as described in Section 4.6 and 4.7 of Chapter 4. Subsequently, this leads to the triangular or trapezoidal membership

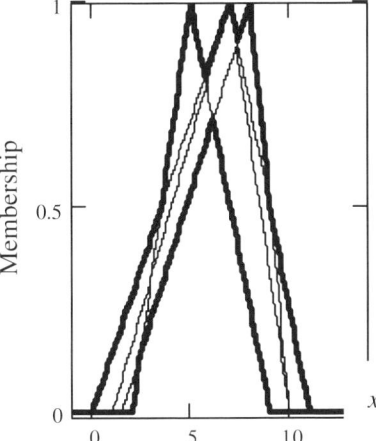

Figure 8.12 Interval-valued fuzzy set resulting from the aggregation of triangular fuzzy sets.

functions defined in [0,1]. In essence, in this way we have constructed a certain type-2 fuzzy set.

There are also a substantial number of developments and augmentations of a generic concept of fuzzy sets that come under different names. Some of them somewhat relate to interval-valued fuzzy sets. One of them is the concept proposed by Atanassov (1998, 2000). The crux of this construct is the following. Let us consider an information granule A defined over some space $\mathbf{X}$ and expressed as a pair of mappings from $\mathbf{X}$ to [0,1], that is, (A^+, A^-) $A^+(x)$ denotes a degree of membership of x to A, whereas $A^-(x)$ stands for a degree of nonmembership of x to A^-. Furthermore, we require that for any $x \in \mathbf{X}$, A^+ and A^- satisfy the relationship

$$A^+(x) + A^-(x) \leq 1 \tag{8.6}$$

meaning that there exists a "gap" between a strength of membership (which is a sort of some "positive" assessment) and a strength of nonmembership (which could be viewed as a type of "negative" assessment). The basic logic operations for (A^-, A^+) and (B^-, B^+) are defined in the form given before in (8.5); note that the result is again a pair of degrees of membership and nonmembership).

8.5 SHADOWED SETS AS A THREE-VALUED LOGIC CHARACTERIZATION OF FUZZY SETS

Fuzzy sets offer a wealth of detailed numeric information conveyed by their detailed numeric membership grades (membership functions). This very detailed conceptualization of information granules can clearly act as a two-edge sword. On the one hand, we may enjoy a very detailed quantification of elements to a given concept (fuzzy set). On the other hand, those membership grades could be somewhat overwhelming and introduce some burden when it comes to a general interpretation. It is also worth noting that numeric processing of membership grades comes sometimes with quite substantial computing overhead. To alleviate these problems, we introduce a certain category of information granules called shadowed sets. Shadowed sets are information granules induced by fuzzy sets so that they capture the essence of fuzzy sets while at the same time reducing the numeric burden because of their limited three-valued characterization of shadowed sets. This nonnumeric character of shadowed sets is also of particular interest when dealing with their interpretation abilities. Given the characteristics of shadowed sets, we may view them as a certain symbolic representation of fuzzy sets.

8.5.1 Defining Shadowed Sets

Formally speaking, a shadowed set A defined in some space $\mathbf{X}$ is a set-valued mapping coming in the following form (Pedrycz, 1998, 1999)

$$A : \mathbf{X} \to \{0, [0,1], 1\} \tag{8.7}$$

8.5 Shadowed Sets as a Three-Valued Logic Characterization of Fuzzy Sets

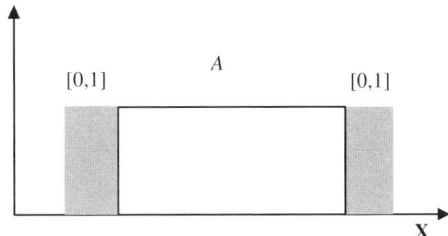

Figure 8.13 An example of a shadowed set *A*; note shadows formed around the core of the construct.

The co-domain of *A* consists of three components, that is, 0, 1, and the unit interval [0,1]. They can be treated as degrees of membership of elements to *A*. These three quantification levels come with an apparent interpretation. All elements for which $A(x)$ assume 1 are called a core of the shadowed set—they embrace all elements that are fully compatible with the concept conveyed by *A*. The elements of **X** for which $A(x)$ attains zero are excluded from *A*. The elements of **X** for which we have assigned the unit interval are completely uncertain—we are not at a position to allocate any numeric membership grade. Therefore, we allow the usage of the unit interval, which reflects uncertainty, meaning that any numeric value could be permitted here. In essence, such element could be excluded (we pick up the lowest possible value from the unit interval), exhibit partial membership (any number within the range from 0 and 1), or could be fully allocated to *A*. Given this extreme level of uncertainty (nothing is known and all values are allowed), we call these elements shadows and hence the name of the shadowed set. An illustration of the underlying concept of a shadowed set is included in Figure 8.13.

One can view this mapping (shadowed set) as an example of a three-valued logic as encountered in the classic model introduced by Lukasiewicz. Having this in mind, we can think of shadowed sets as a *symbolic* representation of *numeric* fuzzy sets. Obviously, the elements of co-domain of *A* could be labeled using symbols (say, certain, shadow, excluded; or *a*, *b*, *c* and alike) endowed with some well-defined semantics.

The operations on shadowed Table 8.1 sets are isomorphic with those encountered in the three-valued logic.

These logic operations are conceptually convincing; we observe an effect of preservation of uncertainty. In the case of the *or* operation, we note that combining a single numeric value of exclusion (0) with the shadow, we arrive at the shadow (as nothing specific could be stated about the result of this logic aggregation). Similar effect occurs for the *and* operator when being applied to the shadow and the logic value of 1.

The simplicity of shadowed sets becomes their advantage. Dealing with three logic values simplifies not only the interpretation but it is also advantageous in all computing, especially when such calculations are contrasted with the calculations completed for fuzzy sets involving detailed numeric membership grades. Let us note that logic operators that are typically realized by means of some t-norms and

Table 8.1 Logic Operations (*and*, *or*, and Complement) on Shadowed Sets; Here a Shadow is Denoted by $S(= [0, 1])$.

Intersection			
A\B	0	S	1
0	0	0	0
S	0	S	S
1	0	S	1
Union			
A\B	0	S	1
0	0	S	1
S	S	S	1
1	1	1	1
Complement			
A			
0	1		
S	S		
1	0		

t-conorms require computing of the numeric values of the membership grades. In contrast, those realized on shadowed sets are based on comparison operations and therefore are far less demanding.

Although shadowed sets could be sought as new and standalone constructs, our objective is to treat them as information granules induced by some fuzzy sets. The bottom line of our approach is straightforward—considering fuzzy sets (or fuzzy relations) as the point of departure and acknowledging computing overhead associated with them, we regard shadowed sets as constructs that capture the essence of fuzzy sets while help reducing the overall computing effort and simplifying ensuing interpretation. In the next section, we concentrate on the development of shadowed sets for given fuzzy sets.

8.5.2 The Development of Shadowed Sets

Accepting the point of view that shadowed sets are algorithmically implied (induced) by some fuzzy sets, we are interested in the transformation mechanisms translating fuzzy sets into the corresponding shadowed sets. The underlying concept is the one of uncertainty condensation or "localization." While in fuzzy sets we encounter intermediate membership grades located in between 0 and 1 and distributed practically across the entire space, in shadowed sets we "localize" the uncertainty effect by building constrained and fairly compact shadows. By doing so we could remove (or better to say, redistribute) uncertainty from the rest of the universe of discourse by bringing the corresponding low and high membership grades to zero and one and

8.5 Shadowed Sets as a Three-Valued Logic Characterization of Fuzzy Sets

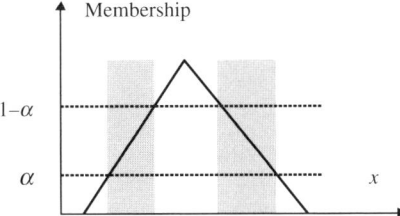

Figure 8.14 The concept of a shadowed set induced by some fuzzy set; note the range of membership grades (located between α and $1 - \alpha$) generating a shadow.

then compensating these changes by allowing for the emergence of uncertainty regions. This transformation could lead to a certain optimization process in which we complete a total balance of uncertainty.

To illustrate this optimization, let us start with a continuous, symmetric, unimodal, and normal membership function A. In this case we can split the problem into two tasks by considering separately the increasing and decreasing portion of the membership function, Figure 8.14.

For the increasing portion of the membership function, we reduce low membership grades to zero, elevate high membership grades to one, and compensate these changes (which in essence lead to an elimination of partial membership grades) by allowing for a region of the shadow where there are no specific membership values assigned, but we admit the entire unit interval as feasible membership grades. Computationally, we form the following balance of uncertainty preservation that could be symbolically expressed as

$$\text{Reduction of membership} + \text{elevation of membership} = \text{Shadow} \tag{8.8}$$

Again referring to Figure 8.14, once given the membership grades below α and above $1 - \alpha$, $\alpha \in (0, \frac{1}{2})$, we express the components of the above relationship in the form (we assume that all integrals do exist)

Reduction of membership (low membership grades are reduced to zero)

$$\int_{x:A(x) \leq \alpha} A(x) dx \tag{8.9}$$

Elevation of membership (high membership grades elevated to 1)

$$\int_{x:A(x) \geq 1-\alpha} (1 - A(x)) dx \tag{8.10}$$

Shadow

$$\int_{x:\alpha < A(x) < 1-\alpha} dx \tag{8.11}$$

The minimization of the absolute difference

$$V(\alpha) = \left| \int_{x:A(x)\leq\alpha} A(x)dx + \int_{x:A(x)\geq 1-\alpha} (1-A(x))dx - \int_{x:\alpha<A(x)<1-\alpha} dx \right| \quad (8.12)$$

completed with respect to α is given in the form of the following optimization problem

$$\alpha_{opt} = \arg\min_\alpha V(\alpha) \quad (8.13)$$

where $\alpha \in (0, \frac{1}{2})$. For instance, when dealing with triangular membership function (and it appears that the result does not require the symmetry requirement), the optimal value of α is equal to $\sqrt{2} - 1 \approx 0.4142$ (Pedrycz, 1999). For the parabolic membership functions, the optimization leads to the value of α equal to 0.405.

For the Gaussian membership function, $A(x) = \exp(-x^2/\sigma^2)$, we get the optimal value of α resulting from the relationship (the calculations here concerns the decreasing part of the membership function defined over $[0, \infty)$

$$V(\alpha) = \left| \int_0^{\sigma\sqrt{-\ln(1-\alpha)}} (1-A(x))dx + \int_{\sigma\sqrt{-\ln(\alpha)}}^{\infty} A(x)dx - \int_{\sigma\sqrt{-\ln(\alpha)}}^{\sigma\sqrt{-\ln(1-\alpha)}} dx \right| \quad (8.14)$$

Then the optimal value of α is equal to 0.3947, and it does not depend upon the spread σ. The plot of V, Figure 8.15, reveals that V exhibits a global minimum.

Let us move on to the most general case in which we do not impose any assumptions as to the form of the membership function. We consider discrete membership values $u_1, u_2, \ldots u_N$. Denote the minimal and maximal value in this set by $u_{\min}$ and $u_{\max}$, respectively. The overall reduction of lower membership grades is expressed in the form of the following sum $\sum_{k\in\Omega} u_k$ where $\Omega = \{k | u_k \leq \alpha\}$. The elevation of higher membership grades to one leads to the expression $\sum_{k\in\Phi}(1-u_k)$ with $\Omega = \{k | u_k \geq u_{\max} - \alpha\}$. For the shadows we consider the cardinality of the set

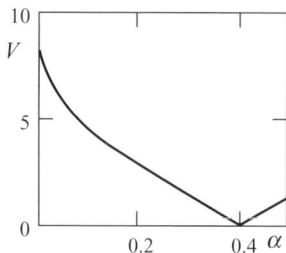

Figure 8.15 Performance index V treated as a function of α ($\sigma = 4$).

8.5 Shadowed Sets as a Three-Valued Logic Characterization of Fuzzy Sets

$\Omega = \{k | u_k \in (\alpha, u_{\max} - \alpha)\}$. Then the above conditions translate into the following optimization problem

$$V(\alpha) = \left| \sum_{k \in \Omega} u_k + \sum_{k \in \Phi} (1 - u_k) - \text{Card}(\Omega) \right| \quad \text{Minimize } V(\alpha) \text{ with respect to } \alpha \tag{8.15}$$

where the range of feasible values of α is given as $[u_{\min}, \frac{u_{\min} + u_{\max}}{2}]$.

Once optimized, the resulting shadowed set can be treated as a concise descriptor of the corresponding fuzzy set. For the original fuzzy set A, we denote by Core(A) and Shadow(A), respectively, the core and shadow of the shadowed set induced by A.

The above design process could be generalized in such a way that we introduce a continuous and increasing functional $\gamma(u) : [0, 1] \to [0, 1]$ that helps quantify the original values of the membership grades when taken into consideration in the balance captured by (8.8). When reducing membership grades we use the expression

$$\int_{x:A(x) \leq \alpha} \gamma(A(x)) dx \tag{8.16}$$

while the elevation of membership is guided by the form

$$\int_{x:A(x) \geq 1-\alpha} (1 - \gamma(A(x))) dx \tag{8.17}$$

The typical form of the functional would be a polynomial $\gamma(u) = u^p, p > 0$.

The shadowed sets are instrumental in fuzzy cluster analysis, especially results produced there. Consider a data set shown in Figure 8.16. The standard FCM run for $c = 2$ clusters returned the partition matrix whose membership grades are then transformed into the shadowed set.

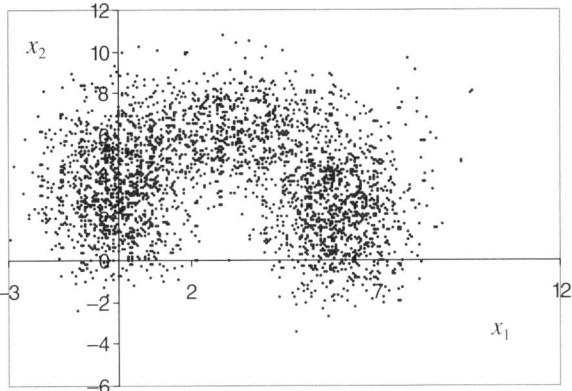

Figure 8.16 Two-dimensional synthetic data set.

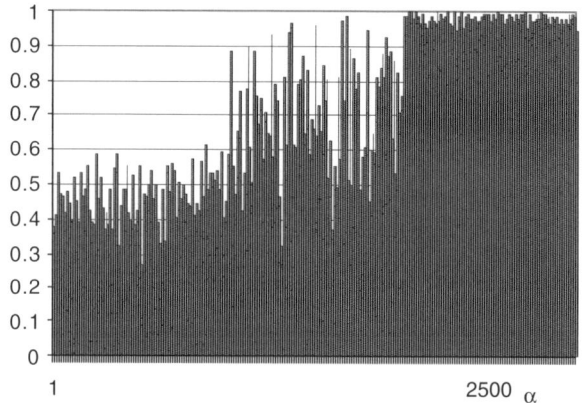

Figure 8.17 Membership grades of one of the clusters developed by the clustering algorithm.

The prototypes are equal to $\mathbf{v}_1 = [5.51, 2.48]$, $\mathbf{v}_2 = [1.05, 4.71]$ and reflect the structure of the data set. The membership function of one cluster (fuzzy relation) is visualized in Figure 8.17.

The optimization of the threshold level (α) inducing the shadowed set is completed through a simple enumeration; following the graph in Figure 8.18, we obtain $\alpha = 0.4322$. This in turn highlights several patterns to be treated as potential candidates for further thorough analysis.

When we complete clustering with $c = 12$ (which is substantially higher than in the first case), the results become quite different, see Figure 8.19. First, the optimal value of α becomes equal to 0.3636. As illustrated in Figure 8.20, the overlap between the clusters has been also reduced quite visibly.

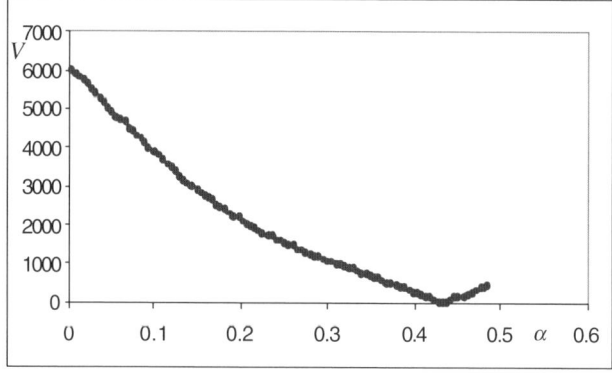

Figure 8.18 V viewed as a function of α.

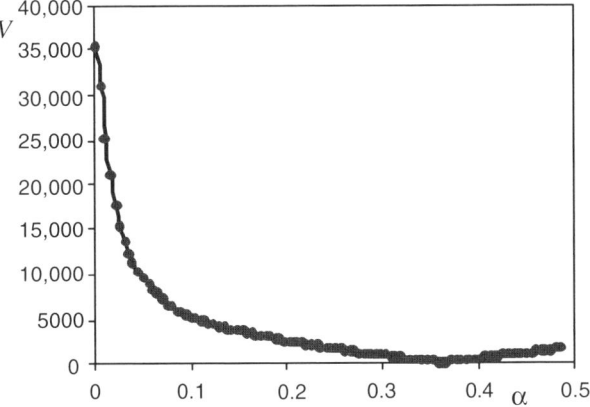

Figure 8.19 V viewed as a function of α for $c = 12$ clusters.

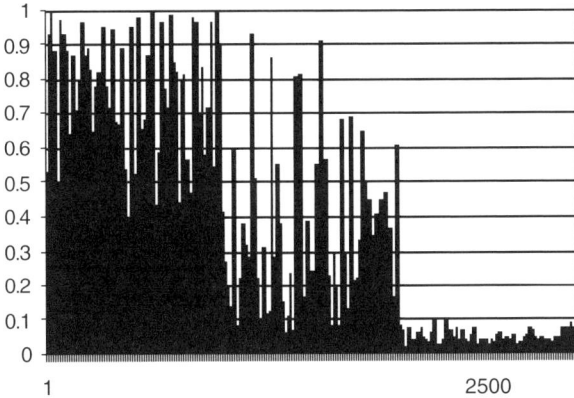

Figure 8.20 Distribution of membership grades of one of the clusters developed by the clustering algorithm.

8.6 PROBABILITY AND FUZZY SETS

Fuzzy sets and probability are different realizations of the fundamental concept of information granules. Given this similarity, they address different facets of information granularity. There exist significant conceptual differences and algorithmic aspects. Let us emphasize the most visible one.

Fuzzy sets are information granules whose existence and a form of boundaries are a result of perception of some phenomena whose comprehension is inherently reflective of the very nature of these concepts and an intension of the observer. Put it briefly, the same experimental evidence can be viewed at from different standpoints

212 Chapter 8 Generalizations and Extensions of Fuzzy Sets

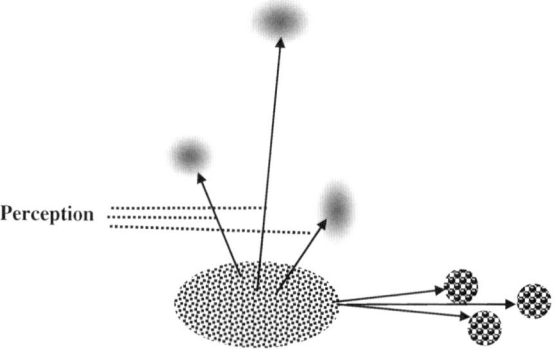

Figure 8.21 Data and their different perception articulated by fuzzy sets and probabilities.

articulated by some fuzzy sets. In this sense, the same data could lead to a fairly diversified manifestation. In

In probability, we capture the distribution of data and transform them into some form of probability density function, probability function, and alike. When dealing with probabilistic information granules, we evidently and directly rely upon the experimental data by capturing their distribution in terms of some probability density function. In spite of possible differences between various techniques used in probability density estimation, it is anticipated that all of the methods will lead to similar results, see Figure 8.21.

The distinction between fuzzy sets and probability can be made even more striking by recalling that fuzzy sets capture the effect of uncertainty associated with perception whereas probability deals with the facet of uncertainty that deals with occurrence or nonoccurrence of some phenomenon. In other words, fuzzy sets and probability theory are orthogonal. Fuzzy sets form a certain view at the concepts and admit that they are inherently elements with partial membership. This has nothing to do with an outcome of experiment. In probability we aim at quantifying opinions about occurrence (chances of occurrence) of some events. Say, we toss a coin. Before this experiment, we cannot say what is going to happen. However, once the experiment has been done, we know the outcome. The uncertainty associated with it has vanished. Instead of expressing our judgment in the language of probability theory, we now know whether the event has occurred or not.

There is a certain point to be made clear: Although fuzzy sets do not directly rely on experimental numeric data and do not capture their characteristics in an explicit manner, they may do so in some loosely defined implicit manner. In other words, when forming a membership function, one might incorporate some experimental evidence and make it being included in the formation of the fuzzy set in some implicit manner. We have witnessed this effect in several methods of membership function determination we discussed in Chapter 4. In particular, this concerns fuzzy equalization and fuzzy clustering.

8.7 PROBABILITY OF FUZZY EVENTS

By identifying the orthogonality of the concepts of fuzzy sets and probability, we can conveniently describe situations in which both of them come together, and it becomes beneficial to consider both of them. Before we move to the formalization of the problems and discuss algorithmic underpinnings, let us offer some illustrative examples.

- What is the probability of *low* temperature tomorrow?
- What is the probability of *high* inflation in a *short* term?
- What is the probability of *small* steady state error in control of pressure of steam boiler?

In all of these statements, we are concerned about quantifying probability of fuzzy events. These events could be simple (such as the one in the first or the third one) or composite (as the one encountered in the second example)

Denote the underlying probability density function (*pdf*) defined in **X** by $p(x)$. The probability of the fuzzy event (Zadeh, 1968, 1975) A defined in the same space **X** is defined as the following integral $\int_X A(x) dP(x) = \int_X A(x) p(x) dx$ (by default we assume that this integral does exist). A careful inspection of the above expression indicates that this integral is nothing but an expected value of the membership function $E(A)$,

$$E(A) = \int_X A(x) p(x) dx \qquad (8.18)$$

Along the same line, we can introduce variance of A (and its standard deviation) as well as higher order moments of the fuzzy event:

- variance $E^2(A) = \int_X [A(x) - E(A)]^2 p(x) dx$
  and standard deviation $\sigma(A) = -\sqrt{E^2(A)}$
- higher order moments $\int_X [A(x) - E(A)]^r p(x) dx$, where $r > 2$

EXAMPLE 8.3

We are interested in the determination of the probability of the fuzzy event in the first example assuming that the distribution of temperatures in this particular season of the year can be described by means of a normal density function with the mean value of 5 and a standard deviation equal to 1, while the concept of *low* temperature comes with the piecewise linear membership function $A(x)$ shown in Figure 8.22. Furthermore, let us find the standard deviation of the fuzzy event.

The integral (8.18) makes sense; however, there is no analytical solution to it, and we need to resort ourselves to numerical integration. The resulting probability of the fuzzy event A is equal to 0.294. The standard deviation of the fuzzy event $\sigma(A)$ is then equal to 3.46×10^{-3}. Noticeably, it is substantially lower in comparison with the probability of the fuzzy event.

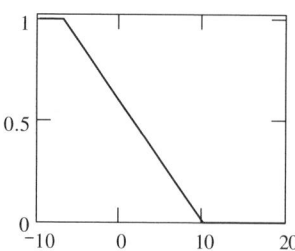 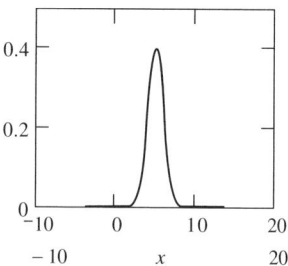

Figure 8.22 Fuzzy event of *low* temperature and the probability density function of temperatures during the season of the year.

In a similar manner we can deal with situations where we encounter computing of probabilities of combinations of fuzzy events A and B such as P(A and B), P(A or B), or in general $P(\phi(A,B))$ where ϕ denotes some logic aggregation of A and B.

Referring to the orthogonality of fuzzy sets and probability, we might be interested in fuzzy sets that are made meaningful in terms of their underlying semantics as well as existing experimental evidence. The semantics issue could be handled by requesting that the energy measure of fuzziness of the fuzzy set does not exceed a certain acceptable level (we want the fuzzy set to be specific enough). The experimental evidence should be above some threshold level, which says that the fuzzy set comes with sufficient data behind it. In other words, we request that the expected value $E(A)$ should exceed the predefined level. Given these two requirements, we can form a feasible region in which such fuzzy sets should be positioned; see Figure 8.23. It includes all fuzzy sets such that their energy measure of fuzziness is less than some threshold λ and the probability of the event is not less than μ.

Another interesting category of expressions deals with situations in which we encounter linguistically quantified statements. For instance, we commonly use common terms such as *low* probability, *high* probability, probability of *about* 70% (that is 0.7), and so on. Given their character, all these probabilities can be referred to as linguistic probabilities (Zadeh, 1975). They are fuzzy sets defined in the unit interval and come with an intuitively appealing membership functions as illustrated in Figure 8.24.

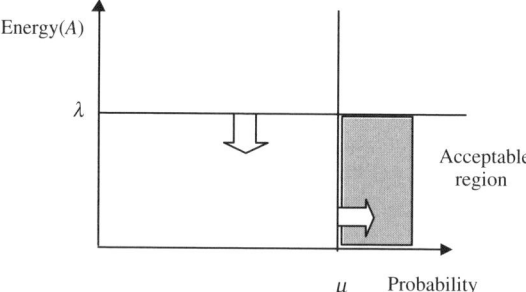

Figure 8.23 The feasible region of fuzzy sets satisfying the requirements of semantic validity and experimental evidence (shaded region); the corresponding threshold levels are denoted by λ and μ, respectively.

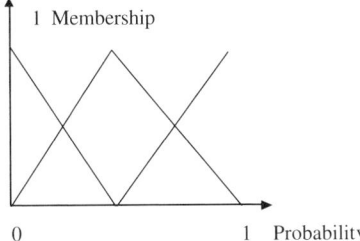

Figure 8.24 Examples of linguistic probabilities of *low*, *medium*, and *high* probability.

As an illustration of processing realized in the presence of linguistic probabilities as shown above, let us compute the expected value of the expression (Zadeh, 1975)

$$Z = \sum_{i=1}^{n} a_i P_i \tag{8.19}$$

where $a_i \in \mathbf{R}$. Recall that the expression $\sum_{i=1}^{n} a_i p_i$ is nothing but an expected value of the random variable assuming values a_i with probabilities p_i. Here as P_i are fuzzy sets, the expected value Z is a fuzzy set as well.

The solution to (8.19) is obtained by applying the extension principle. We have

$$Z(z) = \sup\left[\min(P_1(p_1), P_2(p_2), \ldots, P_n(p_n))\right]$$

$$\text{subject to } z = \sum_{i=1}^{n} a_i p_i \quad \text{and} \quad \sum_{i=1}^{n} p_i = 1 \tag{8.20}$$

The second unity constraint is here to assure that we adhere to the probabilistic nature of the constraints (the probabilities must sum up to 1).

A simple example concerns the following expression

$$Z = a_1 \text{ likely} + a_2 \text{ unlikely}$$

where *likely* and *unlikely* are linguistic probabilities that satisfy the antonym relationship meaning that $likely(p) = unlikely(1 - p), p \in [0, 1]$ a_1 and a_2 are weights in [0,1]. In light of the fuzzy probabilities, the result becomes a fuzzy set. Then we obtain

$$Z(z) = likely\left(\frac{z - a_2}{a_1 - a_2}\right)$$

EXAMPLE 8.4

Assuming the membership function of *likely* to be in the form $likely(u) = u$, the resulting fuzzy set of expected values Z is illustrated in Figure 8.25; here we considered several combinations of values of a_1 and a_2. For the modified form of the membership function $likely(u) = u^2$, the corresponding plots of Z are included as well.

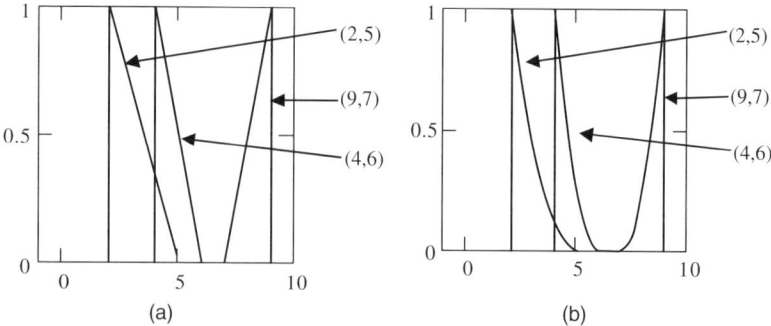

Figure 8.25 Fuzzy sets Z for selected combinations of values of a_1 and a_2: (a) $likely(u) = u$, and (b) $likely(u) = u^2$.

8.8 CONCLUSIONS

Fuzzy sets can be generalized in many quite different ways depending upon the specificity of the problem in which they are being used. We have distinguished between two main categories of generalized constructs.

(a) Generalization at the level of membership grades: Here we generalize from numeric values by moving toward intervals, fuzzy sets defined in the unit interval, probability density functions, and so on.

(b) Generalization at the level of the universe of discourse: The generalization activities are geared toward the construction of more suitable (generalized) universes of discourse. The typical constructs are fuzzy sets of higher order. This feature becomes quite evident given the fact that in second order fuzzy sets, these are defined in elements being fuzzy sets.

(c) Generalization through hybridization: In this way we construct fuzzy sets by expressing them in the language of some other formalisms of information granulation. For instance, we encounter hybrids of fuzzy sets and rough sets (including fuzzy rough sets, rough fuzzy sets).

Since the very inception of fuzzy sets, there were a great deal of discussion as to this concept and probability theory. It is quite commonly accepted today that probability and fuzzy sets are orthogonal concepts that are not in any competition (that may lead to the position "either-or"). On the contrary, they could augment each other as clearly manifested through the constructs such as fuzzy probabilities (probability of fuzzy events), linguistic probabilities, and alike.

Let us emphasize that while the generalizations of the generic concept of fuzzy set can be realized in different ways, one should always exercise some prudence by making sure that the two fundamental questions are satisfactorily addressed, namely, (a) is there a genuine need to expand the generic concept or a careful reformulation of the problem could be equally good? and (b) by proposing the generalization, have we developed a sound and technically viable estimation procedure that helps us effectively compute the details of the generalized construct, say fuzzy sets of higher type or higher order? If there are no clear answers to these questions, we would be better off to resist temptation of moving toward more generalized constructs. They might occur to be less effective than one has originally anticipated.

EXERCISES AND PROBLEMS

1. Given are matrices of pairwise comparison R_1, R_2, R_3, and R_4 provided by four experts and concerning preferences expressed within a finite set of alternatives $\mathbf{X} = \{x_1, x_2, \ldots, x_4\}$. On their basis, determine fuzzy sets and construct a single interval-valued fuzzy set and fuzzy set of type-2.

$$R_1 = \begin{bmatrix} 1 & 9 & 3 & 4 \\ 1/9 & 1 & 5 & 4 \\ 1/3 & 1/5 & 1 & 7 \\ 1/4 & 1/4 & 1/7 & 1 \end{bmatrix} \quad R_2 = \begin{bmatrix} 1 & 9 & 3 & 3 \\ 1/8 & 1 & 7 & 6 \\ 1/3 & 1/7 & 1 & 7 \\ 1/3 & 1/6 & 1/7 & 1 \end{bmatrix}$$

$$R_3 = \begin{bmatrix} 1 & 2 & 6 & 7 \\ 1/2 & 1 & 5 & 4 \\ 1/6 & 1/5 & 1 & 7 \\ 1/7 & 1/4 & 1/7 & 1 \end{bmatrix} \quad R_2 = \begin{bmatrix} 1 & 9 & 3 & 6 \\ 1/9 & 1 & 5 & 2 \\ 1/3 & 1/5 & 1 & 8 \\ 1/6 & 1/2 & 1/8 & 1 \end{bmatrix}$$

2. An error distribution of a given sensor can be modeled by a Gaussian density function

$$p(x) = \frac{1}{\sigma\sqrt{2\pi}} \exp\left(-\frac{x^2}{2\sigma^2}\right)$$

where $\sigma = 2$. The reading of the sensor is modeled by a triangular fuzzy number with spread equal to 0.5. What is the expected value of the reading of this sensor?

3. What should be a spread of the Gaussian membership function of the fuzzy set A so that its expected value is not lower than positive threshold value λ. Assume that the pdf in this problem is uniform and distributed symmetrically around the modal value of the membership function

4. Suggest two or three concepts that could be modeled using order-2 fuzzy sets.

5. We are concerned with an extension of a set (interval) A into a fuzzy set B in a way that the probabilistic evidence of such fuzzy set is equal to the one coming with the set. Refer to Figure 8.26. A is symmetrically distributed around zero.

Assuming that B is defined by a trapezoidal membership function, elaborate on its construction. Next, determine the membership function of B for the following cases:

(a) Uniform *pdf* where $b > a$
(b) Gaussian *pdf* with the zero mean value and a standard deviation σ equal to 3.

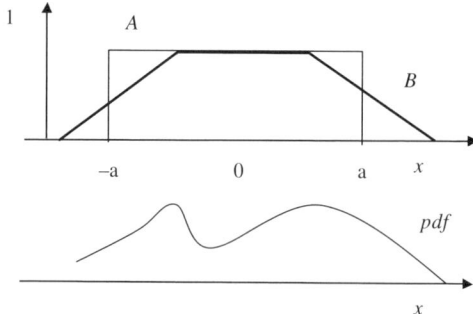

Figure 8.26 Set (interval) A, its extension to a fuzzy set B along with the underlying probability density function $p(x)$.

6. Determine a shadowed set for the fuzzy set described by the cosine-based membership function

$$A(x) = \begin{cases} \cos(\gamma x), & \text{if } x \in \left[-\frac{\pi}{2\gamma}, \frac{\pi}{2\gamma}\right] \\ 0, & \text{otherwise} \end{cases}$$

How does the size (length) the shadow depend upon the values of γ?

7. Consider a finite family of fuzzy sets $\{A_{ij}\}, i,j = -2, -1, 0, 1, 2$ defined in the Cartesian product of real numbers $\mathbf{R}^2$ with the membership functions

$$A_{ij}(x,y) = \exp(-(x-i)^2 + (y-j)^2)$$

Given $\mathbf{X} = [-1, 1] \times [-1, 1]$, describe it in terms of the elements of the family $\{A_{ij}\}$.

HISTORICAL NOTES

Various extensions and generalizations of fuzzy sets have been reported in the literature. Although they follow the main taxonomy, there is a genuine diversity of approaches (Goguen, 1967; Atanassov, 1986, 1999; Mendel, 2001, 2002; Pawlak and Skowron, 2007; Pedrycz, 1999, 2005). An interesting alternative (in the form of so-called balanced fuzzy sets) that addresses an issue of "inhibitory" aspects of membership degrees was presented by Homenda (2006). Probability and fuzzy sets were studied by Zadeh (1968, 2002).

REFERENCES

Atanassov, K. Intuitionistic fuzzy sets, *Fuzzy Set Syst.* **20**, 1986, 87–96.
Atanassov, K. *Intuitionistic Fuzzy Sets*, Physica-Verlag, Heidelberg, 1999.
Gau, W., Buehrer, D. Vague sets, *IEEE Trans. Syst. Man Cybern.* **23**, 1993, 610–614.
Goguen, J. L-fuzzy sets, *J. Math. Anal. Appl.* **18**, 1967, 145–174.
Homenda, W. Balanced fuzzy sets, *Inf. Sci.* **176**, 2006, 2467–2506.
Mendel, J. *Uncertain Rule-Based Fuzzy Logic Systems*, Prentice-Hall, Upper Saddle River, NJ, 2001.
Mendel, J., John, R. Type-2 fuzzy sets made simple, *IEEE Trans. Fuzzy Syst.* **10**, 2002, 117–127.
Pawlak, Z. Rough sets, *Int. J. Comput. Inf. Sci.* **11**, 1982, 341–356.

Pawlak, Z. *Rough Sets: Theoretical Aspects of Reasoning About Data*, Kluwer Academic Publishers, Dordrecht, 1991.

Pawlak, Z., Skowron, A. Rudiments of rough sets, *Inf. Sci.* **177**(1), 2007a, 3–27.

Pawlak, Z., Skowron, A. Rough sets: some extensions, *Inf. Sci.* **177**(1), 2007b, 28–40.

Pawlak, Z., Skowron, A. Rough sets and Boolean reasoning, *Inf. Sci.* **177**(1), 2007c, 41–73.

Pedrycz, W. Relevancy of fuzzy models, *Inf. Sci.* **52**(3), 1990, 285–302.

Pedrycz, W. Shadowed sets: representing and processing fuzzy sets, *IEEE Trans. Syst. Man Cybern. B*, **28**, 1998, 103–109.

Pedrycz, W. Shadowed sets: bridging fuzzy and rough sets, in: S.K. Pal, A. Skowron (eds.), *Rough Fuzzy Hybridization: A New Trend in Decision-Making*, Springer-Verlag, Singapore, 1999, pp. 179–199.

Polkowski, L., Skowron A. (eds.), *Rough Sets in Knowledge Discovery*, Physica-Verlag, Heidelberg, 1998.

Sambuc, R. *Fonctions Φ-floues. Application à l'aide au diagnostic en pathologie thyroidienne*, Ph.D. Thesis, University of Marseille, 1975.

Skowron, A. Rough decision problems in information systems, *Bull. Acad. Polonaise Sci.(Tech)*, **37**, 1989, 59–66.

Wagenknecht, M., Hartmann, K. Application of fuzzy sets of type 2 to the solution of fuzzy equation systems, *Fuzzy Sets Syst.* **25**, 1988, 183–190.

Zadeh, L. A. Probability measures of fuzzy events, *J. Math. Anal. Appl.* **23**, 1968, 421–427.

Zadeh, L. A. The concept of a linguistic variable and its application to approximate reasoning-I, *Inf. Sci.* **8**, 1975, 199–249.

Zadeh, L. A. Toward a perception-based theory of probabilistic reasoning with imprecise probabilities, *J. Stat. Plan. Inf.* **102**, 2002, 233–264.

Chapter 9

Interoperability Aspects of Fuzzy Sets

In this chapter, we introduce an important notion of interoperability with the use of fuzzy sets. Here we formulate and solve various problems dealing with fuzzy sets and other environments of granular computing. We also study interoperability involving fuzzy sets and numeric data that are treated as a special (boundary) case of the environment of granular computing. The fundamental ideas of encoding and decoding (referred commonly in fuzzy sets as fuzzification and defuzzification) are investigated in detail. The linkages between fuzzy sets and sets are revealed and discussed. In this regard a construct of α-cuts plays a pivotal role. We discuss the mechanisms of communication realized both in scalar and multivariable cases. Next, we present shadowed sets and show how they could be viewed as a conceptual and computational vehicle facilitating interpretation of results conveyed by fuzzy sets.

9.1 FUZZY SET AND ITS FAMILY OF α-CUTS

Fuzzy sets offer an important conceptual and operational feature of information granules by endowing their formal models by gradual degrees of membership. We are interested in exploring relationships between fuzzy sets and sets. Although sets come with the binary (yes-no) model of membership, it would be worth investigating whether they are indeed some special cases of fuzzy sets and if so, in what sense a set could be treated as a suitable approximation of some given fuzzy set. This could shed light on some related processing aspects. To gain a detailed insight into this matter, we discuss a concept of an α-cut and a family of α-cuts and show that they relate to fuzzy sets in an intuitive and transparent way.

For a given fuzzy set A expressed in some space $\mathbf{X}$, let us define the following set (Negoita and Ralescu, 1987; Wierman, 1997)

$$A_\alpha = \{x \in \mathbf{X} | A(x) \geq \alpha\} \tag{9.1}$$

Fuzzy Systems Engineering: Toward Human-Centric Computing, by Witold Pedrycz and Fernando Gomide
Copyright © 2007 John Wiley & Sons, Inc.

where $\alpha \in [0, 1]$. Recall from Section 3.1.5 that A_α is referred to as an α-cut or α-level set. Because of the weak inequality, we sometimes refer to the result as a weak α-cut. In contrast, when considering the expression $\{x \in \mathbf{X} | A(x) > \alpha\}$ we refer to it as a strong α-cut. Its semantics of α-cuts is as follows: an α-cut of A embraces all elements of the fuzzy set whose degrees of belongingness (membership) to this fuzzy set are at least equal to α. In this sense, by selecting a sufficiently high value of α, we identify (tag) elements of A that belongs to it to a significant extent, and thus they could be sought as those highly representative of the concept conveyed by A. Those elements of $\mathbf{X}$ exhibiting lower values of the membership grades are suppressed, so this allows us to selectively focus on the elements with the highest degrees of membership while dropping the others.

For α-cuts A_α the following properties hold

(a) $A_0 = \mathbf{X}$
(b) If $\alpha \leq \beta$ then $A_\alpha \supseteq A_\beta$ (9.2)

The first property shows that if we allow for the zero value of α, then all elements of $\mathbf{X}$ are included in this α-cut (0-cut, to be more specific). The second property underlines the monotonic character of the construct: higher values of the threshold imply that more elements are accepted in the resulting α-cuts. In other words, we may say that the level sets (α-cuts) A_α form a nested family of sets indexed by some parameter (α). If we consider the limit value of α, that is, $\alpha = 1$, the corresponding α-cut is nonempty if and only if A is a normal fuzzy set.

It is also worth remembering that α-cuts, in contrast to fuzzy sets, are sets. We showed how for some given fuzzy set, its α-cut could be formed. An interesting question arises as to the construction that could be realized when moving into the opposite direction. Could we "reconstruct" a fuzzy set on the basis of an infinite family of sets? The answer to this problem is offered in what is known as the representation theorem for fuzzy sets (Negoita and Ralescu, 1987).

REPRESENTATION THEOREM
Let $\{A_\alpha\} \alpha \in [0, 1]$ be a family of sets defined in $\mathbf{X}$ such that they satisfy the following properties:

(a) $A_0 = \mathbf{X}$.
(b) If $\alpha \leq \beta$ then $A_\alpha \supseteq A_\beta$.
(c) For the sequence of threshold values $\alpha_1 \leq \alpha_2 \leq \ldots$ such that $\lim \alpha_n = \alpha$ we have $A_\alpha = \bigcap_{n=1}^{\infty} A_{\alpha_n}$.

Then there exists a unique fuzzy set B defined in $\mathbf{X}$ such that $B_\alpha = A_\alpha$ for each $\alpha \in [0, 1]$.

In other words, the representation theorem states that any fuzzy set A can be uniquely represented by an infinite family of its α-cuts. The following reconstruction expression shows how the corresponding α-cuts contribute to the formation of the corresponding fuzzy set

$$A = \bigcup_{\alpha>0} \alpha A_\alpha \qquad (9.3)$$

that is,

$$A(x) = \sup_{\alpha \in (0,1]} [\alpha A_\alpha(x)] \qquad (9.4)$$

where A_α denotes the corresponding α-cut.

The essence of this construct is that any fuzzy set can be uniquely represented by the corresponding family of sets. The illustration of the concept of the α-cut and a way in which the representation of the corresponding fuzzy set becomes realized is shown in Figure 9.1.

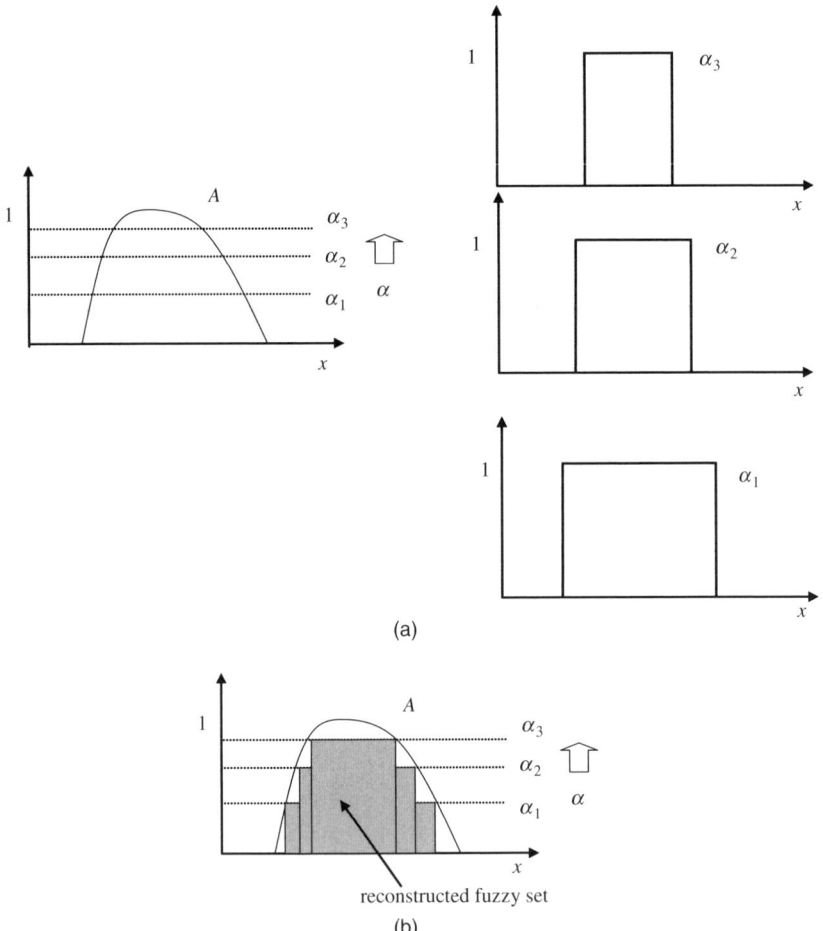

Figure 9.1 Fuzzy set A, examples of some of its α-cuts (a) and a representation of A through the corresponding family of sets (α-cuts) (b).

More descriptively, we may say that fuzzy sets can be reconstructed by a family of sets. Apparently, we need a family of sets (intervals, in particular) to capture the essence of a single fuzzy set. The reconstruction scheme illustrated in Figure 9.1 is self-explanatory in this regard. In more descriptive terms, we may look at the expression offered by (9.3) as a way of decomposing A into a series of layers (indexed sets) being calibrated by the values of the associated levels of α.

For the finite universe of discourse, dim $(\mathbf{X}) = n$, we encounter a finite number of membership grades and subsequently a finite number of α-cuts. This finite family of α-cuts is then sufficient to fully "represent" or reconstruct the original fuzzy set.

EXAMPLE 9.1

To illustrate the essence of α-cuts and the ensuing reconstruction, let us consider a fuzzy set with a finite number of membership grades, $A = [0.8, 1.0, 0.2, 0.5, 0.1, 0.0, 0.0, 0.7]$. The corresponding α-cuts of A are equal to

$$\alpha = 1.0 \quad A_{1.0} = [0, \; 1, \; 0, \; 0, \; 0, \; 0, \; 0, \; 0]$$
$$\alpha = 0.8 \quad A_{0.8} = [1, \; 1, \; 0, \; 0, \; 0, \; 0, \; 0, \; 0]$$
$$\alpha = 0.7 \quad A_{0.7} = [1, \; 1, \; 0, \; 0, \; 0, \; 0, \; 0, \; 1]$$
$$\alpha = 0.5 \quad A_{0.5} = [1, \; 1, \; 0, \; 1, \; 0, \; 0, \; 0, \; 1]$$
$$\alpha = 0.2 \quad A_{0.2} = [1, \; 1, \; 1, \; 1, \; 0, \; 0, \; 0, \; 1]$$
$$\alpha = 0.1 \quad A_{0.1} = [1, \; 1, \; 1, \; 1, \; 1, \; 0, \; 0, \; 1]$$

We clearly see the layered character of the consecutive α-cuts indexed by the sequence of the increasing values of α. Because of the finite number of membership grades, the reconstruction realized in terms of (9.4) returns the original fuzzy set (which is possible given the finite space over which the original fuzzy set has been defined)

$$A(x) = \max\left(1.0 A_{1.0}(x), 0.8 A_{0.8}(x), 0.7 A_{0.7}(x), 0.5 A_{0.5}(x), 0.2 A_{0.2}(x), 0.1 A_{0.1}(x)\right)$$

EXAMPLE 9.2

Let us consider a triangular fuzzy set defined in $\mathbf{R}$. Its reconstruction by a finite family of α-cuts brings some reconstruction error as the result computed by means of (9.3)–(9.4) gives rise to the stepwise character of the reconstructed fuzzy set with the resulting membership function showing a series of jumps see Figure 9.2. The quality of the reconstruction depends upon the number of the assumed levels of α and their distribution in the unit interval (we do not require that they need to be uniformly distributed in [0,1]).

The usefulness of α-cuts can be articulated in several ways. First, α-cuts could be treated as a mechanism supporting a concise interpretation of results conveyed by fuzzy sets. Let A be a fuzzy set of possible solutions (alternatives) to be considered in some decision-making problem. Here we typically encounter a finite universe of discourse where each element represents a certain alternative, and the fuzzy set defined here offers its degree of preference. In this case, the use of α-cuts helps flag and eliminate the weakest alternatives (those described by the lowest membership degrees) and concentrate on the subset of the most promising ones. Obviously, this

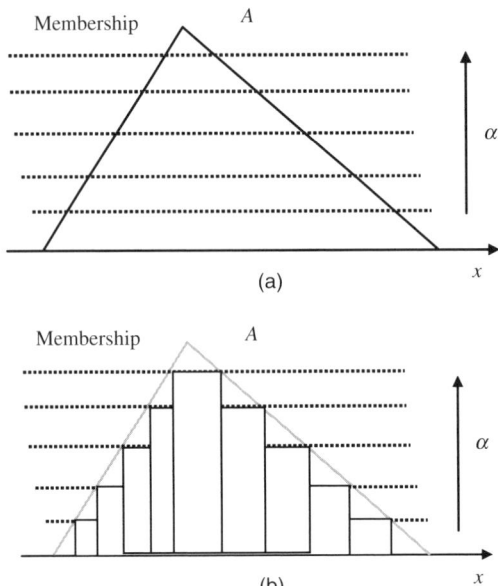

Figure 9.2 A triangular fuzzy set (number) with uniformly distributed values of α-cuts (a) and the resulting reconstruction of this fuzzy set (b). A stairwise character of the "reconstructed" fuzzy set is visible.

selection is implied by the assumed value of the threshold—here the choice of a suitable level of α still remains an open problem. Too low a value of α will leave us with a large number of alternatives, and thus not really leading us to any improvement over the use of the original fuzzy set. Choosing quite high values of α will remove most of the alternatives. However the monotonicity property (9.2) does not offer any constructive guidance as to the selection of the most suitable level of α. Another alternative worth considering is to formulate a problem of representation of the fuzzy set as a certain optimization task. We require that a set approximation B of the fuzzy set A (which is in essence some of its α-cut) is such that it minimizes the error expressed as

$$Q = \int_{x \notin A_\alpha} A(x)dx + \int_{x \in A_\alpha} (1 - A(x))dx \qquad (9.5)$$

where the value of the threshold α (level of the α-cut) is chosen in such a way that it minimizes the value of Q. In the above integral, we account for two components of the error obtained in this manner. The first one results from the reduction of the membership grades down to zero. The second component quantifies the error that appears because of the elevation of higher membership grades to 1. The optimal value of α, say α^{opt}, is the result of the optimization

$$\alpha_{opt} = \arg \min_\alpha Q(\alpha) \qquad (9.6)$$

Nevertheless, in any case of the representation (approximation) of fuzzy sets by sets some prudence is required and eventually a trial-and-error process needs to be put in place. Even if we consider the optimization problem, one should become aware of a conceptual shortcoming coming with the fact that the intermediate membership grades are transformed into 0 or 1 without any provisions that are indicative of the suppression of these values being realized. We will return to this matter a bit later when dealing with shadowed sets where it will be shown how these constructs could help alleviate this problem.

EXAMPLE 9.3

Let us consider a triangular fuzzy set with the membership function $A(x) = \max(1 - x/b, 0)$ defined for positive values of x where b is a cutoff point. The optimal α-cut (set representation) of this fuzzy set is guided by the minimization of (9.5), which in this case reads as

$$Q = \int_{b(1-\alpha)}^{b} \left(1 - \frac{x}{b}\right) dx + \int_{0}^{b(1-\alpha)} \left(1 - 1 + \frac{x}{b}\right) dx$$

After integration, we obtain Q to be a quadratic function of α

$$Q = b - b(1-\alpha) + b(1-\alpha)^2 - b/2$$

The minimum of Q, $\partial Q/\partial \alpha = 0$, yields an intuitively appealing result of the optimal value of α equal to $\frac{1}{2}$.

Second, α-cuts are useful constructs when solving various optimization problems using techniques of interval analysis. Given the long history of interval computing coming along with significant intellectual investments in effective and numerous algorithms of dealing with intervals and interval data, it could be quite tempting to consider those techniques and investigate their potential usefulness in the context of fuzzy sets (as we have already indicated, computing with fuzzy sets not only occurs at a high level of abstraction but also carries some significant price tag). Given this, detailed inquires into the use of already available techniques of interval mathematics could be helpful. α-cuts transform fuzzy sets into intervals, and thus generate the format of data required in interval analysis. The way of proceeding with fuzzy optimization would be to represent fuzzy data (fuzzy sets) by their families of α-cuts (by choosing some values of α), for each specific α-cut solves the problem using some existing techniques of interval analysis and records the corresponding solution that typically comes in the form of some numeric interval. Repeating the optimization process for consecutive (say, increasing) values of α, we end up with a family of solutions—intervals indexed the respective values of α. The organization of the computing processes is visualized in Figure 9.3.

The aggregation of the results, namely intervals being indexed by the values of α, may lead to a fuzzy set of solution being an approximation of the exact solution that could have been obtained when dealing with some technique focused on handling fuzzy sets themselves. Obviously, the result of processing using interval

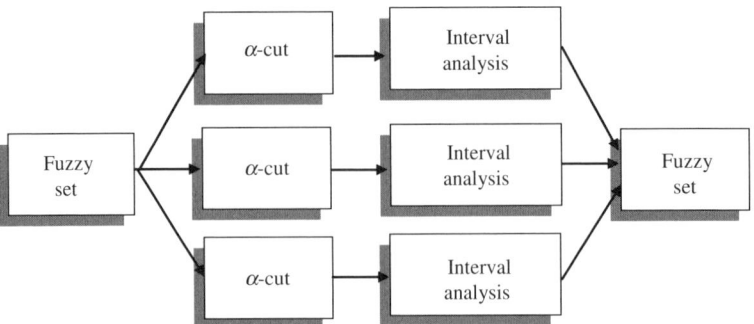

Figure 9.3 Computing and optimization of problems involving fuzzy sets that are represented as families of subproblems indexed by the values of α and handled by techniques of interval analysis.

methods realized for any specific value of α is an interval, yet it remains to be checked whether the sequence of intervals indexed by the corresponding values of α is indeed a realization of a certain fuzzy set. Here the representation theorem becomes useful.

9.2 FUZZY SETS AND THEIR INTERFACING WITH THE EXTERNAL WORLD

Fuzzy set-based models and processes interact with the external world. As we stressed several times, fuzzy sets are abstract constructs that do not exist in our external world as physical entities, but are very much reflective of a way we perceive the world, build its appropriate operational models, and process information at some desired level of abstraction (granularity) that offers the most suitable level of conceptualization. In this sense, any communication and possible interoperability with the external worlds require some well-developed mechanisms of interaction or interfacing. This brings us to the general form of the architecture outlined in Figure 9.4. This figure offers a general schematic view of the interfacing process. The inputs represent information generated by the external word. These could be sensor readings, visual information (provided by camera), audio signals, and so on. The results of decoding are the outputs in the form of some decision variables, control signals, classification results, and alike.

As visualized in Figure 9.4, two important processes come into play. The first concerns encoding—a phase in which any input information is translated into the format acceptable (comprehensible) at the level of processing of fuzzy sets. The phase of decoding focuses on processes of transforming a result available at the level of fuzzy sets into the format acceptable to the external world. In the terminology and tradition of fuzzy sets (especially in such areas as fuzzy controllers) we typically allude to the terms of fuzzification (encoding) and defuzzification (decoding). Although being very much descriptive of what is being

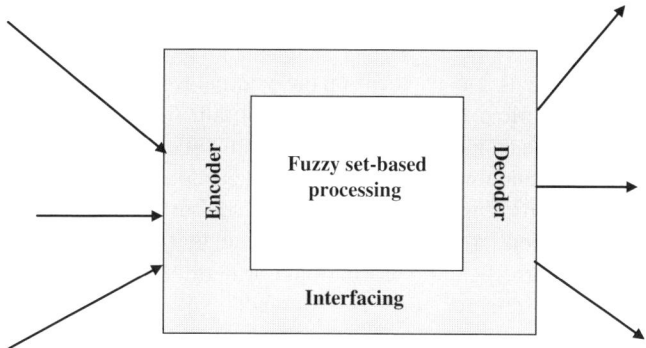

Figure 9.4 Processing at the level of fuzzy sets and mechanisms of interfacing with the external world forming an auxiliary layer of encoding and decoding processes.

realized here, these two terms do not fully capture the technical facet of the processing occurring in such framework.

Before moving on to a detailed discussion about the character of encoding and decoding, one could revisit the general scheme outlined in Figure 9.4 in the context of digital processing that profoundly dominates all our computing pursuits. Here we follow the same general scheme as outlined in Figure 9.5 (with the same interpretation of inputs and outputs of the system).

The computing core concerns digital processing. The external world is inherently analog. The interfacing processes come under the well-known terms of analog-to-digital conversion (encoding) and digital-to-analog conversion (decoding).

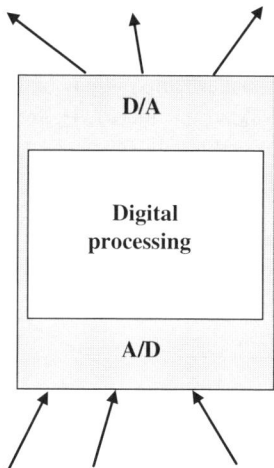

Figure 9.5 Digital processing: interfacing with analog world. Two functional blocks of conversion of analog and digital data (A/D–analog-to-digital; D/A-digital to analog) are shown.

The conceptualization of computing occurs at the level of intervals as opposed to fuzzy sets.

In the detailed discussion, we focus on the external world that generates numeric data and accepts numeric results. This somewhat limits the scope of our discussion, but concentrates on the main and the most commonly encountered developments. In computing fuzzy sets and interfacing problems, we encounter families of fuzzy sets. Typically we are not dealing with a single fuzzy set but a collection of fuzzy sets. Here we use the notation $\{A_1, A_2, \ldots, A_c\}$ to indicate that there are families of fuzzy sets involved in the corresponding processing. Furthermore such families often meet some additional requirements such as coverage of the space (meaning that the membership function of their union is nonzero over the entire space), level of overlap, and alike. Being more strict, one may also require that $A'_i s$ form a fuzzy partition. In what follows, we discuss the algorithmic aspects of the encoding and decoding schemes.

9.2.1 Encoding Mechanisms

In the encoding phase, we accept any numeric input information from the external world and transform it into a format acceptable to the processing carried out within the framework of the formalism of fuzzy sets. Consider an input interface consisting of c fuzzy sets. The numeric input $x \in \mathbf{R}$ is represented in terms of A'_is thus giving rise to a c-dimensional vector located in the unit hypercube $[0,1]^c$. These are the values of the possibility measure of x computed with respect to the corresponding fuzzy sets of the interface, $[A_1(x), A_2(x), \ldots, A_c(x)]$.

We can look at the encoding process as a certain nonlinear *normalization* that converts any real number into its normalized version, usually with the values confined to the unit interval. Here each fuzzy set contributes to this nonlinear normalization: when we compute the membership degree of A_i, we carry out some normalization to the unit interval whereas the nonlinear character is induced by the nonlinear character of the membership function.

Interestingly, a linear normalization that is governed by the expression $(x - a)/(b - a)$ with a and b denoting the lower and upper bound of the range is the one implied by the linear membership function of the form shown in Figure 9.6.

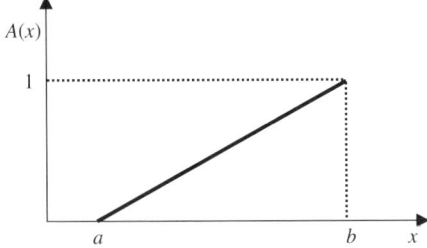

Figure 9.6 Example of a linear normalization implemented by a linearly increasing membership function.

9.2.2 Decoding Mechanisms

The decoding process is complementary to the one realized through the encoding. Given are some fuzzy sets, and our goal is to develop its numeric representative. This transformation is referred to as a decoding mechanism.

There are two main directions that are pursued:

(a) Decoding completed on a basis of a single fuzzy set; this avenue seems to be more vigorously discussed with various techniques being developed here.

(b) Decoding realized on a basis of a certain finite family of fuzzy sets and levels of their activation.

One has to be fully aware that the way the decoding could be realized is by no means unique. This is not surprising at all as membership functions are just continuous functions with infinite number of membership values or when dealing with finite spaces vectors of grades of membership, say $[0, 1]^n$. Associating a single numeric value with the vector of numbers cannot be done in a unique manner. Consider a certain fuzzy set with its membership function $B(x)$. Denote the transformation of B into some numeric representative by $\hat{x} = D(B)$. The most commonly encountered methods include the following (Runkler and Glesner, 1993; Wierman, 1997):

Mean of Maxima. We determine the arguments of **X** for which this membership function achieves its maximal values. Denote them by $\tilde{x}_1 + \tilde{x}_2 + \cdots + \tilde{x}_p$. The result of the decoding is taken as the average of these values, that is,

$$\hat{x} = \frac{\tilde{x}_1 + \tilde{x}_2 + \cdots + \tilde{x}_p}{p} \tag{9.7}$$

If we encounter only a single maximum of B, then this method points at it (which is a highly intuitive choice). There are several possible variations of this technique, such as the first of maxima and the last of the maxima in which we select a particular modal value of the membership function.

Centre of Area. We find a position of $\hat{x}$ such that it results in the equal areas below the membership function positioned on the left and on the right from this representative. In other words we have

$$\int_{-\infty}^{\hat{x}} B(x) dx = \int_{\hat{x}}^{\infty} B(x) dx \tag{9.8}$$

(we assume that the membership function can be integrated).

Centre of Gravity. Here the result of decoding is obtained as follows:

$$\hat{x} = \frac{\int_X B(x) x \, dx}{\int_X B(x) \, dx} \tag{9.9}$$

(we assume that the above integrals do exist).

One of the commonly encountered variations of these methods relies on the use of some threshold level applied to the membership grades of B. Simply, we remove from the considerations all membership grades that are below some threshold $\beta \in [0, 1]$. The rationale being that low membership grades should not significantly impact the decoding results. Given that the centre of gravity formula reads now as

$$\hat{x} = \frac{\int_{x \in \mathbf{X}: B(x) \geq \beta} B(x) x \, dx}{\int_{x \in \mathbf{X}: B(x) \geq \beta} B(x) \, dx} \tag{9.10}$$

We have used the notation showing explicitly the threshold to underline the dependence of the result on the specific value of β. Further generalizations of the decoding are also possible. For instance, one could introduce the following two-parameter decoding scheme

$$\hat{x} = \frac{\int_{x \in \mathbf{X}: B(x) \geq \beta} B^{\gamma}(x) x \, dx}{\int_{x \in \mathbf{X}: B(x) \geq \beta} B^{\gamma}(x) \, dx}$$

We emphasize the parametric character of the decoding by using the notation of the form $\hat{x}(\beta, \gamma)$. Here the positive parameter γ serves as a nonlinear transformation of the original membership function. By doing that, we introduce another level of flexibility to the decoding. Interestingly, when $\gamma > 1$, we encounter the concentration effect. The positive values of γ and lower than 1 yield a dilution of the membership function.

Some of these decoding schemes are governed by the underlying optimization problem. Consider, for instance, the centre of gravity. This scheme becomes a direct result of the minimization of the following performance index:

$$V = \int_X B(x)[x - \hat{x}]^2 \, dx \tag{9.11}$$

that is, min $V(\hat{x})$. By taking the derivative of V with respect to $\hat{x}$ one has

$$\frac{\partial V}{\partial \hat{x}} = 0 \quad \text{and} \quad 2 \int_X B(x)[x - \hat{x}] \, dx = 0$$

which leads to the decoding scheme described by (9.9).

Given the large number of decoding alternatives, it could be helpful to establish some systematic criteria that we can accept and any decoding scheme could satisfy. The example of some axiomatic frameworks has been offered by Runkler and Glesner (1993). The authors proposed a series of requirements that are organized into several groups, namely (a) basic constraints in which issues of specific forms of membership functions (constant and singletons) and monotonicity, (b) graphically motivated requirements including symmetry, translation, scaling, offset, (c) constraints motivated by the use of logic operations and linguistic modifiers (dilation and concentration), and (d) requirements specific to some application domains are discussed.

9.3 ENCODING AND DECODING AS AN OPTIMIZATION PROBLEM OF VECTOR QUANTIZATION

It is of interest and of straightforward practical relevance to consider the mechanisms of encoding and decoding as two dependent and interrelated processes. This allows us to adhere to some well-known principles and algorithms available in data compression as well as come up with some sound optimization criteria. Furthermore, by casting the discussion in this context, we will be able to contrast the features of decoding implemented in fuzzy sets with the properties of decoding realized in terms of sets. Following the general scheme (Fig. 9.7) here we are concerned both with a scalar and multivariable (vector) case.

The key objective is to simultaneously optimize the processes of encoding and decoding, in particular their underlying parameters so that that the result of decoding is made as close as possible to the original numeric entity that has been originally used at the encoding end of the process. In the ideal situation, the encoding–decoding scheme should result in the relationship $\mathbf{x} = \hat{\mathbf{x}}$ to be satisfied for any $\mathbf{x}$ in a multi-dimensional input space forming a subset of $\mathbf{R}^n$. If this equality does not hold, we encounter a decoding error where $\hat{\mathbf{x}}$ becomes a distorted version of $\mathbf{x}$.

It is educational to start with a scalar (one-dimensional) case of quantization as it sheds light on the role of fuzzy sets in the overall optimization process. This is also quite enlightening given the fact that the scalar case with the use of sets is extremely well discussed in the literature, so a thorough comparative analysis could be offered.

9.3.1 Fuzzy Scalar Quantization

The results in the case of interval-driven quantization (in which the codebook is taken as a collection of intervals) is well reported in the literature, cf. Gersho and Gray (1992), Gray (1984), Linde et al. (1988), Lloyd (1982), Patane and Russo (2001), Yair et al. (1992), Campobello et al., 2005; Chang et al., 2005; Cho et al., 1994; Kämpke, 2003; Lin and Yu, 2003; Kim et al., 1998; Laskaris and Fotopoulos, 2004; Liu and Lin, 2002; Moller et al., 1998; Pan et al., 2005; Pregenzer et al., 1996; Sinkkonen and Kaski, 2002; Tsekouras, 2005; Yen and Young, 2004 including a quantization (decoding) error for a uniform distribution of intervals and a distribution of intervals given that the underlying probability density becomes available. Although the

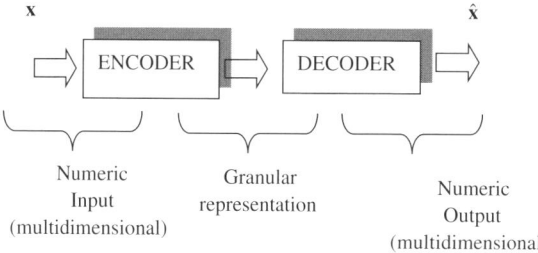

Figure 9.7 Encoding and decoding of numeric data through information granules—fuzzy sets.

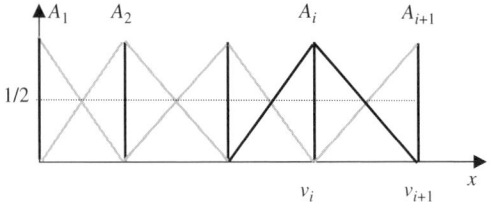

Figure 9.8 An example of the codebook composed of triangular fuzzy sets with an overlap of $\frac{1}{2}$ between each two neighboring elements of the codebook.

quantization error could be minimized, the method is in essence a failure (coming with a nonzero quantization error). In contrast, a codebook formed by fuzzy sets can lead to a zero error of the quantization. In this context, we show a surprisingly simple yet powerful result whose essence could be summarized as follows: a codebook formed of fuzzy triangular fuzzy sets (fuzzy numbers) with an overlap of ½ between two neighboring elements (Fig. 9.8) leads to the lossless compression scheme. The essence of the scheme can be formulated in the form of the following proposition.

PROPOSITION

Let us assume the following:

(a) the fuzzy sets of the codebook $\{A_i\}, i = 1, 2, \ldots, c$ form a fuzzy partition, $\sum_{i=1}^{c} A_i(x) = 1$, and for each x in **X** at least one element of the codebook is "activated", that is, $A_i(x) > 0$

(b) for each x only two neighboring elements of the codebook are "activated", that is, $A_1(x) = 0, \ldots, A_{i-1}(x) = 0, A_i(x) > 0, A_{i+1}(x) > 0, A_{i+2}(x) = \ldots = A_c(x) = 0$

(c) the decoding is realized as a weighted sum of the activation levels and the prototypes of the fuzzy sets v_i, namely $\hat{x} = \sum_{i=1}^{c} A_i(x) v_i$

Then the elements of the codebook described by piecewise linear membership functions

$$A_i(x) = \begin{cases} \dfrac{x - v_{i-1}}{v_i - v_{i-1}}, \text{if } x \in [v_{i-1}, v_i] \\ \dfrac{x - v_{i+1}}{v_i - v_{i+1}}, \text{if } x \in [v_i, v_{i+1}] \end{cases} \qquad (9.12)$$

lead to the zero decoding error (lossless compression) meaning that $\hat{x} = x$.

PROOF

Consider any element x lying in the interval $[v_i, v_{i+1}]$. In virtue of (a) and (b) we can rewrite the decoding formula as follows:

$$\hat{x} = A_i(x) v_i + (1 - A_i(x)) v_{i+1}$$

9.3 Encoding and Decoding as an Optimization Problem of Vector Quantization

We request a lossless compression meaning that $\hat{x} = x$. In other words,

$$x = A_i(x)v_i + (1 - A_i(x))v_{i+1} \qquad (9.13)$$

Then let us write down $A_i(x)$ (which describes the right-hand side of the membership function of A_i spread in-between v_i and v_{i+1} by rearranging the terms in (9.13). This leads to the expression

$$A_i(x)(v_i - v_{i+1}) = x - v_{i+1}$$

and

$$A_i(x) = \frac{x - v_{i+1}}{v_i - v_{i+1}} \qquad (9.14)$$

that reveals the piecewise linear character of the membership function of A_i. In the same fashion, we can deal with the left-hand side of the membership function of A_i by considering the interval $[v_{i-1}, v_i]$. In this case we can demonstrate that for all x in $[v_{i-1}, v_i]$ the membership comes in the form $A_i(x) = \frac{x - v_{i-1}}{v_i - v_{i-1}}$ that completes the proof.

Interestingly enough, triangular fuzzy sets have been commonly used in the development of various fuzzy set constructs as discussed (models, controllers, classifiers, etc.), yet the lossless character of such codebooks is not generally known, perhaps with very few exceptions, cf. Pedrycz (1994, 2001). The above proposition does not make any assumption on the distribution of fuzzy sets, just only imposing a requirement of the 0.5 overlap that need to be met between the elements of the codebook. The results reported in Pedrycz (2001) underline that although fuzzy sets and probability are two orthogonal concepts, one could expect that each fuzzy set comes with some experimental evidence meaning that its probability is sufficiently high. This rationale was introduced and discussed in Pedrycz (2001), and in the sequel it has resulted in an algorithmic realization of the so-called equalization of fuzzy sets of the codebook. The individual fuzzy sets are distributed across the space in such a way that their probability $E(A_i) = \int_X A(x)p(x)dx$ is made equal to $1/c$, where $p(x)$ denotes a probability density function (*pdf*) of the underlying experimental evidence. If the *pdf* is quite low in some regions of universe of discourse, to compensate for that and produce the required value of the integral, the corresponding fuzzy set has to be "spread" more across the space. It is worth noting that if the data are governed by the uniform pfd, then the fuzzy sets of the codebook are uniformly distributed across the universe of discourse.

Unfortunately the lossless property of this fuzzy quantization does not hold to the multivariable case. However, the successful performance of the codebook of fuzzy sets has prompted our interest into the investigations of the fuzzy vector quantization.

As already emphasized (see Chapter 4), fuzzy clusters are multidimensional information granules that reflect upon the experimental data and lead to their conceptual compression. In general, we could regard the encoding–decoding problem as an optimization of the clustering mechanisms guided by a minimization of some assumed performance index that quantifies a departure (distance) of $\hat{\mathbf{x}}$ from the original numeric entry processed by the encoder ($\mathbf{x}$), that is $\| \mathbf{x} - \hat{\mathbf{x}} \|$ with $\| \cdot \|$ being a distance function. The formulation of the problem captures the essence of the numeric–granular–numeric transformation as being encountered along the overall encoding–decoding processing path.

The underlying architectures of these two phases of encoding and decoding have to be specified before moving forward with any further detailed analysis and possible design guidelines. As already stressed, to focus now on the discussion, we confine ourselves to the standard version of the FCM.

9.3.2 Forming the Mechanisms of the Fuzzy Quantization: Beyond a Winner-Takes-All Scheme

The crux of our considerations dwells upon the following conjecture. Although the set-based codebook inherently forms a decoding scheme that decodes the result using a single element of the codebook (which in essence becomes a manifestation of the well-known concept of the winner-takes-all strategy), here we are interested in the exploitation of the nonzero degrees of membership of several elements (fuzzy sets) of the codebook while representing the input datum.

In other words, rather than using a single prototype as a sole representative of a collection of neighboring data, our opinion is that by involving several prototypes at different degrees of activation (weights) could be beneficial to the ensuing decoding. In the spirit of abandoning the winner-takes-all principle, let us start with a collection of weights $u_i(\mathbf{x}), i = 1, 2, \ldots, c$. Adhering to the vector notation, we denote $\mathbf{u}(\mathbf{x}) = [u_1(\mathbf{x}), u_2(\mathbf{x}), \ldots, u_c(\mathbf{x})]$ to be a vector of membership grades. These weights (membership degrees) express an extent to which the corresponding datum $\mathbf{x}$ is encoded in the language of the given prototypes (elements of the codebook) should be involved in the decoding (decompression) scheme. We require that these membership degrees are positioned in-between 0 and 1 and sum up to 1. Thus at the encoding end of the overall scheme, we represent each vector $\mathbf{x}$ by $c - 1$ values of $u_i(\mathbf{x})$. The decoding is then based upon a suitable aggregation of the degrees of membership and the prototypes. Denote this operation by $\hat{\mathbf{x}} = D(\mathbf{u}(\mathbf{x}), \mathbf{v}_1, \mathbf{v}_2, \ldots, \mathbf{v}_c)$ where $\hat{\mathbf{x}}$ denotes the result of the decoding. On the contrary, the formation of the membership degrees (encoding) can be succinctly described in the form of the encoding mapping $E(\mathbf{x}, \mathbf{v}_1, \mathbf{v}_2, \ldots, \mathbf{v}_c)$.

The overall development process is split into two fundamental phases, namely:

(a) Local activities confined to the individual datum $\mathbf{x}$ that involve (i) encoding each $\mathbf{x}_k$ leading to the vector of the membership degrees (u), (ii) decoding being realized in terms of the membership degrees.

(b) Global activities concerning the formation of the codebook in which case we take all data into consideration, thus bringing the design to the global level.

The overall process described above will be referred to as fuzzy vector quantization (FVQ, for brief) (Karayiannis and Pai, 1995; Wu and Yang, 2003). Let us note that the "winner-takes-all" strategy is a cornerstone of the vector quantization (VQ). To contrast the underlying computing in the VQ and FVQ schemes, we portray the essential computational facets in Figure 9.9. As becomes apparent in Figure 9.9,

9.3 Encoding and Decoding as an Optimization Problem of Vector Quantization

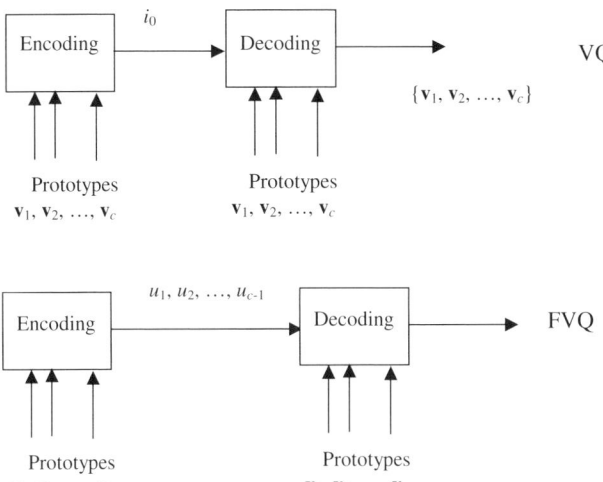

Figure 9.9 VQ and FVQ—a view contrasting the essence of the process and showing the key elements of the ensuing encoding and decoding.

the decoding relies only on a single prototype whose index (i_0) becomes available at the decoding scheme.

In what follows, we discuss the algorithmic details of the optimization of the FVQ. They lead to some constrained and constraint-free optimization.

9.3.3 Coding and Decoding with the Use of Fuzzy Codebooks

We focus on the detailed formulas realizing the coding and decoding schemes. Let us emphasize that both coding and decoding emerge as solutions to the well-defined optimization problems. At this point we assume that the codebook—denoted here as $\{v_1, v_2, \ldots, v_c\}$ has been already formed with the use of the FCM clustering (Bezdek, 1981).

9.3.3.1 Encoding Mechanism

A way of encoding (representing) the original datum **x** is done through the collection of the degrees of activation of the elements of the codebook. We require that the membership degrees are confined to the unit interval and sum up to 1. Let us determine their values by minimizing the following performance index:

$$Q_1(\mathbf{x}) = \sum_{i=1}^{c} u_i^m \|\mathbf{x} - \mathbf{v}_i\|^2 \qquad (9.15)$$

subject to the following constraints already stated above, that is,

$$u_i(\mathbf{x}) \in [0,1], \sum_{i=1}^{c} u_i(x) = 1 \qquad (9.16)$$

The distance function is denoted by $\|\cdot\|$. The fuzzification coefficient $(m, m > 1)$, standing in the above expression is used to adjust the level of contribution of the impact of the prototypes on the result of the encoding. The collection of c weights $\{u_i(\mathbf{x})\}$ is then used to encode the input datum $\mathbf{x}$. These membership degrees along with the corresponding prototypes are afterward used in the decoding scheme.

The minimization of (9.15)-(9.16) follows the standard way of transforming the problem to unconstrained optimization using Lagrangre multipliers. Once solved, the resulting weights (membership degrees) read as

$$u_i(\mathbf{x}) = \frac{1}{\sum \left(\frac{\|\mathbf{x} - \mathbf{v}_i\|}{\|\mathbf{x} - \mathbf{v}_j\|}\right)^{\frac{2}{m-1}}} \qquad (9.17)$$

9.3.3.2 The Decoding Mechanism of Fuzzy Quantization

The decoding is concerned with the "reconstruction" of $\mathbf{x}$, denoted here by $\hat{\mathbf{x}}$. It is based on some aggregation of the elements of the codebook and the associated membership grades $\mathbf{u}(\mathbf{x})$. The proposed way of forming $\hat{\mathbf{x}}$ is through the minimization of the following expression:

$$Q_2(\hat{\mathbf{x}}) = \sum_{i=1}^{c} u_i^m \|\hat{\mathbf{x}} - \mathbf{v}_i\|^2 \qquad (9.18)$$

Given the Euclidean distance, the problem of unconstrained optimization leads to a straightforward solution expressed as a combination of the prototypes weighted by the membership degrees that

$$\hat{\mathbf{x}} = \frac{\sum_{i=1}^{c} u_i^m \mathbf{v}_i}{\sum_{i=1}^{c} u_i^m} \qquad (9.19)$$

Note that again all prototypes contribute to the decoding process that stands in a sharp contrast with the winner-takes-all decoding scheme encountered in the VQ where $\hat{\mathbf{x}} = \mathbf{v}_l$ where l stands for the index of the prototype identified during the decoding phase. The quality of the reconstruction depends on a number of essential parameters, including the size of the codebook as well as the value of the fuzzification coefficient. The impact of this particular parameters is illustrated in Figure 9.10 in which we quantify the distribution of the decoding error by showing values of the Hamming distance between $\mathbf{x}$ and $\hat{\mathbf{x}}$. Evidently, what could have been anticipated, all $\mathbf{x}_s$ situated in a close vicinity of the prototypes exhibit low values of the decoding error. Low values of m that position FVQ close to its set-based counterpart show significant jumps of the error resulting from the abrupt switching in-between the prototypes used in the decoding process. For the higher value of the fuzzification

9.3 Encoding and Decoding as an Optimization Problem of Vector Quantization

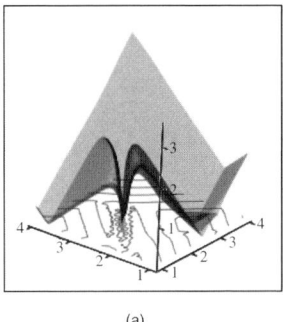
(a)
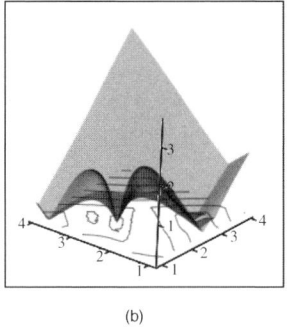
(b)
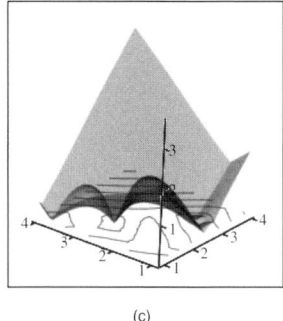
(c)

Figure 9.10 Plots of the decoding error expressed in the form of the Hamming distance between **x** and $\hat{\mathbf{x}}$ (3D and contour plots) for selected values of the fuzzification coefficient: (a) $m = 1.2$ (b) $m = 2.0$, (c) $m = 3.5$.

coefficient, say $m = 2.0$, the jumps are significantly reduced. Further reduction is achieved with the larger values of m; see Figure 9.10 (c).

In the regions that are quite remote from the location of the prototypes, the error increases quite quickly. For instance, this effect is quite visible for the values of **x** close to (4.0 4.0).

In the following illustrative example, we quantify the decoding error in the one-dimensional case where the data come as a mixture of two Gaussian *pdf*'s of the form $p(x) = 0.3N(2, \sigma) + 0.7N(4, \sigma)$ with some standard deviation σ. The prototypes are equal to the mean values of the two components of the mixture of the *pdf* densities, which are 2 and 4. The error of the decoding is then expressed as the integral $\varepsilon(x) = \int_2^4 (x - \hat{x})^2 p(x) dx$. We consider several scenarios by varying the values of the fuzzification coefficient and changing the standard deviation of the data. For the VQ, the decoding error shown as a function of σ is illustrated in Figure 9.11; we consider this to be a reference point. The FVQ comes with the lower values of the error as illustrated in Figure 9.12.

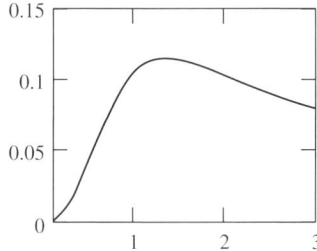

Figure 9.11 Decoding error of the VQ versus σ.

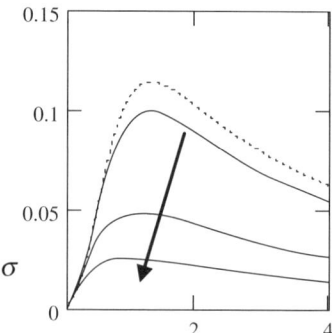

Figure 9.12 Decoding error of the FVQ versus σ for selected values (1.1, 2.0, and 5.0) of the fuzzification coefficient; results for the VQ (dotted line) are also shown.

9.4 DECODING OF A FUZZY SET THROUGH A FAMILY OF FUZZY SETS

Now we further expand the investigations along the line of encoding and decoding by considering here the granular data represented in the form of fuzzy sets. The use of fuzzy sets in place of numeric data used so far makes a substantial difference in comparison with the encoding and decoding schemes discussed so far (Hirota and Pedrycz, 1999; Nobuhara et al., 2000). The representation of a certain fuzzy sets in terms of some fixed and predefined family of fuzzy sets becomes central to all further encoding and decoding pursuits.

9.4.1 Possibility and Necessity Measures in the Encoding of Fuzzy Data

As before, we consider here a finite codebook consisting of c fuzzy sets or fuzzy relations (in multivariable case) $A_1, A_2, \ldots, A_c$. Now the input datum is a fuzzy set (or fuzzy relation) denoted here by X. As the underlying constructs are described by membership functions, capturing the level of matching or coincidence between X and A_i cannot be realized in a unique manner. A viable alternative is to consider possibility and necessity measures. We will be concerned with the probing X by the corresponding elements—fuzzy sets of the codebook.

Let us briefly recall that the possibility measure, $\text{Poss}(A_i, X)$, describes a level of overlap between the two fuzzy sets. The necessity measure, $\text{Nec}(A_i, X)$ captures a level of inclusion of A_i in X. The corresponding formulas read as

$$\text{Poss}(A_i, X) = \sup_{x \in \mathbf{X}}[X(x) t A_i(x)] \tag{9.20}$$
$$\text{Nec}(A_i, X) = \inf_{x \in \mathbf{X}}[(1 - A_i(x)) s X(x)] \tag{9.21}$$

The plot visualizing the computations of the possibility and necessity measures is shown in Figure 9.13. Let us recall that the possibility measure reflects the intersection

9.4 Decoding of a Fuzzy Set through a Family of Fuzzy Sets

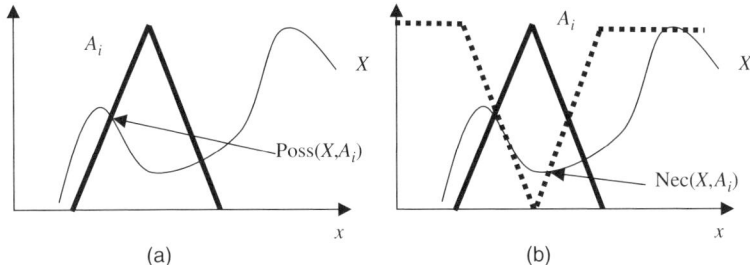

Figure 9.13 Computing of the possibility (a) and necessity (b) measures treated as a vehicle of encoding X with the aid of A_i.

between A_i and X and then takes an optimistic aggregation of the intersection operation by picking up the highest values among the intersection grades of X and A_i that are taken over all elements of the universe of discourse **X**. The necessity measure expresses a pessimistic degree of inclusion of A_i in X. The elements of the codebook operate as a collection of "probes" that probe the input X and build its some internal representation by the

EXAMPLE 9.4

Given are two fuzzy sets $X = [0.0, \ 0.2, \ 0.8, \ 1.0, \ 0.9, \ 0.5, \ 0.1, \ 0.0]$ and

$A_i = [0.6, \ 0.5, \ 0.4, \ 0.5, \ 0.6, \ 0.9, \ 1.0, \ 1.0]$. The values of the possibility and necessity measures become then equal to

$$\text{Poss}(A_i, X) = \max(0.0, 0.2, 0.4, 0.5, 0.6, 0.5, 0.1, 0.0) = 0.6$$
$$\text{Nec}(A_i, X) = \min(0.4, 0.5, 0.8, 1.0, 0.9, 0.5, 0.1, 0.0) = 0.0$$

Now let us consider A_i to be in the following form: $A_i = [0\ 0\ 0\ 0\ 1\ 1\ 1\ 0]$ so it is a set-based probe of X. The computed values of the possibility and necessity measures are equal to $\text{Poss}(A_i, X) = 0.9$, $\text{Nec}(A_i, X) = 0.1$. If A_i is made even more specific and reduced to a single numeric value (singleton) with the membership function of the form $A_i = [0\ 1\ 0\ldots 0\ 0]$, then we obtain $\text{Poss}(A_i, X) = \text{Nec}(A_i, X) = 0.2$.

In general, it becomes apparent that the possibility and necessity measures may not be equal (and usually are not; the differences between the values are implied by the granularity of A_i). If the granularity of the probe increases, the values of these two measures are getting closer to each other. In the limit (as also illustrated in the example shown above) when the probe A_i is a singleton, we end up with the same values of the possibility and necessity.

Considering now the finite family of fuzzy sets of the codebook, we compute the possibility and necessity measures with respect to X for each of them, thus ending up with its $2c$-tuple representation

$$\lambda_i = \text{Poss}(A_i, X), \mu_i = \text{Nec}(A_i, X) \tag{9.22}$$

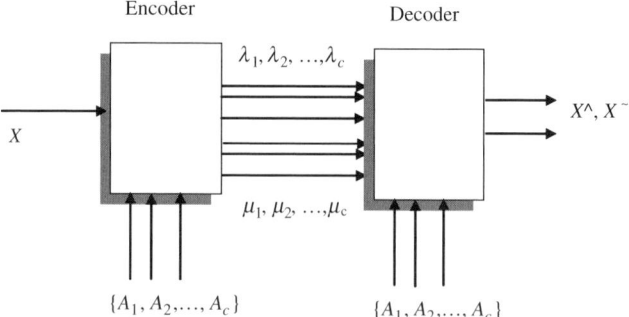

Figure 9.14 A schematic view of the encoding–decoding processes in presence of fuzzy input datum X. The results of decoding come in the form of the interval-valued fuzzy set. For details refer to the text.

which is a manifestation of X in the "language" of the codebook. As a matter of fact, (9.22) is the result of the encoding of X. In parallel to the scheme shown in Figure 9.9, the functional presentation of the encoding–decoding architecture involving granular inputs is presented in Figure 9.14. Note the format of the encoding information to be used by the decoder.

9.4.2 The Design of the Decoder of Fuzzy Data

The decoding of X is inherently related to $A_i's$ and might be regarded as an outcome of reconstruction guided by the values of $(\lambda_i, \mu_i), i = 1, 2, \ldots, c$. From the conceptual standpoint, we can view the decoding as a sort of inverse problem to the encoding task. As X is a fuzzy set itself, the decoder has to produce a fuzzy set, ideally equal to X. This however, does require more attention and a careful analysis of the problem. We proceed in a stepwise manner and start with a single element of the codebook (to come up with a concise notation, let us drop the index of the element of the codebook and simply denote it by A) for which the values of encoding of λ and μ are known. The problem reads as follows: given A, λ, and μ, decode X. One can look at (9.20) and (9.21) and treat them as equations with respect to X. There is no unique solution to neither the first nor the second one. There are, however, extreme solutions that are unique to the problem in hand. For the possibility part of the encoder/decoder, we encounter the maximal solution. Its construction is supported by the theory of fuzzy relational equations (as a matter of fact, (9.20) as a sup-t composition of X and A). In light of the fundamental results available in the theory, the membership function of this maximal fuzzy set (mapping) induced by the A reads as

$$\hat{X}(x) = A(x)\varphi\lambda = \begin{cases} 1, & \text{if } A(x) \leq \lambda \\ \lambda, & \text{otherwise} \end{cases} \quad (9.23)$$

The above formula applies for t-norm realized as a minimum operator. In general, (9.23) reads in the form

$$\hat{X}(x) = A(x)\varphi\lambda = \sup[c \in [0, 1] | ct\, A(x) \leq \lambda] \quad (9.24)$$

9.4 Decoding of a Fuzzy Set through a Family of Fuzzy Sets

When using the entire family of $A'_i s$ (that leads to the intersection of $\hat{X}'_i s$) we obtain

$$\hat{X} = \bigcap_{i=1}^{c} \hat{X}_i \qquad (9.25)$$

From the theoretical point of view that arise in the setting of fuzzy relational equations, we note that we are dealing here with a system of equations $\lambda_i = \text{Poss}(X, A_i), i = 1, 2, \ldots, c$ to be solved with respect to X for λ_i and A_i provided.

The theory of fuzzy relational equations plays the same dominant role in the case of the necessity computations. It is worth noting that we are faced with the so-called dual fuzzy relational equations. Here the minimal solution to (9.21) for A and μ_i given reads in the form

$$\tilde{X}(x) = (1 - A(x))\beta\mu = \begin{cases} \mu, & \text{if } 1 - A(x) < \mu \\ 0, & \text{otherwise} \end{cases} \qquad (9.26)$$

Again the above formula applies to the maximum realization of the s-norm. The general formula takes on the form

$$\tilde{X}(x) = (1 - A_i(x))\beta\mu = \inf\{c \in [0,1] | cs(1 - A(x)) \geq \mu\} \qquad (9.27)$$

Because of the minimal solution, the collection of the probes A_i leads us to the partial results that are afterward combined through the union of a partial solution

$$\tilde{X} = \bigcup_{i=1}^{c} \tilde{X}_i \qquad (9.28)$$

In conclusion, (9.25) and (9.28) become the granular representation of the input datum (X) arising in the context of the collection of the elements of the codebook. The following containment relationship holds

$$\tilde{X} \subseteq X \subseteq \hat{X} \qquad (9.29)$$

As a simple yet highly illustrative example, consider a collection of sets (intervals) regarded as the elements of the codebook; refer to Figure 9.15. A single information granule produces a result shown in Figure 9.16(a), which in fact constructs

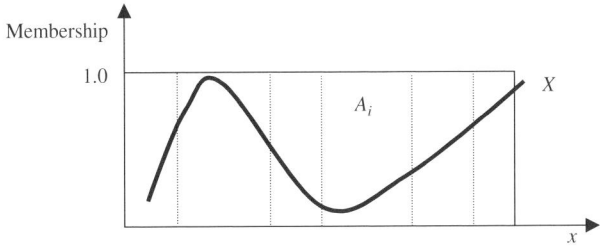

Figure 9.15 Input datum X with a collection of superimposed sets (intervals) A_i.

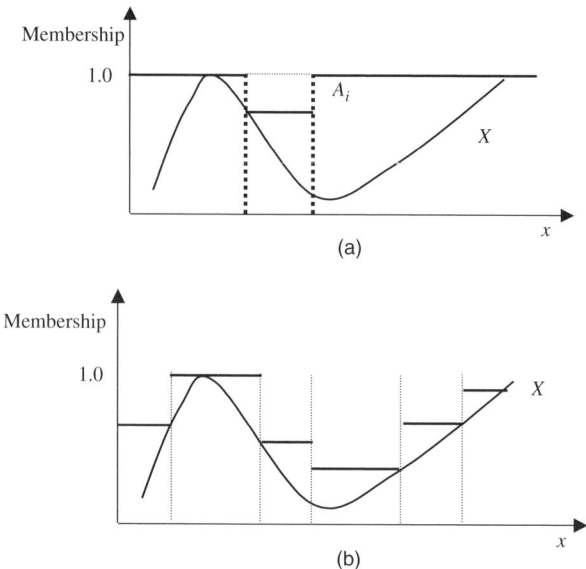

Figure 9.16 Computing of the upper bound of X (solid staircase line) with the use of a single set (A) (a) and the family of sets $\{A_i\}$ (b).

membership values equal to 1 over any argument not belonging to A_i. The aggregation of all partial results gives more specific outcome; see again Figure 9.16(b).

Interestingly, the reconstructed (decoded) fuzzy set exhibits a stepwise type of membership function where the height of the individual jumps and their distribution across the space depends on the distribution of the elements of the codebook $A'_i s$. The same effect that concerns the lower bound of X is present in Figure 9.17. When combined together, the result is a type-2 fuzzy set (or an interval-valued fuzzy set, to be more precise), Figure 9.18. It is worth noting that by changing the position of the cutoff points (intervals), we end up with different granular mappings. Eventually the mapping can be subject to some optimization in which we develop the collection of $A'_i s$ in such a way that the granular mapping is as specific as possible (so that the bounds are made tight).

9.5 TAXONOMY OF DATA IN STRUCTURE DESCRIPTION WITH SHADOWED SETS

The three-valued evaluation offered by shadowed sets (Chapter 8) is helpful in the interpretation of results of fuzzy clustering. In what follows, we will be referring to A_i as a shadowed set (which in essence does not lead to any misunderstanding as it has been induced by the corresponding fuzzy sets). We introduce the following sets of patterns based on their allocation to the components of the shadowed sets of the clusters (Pedrycz, 2005).

9.5 Taxonomy of Data in Structure Description with Shadowed Sets

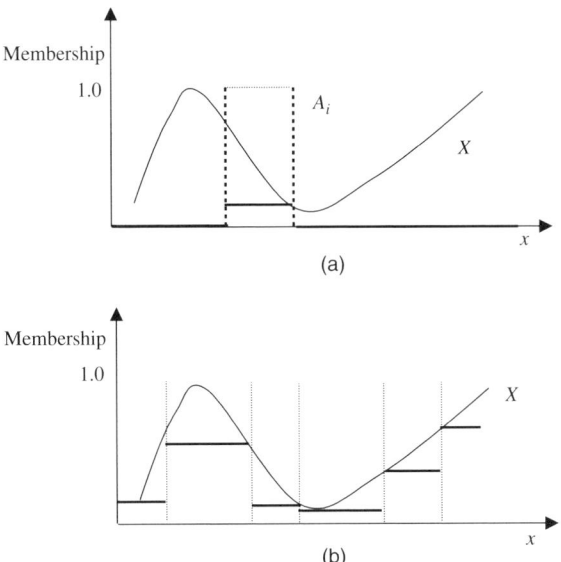

Figure 9.17 The mechanism of decoding realized with the use of the necessity measure.

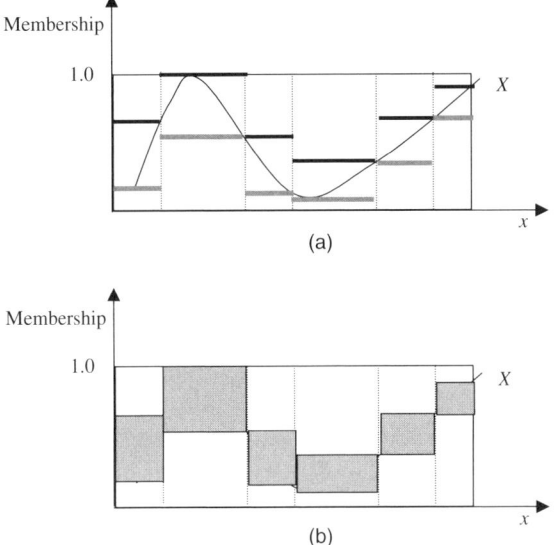

Figure 9.18 Decoding of X: upper and lower bounds of the decoding (a) and interval-valued realization of the decoding presented in the output space (b).

9.5.1 Core Data Structure

Those are the patterns that belong to a core of at least one or more shadowed sets

$$\text{Core data structure} = \{\mathbf{x} | \exists_i \ \mathbf{x} \in \text{Core}(A_i)\} \quad (9.30)$$

The core is composed of the data points that form the backbone of the structure revealed through the clustering mechanisms. They clearly belong to a single cluster or could be shared between several clusters (in such case they overlap).

9.5.2 Shadowed Data Structure

This structure is formed by patterns that do not belong to the core of any of the shadowed sets, but fall within the shadow of one or more shadowed sets. Formally, we write this down in the following form:

$$\text{Shadowed data structure} = \{\mathbf{x} | \exists_i \ \mathbf{x} \in \text{Shadow}(A_i) \text{ and } \forall_i \mathbf{x} \notin \text{Core}(A_i)\} \quad (9.31)$$

Noticeably, this structure embraces patterns that raise some hesitation as to their possible interpretation. The pattern falling within this region requires more attention as to its possible membership and final quantification

9.5.3 Uncertain Data Structure

The patterns belonging to this structure are those that are left out from all shadows meaning that

$$\text{Uncertain data structure} = \{\mathbf{x} | \forall_i \ \mathbf{x} \notin \text{Shadow}(A_i) \text{ and } \forall_i \ \mathbf{x} \notin \text{Core}(A_i)\} \quad (9.32)$$

This structure consists of patterns that could be practically regarded as peripheral to the clusters revealed in the data set. It is likely that most of them could be the outliers or highly atypical data points quite distinct from the primary structure (which is the core and shadowed structure) that require more attention. In this sense we have formed a mechanism attracting attention to those patterns that may potentially trigger some action.

The illustration of these three concepts describing the data structure is included in Figure 9.19. It is worth noting that such data categorization forms an obvious hierarchy of structures revealed by the clustering procedure. We start with the core data structure (which is the most central to the structure description), move down to the shadowed structure, and finally flag the uncertain structure.

We illustrate the performance of the development of the shadowed sets by presenting a two-dimensional synthetic data set (which could be supported by a detailed visualization). The objective function-based clustering used here is the well-known FCM we studied earlier. The data set consisting of 15 patterns is presented in Figure 9.20. There is some structure in this dataset, yet the number of well-delineated clusters is not evident.

The experiment was carried out for $m = 2$ while varying the number of clusters in-between 2 and 4. The successive results are shown in terms of the prototypes,

9.5 Taxonomy of Data in Structure Description with Shadowed Sets

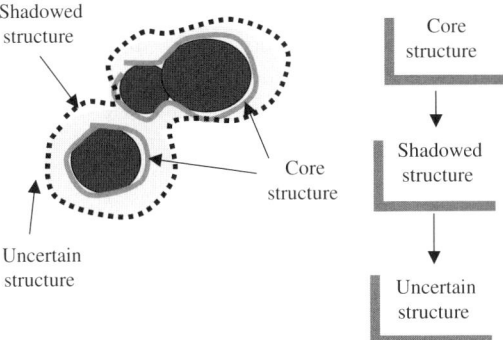

Figure 9.19 Interpretation of data structure revealed in the clustering process along with the hierarchy of concepts describing the data.

ranges of membership grades forming the shadows, and the number of patterns falling under various categories of data structures being formed.

$c = 2$

$\mathbf{v}_1 = [1.32, 1.79], \mathbf{v}_2 = [2.71, 4.19]$

Shadows [0.42, 0.61] [0.38, 0.57]

Here the ranges of the membership grades of both clusters are quite similar and all elements fall within the core of the clusters.

$c = 3$

$\mathbf{v}_1 = [1.69, 3.83], \mathbf{v}_2 = [3.56, 4.21], \mathbf{v}_3 = [1.31, 1.54]$

shadows [0.32, 0.45] [0.25, 0.74] [0.29, 0.68]

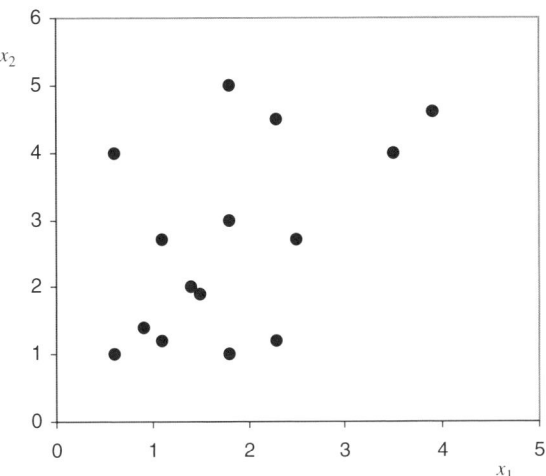

Figure 9.20 Two-dimensional synthetic data set.

246 Chapter 9 Interoperability Aspects of Fuzzy Sets

As more detailed structure has been revealed, the clusters are more diversified with respect to the membership ranges forming the shadows. One pattern has been categorized as the element of the shadowed structure, which is reflective of the increasing specificity of the description of the data. Here, the plots of the Cores and Shadows of the shadowed sets for $c = 3$ are included in the series of figures (Fig. 9.21).

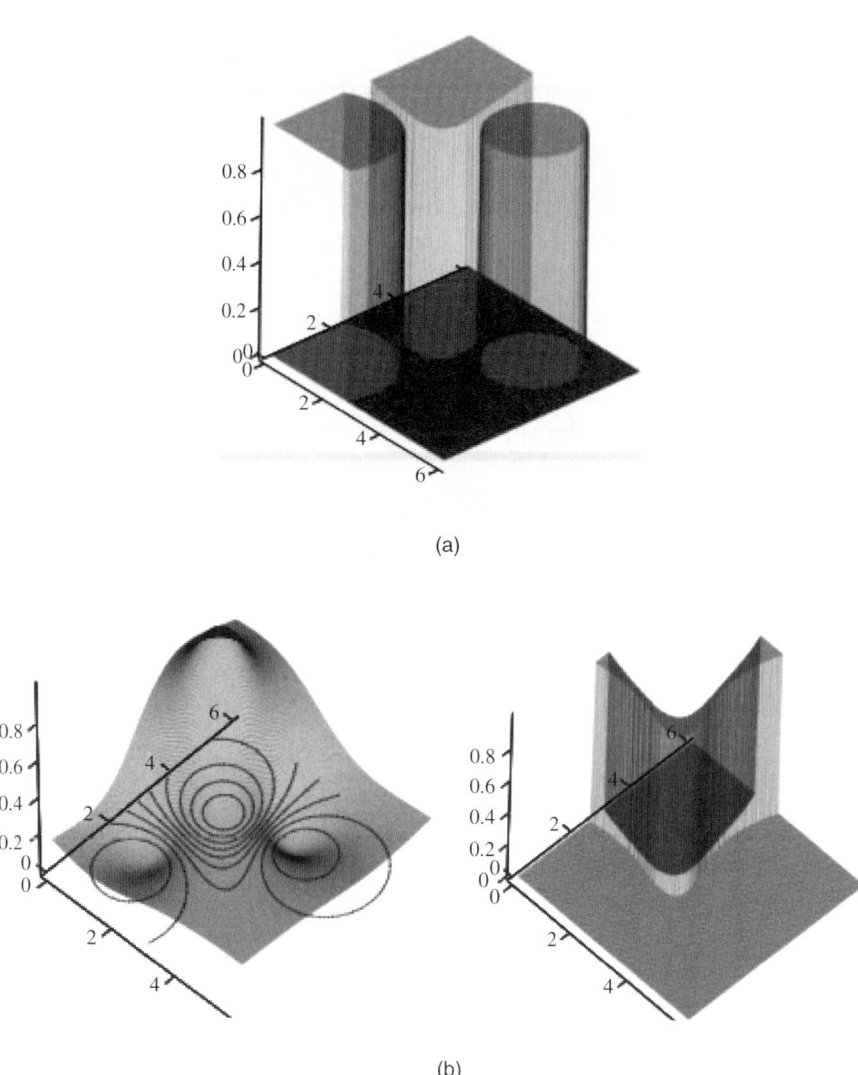

Figure 9.21 Structure representation of data for $c = 3$ clusters: (a) cores of the clusters and original fuzzy clusters and associated shadows of the induced shadowed sets (b)–(d).

9.5 Taxonomy of Data in Structure Description with Shadowed Sets

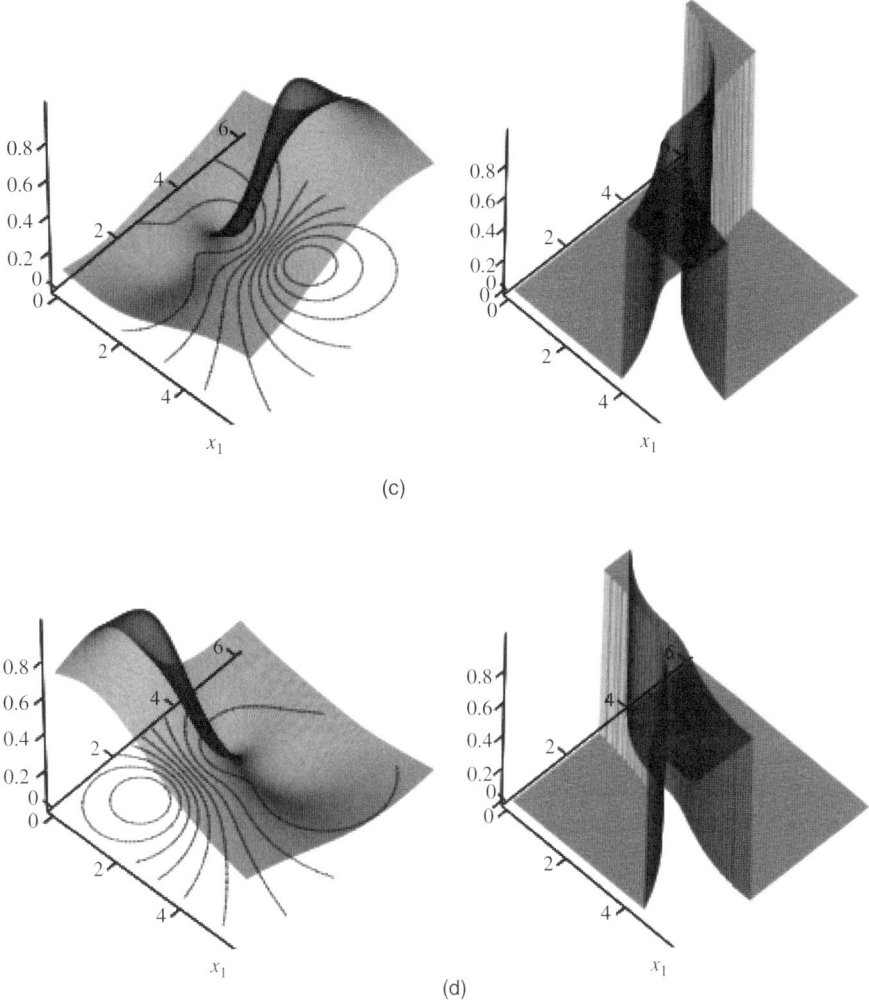

Figure 9.21 (*Continued*)

It helps visualize the location of the shadows vis-à-vis the experimental data and raise awareness as to their role.

$c = 4$

$\mathbf{v}_1 = [1.93, \ 4.61], \mathbf{v}_2 = [3.65, \ 4.25], \mathbf{v}_3 = [1.10, \ 1.29], \mathbf{v}_4 = [1.55, \ 2.38]$

shadows [0.22, 0.72] [0.18, 0.79] [0.38, 0.62] [0.33, 0.43]

The more detailed description of the structure (which comes with more clusters) positions more patterns in the shadowed structure, which is quite understood.

9.6 CONCLUSIONS

The concept of interoperability between fuzzy sets and other types of granular information including numeric data is fundamental to fuzzy modeling and any other interaction including mechanisms of user feedback. We formulated the two key issues of encoding and decoding and showed that a way in which their mechanisms are realized become of paramount relevance to the effectiveness of the processing carried out at the level of fuzzy sets. Along this line, we showed that the interoperability between fuzzy sets and numeric data (being treated as a boundary case of information granules) comes with the specific fuzzification and defuzzification scheme whose suitability could be assessed through the use of the reconstruction criterion. Interestingly, fuzzy sets come with an important feature of lossless reconstruction if selected properly. Here the commonly encountered family of triangular fuzzy sets with an overlap of $1/2$ between successive elements of this family offers the zero value of the decoding error. The linkages between fuzzy sets and sets through the construct of α-cuts are helpful for representation purposes. They also show how the well-established techniques of interval mathematics could be found useful in solving optimization problems involving fuzzy sets. We introduced the concept of shadowed sets and demonstrated how it could be viewed as a conceptual and computational vehicle aimed at the interpretation of fuzzy sets.

EXERCISES AND PROBLEMS

1. Express the concepts of support and core in terms of α-cuts and strong α-cuts.
2. Given is a fuzzy set $B = [0.3, 0.9, 1.0, 1.0, 0.5]$. What would be its best set-based approximation?
3. We have learned that the triangular fuzzy sets with two neighboring fuzzy sets with an overlap of $\frac{1}{2}$ guarantee a zero decoding error irrespective of the number of the fuzzy sets being used. Given this, we could think of using only two fuzzy sets. Is this a wise choice? Why yes? Why not? Offer a detailed explanation.
4. Plot a decoding (reconstruction) error resulting from the use of intervals instead of fuzzy sets. This error is also known as a quantization error.
5. Given a Gaussian fuzzy set, $A(x) = \exp(-x^2/10)$, find its α-cut. How does its length depend upon the values of α? Draw the corresponding graph of length = length (α). By the length we mean the length of the resulting interval of the α–cute.
6. You are provided with a unimodal fuzzy set A. If you are allowed to use only two levels of α-cuts (say β and γ) how would you choose their values?
7. For the nonsymmetrical fuzzy set described by the parabolic membership function with its modal value positioned at zero

$$A(x) = \begin{cases} 1 - \left(\frac{x}{a}\right)^2, & \text{if } x \in [0, a] \\ 1 - \left(\frac{x}{b}\right)^2, & \text{if } x \in [-b, 0] \\ 0, & \text{otherwise} \end{cases}$$

determine the corresponding shadowed set.

8. Derive the expression for the reconstruction error if we consider a collection of Gaussian functions $A_i(x) = \exp(-(x-m_i)^2/2), i = 1, 2, \ldots, c$ that are uniformly distributed across the space $\mathbf{X} = [-10, 10]$. How does the total reconstruction error depend upon the number c of fuzzy sets being used? Plot this relationship. Could you offer any practical insight into the choice of the number of these fuzzy sets?

9. Given is the fuzzy set shown in Figure 9.22.

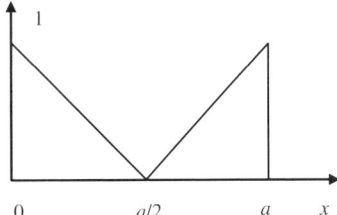

Figure 9.22 Fuzzy set to be represented by a single numeric quantity.

Find its numeric representative using the following methods: (a) center of gravity, (b) mean of maxima, (c) center of area approach. Compare the results; what conclusions could you derive? Are the findings intuitive? Justify your judgment.

10. Derive the encoding and decoding formulas (9.17) and (9.19).

11. Consider the encoding-decoding scheme in the presence of fuzzy input data when the codebook is formed by triangular fuzzy sets with an overlap of 1/2 between the neighboring fuzzy sets.

12. In fuzzy decision-making we are usually provided with a fuzzy set of decision and need to choose several or a single alternative. Describe in detail and contrast several paths you could follow here. What are their advantages? What difficulties could you encounter? To illustrate your considerations, use two fuzzy sets defined in the decision space, which are $B = [0.1, 0.4, 0.5, 0.8, 0.9, 1.0, 1.0, 0.3, 0.1, 0.0, 0.0]$ and $C = [1.0, 0.5, 0.2, 0.1, 0.1, 0.1, 0.0, 0.0, 0.0, 0.0, 0.0]$.

HISTORICAL NOTES

Almost since the very inception of fuzzy sets, it was a strong interest in the characterization of fuzzy sets in the language of sets; α-cuts and representation theorem are essential to these investigations (Negoita and Ralescu, 1987). The concepts of fuzzification and defuzzification arose in the context of various studies on fuzzy controllers and found their own research niche in fuzzy sets. Given the evident practicality of the problem itself, it is not surprising that we see a suite of various decoding (defuzzification) algorithms. There have been some formal requirements formulated (as those offered by Runkler and Glenser, 1993). In general one should become alerted that there is no unique and universal defuzzification scheme (as each of them focuses on a specific form of transformation). Given the very nature of the problem, there will not be a unique method in the future. Any selection of the defuzzification algorithm needs to be guided by the computational constraints (which we encounter in any problem), character of anticipated interaction, and allowed level of information losses. The mechanism of fuzzification is far more obvious in the case of numeric data, however, various alternatives exist when dealing with granular data. The possibility-necessity encoding is one among viable alternatives.

The linkages with the issues of quantization and decoding as encountered in information (signal, image) compression (for such problems we have a genuine abundance of literature with some classic readings such as Gersho and Gray (1992); Lloyd (1982); Linde et al. (1988)) have not been vividly investigated in fuzzy sets. Some studies along this line were reported in Pedrycz (2001) and Pedrycz and de Oliveira (1996).

Shadowed sets formed as constructs induced by fuzzy sets were introduced and studied by Pedrycz (1998, 1999); refer also to for further results and applications (Cattaneo and Ciucci, 2001; Gürkan et al., 2001).

REFERENCES

Aiyer, A., Pyun, K., Huang, Y., O'Brien, D., Lloyd, R. Clustering of Gauss mixture models for image compression and classification, *Signal Process-Image* **20**, 2005, 459–485.

Bezdek, J. *Pattern Recognition with Fuzzy Objective Function Algorithms*, Plenum Press, New York, 1981.

Cattaneo, G., Ciucci, D. An algebraic approach to shadowed sets, *Electronic Notes in Theor. Comp.* **82**(4), 2001, 1–12.

Campobello, G., Mantineo, M., Patane, G., Russo, M. LBGS: a smart approach for very large data sets vector quantization, *Signal Process-Image* **20**, 2005, 91–114.

Chang, C., Chou, Y., Yu, Y., Shih, K. An image zooming technique based on vector quantization approximation, *Image Vision Comput.* **23**, 2005, 1214–1225.

Cho, C., Lee, K., Lee-Kwang, H. Ranking fuzzy values with satisfaction function, *Fuzzy Set Syst.* **64**, 1994, 295–309.

Gath, I., Geva, A. Unsupervised optimal fuzzy clustering. *IEEE Trans. Pattern Anal.* **11**, 1989, 773–791.

Gersho, A., Gray, R. M. *Vector Quantization and Signal Compression*, Kluwer Academic Publishers, Boston, MA, 1992.

Gray, R. Vector quantization, *IEEE Acoust. Speech, Signal Proc. Mag.* **1**, 1984, 4–29.

Gürkan, E., Erkmen, I., Erkmen, A. Two-way fuzzy adaptive identification and control of a flexible-joint robot arm, *Inf. Sci.* **145**(1–2), 2001, 13–43.

Hirota, K., Pedrycz, W. Fuzzy relational compression, *IEEE Trans. Syst. Man Cyb.* **29**(3), 1999, 407–415.

Jee, I., Haddad, R. Optimum design of vector-quantized subband codecs, *IEEE T. Signal Process.* **46**(8), 1998, 2239–2243.

Kämpke, T. Constrained quantization, *Signal Process.* **83**(9), 2003, 1839–1858.

Lin, T., and Yu, P. Centroid neural network adaptive resonance theory for vector quantization, *Signal Process.* **83**(3), 2003, 649–654.

Karayiannis, N., Pai, P. I. Fuzzy vector quantization algorithms and their application in image compression, *IEEE Trans. Image Process.* **4**(9), 1995, 1193–1201.

Kim, J. K., Cho, C. H., Lee-Kwang, H. A note on the set-theoretical defuzzification, *Fuzzy Set Syst.* **98**, 1998, 337–341.

Laskaris, N., Fotopoulos, S. A novel training scheme for neural-network-based vector quantizers and its application in image compression, *Neurocomputing* **61**, 2004, 421–427.

Lendasse, A., Francois, D., Wertz, V., Verleysen, M. Vector quantization: a weighted version for time-series forecasting, *Future Gener. Comp. Sy.* **21**, 2005, 1056–1067.

Linde, Y., Buzo, A., Gray, R. An algorithm for vector quantizer design, *IEEE Trans. Commun.* **COM-28**(1), 2, 1988, 84–95.

Liu, S., Lin, J. Vector quantization in DCT domain using fuzzy possibilistic c-means based on penalized and compensated constraints, *Pattern Recogn.* **35**, 2002, 2201–2211.

Lloyd, S. Least squares quantization in PCM, *IEEE Trans. Inform. Theor.* **28**, 1982, 129–137.

Moller, U., Galicki, M., Baresova, E., Witte, H. An efficient vector quantizer providing globally optimal solutions, *IEEE Trans. Signal Process.* **46**(9), 1998, 2515–2529.

Negoita, C., Ralescu, D. *Simulation, Knowledge-Based Computing, and Fuzzy Statistics*, Van Nostrand Reinhold, New York, USA, 1987.

Nobuhara, H., Pedrycz, W., Hirota, K. Fast solving method of fuzzy relational equation and its application to lossy image compression/reconstruction, *IEEE Trans. Fuzzy Syst.* 2000, **8**(3), 325–335.

Pan, Z., Kotani, K., Ohmi, T. A generalized multiple projection axes method for fast encoding of vector quantization, *Pattern Recogn. Lett.* **26**, 2005, 1316–1326.

Patane, G., Russo, M. The enhanced LGB algorithm, *Neural Networks* **14**, 2001, 1219–1237.

Pedrycz, W. Why triangular membership functions? *Fuzzy Set Syst.* **64**, 1994, 21–30.

Pedrycz, W., Valente de Oliveira, J. Optimization of fuzzy models, *IEEE Trans. Syst. Man Cyb. B*, **26(4)**, 1996, 627–636.

Pedrycz, W. Shadowed sets: representing and processing fuzzy sets, *IEEE Trans. Syst. Man Cyb. B*, **28**, 1998, 103–109.

Pedrycz, W. Shadowed sets: bridging fuzzy and rough sets, in: Pal, S. K., Skowron, A. (eds.), *Rough Fuzzy Hybridization. A New Trend in Decision-Making*, Springer-Verlag, Singapore, 1999, pp. 179–199.

Pedrycz, W. Fuzzy equalization in the construction of fuzzy sets, *Fuzzy Set Syst.* **119**, 2001, 329–335.

Pedrycz, W. Interpretation of clusters in the framework of shadowed sets, *Pattern Recog. Lett.* **26,** 2005, 2439–2449.

Pregenzer, M., Pfurtscheller, G., Flotzinger, D. Automated feature selection with distinction sensitive learning vector quantization, *Neurocomputing* **11**, 1996, 19–29.

Runkler, T., Glesner, M. A set of axioms for defuzzification strategies toward a theory of rational defuzzification operators, *Proceedings of the 2nd IEEE International Conference on Fuzzy Systems*, San Francisco, IEEE Press 1993, pp. 1161–1166.

Sencara, H., Ramkumarb, M., Akansua, A. An overview of scalar quantization based data hiding methods, *Signal Process.* 2005 (available online).

Sinkkonen, J., Kaski, S. Clustering based on conditional distribution in an auxiliary space. *Neural Comput.* **14**, 2002, 217–239.

Tsekouras, G. A fuzzy vector quantization approach to image compression, *Appl. Math. Comput.* **167**, 2005, 539–560.

Wierman, M. Central values of fuzzy numbers-defuzzification, *Inf. Sc.* **100**, 1997, 207–215.

Yair, E., Zeger, K., Gersho, A. Competitive learning and soft competition for vector quantizer design, *IEEE Trans. Signal Process.* **40**(2), 1992, 294–309.

Yen, C., Young, C., Nagurka, M. A vector quantization method for nearest neighbor classifier design, *Pattern Recogn. Lett.* **25**, 2004, 725–731.

Wu, K., Yang, M. A fuzzy-soft learning vector quantization, *Neurocomputing* **55**, 2003, 681–697.

Chapter 10

Fuzzy Modeling: Principles and Methodology

This chapter offers an in-depth discussion on the principles of fuzzy modeling (Bezdek, 1993; Kacprzyk, 1983, 1997; Pedrycz, 1996; Zadeh, 1973, 2005), their design objectives, including accuracy, interpretability, stability, an overall design process, and related verification and validation mechanisms. We also present some general categories of the fuzzy models, and for each of them elaborate on the satisfaction of the already identified design objectives.

10.1 THE ARCHITECTURAL BLUEPRINT OF FUZZY MODELS

In general, fuzzy models operate at a level of information granules—fuzzy sets, and in this way they constitute highly abstract and flexible constructs (Pedrycz and Vukovich, 2001). Given the environment of physical variables describing the surrounding world and an abstract view of the system under modeling, a very general view of the architecture of the fuzzy model can be portrayed, as presented in Figure 10.1 (Pedrycz, 1996; Pedrycz and Gomide, 1998; Oh and Pedrycz, 2005; Pedrycz and Reformat, 2005).

We clearly distinguish between three functional components of the model where each of them comes with well-defined objectives. The input interface builds a collection of modalities (fuzzy sets and fuzzy relations) that are required to link the fuzzy model and its processing core with the external world. This processing core realizes all computing being carried out at the level of fuzzy sets (membership functions) already used in the interfaces. The output interface converts the results of granular (fuzzy) processing into the format acceptable by the modeling environment. In particular, this transformation may involve numeric values that are the representatives of the fuzzy sets produced by the processing core. The interfaces could be present in different categories of the models, yet they may show up to a significant

---

Fuzzy Systems Engineering: Toward Human-Centric Computing, by Witold Pedrycz and Fernando Gomide
Copyright © 2007 John Wiley & Sons, Inc.

10.1 The Architectural Blueprint of Fuzzy Models

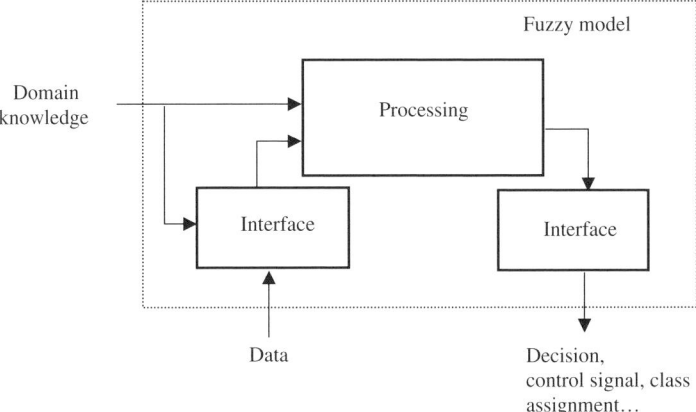

Figure 10.1 A general view at the underlying architecture of fuzzy models along with its three fundamental functional modules.

extent. Their presence and relevance of the pertinent functionality depends upon the architecture of the specific fuzzy model and the way in which the model is utilized. The interfaces are also essential when the models are developed on the basis of available numeric experimental evidence as well as some prior knowledge provided by designers and experts.

For instance, for rule-based topology of the model that is based upon fuzzy sets in input and output variables we require well-developed interfaces. The generic models in this category are formulated as follows:

$$\text{if } X_1 \text{ is } A \text{ and } X_2 \text{ is } B \text{ and } \ldots \text{then } Y \text{ is } C \qquad (10.1)$$

where $X_1, X_2, \ldots$ are input variables and Y is the output variable, whereas $A, B, C, \ldots$ are the fuzzy sets defined in the corresponding spaces (universe). Any logic processing carried out by the rule-based inference mechanism requires that any input is transformed, that is, expressed in terms of fuzzy sets and the results of reasoning is offered in its numeric format (at which stage we require that the result is produced through the transformation of the fuzzy set of conclusion).

The rule-based models endowed with local models forming their consequents (conclusion parts), commonly referred to as fuzzy functional or Takagi–Sugeno fuzzy models (Takagi and Sugeno, 1985; Chen and Linkens, 2004; Babuska, 1998), are governed by the formula

$$\text{if } X_1 \text{ is } A_i \text{ and } X_2 \text{ is } B_i \text{ and} \ldots \text{then } y \text{ is } f_i(\mathbf{x}, \mathbf{a}_i) \qquad (10.2)$$

where $f_i(\mathbf{x}, \mathbf{a}_i)$ denotes a multivariable local model, $\mathbf{x} = [x_1, x_2, \ldots, x_n]^T$ is the vector of base variables of $X_1, X_2, \ldots, X_n$ and $\mathbf{a}_i$ is a vector of parameters $\mathbf{a}_i = [a_{i1}, a_{i2}, \ldots, a_{in}]^T$. In particular, one can envision a linear form of the model in which f_i becomes a linear function of its parameters, namely $f_i(\mathbf{x}, \mathbf{a}_i) = \mathbf{a}_i^T \mathbf{x}$. Depending upon the specificity of the problem and the structure of available data, these regression models could be made nonlinear. Given the character dictated by the problem at

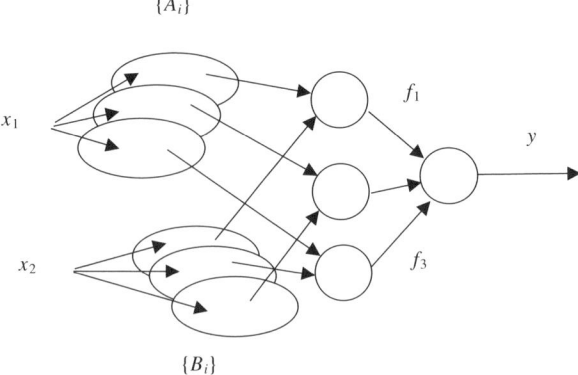

Figure 10.2 A schematic view of the two-input (x_1 and x_2) Takagi-Sugeno model with local regression models; the connections of the output unit realize processing through the local model f_i.

hand, we may be concerned with polynomial regression models (say, quadratic, cubic, etc.), trigonometric, and so on. The region of operation (viz. the area where the rule is relevant) of the rule is determined by the form and the location of the fuzzy sets located in the input space, which then occur in this particular rule; see Figure 10.2.

In this case the output interface is not required as the output of the model is numeric. Furthermore, we still have to use a well-defined input interface as its components (fuzzy sets) form condition parts of the rules. Any input has to be transformed and communicated to the inference procedure making use of the fuzzy sets of the interface. Rule-based models are central architectures of fuzzy models. We will devote a separate chapter to cover the fundamentals and algorithmic developments of fuzzy rule-based computing.

10.2 KEY PHASES OF THE DEVELOPMENT AND USE OF FUZZY MODELS

There are several fundamental schemes that support the design and the use of fuzzy models. Referring to Figure 10.3, we encounter four essential modes of their usage:

(a) The use of numeric data and generation of results in the numeric form is shown in Figure 10.3(a). This mode reflects a large spectrum of modeling scenarios we typically encounter in system modeling. Numeric data available in the problem are transformed through the interfaces and used to construct the processing core of the model. Once developed, the model is then used in a numeric fashion: It accepts numeric entries and produces numeric values of the corresponding output. From the perspective of the external "numeric" world, the fuzzy model manifests itself as a multivariable nonlinear input–output mapping. Later on, we discuss the nonlinear character of the mapping in the context of rule-based systems. It will be demonstrated how the form of the mapping depends directly upon the number of the rules, membership

10.2 Key Phases of the Development and Use of Fuzzy Models

functions of fuzzy sets used there, inference scheme, and other design parameters. Owing to the number of design parameters, rule-based systems bring in a substantial level of modeling flexibility, and this becomes highly advantageous to the design of fuzzy models.

(b) The use of numeric data and the presentation of results in a granular format (through some fuzzy sets) is shown in Figure 10.3 (b). This mode makes the model highly user-centric. The result of modeling comes as a collection of elements with the corresponding degrees of membership, and in this way it becomes more informative and comprehensive than a single numeric quantity. The user/decision-maker is provided with preferences (membership degrees) associated with a collection of possible outcomes.

(c) The use of granular data as inputs and the presentation of fuzzy sets as outcomes of the models is shown in Figure 10.3(c). This scenario is typical for granular modeling in which instead of numeric data we encounter a collection of linguistic observations such as expert's judgments, readings coming from unreliable sensors, outcomes of sensors summarized over some time horizons, and so on. The results presented in the form of fuzzy sets are beneficial for the interpretation purposes and support the user-centric facet of fuzzy modeling.

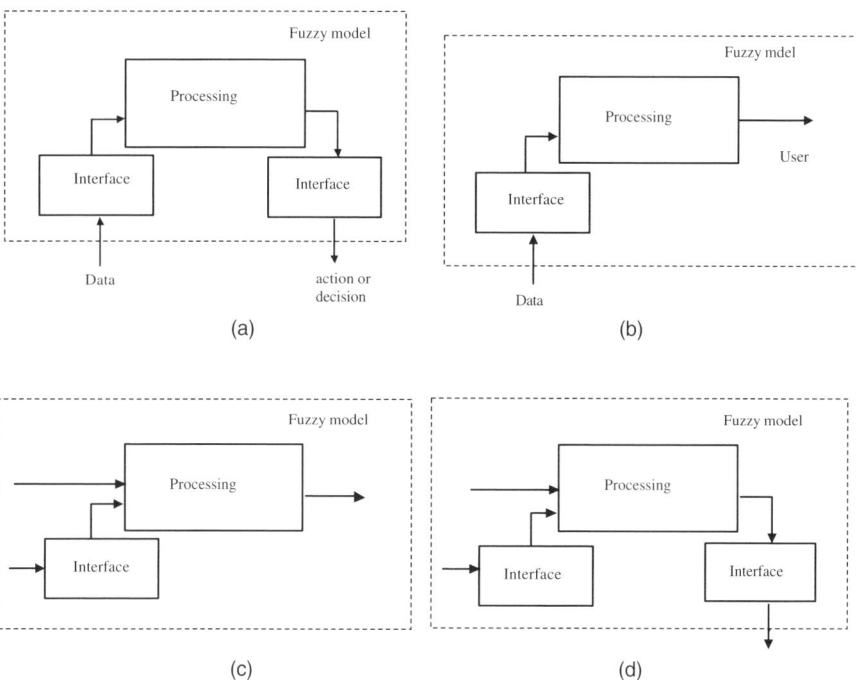

Figure 10.3 Four fundamental modes of the use of fuzzy models; note a role of input and output interfaces played in each of them. See the details in the text.

(d) The use of fuzzy sets as inputs of the model and a generation of numeric outputs of modeling is shown in Figure 10.3(d). Here we rely on expert opinions as well as granular data forming aggregates of detailed numeric data. The results of the fuzzy model are then conveyed (through the interface) to the numeric environment in the form of the corresponding numeric output values. Although this becomes feasible, we should be cognizant that the nature of the numeric output is not fully reflective of the character of the granular input.

10.3 MAIN CATEGORIES OF FUZZY MODELS: AN OVERVIEW

The landscape of fuzzy models is highly diversified. There are several categories of models where each class of the constructs comes with interesting topologies, functional characteristics, learning capabilities, and the mechanisms of knowledge representation. In what follows, we offer a general glimpse at some of the architectures that are most visible and commonly envisioned in the area of fuzzy modeling.

10.3.1 Tabular Fuzzy Models

Tabular fuzzy models are formed as some tables of relationships between the variables of the system granulated by some fuzzy sets (Fig. 10.4). For instance, given two input variables with fuzzy sets $A_1, A_2, A_3, B_1 - B_5$ and the output fuzzy sets C_1, C_2, and C_3, the relationships are articulated by filling in the entries of the table; for each combination of the inputs quantified by fuzzy sets, say A_i and B_j, we associate the corresponding fuzzy set C_k formed in the output space.

The tabular models produce a fairly compact suite of transparent relationships represented at the level of information granules. In the case of many input variables, we end up with multidimensional tables (relations). The evident advantage of the tabular fuzzy models resides with their evident readability. The shortcoming comes with the lack of existence of the direct mapping mechanisms. This means that we do not have any machinery of transforming input (either numeric or granular) into the respective output. Furthermore, the readability of the model could be substantially hampered when dealing with the growing number of variables we consider in this model.

	B_1	B_2	B_3	B_4	B_5
A_1					
A_2			C_3		
A_3					C_1

Figure 10.4 An illustrative example of the two-input tabular fuzzy model (fuzzy decision table).

10.3.2 Rule-Based Systems

Rule-based systems are highly modular and easily expandable fuzzy models composed of a family of conditional "if – then" statements (rules) where fuzzy sets occur in their conditions and conclusions. The standard format of the rule with many inputs (conditions) comes in the form

if $condition_1$ is A *and* $condition_2$ is B *and* ... *and* $condition_n$ is W then conclusion is Z

$$(10.3)$$

where $A, B, \ldots W, Z$ are fuzzy sets defined in the corresponding input and output spaces. The conditions of the *if* part are generically called rule autecedent whereas conclusions, (decisions) of the *then* part is called rule.

The models support the principle of locality and the distributed nature of modeling as each rule can be interpreted as an individual local descriptor of the data (problem) that is invoked by the fuzzy sets defined in the space of conditions (inputs). The local nature of the rule is directly expressed through the support of the corresponding fuzzy sets standing in its condition part. The level of generality of the rule depends upon many aspects that could be easily adjusted making use of the available design components associated with the rules. In particular, we could consider fuzzy sets of condition and conclusion whose granularity could be adjusted so that we could easily capture the specificity of the problem. By making the fuzzy sets in the condition part very specific (that is being of high granularity), we come up with the rule that is very limited and confined to some small region in the input space. When the granularity of fuzzy sets in the condition part is decreased, the generality of the rule increases. In this way the rule could be applied to more situations. To emphasize a broad spectrum of possibilities emerging in this way refer to Figure 10.5, which underlines the very nature of the cases discussed above.

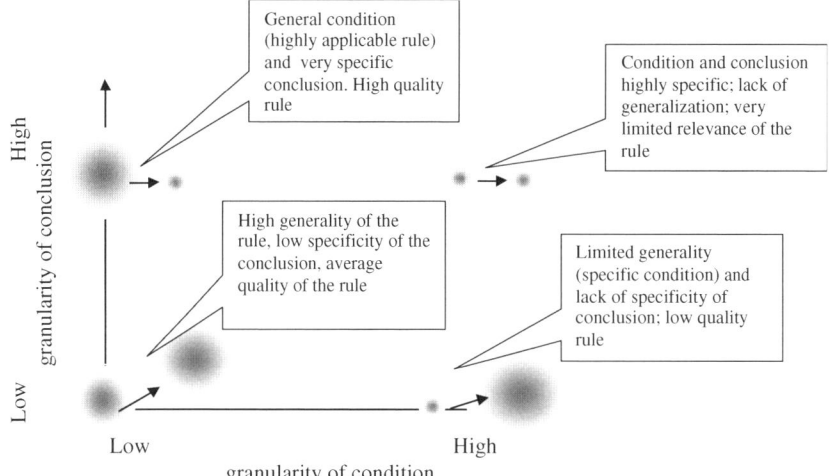

Figure 10.5 Examples of rules and their characterization with respect to the level of granularity of condition and conclusion parts.

258 Chapter 10 Fuzzy Modeling: Principles and Methodology

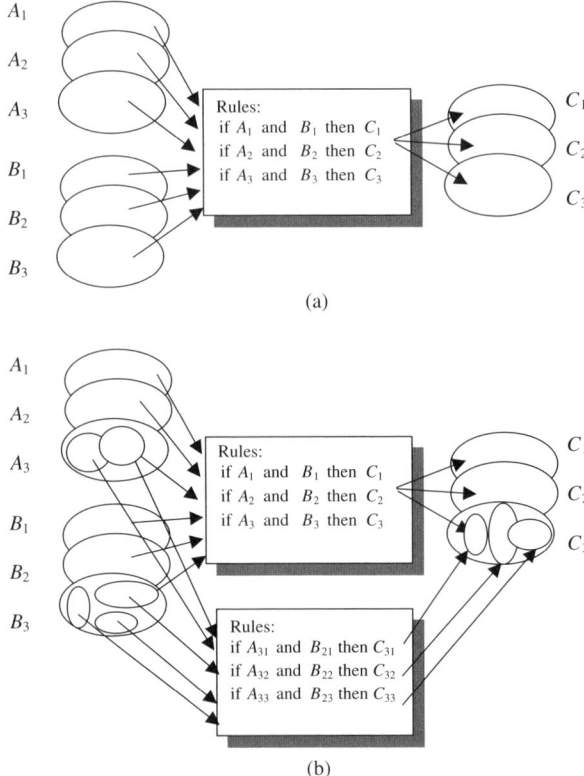

Figure 10.6 Examples of rule-based systems: (a) single-level architecture with all rules expressed at the same level of generality, (b) rules formed at several levels of granularity (specificity) of fuzzy sets standing in the condition parts of the rules. A_i and B_j stand for fuzzy sets forming the condition parts of the rules.

While the rules represented in (10.3) form a single level structure (the rules are built at the same level), there are also hierarchical architectures composed of several levels of knowledge representation where there are collections of rules formed at a few very distinct levels of granularity (generality); refer to Figure 10.6. The level of generality of the rules is directly implied by the information granules forming the input and output interfaces. As we have already emphasized, given the importance of rule-based systems, we cover their design in a separate chapter.

10.3.3 Fuzzy Relational Models and Associative Memories

Fuzzy relational models and associative memories are examples of constructs whose computing dwells upon logic processing of information granules. The spirit of the

10.3 Main Categories of Fuzzy Models: An Overview

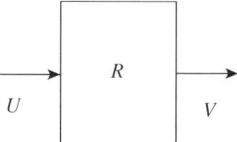

Figure 10.7 A schematic view of the relational models and associative memories.

underlying architectures is that the mapping between input and output information granules is realized in the form of some relational transformation (Fig. 10.7). The mapping itself comes in the form of a certain composition operator (say, max–min or being more general, s-t composition). The same development scheme applies to fuzzy associative memories. The quality of recall carried out in the presence of incomplete or distorted (noisy) inputs is regarded as one of the leading indicators describing the performance of the memories. Given the input and associated output items—fuzzy sets $A_1, A_2, \ldots, A_c$ and $B_1, B_2, \ldots, B_c$, respectively, we construct a fuzzy memory (relation) storing all pairs of items by or-wise aggregating the Cartesian products of the input–output pairs, $R = \cup_{k=1}^{N}(A_k \times B_k)$. Next any input item A leads to the recall of the corresponding output B through some relational operator (say, max–min composition), $V = U \circ R$.

In fuzzy relational equations (that constitute an operational framework of associative memories) we encounter a wealth of architectures that is driven by the variety of composition operators. Although the sup-t (max–min) composition is commonly used, there are also other alternatives available; some of them are presented in Figure 10.8. The selection of the composition operation (hence the form of the equation) depends upon the problem at hand as each composition operator comes with its own well-defined semantics (with the underlying logic

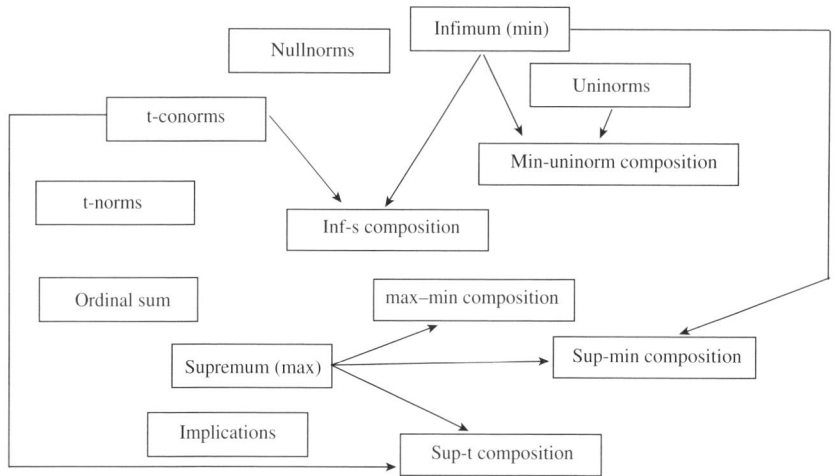

Figure 10.8 A general taxonomy of fuzzy relational equations (modeling structures) presented with respect to the combination of composition operators used in their realization.

underpinnings and interpretation abilities). From the algorithmic standpoint, let us note that some of the composition operators lead to fuzzy relational equations for which we could derive analytical solutions. In other more advanced cases, one has to proceed with some numeric optimization and develop pertinent learning schemes.

10.3.4 Fuzzy Decision Trees

Fuzzy decision trees are generalizations of well-known and commonly used decision trees (Apolloni et al., 1998; Chang and Pavlidis, 1977; Janikow, 1998; Pedrycz and Sosnowski, 2001; Qin and Lawry, 2005). In essence, a decision tree is a directed acyclic graph (DAG) whose nodes are marked by the attributes (input variables of the model) and links are associated with the discrete (finite) values of the attributes associated with the corresponding nodes (Fig. 10.9). The terminal nodes concern the values of the output variable (which depending upon the nature of the problem could assume discrete or continuous values).

By traversing the tree starting from the root node, we arrive at one of its final nodes. In decision tree only one terminal node can be reached as the values of the inputs uniquely determine the path one traverses through the tree. In contrast, in fuzzy decision trees several paths could be traversed in parallel. When moving down a certain path, several alternative edges originating from a given node are explored where each of them comes with the degree of matching of the current data (more specifically the value of the attribute that is associated with the node) and the fuzzy sets representing the values of the attribute coming with each node. The reachability of the node is computed by aggregating the degrees of matching along the path that has been traversed to reach it. Typically, we use here some t-norm as we adhere to the *and*-like aggregation of the activation levels reported along the edges of the tree visited so far. In this way, several terminal nodes are reached and each of them comes with its own value of the reachability index computed by an *and* aggregation (using some t-norm) of the activation (matching) degrees between the data and the values of the attributes represented as fuzzy sets. The pertinent details are illustrated in Figure 10.9 (c).

10.3.5 Fuzzy Neural Networks

Fuzzy neural networks are fuzzy set-driven models composed with the aid of some logic processing units—fuzzy neurons (cf. Pedrycz and Gomide, 1998; Pedrycz, 2004; Ciaramella et al., 2006). These neurons realize a suite of generic logic operators (such as *and, or, inclusion, dominance, similarity, difference*, etc.). Each neuron comes with a collection of the connections (weights). These weights bring a highly required flexibility to the processing units that could be exploited during the learning of the network. From the perspective of the topology of the network, we can envision several well-delineated layers of the processing units; see Figure 10.10.

There are some interesting linkages between the fuzzy neural networks and the relational structures (fuzzy relational equations) we have discussed earlier. Both of

10.3 Main Categories of Fuzzy Models: An Overview

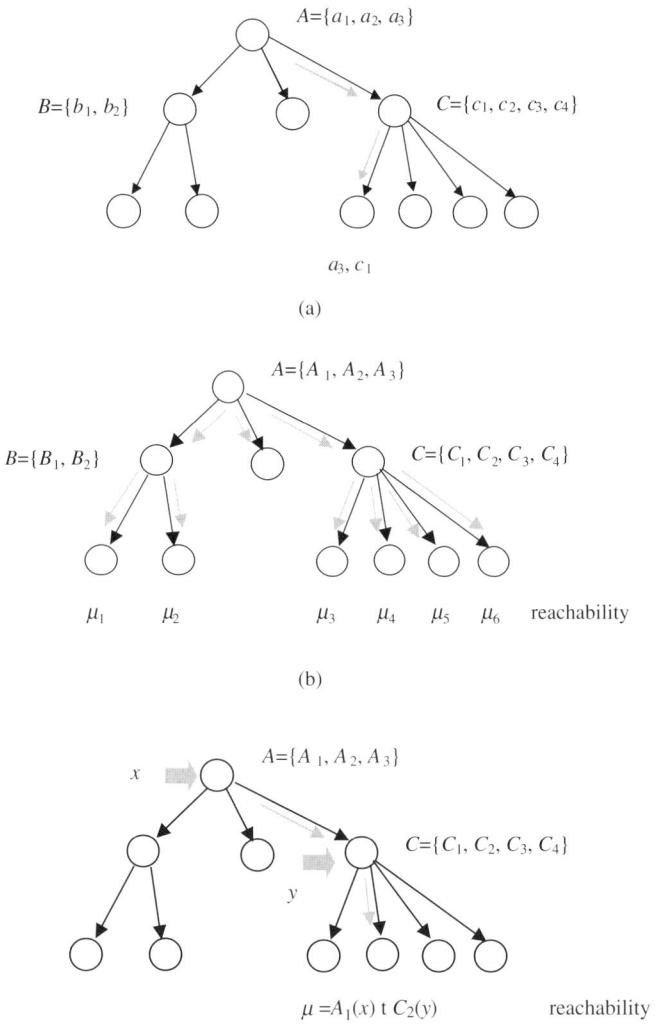

Figure 10.9 An example of the decision tree (a) and the fuzzy decision tree (b); in this case one can reach several terminal nodes at different levels of reachability (μ_i). The level of reachability is determined by aggregating activation levels along the path leading to the terminal node (c).

them rely on the same pool of composition operators (logic mappings); however, when it comes to the networks, those are typically multilayer architectures.

10.3.6 Network of Fuzzy Processing Units

The essence of these modeling architectures is to allow for a higher level of autonomy and flexibility. In contrast to the fuzzy neural networks, there is no layered

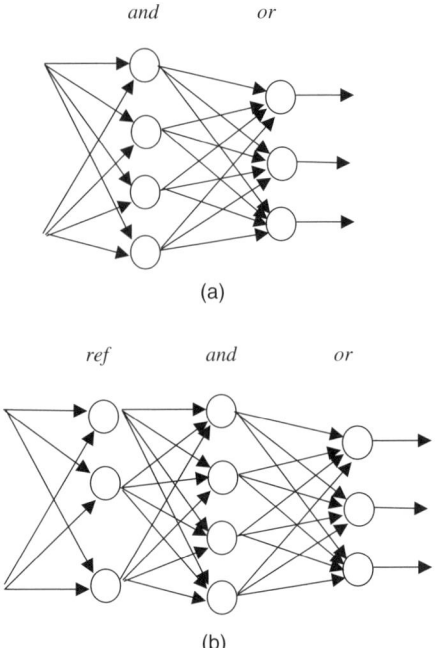

Figure 10.10 Examples of architectures of fuzzy neural networks: generalized (multivalued) logic functions realized in terms of *and* and *or* neurons (a), and the network with an auxiliary referential layer consisting of referential neurons (b).

structure. Rather than that, we allow for loosely connected processing units that can operate individually and communicate with the others. Furthermore, when dealing with dynamic systems, the network has to exhibit some recurrent links.

One of the interesting and representative architectures in this category are fuzzy cognitive maps (Kosko, 1986, 1992; Papageorgiou and Groumpos, 2005; Papageorgiou et al., 2006, Carvalho and Torne, 2007). These maps, being the generalization of the binary concepts introduced by Axelrod (1976) represent concepts and show linkages between them. A collection of basic concepts is represented as nodes of the graph that are interrelated through a web of links (edges of the graph). The links could be excitatory (so the increase of intensity of one concept triggers the increased level of manifestation of the related one) or inhibitory (in which case we see an opposite effect: the increase of intensity of one concept triggers the decline of intensity of the other one). Traditionally, the connections (links) assume numeric values from -1 to 1. An example of the fuzzy cognitive map is shown in Figure 10.11.

The detailed computing realized at the level of the individual node is governed by the expression $x_j = f(\sum_{\substack{j=1 \\ j \neq i}}^{n} w_{ji} x_j)$, where x_j denotes a resulting level of activity (intensity) at the node of interest and x_i is the intensity level associated with the ith node. The connections (linkages) between the two nodes are denoted by w_{ij}. The

10.3 Main Categories of Fuzzy Models: An Overview

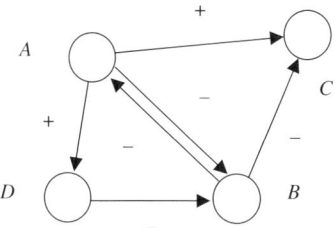

Figure 10.11 An example of a fuzzy cognitive map composed of four concepts (A, B, C, and D). The sign of the corresponding connection identifies the effect of inhibition (−) or excitation (+) between the concepts (nodes). For instance, A excites C.

nonlinear mapping f is typically a monotonically increasing function, say a sigmoid one, $f(u) = 1/(1 + \exp(-u))$. A node could be equipped with its own dynamics (internal feedback loop), and in this case we consider a nonzero link w_{jj} for this particular node. Given this, we arrive at a recurrent expression of the form $x_i = f(\sum_{j=1}^{n} w_{ji} x_j)$. The interfaces of this fuzzy model is not shown explicitly; however, we should have in mind that the inputs to the nodes are the inputs from the modeled world that were subject to a certain transformation realized through some fuzzy sets defined in the corresponding variables or the Cartesian products of these variables.

The structure of the network offers a great deal of flexibility and is far less rigid than the fuzzy neural networks where typically the nodes (neurons) are organized into some layers. The individual nodes of the fuzzy cognitive maps could be realized as some logic expressions and implemented as *and* or *or* neurons. The connections could assume values in [0,1], and the inhibitory effect can be realized by taking the complement of the activation level of the node linked to the one under consideration. An example of the logic-based fuzzy cognitive map is presented in Figure 10.12.

In all these categories of fuzzy models, we can envision a hierarchy of the structures that could be formed at each level of the hierarchy (Cordón et al., 2003). We start from the highest, the most general level and expand it by moving down to

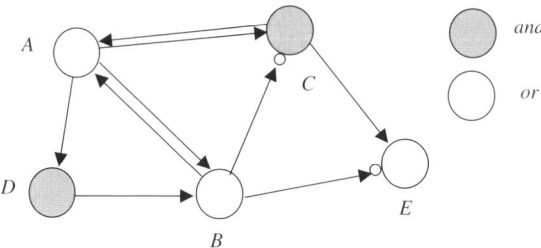

Figure 10.12 An example of a fuzzy cognitive map whose nodes are realized as logic expressions (*and* and *or* neurons). The inhibitory effect is realized by taking the complement of the activation level of the interacting node (here indicated symbolically by a small dot).

264 Chapter 10 Fuzzy Modeling: Principles and Methodology

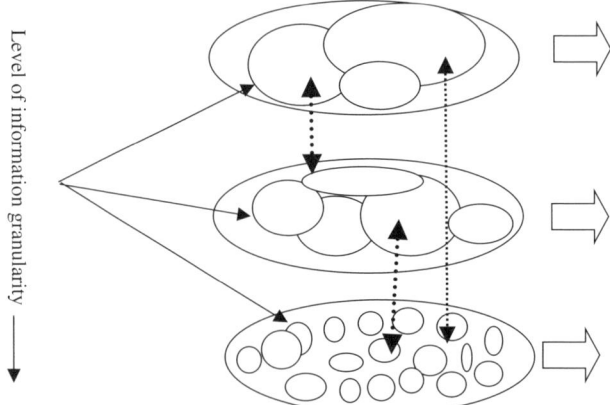

Figure 10.13 A general concept of hierarchy in fuzzy modeling; depending upon a certain level of specificity, various sources of data could be accommodated and processed, and afterward the results communicated to the modeling environment.

capture more details. In general, we can envision a truly hierarchical structure shown in Figure 10.13.

A more specific and detailed visualization of the hierarchy of the model is shown in Figure 10.14 where we are concerned with fuzzy cognitive maps. Here, a certain concept present at the higher level and represented as one of the nodes of the map unfolds into several subconcepts present at the lower level. With the computing

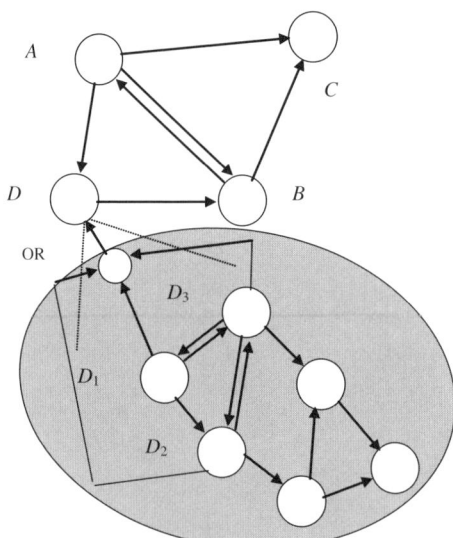

Figure 10.14 An example of a hierarchy of fuzzy cognitive maps; a certain concept at the higher level of generality is constructed as a logic *or*-type of aggregation of the more detailed ones used in the

occurring at the lower level, more detailed level produces some level of activation of the more detailed nodes (sub-concepts), and these levels aggregated *or*-wise are then used in computing realized at the higher level of generality.

10.4 VERIFICATION AND VALIDATION OF FUZZY MODELS

The processes of verification and validation (referred to as V&V) are concerned with the fundamental issues of the development of the model and assessment of its usefulness. Following the standard terminology (which is well-established in many disciplines, such as, e.g., software engineering), verification is concerned with the analysis of the underlying processes of constructing the fuzzy model. Are the design principles guiding the systematic way the model is built fully adhered to? In other words, rephrasing the concept in the setting of software engineering, we are focused on the following question, "Are we building the product *right*?" Validation, on the other hand, is concerned with ensuring that the model (product) meets the requirements of the customer. Here we concentrate on the question, "Are we building the *right* product?" Put it differently: Is the resulting model in compliance with the expectations (requirements) of the users or groups of users of the model?

Let us elaborate on the verification and validation in more detail.

10.4.1 Verification of Fuzzy Models

Fuzzy models and fuzzy modeling are the pursuits that in spite of their specific requirements are still adhere to by the fundamentals of system modeling. In this sense, they also have to follow the same principles of model verification. There are several fundamental guidelines in this respect. Let us highlight the essence of them.

(a) An iterative process of constructing a model in which we successively develop a structure of the model and estimate its parameters. There are well-established estimation procedures that come with a suite of optimization algorithms. It is quite rare that the model is completely built through a single pass through these two main phases.

(b) Thorough assessment of accuracy of the developed model. The underlying practice is that one should avoid any bias in the assessment of this quality, especially by developing a false impression about the high accuracy of the model. To avoid this and gain a significant level of objective evaluation, we split the data into training and testing data. While the model is constructed, we use the training data.

(c) Generalization capabilities of the resulting model. Although the accuracy is evaluated for the experimental data being used to construct the model, the accuracy obtained quantified in this way could lead to highly and optimistically biased evaluation. The assessment of the performance of the model on the testing data helps to eliminate this shortcoming.

(d) The lowest possible complexity of the model. This is usually referred to as an Occam's razor principle. The principle states that among several models of a very similar accuracy, we always prefer a model that is the simplest. The concept of simplicity requires some clarification as the concept itself is not straightforward as one might have envisioned. If you consider a collection of polynomial models, linear models are obviously simpler than those involving second or higher order polynomials. On the other hand, the notion of complexity could also carry a subjective component. For instance, it could significantly depend on the preferences of designers and users of the model. In a certain environment where neurocomputing is dominant, models of neural networks are far more acceptable and, therefore, perceived as being simpler than polynomial models. One should stress, however, that this type of assessment comes with a substantial level of subjectivity.

(e) High level of design autonomy of the model. Given that usually we encounter a significant number of design alternatives as to the architecture of the model, various parameters one can choose from to construct the detailed topology of the model, it is highly desirable to endow the development environment of the model with a significant level of design autonomy, that is exploit suitable optimization techniques that offer a variety of capabilities aimed at the structural development of the model. In this regard, evolutionary techniques of optimization play a pivotal role. Their dominant features such as population-based search, minimal level of guidance (a suitable fitness function is just a suitable mechanism guiding optimization efforts), collaborative search efforts (through some mechanisms of communication between the individual solutions) are of particular interest in this setting. The design of fuzzy models in the presence of the number of objectives is an example of multiobjective optimization in which the objectives are highly conflicting. The set of efficient solutions, called nondominated Pareto optimal, is formed by all elements in the solution space for which there is no further improvement without degradation in other design objectives. Hence, the machinery of genetic optimization becomes a highly viable and promising alternative.

When constructing fuzzy models, we also adhere to the same principles (viz. iterative development and successive refinements, accuracy assessment through training and testing sets, and striving for the lowest possible complexity of the construct). When it comes to the evaluation of accuracy of the fuzzy models, it is worth stressing that given the topology of these models in which the interface module constitutes an integral part (Pedrycz and Valente de Oliveira, 1996; Bortolan and Pedrycz, 2002; Mencar et al., 2006), there are two levels at which the accuracy of the models can be expressed. We may refer to them as internal and external levels of accuracy characterization. Their essence is schematically visualized in Figure 10.15. At the external (viz. numeric) level of accuracy quantification, we compute the distance between the numeric data and the numeric output of the model resulting from the transformation realized by the interface of the model.

10.4 Verification and Validation of Fuzzy Models

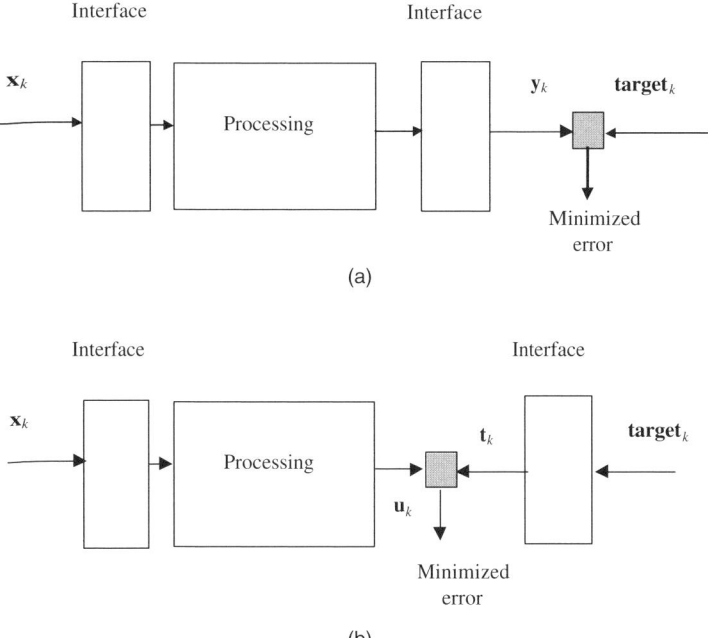

Figure 10.15 Two fundamental ways of expressing the accuracy of fuzzy models: (a) at the numeric level of experimental data and results of mapping through the interface, and (b) at the internal level of processing after the transformation through the interfaces.

In other words, the performance index expressing the (external) accuracy of the model reads in the form

$$Q = \sum_{k=1}^{N} \| \mathbf{y}_k - \mathbf{target}_k \|^2 \qquad (10.4)$$

where the summation is carried out over the numeric data available in the training, validation, or testing set. The form of the specific distance function (Euclidean, Hamming, Tchebyschev, or more generally, Minkowski distance) could be selected when dealing with the detailed quantification of the proposed performance index.

At the internal level of assessment of the quality of the fuzzy model, we transform the output data through the output interface so now they become vectors in the $[0, 1]^m$ hypercube and calculate distance at the internal level by dealing with two vectors with the [0,1] entries. As before, the calculations may involve training, validation, or testing data. More specifically, we have

$$Q = \sum_{k=1}^{N} \| \mathbf{u}_k - \mathbf{t}_k \|^2 \qquad (10.5)$$

refer also to Figure 10.15 (b). These two ways of quantifying the accuracy are conceptually different, and there is no equivalence between them unless their granular to numeric interface does not produce any additional error. We have already elaborated on this issue in Chapter 9 when dealing with the matter of interoperability. Quite often the interface itself could introduce an additional error. In other words, we may have a zero error at the level of granular information, however once we transform these results through the interface they become associated with the nonzero error. The performance index in the form shown above (3)–(4) is computed at either the numeric or the granular level. In the first case, refer to Figure 10.15 (a), it concerns the real numbers. At the level of information granules, the distances are determined at the level of the elements located in the unit hypercubes $[0, 1]^m$; see Figure 10.15 (b).

10.4.2 Training, Validation, and Testing Data in the Development of Fuzzy Models

When assessing the quality of the fuzzy models in terms of accuracy, stability, and transparency, it is important to quantify these features in an environment in which we could gain a high confidence as to the produced findings. Having overly optimistic evaluations is not advisable. Likewise, producing some pessimistic bias is not helpful as well. In order to strive for high reliability of the evaluation process of the model, one should release results of assessment based on the prudent use of available data. Here we follow the general guidelines and evaluation procedures encountered in system modeling (and these guidelines hold in spite of the diversity of the models and their underlying fundamentals).

Split of Data into Training and Testing Subsets To avoid any potential bias, the available data are split randomly into two disjoint subsets of training and testing data (with the split of 60–40% where 60% is used in the training set). The model is built using the training data. Next its performance is evaluated on the testing data. As this data set has not been used in the development of the model, we avoid any potential bias in its evaluation.

10 Fold Cross-Validation Although the use of the training and testing data helps gain some objectivity in the assessment, there still could be some variability in the evaluation, which could be the result of the random split. To reduce this effect, we randomly split the data into training–testing subsets, evaluate the performance of the model, and repeat the split and evaluation 10 times, in each case producing a random split of the data. In this sense, the obtained results could help reduce variability. Both the mean value of the performance and the related standard deviation are reported. When preparing the data, the split is typically carried out at the level of 90–10%.

Leave-One-Out Evaluation Strategy This strategy is of particular relevance when dealing with small data sets in which case the 60–40% split is not justifiable. Consider, for instance, 20 data points (which could be quite typical when dealing with data coming from some software projects; we do not have hundreds of those). In this case, the use of the approaches presented above could be quite unstable—as a matter of fact the number of data points is quite low, say 12 data for the training purposes; hence, the development of the fuzzy model could be affected by the reduced size of the data. In this case we consider a leave-one-out strategy. Here we use all but one data point in the training data, construct the model, and evaluate its performance on the one left out from the training data. The process is repeated for all data points starting from the first one left out, building the model and testing it on the single data point. Thus for N data, this strategy produces results of performance on all points being left out. Then average and standard deviation of the results could serve as a sound assessment of the quality of the fuzzy model.

So far, we have indicated that the available data set is split into its training and testing part. Quite often we also use a so-called validation set. The role of the validation set is to guide the development of the model with respect to its structural optimization. For instance, consider that we are at a position to adjust the number of fuzzy sets defined for individual variables. It is quite anticipated that when we start increasing the number of fuzzy sets, the accuracy of the model on the training set is going to become better. It is very likely that the tendency on the testing set is going to be quite different. The question as to the "optimal" number of fuzzy sets cannot be answered on the basis of the training data. The testing set is supposed not to be used at all in the construction of the model. To solve the problem, in addition to the training data set, we set aside a portion of the data—validation set, which is used to validate the model constructed on the basis of the training data. The development process proceeds as follows: We construct a model (estimate its parameters, in particular) on the basis of the training set. When it comes to the structural development (say, the number of fuzzy sets and alike) where there is a strong monotonic tendency, we have to resort to the validation set: choose the value of the structural parameter (say, the number of nodes, processing units, fuzzy sets, etc.), optimize the model on the training set and check its performance on the validation set, and select the value for which we get the best results on the validation set.

10.4.3 Validation of Fuzzy Models

As already indicated, validation is focused on the issues related to the question, "Are we building the *right* system?" In essence, the term of validation is inherently multifaceted. It embraces several important aspects, in particular transparency and stability of the model. Let us discuss them in more detail.

Transparency of the Model The interpretation of transparency or "readability" of the model is directly associated with the form of the fuzzy model (Casillas et al. 2003; Paiva and Dourado, 2004). The essence of this feature of the model is associated with the ability to easily comprehend the model, namely pick up the key relationships captured by it. There is also a substantial level of flexibility in the formalization of the concept of transparency. For instance, consider a rule-based model that is composed of a series of rules "if condition$_1$ *and* condition$_2$ *and* .... *and* condition$_n$ then conclusion" where the conditions and conclusions are quantified in terms of some fuzzy sets. The transparency of the model can be quantified by counting the number of rules and taking into consideration the complexity of each rule. This complexity can be expressed by counting the number of the conditions standing in the rule. The larger the number of the rules and the longer they are, they lower the readability (transparency) of the model. When dealing with network type of fuzzy models such as, for example fuzzy cognitive maps, the immediate criterion we may take into consideration is the number of nodes or the number of connections between them (or alternatively the density of connections, which is determined by counting the number of connections and dividing them by the number of nodes). The higher these values, the more difficult it becomes to "read" the model and interpret it in a meaningful manner. The transparency of the model is also essential when dealing with an ability to accommodate any prior domain knowledge that is available in any problem solving. The existing components of such domain knowledge are highly instrumental in the development of the models. For instance, one could easily reduce learning effort going toward the estimation of the parameters of the model (say, a fuzzy neural network) once the learning starts from a certain promising point in the usually huge search space.

Stability of the Model The substantial value of fuzzy models comes with their stability. We always prefer a model that is "stable" so it does not change over some minor variations of the environment, and experimental data in particular. Practically, if we take some subsets of training data, we anticipate that the resulting fuzzy model does not radically change and retains its conceptual core, say a subset of rules that are essential descriptors of the phenomenon or the process of interest. Some minor variations of other less essential rules cannot be avoided and are less detrimental to the overall stability of the model. There could also be some changes in the numeric values of the parameters of the model, yet their limited changes could be secondary to the stability of the model. The aspect of the model's stability is somewhat associated with the transparency we have considered so far: Once provided with the model, we expect that it concentrates on these aspects of reality that repeats all the time in spite of the variations of the environment. By the same token we intend to avoid some highly variable components of the model as not contributing to the essence of the underlying phenomenon. Intuitively, one could conclude that the stability is inherently associated with the level of granularity we establish for the description. This is not surprising at all: the higher the generality, the higher the stability of the model.

It should be stressed that these fundamental features of the fuzzy models could be in competition. High accuracy could reduce readability. High transparency could come at the cost of reduced accuracy. Always a sound compromise should be strived for. A suitable choice depends upon the relationships between these characteristics. Some examples are illustrated in Figure 10.16. Reaching a compromise should position us at a point where abrupt changes are avoided.

10.5 CONCLUSIONS

We have discussed the paradigm of fuzzy modeling and identified the main features of this modeling environment. While a number of fundamental requirements we

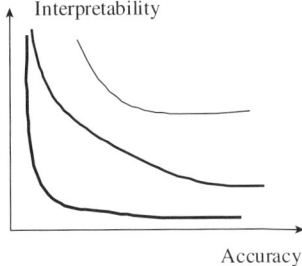

Figure 10.16 Examples of relationships between interpretability and accuracy of fuzzy models; the character of these dependencies are specific to the type of fuzzy model under consideration.

encounter in general system modeling are pertinent here, we also have new aspects to fully deal with, which come because of the processing of granular information realized in such models. In particular, the three module topology of the fuzzy models with clearly delineated interfaces is definitely worth noting here. There are also several essential modes of use of fuzzy models, which definitely address the point of human-centricity that fuzzy modeling brings into the picture. The review of the main architectural categories of the fuzzy models is beneficial in asserting their capabilities as far as the concepts of interpretability, stability, and accuracy are concerned.

This chapter provides a general overview and focuses on the methodological issues. The detailed discussion on the main categories of the models such as fuzzy rule-based systems and fuzzy neural networks will be studied in detail in the consecutive chapters.

EXERCISES AND PROBLEMS

1. Suggest a fuzzy model of decision-making concerning a purchase of a car. In particular, consider the following components that are essential to the model:

   (a) variables and their possible granulation realized in the form of some fuzzy sets
   (b) optimization criteria that could be viewed meaningful in the problem formulation
   (c) type of the fuzzy model; justify your choice

2. Transform the following decision tree into a collection of rules.

   The information granules of the attributes are given in the form of fuzzy sets or intervals as illustrated above. What would be the result (class membership) for the input $x = 1.5$ and $y = 2.0$?

3. Offer a representation of the following problem in the form of a suitable fuzzy cognitive map.

   Consider a process of weed control through spraying. The herbicides increase the mortality of weeds. Both the crop and weed depend upon the nutrients and water in the soil. There is also another component in the process that is herbivore. The growth of herbivore population depends on the amount of weed eaten; also each individual has a natural rate of mortality.

272 Chapter 10 Fuzzy Modeling: Principles and Methodology

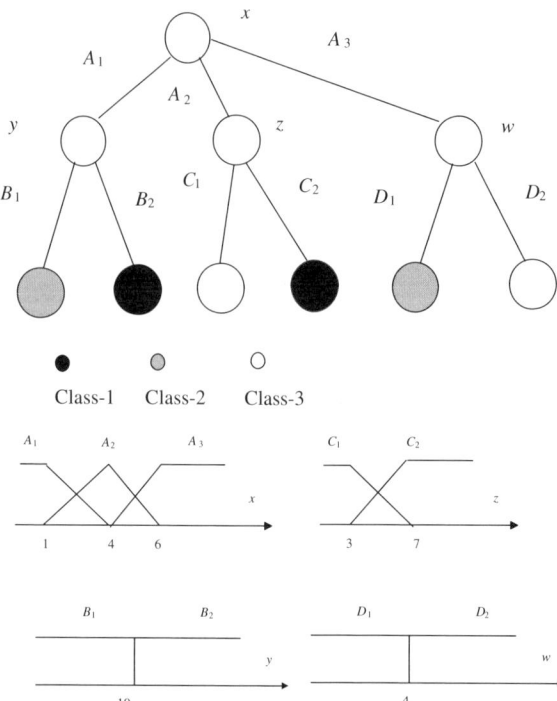

Justify the choice of the structure. Identify the pertinent variables describing the system and run the map for several iterations. Start with a single concept being fully activated; consider that the others are equal to zero.

4. The experimental data for a single-input–single-output relationship are shown below.

What form of rules and how many would you suggest in each case; justify your choice.

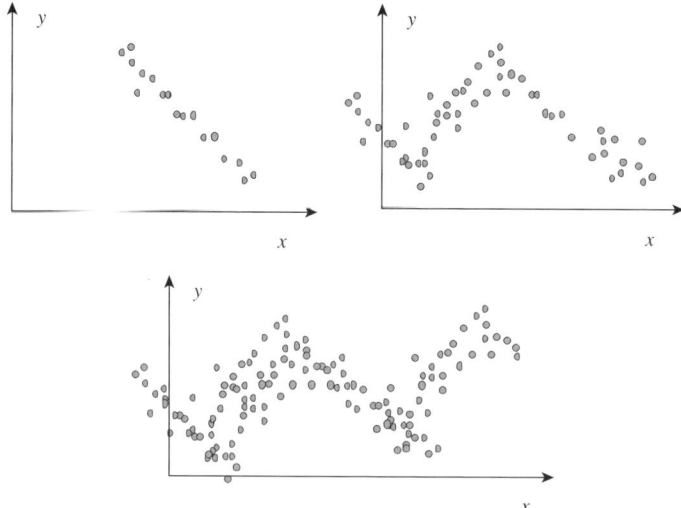

5. Is there any relationship between the 10-fold cross-validation strategy and the leave-one-out strategy?
6. Pick up a manual of any appliance, electronic device, cellular phone and alike. Could you develop a fuzzy model on its basis (say, a collection of fuzzy rules).
7. The output of the fuzzy model produces the membership functions shown below.

    How could these fuzzy sets help you interpret the result produced by the fuzzy model? Assume that the output y can assume values in the range $[a, b]$

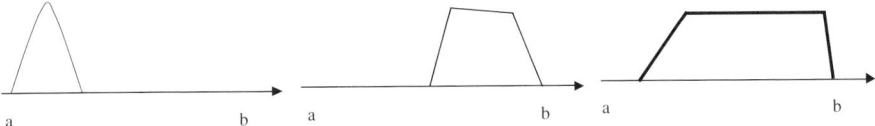

8. Give several reasons why we should not go for too many fuzzy sets defined in the system's variables.

HISTORICAL NOTES

In system modeling we are commonly after simplicity of the resulting constructs. Here the concept of the Occam's razor principle fully applies. This principle states that the explanation of any phenomenon (system) should be realized in the simplest possible way using a few assumptions (i.e., entities should not be multiplied beyond necessity).

The 14th-century English logician and Franciscan friar William of Ockham.(1295–1349) is attributed to law of parsimony (simplicity or succinctness) in scientific theories. Occam's razor states that the explanation of any phenomenon should make as few assumptions as possible.

The principle of incompatibility lucidly formulated by Zadeh (1973) underlines the fundamental conflict between numeric accuracy and overall meaningfulness of the models.

> *As the complexity of a system increases, our ability to make precise and yet significant statements about its behavior diminishes until a threshold is reached beyond which precision and significance (or relevance) become almost mutually exclusive characteristics*

Fuzzy modeling is an example of granular modeling, namely, modeling that is inherently associated with information granules. In this regard, it is instructive to refer to qualitative modeling (Kuipers,

1986; Forbus, 1984, 1993). Qualitative modeling is based predominantly on symbols. Very often these symbols do not come with any numeric characterization, so when dealing with the terms such as low, medium, high used there, one has to view them as some entities for which there is some simple calculus of algebraic manipulation. Note, however, that the concepts are symbolic and do not carry any numeric quantification (such as the one being formed through membership functions). The relevance of qualitative modeling was also stressed by Puccia and Levins in their book published in 1985 (Puccia and Levins, 1985).

The methodology, algorithms, and applications of fuzzy modeling have been vigorously pursued since the inception of fuzzy sets. Most of the models were those of a rule-based format. Owing to this architecture, the models supported their high readability. Starting from the collections of associations between fuzzy sets in input and output space, the rule-based models started to become very much dominated by the rules with local regression models (as originally introduced by Takagi and Sugeno in 1985). With the renaissance of neurofuzzy systems where neural networks came in a variety of synergistic constructs with fuzzy sets, the accuracy of fuzzy models started to dominate the development agenda. For some period of time, the issues of transparency and readability as well as simplicity of the fuzzy models were not predominantly visible. Only recently (Casillas et al., 2003), they have started growing in importance.

REFERENCES

Apolloni, B., Zamponi, G., Zanaboni, A. Learning fuzzy decision trees, *Neural Netw.* **11**, 1998, 885–895.

Axelrod, R. *Structure of Decision: The Cognitive Maps of Political Elites*, Princeton University Press, Princeton, NJ, 1976.

Babuska, R. *Fuzzy Modeling for Control*, Kluwer Academic Publishers, Dordrecht, 1998.

Bezdek, J. Fuzzy models—what are they and why? *IEEE Trans. Fuzzy Syst.* **1**, 1993, 1–6.

Bortolan, G., Pedrycz, W. Linguistic neurocomputing: the design of neural networks in the framework of fuzzy sets, *Fuzzy Set Syst.* **128**(3), 2002, 389–412.

J. Casillas et al. (eds.), *Interpretability Issues in Fuzzy Modeling*, Springer Verlag, Berlin, 2003.

Chang, R., Pavlidis, T. Fuzzy decision tree algorithms, *IEEE Trans. Syst. Man Cybern.* SMC-7, **1**, 1977, 28–35.

Chen, M., Linkens, D. Rule-base self-generation and simplification for data-driven fuzzy models, *Fuzzy Set Syst.* **142**(2), 2004, 243–265.

Ciaramella, A., Tagliaferri, R., Pedrycz, W., Di Nola, A. Fuzzy relational neural network, *Int. J. Approx. Reason.* **41**(2), 2006, 146–163.

Cordón, O., Herrera, F., Zwir, I. A hierarchical knowledge-based environment for linguistic modeling: models and iterative methodology, *Fuzzy Set Syst.* **138**(2), 2003, 307–341.

Forbus, K. Qualitative process theory, *Artif. Intell.* **24**(1–3), 1984, 85–168.

Forbus, K. Qualitative process theory: twelve years after, *Artif. Intell.* **59**(1–2), 1993, 115–123.

Janikow, C. Fuzzy decision trees: issues and methods, *IEEE Trans. Syst. Man Cybern.—Part B*, **28**(1), 1998, 1–14.

Kacprzyk, J. *Multistage Decision-Making under Fuzziness*, TUV Verlag, Rheinland, Cologne, 1983.

Kacprzyk, J. *Multistage Fuzzy Control*, John Wiley & Sons, Ltd, Chichester, 1997.

Kosko, B. Fuzzy cognitive maps, *Int. J. Man-Mach. Stud.* **24**, 1986, 65–75.

Kosko, B. *Neural Networks and Fuzzy Systems*, Prentice Hall, Englewood Cliffs, NJ, 1992.

Kuipers, B. Qualitative simulation, *Artif. Intell.* **29**, 1986, 289–338.

Mencar, C., Castellano, G., Fanelli, A. Interface optimality in fuzzy inference systems, *Int. J. Approx. Reason.* **41**(2), 2006, 128–145.

Oh, S., Pedrycz, W. A new approach to self-organizing fuzzy polynomial neural networks guided by genetic optimization, *Phys. Lett. A* **345**(1–3), 2005, 88–100.

Paiva, R., Dourado, A. Interpretability and learning in neuro-fuzzy systems, *Fuzzy Set Syst.* **147**(1), 2004, 17–38.

Papageorgiou, E. I., Stylios, C., Groumpos, P. P. Unsupervised learning techniques for fine-tuning fuzzy cognitive map causal links, *Int. J. Hum. Comput. Stud.* **64**(8), 2006, 727–743.

Papageorgiou, E., Groumpos, P. A new hybrid method using evolutionary algorithms to train fuzzy cognitive maps, *Appl. Soft Comput.* **5**(4), 2005, 409–431.

Pedrycz, W., Valente de Oliveira, J. Optimization of fuzzy models, *IEEE Trans. Syst. Man Cybern.—Part B*, **26**(4), 1996, 627–636.

Pedrycz W. (Ed.), *Fuzzy Modelling: Paradigms and Practice*, Kluwer Academic Press, Dordrecht, 1996.

Pedrycz, W., Gomide, F. *An Introduction to Fuzzy Sets*, MIT Press, Cambridge, MA, 1998.

Pedrycz, W., Sosnowski, Z. The design of decision trees in the framework of granular data and their application to software quality models, *Fuzzy Set Syst.* **123**(3), 2001, 271–290.

Pedrycz, W., Vukovich, G. Granular neural networks, *Neurocomput.* **36**(1–4), 2001, 205–224.

Pedrycz, W. Logic-driven fuzzy modeling with fuzzy multiplexers, *Eng. Appl. Artif. Intell.* **17**(4), 2004, 383–391.

Pedrycz, W., Reformat, M. Genetically optimized logic models, *Fuzzy Set Syst.* **150**(2), 2005, 351–371.

Puccia, C., Levins, R. *Qualitative Modeling of Complex Systems*, Harvard University Press, Cambridge, MA, 1985.

Qin, Z., Lawry, J. Decision tree learning with fuzzy labels, *Inf. Sci.* **172**(1–2), 2005, 91–129.

Takagi, T., Sugeno, M. Fuzzy identification of systems and its application to modelling and control, *IEEE Trans. Syst. Man Cybern.* **15** 1985, 116–132.

Zadeh, L. A. Outline of a new approach to the analysis of complex system and decision process, *IEEE Trans. Syst. Man Cybern.* **3**, 1973, 28–44.

Zadeh, L. A. Toward a generalized theory of uncertainty (GTU): An outline, *Inf. Sci.* **172**(1–2), 2005, 1–40.

Chapter 11

Rule-Based Fuzzy Models

Rule-based models play a central role in fuzzy modeling. Fuzzy rules capture relationships among fuzzy variables and provide a mechanism to link linguistic descriptions of systems with their computational realizations. Fuzzy rules can be formalized via collections of fuzzy relations. As such, they naturally provide a way to construct models of systems involving domain knowledge, experience, and experimental data. Interpreting fuzzy rules as fuzzy relations allows the use of relational calculus to process information and carry out efficient computing with fuzzy rules in numerous applications. This chapter addresses the concepts of fuzzy rule-based systems and architectures, discusses their structural representations and interpretations, and offers procedures to carry out rule-based computing. Design issues, adaptation, and parametric optimization mechanisms are addressed as well.

11.1 FUZZY RULES AS A VEHICLE OF KNOWLEDGE REPRESENTATION

In their generic format, rules forming a core of rule-based systems come in the form of conditional statements.

$$\text{If input variable is } A \text{ then output variable is } B \qquad (11.1)$$

where A and B standing in the "If" and "then" parts of the rules are descriptors of some pieces of knowledge about the domain (problem to be represented). The rule itself expresses a certain relationship between these input and output variables (descriptors). For instance, the rule "if the temperature is *high* then the electricity demand is *high*" captures a piece of domain knowledge that is essential to some specific planning activities exercised by an electric company. Notably, the rule of this character is quite qualitative yet at the same time highly expressive. We are perhaps not so much concerned about the detailed numeric quantification of the descriptors standing in the rules; however, we appreciate the fact that the rule presents some interesting and transparent relationship pertinent to the problem. We may note that both the "If" and "then" parts are formed as information granules—conceptual

---

Fuzzy Systems Engineering: Toward Human-Centric Computing, by Witold Pedrycz and Fernando Gomide
Copyright © 2007 John Wiley & Sons, Inc.

entities that are semantically sound abstractions. The operational context within which information granules are formalized and used afterward could be established by considering one among the available formal frameworks, say sets, fuzzy sets, rough sets, and others.

In practical cases, the domain knowledge is typically structured into a family of c rules with each of them assuming the format

$$\text{If input variable is } A_i \text{ then output is } B_i \qquad (11.2)$$

$i = 1, 2, \ldots, c$ where A_i and B_i are information granules. The rules articulate a collection of meaningful relationships existing among the variables of the problem.

We may envision more complex rules whose left-hand sides may include several conditions. Such rules read as

If input variable$_1$ is A *and* input variable$_2$ is B *and* ... *and* input variable$_n$ is W
then output is Z (11.3)

with the multidimensional input space formed as a Cartesian product of the input variables. Note that the individual conditions are aggregated together by the *and* connective. From the system modeling standpoint, rules give rise to a highly parallel, modular form of a granular model; an expansion of the model usually requires an addition of some new rules while the existing ones are left intact.

11.2 GENERAL CATEGORIES OF FUZZY RULES AND THEIR SEMANTICS

Typically, multi-input multi-output fuzzy rules come in the following form:

If X_1 is A_1 *and* X_2 is A_2 *and* ... *and* X_n is A_n
then Y_1 is B_1 *and* Y_2 is B_2 *and* ... *and* Y_m is B_m

where X_i and Y_j are viewed as variables whose values are fuzzy sets A_i and B_j such as for example, *high*, *medium*, and *low*, defined in the corresponding spaces $\mathbf{X}_i$ and $\mathbf{Y}_j$, $i = 1, 2, \ldots, n$ and $j = 1, 2, \ldots, m$. The left-hand side of the rule is referred to as antecedent and the right-hand side is known as consequent. In the simplest case, $n = m = 1$. Quite commonly we encounter a case in which $n = 2$ and $m = 1$. Although the rule can be easily generalized to higher values of n and m, it is sufficient to discuss the computing details of the inference scheme. In this two-input single-output case, the rules assume the following format:

If X is A *and* Y is B then Z is C

which could be treated as the generic one.

Rules may come in different formats depending upon the nature of the problem and a character of domain knowledge being available. For instance, rules could be uncertain, gradual, relational, and others.

11.2.1 Certainty-Qualified Rules

Instead of allocating full confidence in the validity of the rules, we allow to treat them as being satisfied (valid) at some level of confidence. The degree of uncertainty leads to certainty-qualified expressions of the following form:

If X is A and Y is B then Z is C with certainty μ

where $\mu \in [0, 1]$ denotes the degree of certainty of this rule. If $\mu = 1$, we say that the rule is certain.

11.2.2 Gradual Rules

Rules may also involve gradual relationships between objects, properties, or concepts. For example, the rule

the *more* X is A, the *more* Y is B

expresses a relationship between changes in Y triggered by the changes in X. In these rules, rather than expressing some association between antecedents and consequents, we capture the tendency between the information granules; hence the term of *graduality* occurring in the condition and conclusion part. For instance, the rules of the form

the *higher* the values of condition, the *higher* the values of conclusion (11.4)

or

the *lower* the values of condition, the *higher* the values of conclusion (11.5)

represent the knowledge about the relationships between the changes of the condition and conclusion. For instance, the graduality occurs in the rule "the *higher* the income, the *higher* the taxes" or "typically the *higher* the horsepower, the *higher* the fuel consumption." Gradual rules are frequently encountered in commonsense reasoning.

11.2.3 Functional Fuzzy Rules

In this category of rules the consequent comes in the form of some function of the input variables whose scope is narrowed down (focused) to the condition part of the rule, that is,

$$\text{If } \mathbf{x} \text{ is } A_i, \text{ then } y = f_i(\mathbf{x}, \mathbf{a}_i) \quad (11.6)$$

The function f_i itself could be linear or nonlinear and applies only to the inputs $\mathbf{x} \in \mathbf{R}^n$. Being more descriptive, we may say that the scope of this function is confined to the support of the information granule represented by A_i.

Let us highlight two general and essential observations that hold in spite of the significant diversity of the formats of the rules themselves and their ensuing applications. First, the rules give rise to highly modular architectures and such an organization becomes crucial to the efficient realization of the mechanisms of model formation, reasoning, and future maintenance and update. Second, rules are always formulated in the language of information granules; hence they constitute an abstract reflection of the problem or problem solving.

In rule-based models, rules could be used in a mixed-mode format. This means that the collection of rules can include a combination of rules in which both conditions and conclusions are represented in the form of fuzzy sets as well as rules. This could be a reflection of the character of the available data. For instance, we could have data of different levels of distribution, and this may easily lead to different ways of system modeling.

In general, fuzzy rules can be classified into two broad categories: categorical rules and dispositional rules. Categorical rules are statements with no quantification. They do not contain any fuzzy quantifiers, fuzzy probabilities, or alike. Dispositional expressions are rules with fuzzy quantifiers or probabilities. We say that they may be preponderantly, but not necessarily always, true (Zadeh, 1989). Dispositional rules will not be considered here, but details can be found in Pedrycz and Gomide (1998).

11.3 SYNTAX OF FUZZY RULES

A syntax of fuzzy rules can be summarized with the use of the BNF (Backus-Nauer Form) notation, which offers a higher level of formalism and rigor. An example of the syntax of the rules is shown below:

$$\langle \text{if\_then\_rule} \rangle ::= \text{if} \langle \text{antecedent} \rangle \text{then} \langle \text{consequent} \rangle \{ \langle \text{certainty} \rangle \}$$
$$\langle \text{gradual\_rule} \rangle ::= \langle \text{word} \rangle \langle \text{antecedent} \rangle \langle \text{word} \rangle \langle \text{consequent} \rangle$$
$$\langle \text{word} \rangle ::= \langle \text{more} \rangle \{ \langle \text{less} \rangle \}$$
$$\langle \text{antecedent} \rangle ::= \langle \text{expression} \rangle$$
$$\langle \text{consequent} \rangle ::= \langle \text{expression} \rangle$$
$$\langle \text{expression} \rangle ::= \langle \text{disjunction} \rangle \{ and \langle \text{disjunction} \rangle \}$$
$$\langle \text{disjunction} \rangle ::= \langle \text{variable} \rangle \{ or \langle \text{variable} \rangle \}$$
$$\langle \text{variable} \rangle ::= \langle \text{attribute} \rangle \text{ is } \langle \text{value} \rangle$$
$$\langle \text{certainty} \rangle ::= \langle \text{none} \rangle \{ \text{certainty } \mu \in [0, 1] \}$$

Fuzzy rules can be regarded as fuzzy relations constructed with the values of the fuzzy variables, and the fuzzy sets of rule antecedent and consequent. The membership functions of the fuzzy relations depend on the operators chosen to combine the fuzzy sets. In turn, the choice of the logic operators depends on the meaning of the rules and the inference procedure under consideration.

The construction of computable (operational) representations of rule-based systems involves the following steps:

1. specification of the fuzzy variables to be used in the rule;
2. association of the fuzzy variables with the use of fuzzy sets;

3. computational formalization of each rule by means of the corresponding fuzzy relation and development of aggregation mechanisms determining a way in which individual rules are put together (Dubois and Prade, 1996).

11.4 BASIC FUNCTIONAL MODULES: RULE BASE, DATABASE, AND INFERENCE SCHEME

A general architecture of a rule-based fuzzy model is shown in Figure 11.1. It is composed of five generic modules: the input interface, rule base, database, fuzzy inference, and output interface.

The *input interface* accepts inputs X defined in universe $\mathbf{X}$ and converts the inputs into the format of some propositions that the fuzzy inference can use to activate and process the fuzzy rules. In general, the input X is a fuzzy set on $\mathbf{X}$, but in certain application domains such as signal processing, system control, forecasting, and diagnosis, it could be a single numeric entity. In other words, X is a single-element set (singleton).

The *rule base* is composed of a set of fuzzy "if–then" rules describing an input–output relationship being perceived at the level of information granules.

The *database* stores the values of the parameters of the rule-based model. These parameters concern values of the scaling factors, detailed definitions of the universes of the input and output variables, and details of the membership functions (including their types and corresponding parameters). The values of these parameters are highly problem dependent.

The *fuzzy inference* module processes the inputs using the rule base and exploiting the mechanisms of fuzzy inference and approximate reasoning. Because the inputs and rules are fuzzy expressions, fuzzy inference transforms the inputs using fuzzy rules as a means to develop the outputs. Quite commonly, these transformations are treated compositions of fuzzy sets and fuzzy relations.

The *output interface* translates the result of fuzzy inference into a suitable format required by the application environment. Typically, the result of approximate reasoning is a certain fuzzy set. In some categories of applications, say, in control engineering, we require a single numeric value of control. In this case, the result

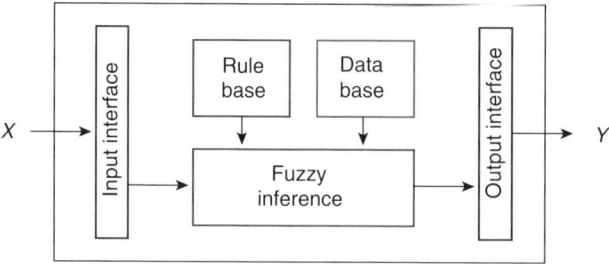

Figure 11.1 General architecture of fuzzy rule-based models outlining main modules of the system.

of reasoning has to be transformed (decoded) into a number. The decoding mechanism involved here is commonly referred to as defuzzification.

11.4.1 Input Interface

The generic component used to represent inputs are expressions of the following form:

⟨attribute⟩ of ⟨input⟩ is ⟨value⟩

sometimes referred to as triples ⟨attribute, object, value⟩. For instance, the proposition

the temperature of the motor is *high*

states that the object under discussion is motor whose attribute (variable) is temperature, which is high. The basic information units are atomic expressions whose canonical form is

$$p : X \text{ is } A \tag{11.7}$$

where X is a linguistic variable representing the pair attribute (input) and A is its value, a fuzzy set on an universe **X**. For instance, the proposition shown above can be written down as

temperature (motor) is *high*

where temperature (motor) is the variable X and *high* is its value quantified in the form of the corresponding fuzzy set. Therefore, variables in fuzzy expressions are viewed as linguistic variables whose linguistic values are fuzzy sets. Fuzzy sets provide a computable representation of a proposition as illustrated in Figure 11.2.

Multiple fuzzy inputs can be handled analogously through forming an atomic fuzzy expression for each input and then using conjunctions or disjunctions to form the compound expression. The aggregations into the corresponding conjunctions and disjunctions are realized with the use of t-norms and t-conorms, respectively. We have

$$p : X_1 \text{ is } A_1 \text{ and } X_2 \text{ is } A_2 \text{ and } \ldots \text{ and } X_n \text{ is } A_n \quad \text{conjunctive canonical form} \tag{11.8}$$

$$q : X_1 \text{ is } A_1 \text{ or } X_2 \text{ is } A_2 \text{ or } \ldots \text{ or } X_n \text{ is } A_n \quad \text{disjunctive canonical form} \tag{11.9}$$

where A_i is a fuzzy set defined in $\mathbf{X}_i$, and X_i stands for linguistic variables, $i = 1, 2, \ldots, n$.

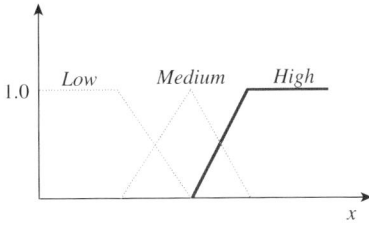

Figure 11.2 Representation of fuzzy proposition X is *high*.

The compound proposition induces fuzzy relations P and Q on $\mathbf{X}_1 \times \mathbf{X}_2 \times \cdots \times \mathbf{X}_n$ whose membership functions are given as

$$P(x_1, x_2, \ldots, x_n) = A_1(x_1) t A_2(x_2) t \ldots t A_n(x_n) = \mathop{T}_{i=1}^{n} A_i(x_i) \qquad (11.10)$$

$$Q(x_1, x_2, \ldots, x_n) = A_1(x_1) s A_2(x_2) s \ldots s A_n(x_n) = \mathop{S}_{i=1}^{n} A_i(x_i) \qquad (11.11)$$

Figure 11.3 illustrates the case with $n = 2$ when using triangular membership functions $A_1(x, 4, 5, 6)$ and $A_2(y, 8, 10, 12)$, and the algebraic product t-norm ($p = ab$) to build P, and the probabilistic sum ($q = a + b - ab$) to form Q,

$$P(x, y) = A(x) \cdot B(y), \quad \forall (x, y) \in \mathbf{X} \times \mathbf{Y}$$
$$Q(x, y) = [A(x) + B(y) - A(x) \cdot B(y)], \quad \forall (x, y) \in \mathbf{X} \times \mathbf{Y}$$

The compound expressions (11.8) and (11.9) can be rewritten, using (11.10) and (11.11), as follows:

$$\mathrm{p} : (X_1, X_2, \ldots, X_n) \text{ is } P$$
$$\mathrm{q} : (X_1, X_2, \ldots, X_n) \text{ is } Q$$

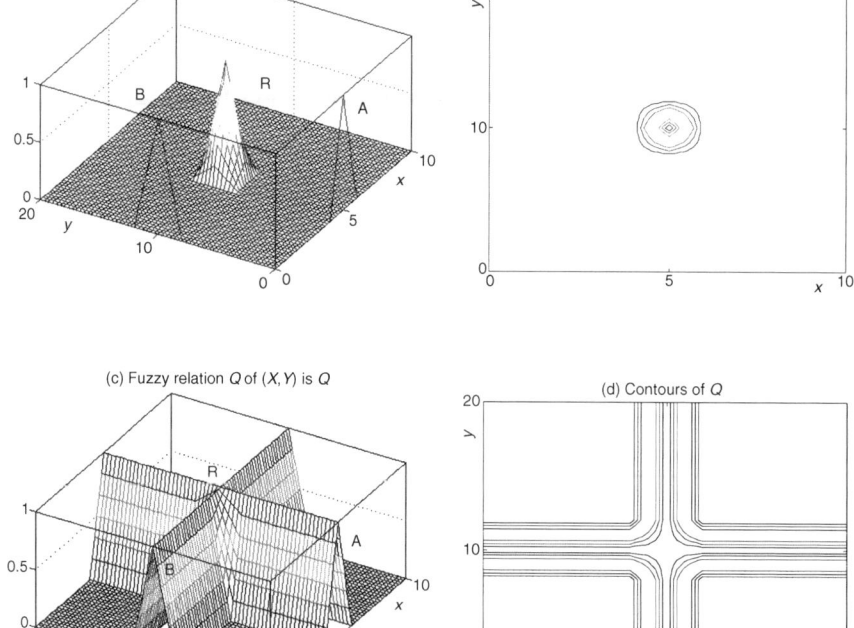

Figure 11.3 Fuzzy relations associated with propositions (a) (X, Y) is P and (c) is (X,Y) is Q t-norm: product, t-conorm: probabilistic sum.

11.4 Basic Functional Modules: Rule Base, Database, and Inference Scheme

Analogously, introducing a compound variable $X = (X_1, X_2, \ldots, X_n)$ and recalling the multivariable nature of the membership functions of fuzzy relations P and Q, expressions (11.8) and (11.9) take on the form equivalent to (11.7)

$$p : X \text{ is } P \qquad (11.12)$$
$$q : X \text{ is } Q \qquad (11.13)$$

11.4.2 Rule Base

Intuitively, expressions coming in the form of the rule "If X is A then Y is B" describe some relationship between fuzzy variables X and Y. The semantics of this dependency is formalized in terms of some fuzzy relation R. Here the membership function $R(x,y)$ represents the degree to which a pair $(x, y) \in \mathbf{X} \times \mathbf{Y}$ is compatible with the relation zbetween the variables X and Y involved in the rule. Because A and B are fuzzy sets on X and Y, the relation R on $\mathbf{X} \times \mathbf{Y}$ can be determined by an relational assignment of the form

$$R(x, y) = f(A(x), B(y)) \quad \forall (x, y) \in \mathbf{X} \times \mathbf{Y}$$
where $f : [0, 1]^2 \to [0, 1]$.

In general, the fuzzy relation capturing the relationship between A and B can fall under one of the three general categories such as

1. fuzzy conjunction
2. fuzzy disjunction
3. fuzzy implication

Fuzzy conjunction and implication are the two models being encountered most often. Fuzzy conjunction, denoted here by f_t, and fuzzy disjunction, f_s, may be viewed as two dual generalizations of the realization of some sort of Cartesian product of the fuzzy sets via triangular norms and conorms. On the other hand, fuzzy implications, denoted by f_i, are generalizations of implications encountered in multiple-valued logic.

Let us briefly discuss these three realizations.

11.4.2.1 Fuzzy Conjunction

A fuzzy conjunction $f_t : [0, 1]^2 \to [0, 1]$ is treated as a function such that

$$f_t(A(x), B(y)) = A(x) t B(y), \quad \forall (x, y) \in \mathbf{X} \times \mathbf{Y} \qquad (11.14)$$

In other words, the fuzzy rule formed in this way is a fuzzy relation R_t that induces a constraint on the joint variable (X, Y) and comes in the form (X, Y) is R_t. For example, consider the following rule:

If X is A then Y is B.

If we choose the minimum operation, then the membership function f_c of the corresponding fuzzy relation, denoted by R_c, is the fuzzy Cartesian product, namely,

$$R_c(x,y) = f_c(A(x), B(y)) = \min[A(x), B(y)], \quad \forall (x,y) \in \mathbf{X} \times \mathbf{Y} \quad \text{(Mamdani)}$$

Typically, in the context of rule-based computing, we refer to this construct as a Mamdani relation, or Mamdani, for short. Figure 11.4(a) shows the function f_c and Figure 11.4(c) an example of the relation R_c in the case when A and B are triangular fuzzy sets.

Similarly, if we consider the realization of t-norm by means of the algebraic product, then the membership function f_p of the respective fuzzy relation R_p reads as

$$R_p(x,y) = f_p(A(x), B(y)) = A(x) \cdot B(y), \quad \forall (x,y) \in \mathbf{X} \times \mathbf{Y} \quad \text{(Larsen)}$$

The plot of this fuzzy relation is included in Figure 11.4(b) and (d), respectively.

The certainty-qualified rules coming in the form

If X is A then Y is B with certainty μ

give rise to the conjunctive type of representation $f_e : [0,1]^2 \to [0,1]$ such that (Yager, 1984)

$$f_e(A(x), B(y)) = (f_t(A(x), B(y))t\mu) + (1-\mu), \quad \forall (x,y) \in \mathbf{X} \times \mathbf{Y}$$

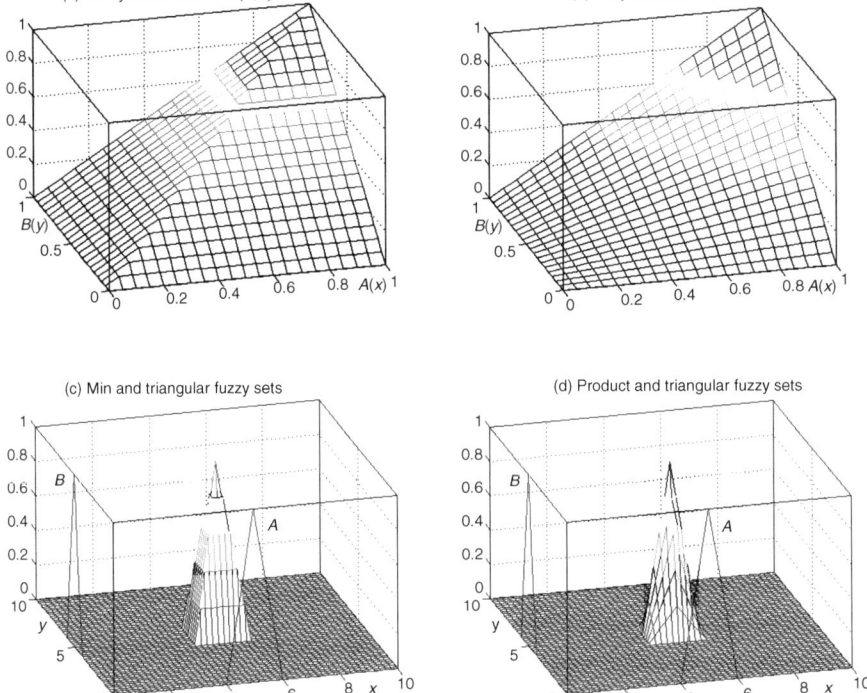

Figure 11.4 Fuzzy rule "If X is A then Y is B" interpreted as conjunction: the functions (a) f_c and (b) f_p, and examples of the relations (c) R_c and (d) Rp induced by f_c and fp, whereas A and B are triangular fuzzy sets $A(x, 4, 5, 6)$ and $B(x, 4, 5, 6)$.

11.4 Basic Functional Modules: Rule Base, Database, and Inference Scheme

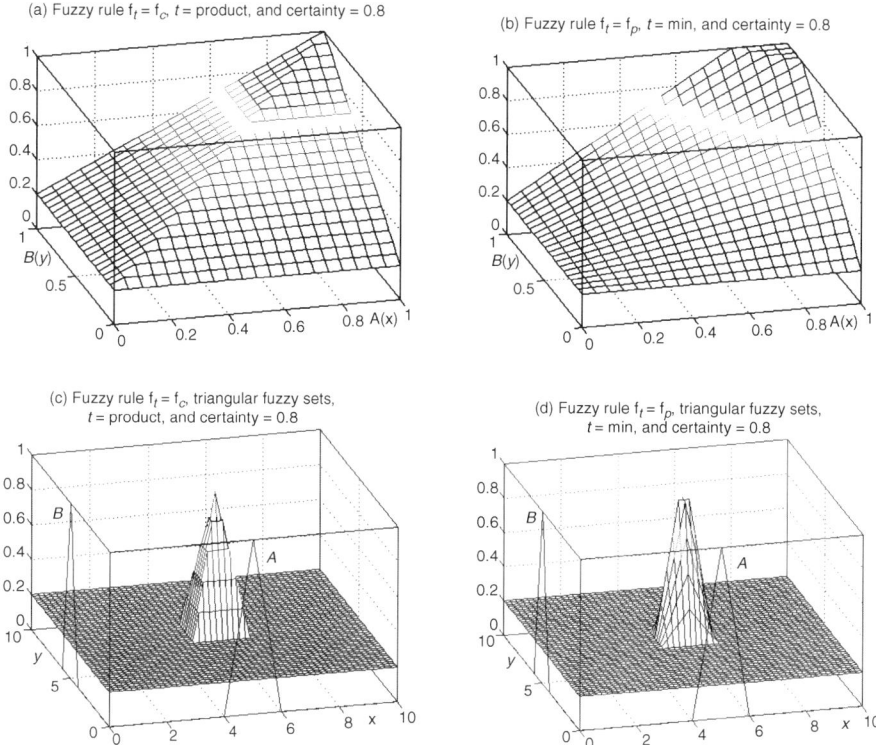

Figure 11.5 Fuzzy rule "If X is A then Y is B" with certainty factor: (a) with minimum f_c and (b) algebraic product f_c conjunctions, considering the algebraic (c) product and (d) minimum t-norm, respectively, to handle the certainty factor $\mu = 0.8$ and triangular fuzzy sets.

In other words, the certainty-qualified rule is interpreted as a fuzzy relation R_e scaled by the certainty factor μ, that is,

$$R_e(x,y) = (f_t(A(x), B(y))t\mu) + (1-\mu), \quad \forall (x,y) \in \mathbf{X} \times \mathbf{Y}$$

If $\mu = 1$, then such certainty-qualified fuzzy rule becomes a conjunction as shown in Figure 11.4. Otherwise, the membership function becomes scaled down by the value of the certainty factor. Figure 11.5 shows an example of the rule.

11.4.2.2 Fuzzy Disjunction

A fuzzy disjunction is a function $f_s : [0,1]^2 \to [0,1]$ such that

$$f_s(A(x), B(y)) = A(x) s B(y), \quad \forall (x,y) \in \mathbf{X} \times \mathbf{Y} \quad (11.15)$$

This fuzzy rule can also be viewed as a fuzzy relation R_s that induces a constraint on the joint variable (X, Y) in the form (X, Y) is R_s. For instance, considering the same rule as before

If X is A then Y is B

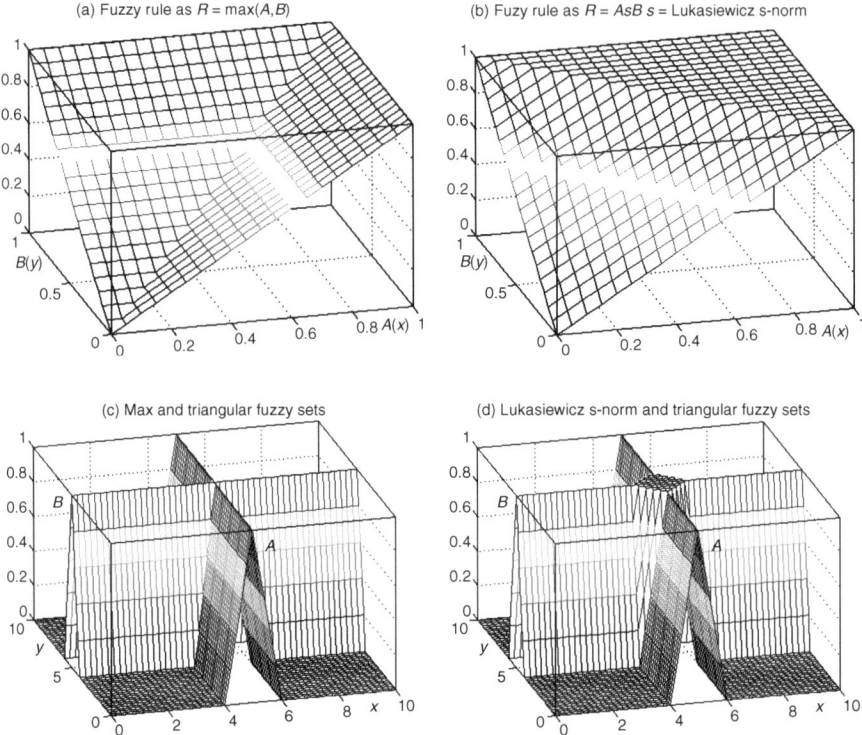

Figure 11.6 Fuzzy rule "If X is A then Y is B" interpreted as disjunction: function f_m as (a) max and (b) Lukasiewicz t-conorm and an example of the relations R_m induced by f_m when A and B are triangular fuzzy sets $A(x, 4, 5, 6)$ and $B(x, 4, 5, 6)$ using (c) max and (d) Lukasiewicz, respectively.

and selecting the maximum as the realization of the t-conorm, the relational assignment leads to the fuzzy relation R_m of the following form:

$$R_m(x, y) = f_m(A(x), B(y)) = \max[A(x), B(y)], \quad \forall (x, y) \in \mathbf{X} \times \mathbf{Y}$$

An illustration of this fuzzy relation is shown in Figure 11.6.

Given the Lukasiewicz t-conorm, we obtain

$$R_\ell(x, y) = f_\ell(A(x), B(y)) = \min[1, A(x) + B(y)], \quad \forall (x, y) \in \mathbf{X} \times \mathbf{Y}$$

See Figure 11.6.

11.4.2.3 Fuzzy Implication

A fuzzy implication is a function $f_i : [0, 1]^2 \to [0, 1]$ with the following properties:

1. $B(y_1) \leq B(y_2) \Rightarrow f_i(A(x), B(y_1)) \leq f_i(A(x), B(y_2))$ monotonicity

11.4 Basic Functional Modules: Rule Base, Database, and Inference Scheme

2. $f_i(0, B(y)) = 1$ dominance of falsity
3. $f_i(1, B(y)) = B(y)$ neutrality of truth

In addition to the list shown above, two properties are usually added:

4. $A(x_1) \leq A(x_2) \Rightarrow f_i(A(x_1), B(y)) \geq f_i(A(x_2), B(y))$ monotonicity
5. $f_i(A(x_1), f_i(A(x_2), B(y))) = f_i(A(x_2), f_i(A(x_1), B(y)))$ exchange

Further properties that can also be considered (Klir and Yuan, 1995) are as follows:

6. $f_i(A(x), A(x)) = 1$ identity
7. $f_i(A(x), B(y)) = 1$ if and only if $A(x) \leq B(y)$ boundary condition
8. $f_i(A(x), B(y)) = f_i(\overline{B}(y), \overline{A}(x))$ contraposition
9. f_i is a continuous function   continuity

There are a number of possible realizations of implication operators. Several of the commonly encountered models are summarized in Table 11.1. The corresponding plots of the implications are contained in Figure 11.7.

Figure 11.7(a) provides a graphical visualization of the Lukasiewicz implication f_ℓ and (c) provides an example of the relation R_ℓ generated by f_ℓ when A and B are triangular fuzzy sets $A(x, 4, 5, 6)$ and $B(x, 4, 5, 6)$.

Figure 11.7(b) illustrates the Klir–Yuan implication f_k and (d) illustrates the relation R_k generated by f_k when A and B are triangular fuzzy sets $A(x, 4, 5, 6)$ and $B(x, 4, 5, 6)$.

Table 11.1 Examples of Implications.

Name	Definition	Comment
Lukasiewicz	$f_\ell(A(x), B(y)) = \min[1, 1 - A(x) + B(y)]$	
Pseudo-Lukasiewicz	$f_\lambda(A(x), B(y)) = \min\left[1, \dfrac{1 - A(x) + (\lambda + 1)B(y)}{1 + \lambda A(x)}\right]$	$\lambda > -1$
Pseudo-Lukasiewicz	$f_w(A(x), B(y)) = \min[1, (1 - A(x)^w + B(y)^w)^{1/w}]$	$w > 0$
Gaines	$f_a(A(x), B(y)) = \begin{cases} 1, & \text{if } A(x) \leq B(y) \\ 0, & \text{otherwise} \end{cases}$	
Gödel	$f_g(A(x), B(y)) = \begin{cases} 1, & \text{if } A(x) \leq B(y) \\ B(y), & \text{otherwise} \end{cases}$	
Goguen	$f_e(A(x), B(y)) = \begin{cases} 1, & \text{if } A(x) \leq B(y) \\ \dfrac{B(y)}{A(x)}, & \text{otherwise} \end{cases}$	
Kleene	$f_b(A(x), B(y)) = \max[1 - A(x), B(y)]$	
Reichenbach	$f_r(A(x), B(y)) = 1 - A(x) + A(x)B(y)$	
Zadeh	$f_z(A(x), B(y)) = \max[1 - A(x), \min(A(x), B(y))]$	
Klir–Yuan	$f_k(A(x), B(y)) = 1 - A(x) + A(x)^2 B(y)$	

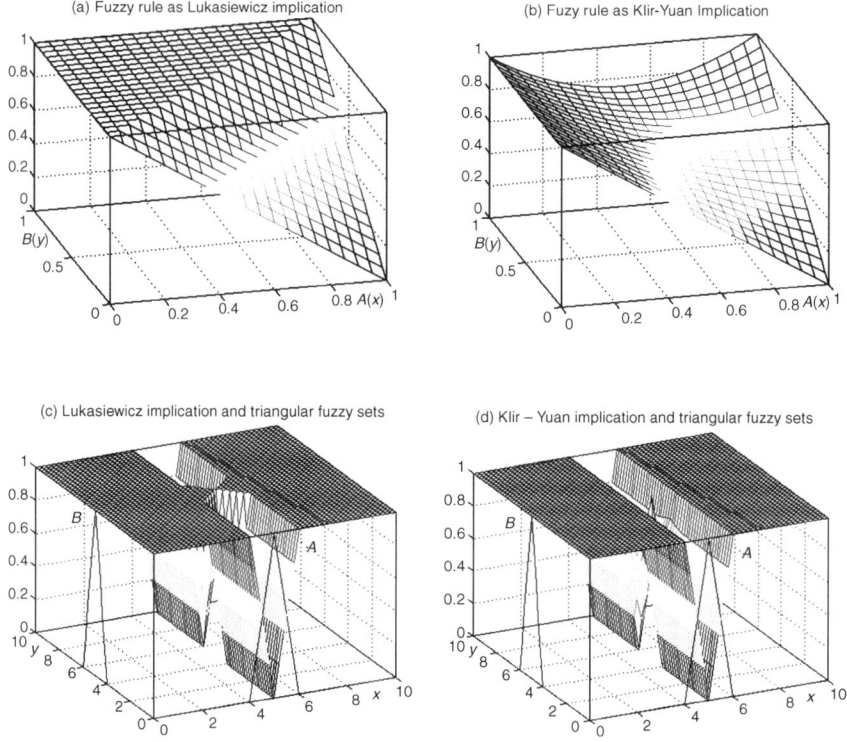

Figure 11.7 Fuzzy rule "If X is A then Y is B" interpreted as implication.

It is worth noting that fuzzy implications can be arranged into two main categories, namely, s-implications and r-implications. The class of s-implications arises from the formalism of classical logic when an implication is defined in terms of negation and disjunction, that is,

$$p \Rightarrow q \equiv \bar{p} \vee q$$

In this category, the implications possesses properties (1)–(5) of fuzzy implications. Formally, an s-implication is defined in the following way:

$$f_{is}(A(x), B(y)) = \bar{A}(x) \vee B(y), \quad \forall\, (x, y) \in \mathbf{X} \times \mathbf{Y}$$

The Lukasiewicz, Kleene, and Reichenbach implications are typical examples of s-implications.

The class of r-implications originates from the formalism of intuitionistic logic and is formed by taking a residuation of a continuous t-norm. Formally, an r-implication reads as follows:

$$f_{ir}(A(x), B(y)) = \sup[c \in [0,1] | A(x) t c \leq B(y)], \quad \forall\, (x, y) \in \mathbf{X} \times \mathbf{Y}$$

One can verify that for these implications, the list of properties (11.1)–(11.7) holds. Typical examples of implications coming from this category include Lukasiewicz, Gödel, and Goguen implications.

11.4 Basic Functional Modules: Rule Base, Database, and Inference Scheme

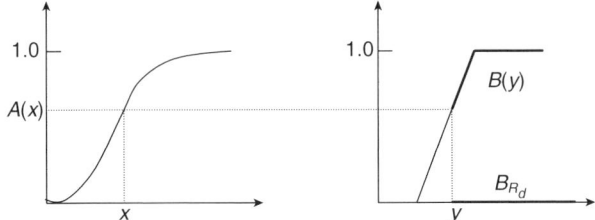

Figure 11.8 Constraint associated with a gradual rule.

Gradual rules of the form

$$\text{the } \textit{more } X \text{ is } A, \text{ the } \textit{more } Y \text{ is } B$$

or rules having a similar format that appear when "less" is substituted for one or more occurrence of *more* are examples of rules closely related with implications. Given fuzzy sets A and B, a gradual rule "the *more* X is A the *more* Y is B" translates into the following constraint (Dubois and Prade, 1996, 1992):

$$B(y) \geq A(x), \quad \forall x \in \mathbf{X} \text{ and } \forall y \in \mathbf{Y} \tag{11.16}$$

which defines a relation R_d between the variables x and y. The constraint expresses the idea conveyed by the linguistic form of the rule: Whenever the degree of membership of x in A increases, the degree of membership of y in B also increases. The constraint (11.16) associates to each value of $x \in \mathbf{X}$ a subset $B_{R_d} = \{y \in \mathbf{Y} | B(y) \geq A(x)\}$ of values of y viewed as possible when x is given, refer to Figure 11.8.

For instance, we can define the fuzzy relation R_d using the Gaines implication f_a, that is, $R_d = f_a$ and

$$R_d(x, y) = \begin{cases} 1, & \text{if } B(y) \geq A(x) \\ 0, & \text{otherwise} \end{cases}$$

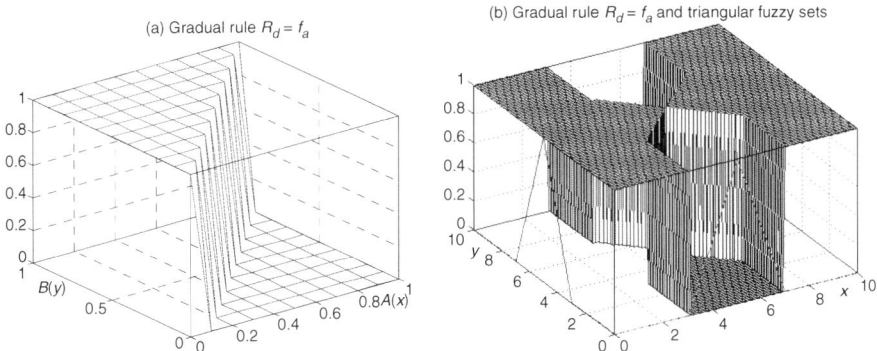

Figure 11.9 Gradual rule "the more X is A the more Y is B" as (a) Gaines implication and (b) Gaines implication when A and B are triangular fuzzy sets $A(x, 3, 5, 7)$ and $B(y, 3, 5, 7)$.

The relation is illustrated in Figure 11.9. Other choices of implications can be made, including Gödel (f_g) and Goguen (f_e).

11.4.3 Main Types of Rule Bases

A rule base is a collection of "If–then" rules encoding a description of how inputs of the system are transformed to produce the corresponding outputs. A fundamental question is about a way in which two or more rules of a rule base are to be combined and how the rule base itself has to be interpreted. There are several alternative ways of proceeding with the aggregation of the rules.

11.4.3.1 Fuzzy Graphs

This is the most common and frequently used interpretation and considers a fuzzy rule of the form "If X is A then Y is B" as a single information granule. In this setting, let us emphasize that fuzzy granules are generalizations of points and can be treated as *fuzzy* points. To underline this effect, recall that a point P in $\mathbf{X} \times \mathbf{Y}$ is the set $P = \{(x, y) | x \in A, y \in B\}$, that is, P is the Cartesian product of the degenerated, one-element sets $A = \{x\}$ and $B = \{y\}$ (Fig. 11.10).

If now A is an interval on $\mathbf{X}$ and B an interval on $\mathbf{Y}$, then an information granule G on $\mathbf{X} \times \mathbf{Y}$ comes as the set $G = \{(x, y) | x \in A, y \in B\}$. Figure 11.11 shows an example of granule G where $A = \{x \in \mathbf{R}, 3 \leq x \leq 6\}$ and $B = \{y \in \mathbf{R}, 4 \leq x \leq 6\}$.

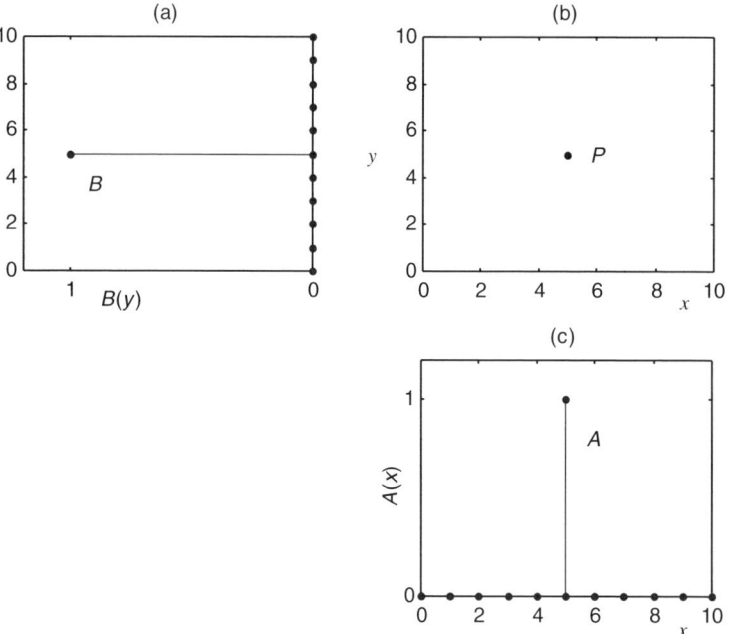

Figure 11.10 A point P in $\mathbf{X} \times \mathbf{Y}$ as a Cartesian product of two degenerated sets A and B.

11.4 Basic Functional Modules: Rule Base, Database, and Inference Scheme

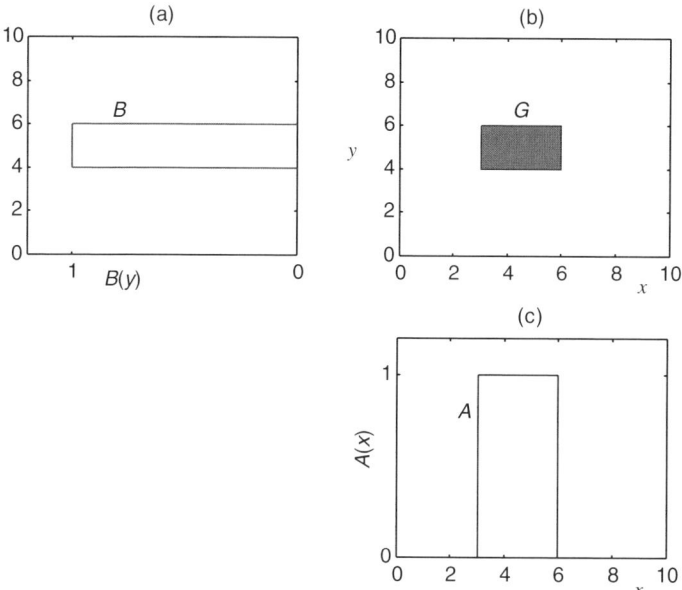

Figure 11.11 An information granule G in $\mathbf{X} \times \mathbf{Y}$ for A and B represented as two numeric intervals.

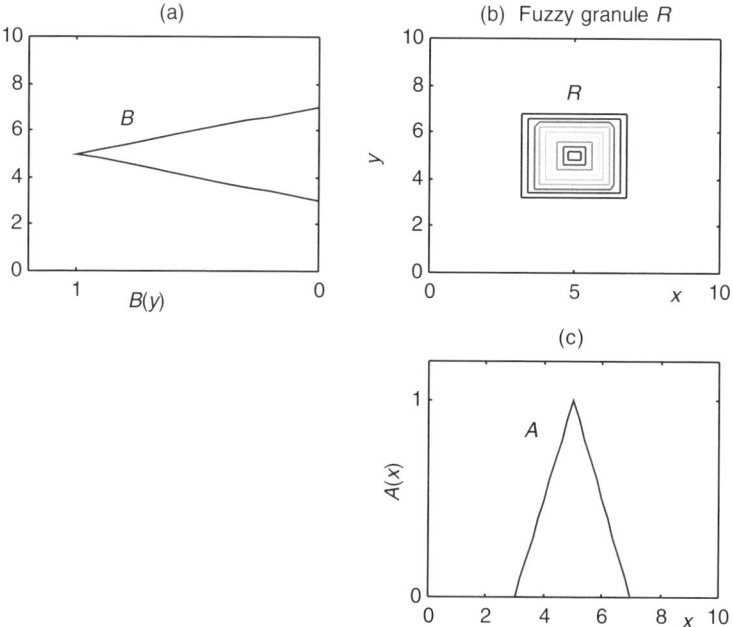

Figure 11.12 A fuzzy granule R in $\mathbf{X} \times \mathbf{Y}$ coming as the Cartesian product of two triangular fuzzy sets A and B.

292 Chapter 11 Rule-Based Fuzzy Models

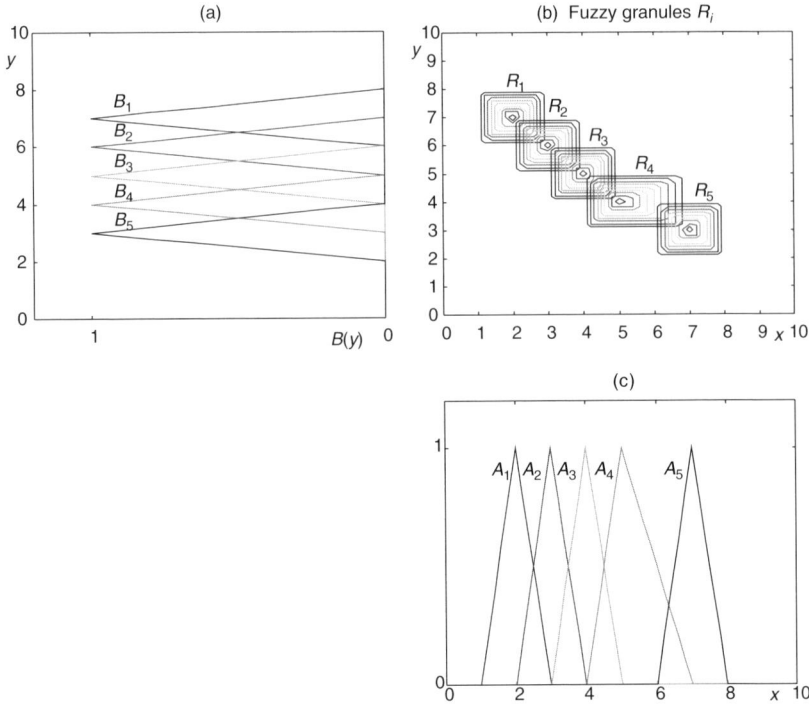

Figure 11.13 Rule base as a set of fuzzy granules.

In the sequel, a fuzzy granule, alternatively, fuzzy point, R is the fuzzy set $R = AtB$, where A and B are normal, convex fuzzy sets in **X** and **Y**, respectively. If the t-norm is the minimum, then the fuzzy granule is the Cartesian product of A and B. See Figure 11.12 for an illustrative example.

For a rule base formed by a collection of fuzzy rules

$$\text{If } X \text{ is } A_i \text{ then } Y \text{ is } B_i$$

where $i = 1, \ldots, N$; these N fuzzy rules correspond to N fuzzy granules as illustrated in Figure 11.13. Here $N = 5$ while the fuzzy sets are described by triangular fuzzy sets A_i and B_i. The t-norm is implemented as the minimum operation. Each rule is implemented as a fuzzy granule (the Cartesian product of triangular fuzzy sets A_i and B_i), namely, $R_i = A_i \times B_i$, or in terms of the following membership function:

$$R_i(x, y) = \min[A_i(x), B_i(y)], \quad \forall x \in \mathbf{X} \text{ and } \forall y \in \mathbf{Y}$$

The rule base is interpreted as a fuzzy graph, that is, an approximate description of the relationship between X and Y. As a whole, it results in a union of the fuzzy rules. This underlines an observation that the rule base becomes a fuzzy graph in the same manner as the set $f = \{(x, y), y = f(x), x \in \mathbf{X}, y \in \mathbf{Y}\}$ is the graph of the

11.4 Basic Functional Modules: Rule Base, Database, and Inference Scheme

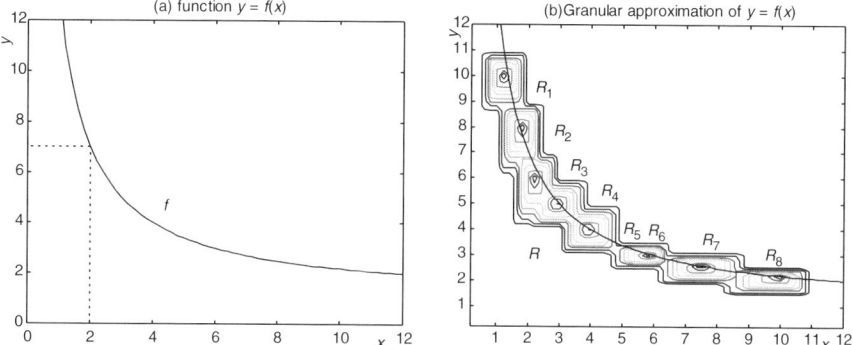

Figure 11.14 Graph of a function f as its granular approximation R.

function f and can be viewed as the union of degenerated sets (singletons) (x, y) in $\mathbf{X} \times \mathbf{Y}$ such that $y = f(x)$, refer to Figure 11.14(a). As an immediate generalization of this concept, a fuzzy graph R of a functional dependency $f : \mathbf{X} \to \mathbf{Y}$ between the fuzzy variables X and Y is defined (Zadeh, 1994) as an approximate, granular representation of f:

$$R = \bigcup_{i=1}^{N} R_i = \bigcup_{i=1}^{N} (A_i \times B_i)$$

Figure 11.14 (b) shows an example of such a granular approximation of a function shown in (a), whose fuzzy graph is composed by the union of $N = 8$ fuzzy granules, each of which is defined by the Cartesian product of the triangular fuzzy sets.

Overall, a collection of rules "If X is A_i then Y is B_i" can be aggregated to form a fuzzy rule base using operations such as fuzzy unions, or fuzzy disjunctions of the fuzzy granules (fuzzy relations R_i) associated with each rule. The result of the aggregation of R_is, where $i = 1, \ldots, N$, is a global fuzzy relation R.

$$R = R_1 \, or \, R_2 \, or \, \ldots R_N$$

Given the fuzzy disjunction (which is realized by any t-conorms), we rewrite the above expression for the fuzzy graph as follows:

$$R(x, y) = \underset{i=1}{\overset{N}{S}} [A_i(x) t B_i(y)], \, \forall \, (x, y) \in \mathbf{X} \times \mathbf{Y}$$

Figure 11.14 illustrates the underlying concept of the granular representation of input–output relationships. Furthermore, Figure 11.15 shows some typical examples of fuzzy rule bases of $N = 5$ fuzzy rules with triangular membership functions in their antecedents and consequents.

11.4.3.2 Fuzzy Implication Rule Bases

In contrast to fuzzy graphs, which use conjunction as a form of knowledge accumulation, the implication-based models of fuzzy rules and fuzzy rule bases view rules as

294 Chapter 11 Rule-Based Fuzzy Models

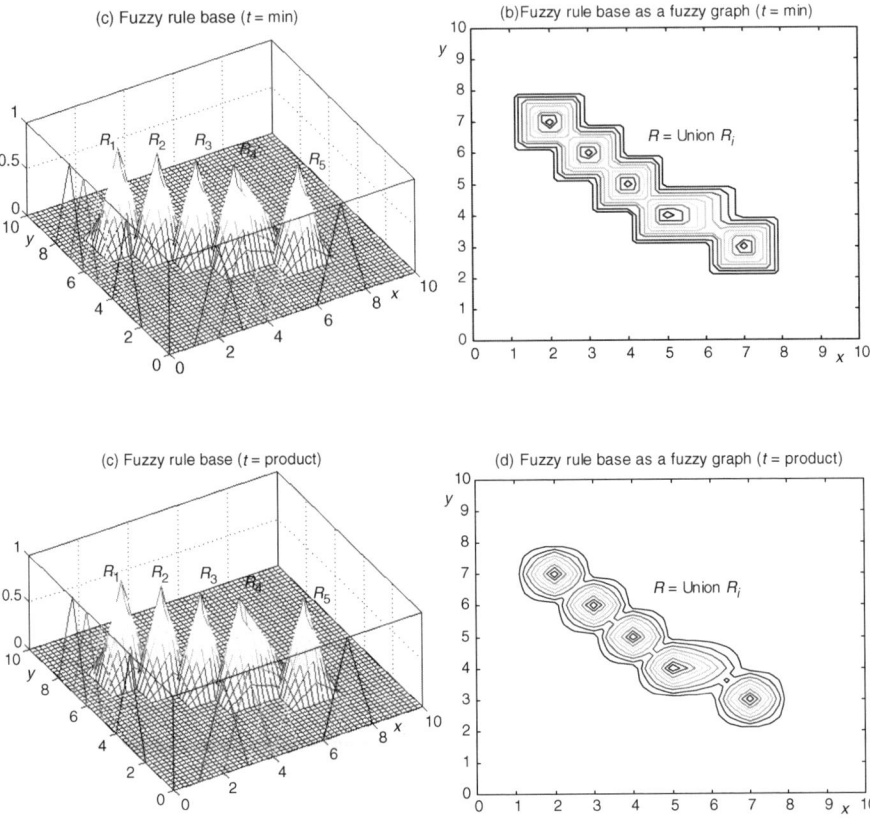

Figure 11.15 Rule base and the resulting fuzzy graphs: (a) and (b) use the minimum t-norm while (c) and (d) the algebraic product t-norm. (a), (b), (c), and (d) use the maximum t-conorm as a *or*.

a constraint that restricts a set of possible solutions. In this case, rule combination is an aggregation acting as a refinement mechanism to enhance the precision via completing a conjunction of the individual rules.

As before, consider the rules of the form

$$\text{If } X \text{ is } A_i \text{ then } Y \text{ is } B_i, \text{with } i = 1, 2, \ldots, N$$

where now each rule is represented as some fuzzy relation R_i whose membership function is constructed using a fuzzy implication. For instance, Figure 11.16 shows an example of triangular fuzzy sets in rule antecedent and consequent where we use the Lukasiewicz implication to construct R_i, that is, $R_i(x, y) = f_\ell(A_i(x), B_i(y))$.

The result of aggregating the relations $R_i, i = 1, .., N$, rules is a overall fuzzy relation R

$$R = R_1 \text{ and } R_2 \text{ and } \ldots R_N$$

11.4 Basic Functional Modules: Rule Base, Database, and Inference Scheme

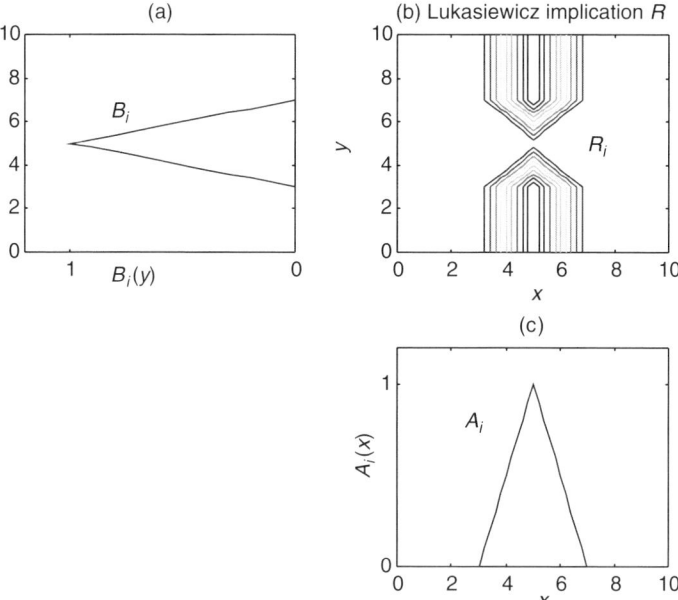

Figure 11.16 A fuzzy rule formed with the use of Lukasiewicz implication.

in which the aggregation is carried out in the form of some conjunction and therefore implemented trough some t-norm,

$$R(x,y) = \mathop{T}_{i=1}^{N} [f_i(A_i(x), B_i(y))], \; \forall \, (x,y) \in \mathbf{X} \times \mathbf{Y}$$

Figures 11.17 and 11.18 illustrate typical examples of fuzzy rule bases composed by $N = 5$ fuzzy rules with triangular membership functions in their antecedents and consequents. Figures 11.17 and 11.18(a,b) adopt the minimum t-norm aggregation while Figures 11.17 and 11.18(c,d) use the Lukasiewicz t-norm aggregation.

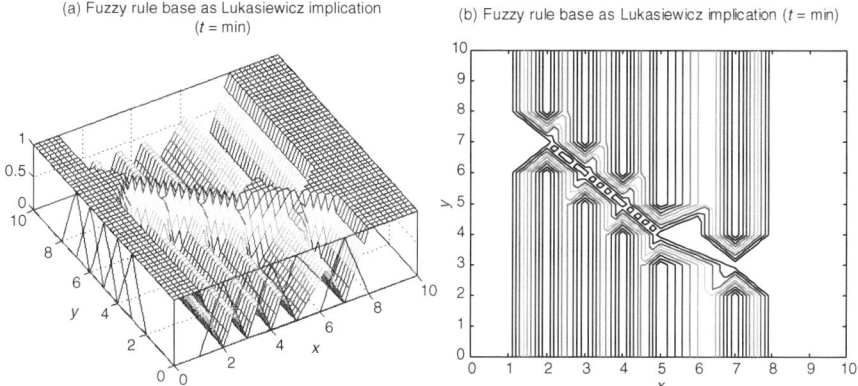

Figure 11.17 A fuzzy rule base resulting from the use of the Lukasiewicz implication.

296 Chapter 11 Rule-Based Fuzzy Models

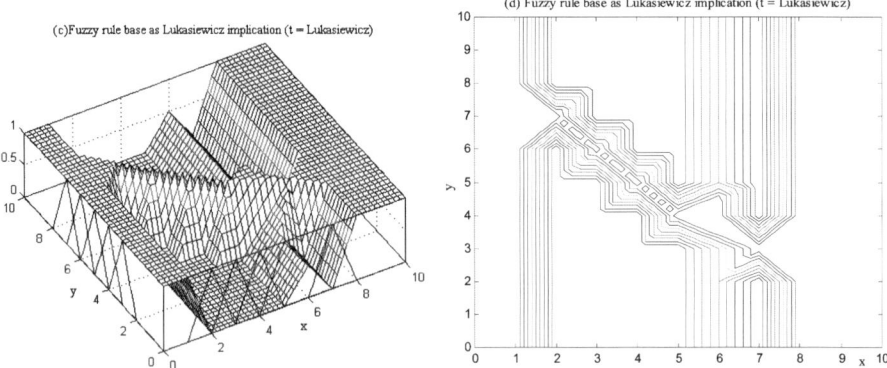

Figure 11.17 (*Continued*)

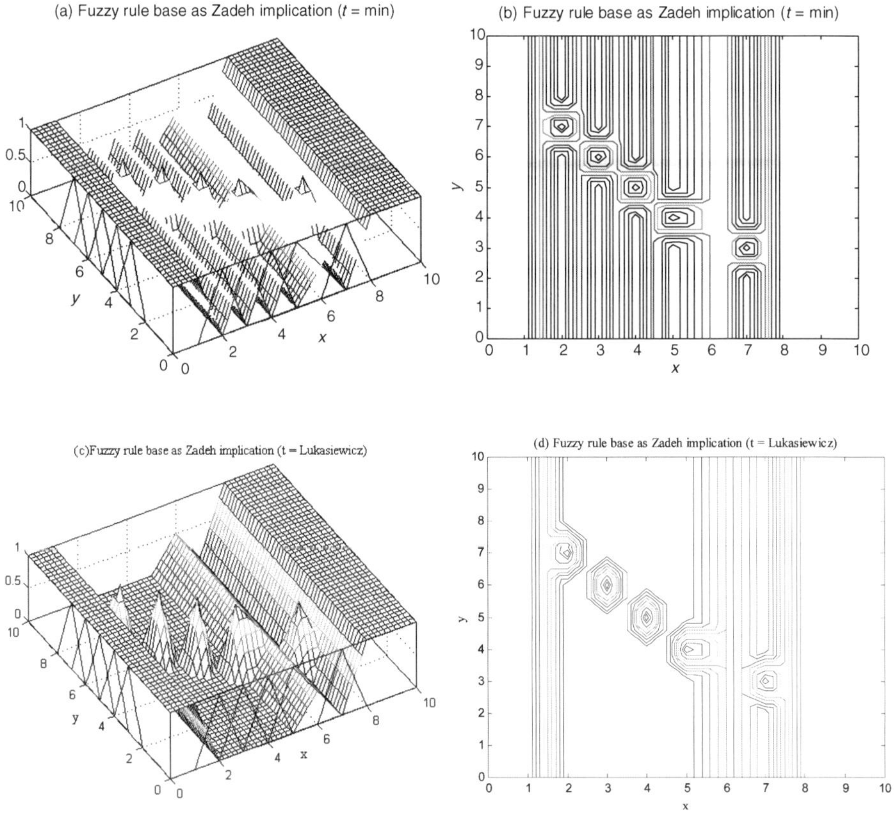

Figure 11.18 A fuzzy rule base as Zadeh implication.

Figure 11.17 illustrates the Lukasiewicz implication rule base and Figure 11.18 shows the Zadeh implication rule base. As both the figures clearly indicate, the choice of the implication and aggregation operations leads to quite distinct form of relationships. The diversity available in this manner becomes essential when designing and developing applications in areas such as system diagnosis, control, and pattern recognition.

Notice that in Figure 11.17(a,b) the overall fuzzy relation R is $R = \bigcap_{i=1} R_i$. It uses the Lukasiewicz implication and the minimum t-norm aggregation. Thus, the membership function of R is

$$R(x,y) = \max\{0, \sum_{i=1}^{5} \min[1, 1 - A_i(x) + B_i(y)] - 4\}, \forall (x,y) \in \mathbf{X} \times \mathbf{Y}$$

If aggregation is performed using the Lukasiewicz t-norm instead of minimum t-norm (Fig. 11.17(c,d)), the membership of R becomes

$$R(x,y) = \max\{0, \sum_{i=1}^{5} \min[1, 1 - A_i(x) + B_i(y)] - 4\}, \forall (x,y) \in \mathbf{X} \times \mathbf{Y}$$

Similarly, as illustrated in Figure 11.18, the membership functions of fuzzy relation R using the case of Zadeh implication and the minimum and Lukasiewicz t-norms as aggregations are, respectively,

$$R(x,y) = \min\{\max[1 - A(x), \min(A_i(x), B_i(y))], i = 1, 2, \cdots, 5\}, \forall (x,y) \in \mathbf{X} \times \mathbf{Y}$$

and

$$R(x,y) = \max\{0, \sum_{i=1}^{5} \max[1 - A_i(x), \min(A_i(x), B_i(y))] - 4\}, \forall (x,y) \in \mathbf{X} \times \mathbf{Y}$$

11.4.4 Data Base

The database contains definitions of the universes, description of scaling functions of the input and output variables, granulation of the universes, and membership functions of the fuzzy sets assigned to each linguistic value of linguistic variables, that is, the term sets. Granulation of the input and output universes determines how many terms should exist in a term set. This is equivalent to definition the primary fuzzy sets.

There are two basic schemes to granulate a given universe $\mathbf{X}$. The first explores the view of fuzzy expressions as granular constructs in the form of fuzzy points, refer to Figure 11.19(a). The second one assumes granules uniquely specified along

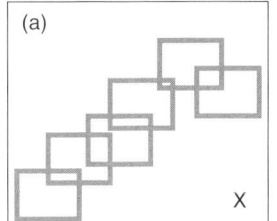
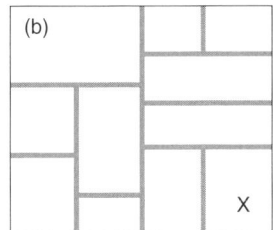

Figure 11.19 Examples of granulation of input and output universes.

298 Chapter 11 Rule-Based Fuzzy Models

different regions of the universe, refer to Figure 11.19 (b). These schemes are based on an assumption that the input variables are independent, which is useful and pragmatic in many practical situations. If this assumption does not hold due to the presence of correlated variables, then partitions incorporating the interactions must be pursued such as product space clustering (Setnes, 1998) and relational antecedents (Gaweda and Zurada, 2003).

There are basically two mechanisms to construct membership functions, namely, expert knowledge and learning from experimental data (we discuss these approaches later in Section 11.7).

11.4.5 Fuzzy Inference

The fundamental concept that supports fuzzy inference concerns a way of carrying out composition of fuzzy sets and fuzzy relations. A way of completing this composition is commonly referred to as a compositional rule of inference. In a nutshell, it is about a generalization of a well-known mapping problem: Given a function $y = f(x)$ and a certain value of its argument, say $x = a$, determine the corresponding value of y, that is, $y = f(a) (= b)$. Schematically, we can describe the situation as follows:

$$x = a$$
$$y = f(x)$$
$$y = b$$

as illustrated in Figure 11.20(a). The procedure to find the value of y comprises the following steps:

1. Draw a vertical line a_c in the Cartesian product of space $\mathbf{X} \times \mathbf{Y}$, such that $x = a$. Formally this construct is the same as the cylindrical extension a_c of a set that is a single point x with base a.

2. Find the intersection point I between the cylindrical extension a_c and the graph of the function. Notice that both, the graph of the function and the cylindrical extension, are subsets of the same universe, that is, $\mathbf{X} \times \mathbf{Y}$.

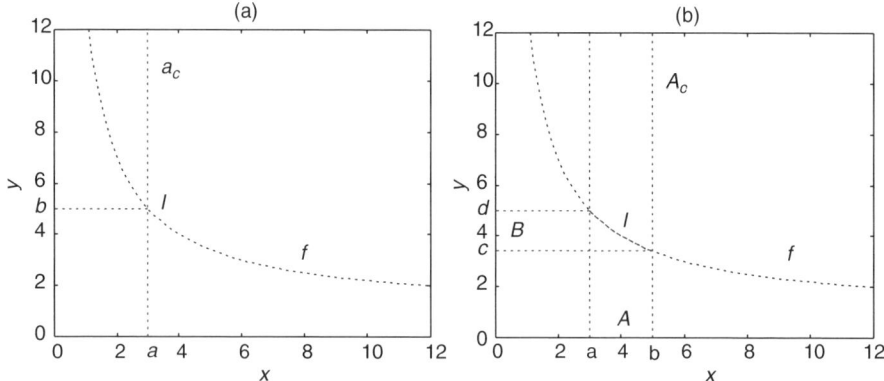

Figure 11.20 Computing function values with operations on sets.

11.4 Basic Functional Modules: Rule Base, Database, and Inference Scheme

3. Draw a horizontal line crossing I. Formally, this is the same as to project the intersection I, which comes as the corresponding point in **Y**.

Now, if instead of a single point we consider set $A \subset \mathbf{X}$, for example, an interval $A = [a, b]$, the steps 1–3 can be repeated for each point $x \in [a, b]$. By assembling the results, we form a set $B = f(A) = \{f(x), x \in A\}$ as depicted in Figure 11.20(b). Here A_c is the cylindrical extension of A, and I is a set being the intersection of the graph of f and A_c.

Moving one step further, given that x is now A and instead of a function with graph f, we have a relation $R \subset \mathbf{X} \times \mathbf{Y}$, we can compute the set $B \subset \mathbf{Y}$ following the same procedure as before. Figure 11.21 illustrates the underlying concept. As we soon realize, B is the result of the composition of A with the relation R.

In the most general case, we treat A as a fuzzy set in **X** and view R to be a fuzzy relation on $\mathbf{X} \times \mathbf{Y}$. The problem of determining the fuzzy set B in **Y** can be expressed in the form

$$\frac{\begin{array}{c} x \text{ is } A \\ (x, y) \text{ is } R \end{array}}{y \text{ is } B}$$

Now the set operations are replaced by their fuzzy set counterparts. Thus, if cylindrical extension, intersection, and projection are understood as constructs of fuzzy sets, fuzzy inference follows the same sequence of steps as in the previous cases, refer to Figure 11.21(b).

Fuzzy set B is determined by projecting $I = A_c \cap R$ onto **Y**:

$$B = \text{Proj}_Y (A_c \cap R)$$

Because intersection is realized by some t-norm while projection uses the supremum operation, the membership function of B reads as

$$B(y) = \sup_{x \in X} \{A_c(x, y) t R(x, y)\}$$

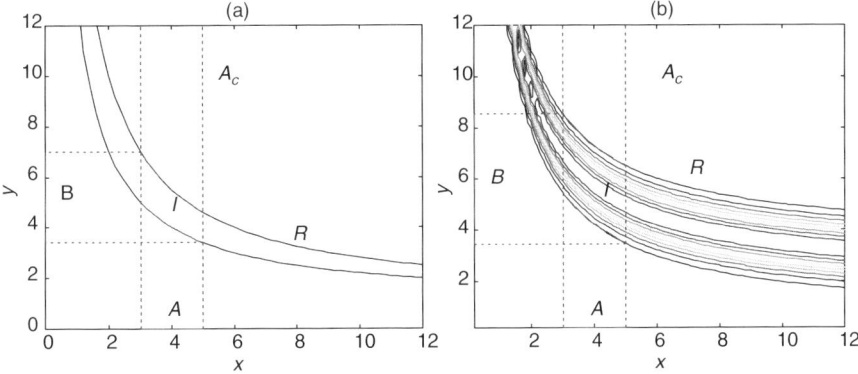

Figure 11.21 Composition of sets with relations.

and recalling that, by definition, $A_c(x,y) = A(x)$ we get

$$B(y) = \sup_{x \in X}\{A(x) t R(x,y)\}$$

Summarizing, fuzzy set B results from the sup-t composition of fuzzy set A and fuzzy relation R

$$B = A \circ R$$

The general scheme of fuzzy inference becomes

$$X \text{ is } A$$
$$(X,Y) \text{ is } R$$
$$\overline{Y \text{ is } A \circ R}$$

The procedure of fuzzy inference is summarized as follows:

procedure FUZZY-INFERENCE (A, R) **returns** a fuzzy set
input: fuzzy relation: R
       fuzzy set: A
local: (x, y): elements of **X** and **Y**
   t: t-norm
 for all x and y **do**
   $A_c(x,y) \leftarrow A(x)$
 for all x and y **do**
   $I(x,y) \leftarrow A_c(x,y) t R(x,y)$
   $B(y) \leftarrow \sup_x I(x,y)$
 return B

Figure 11.22 illustrates successive steps to perform fuzzy inference using the sup–min composition. The fuzzy relation R has a Gaussian membership function slices with unity dispersion and modal values positioned along the function

$$y = f(x) = \frac{k}{x} + c \qquad (11.17)$$

where $k = 15$ and $c = 2$. The input is a triangular fuzzy set $A(x, 3, 4, 5)$. The fuzzy relation R is the same as the one whose contour plot is given in Figure 11.21(b).

The fuzzy inference of Figure 11.21 is the compositional rule of inference introduced by Zadeh (1975, 1988). The fuzzy inference procedure is general and can naturally handle the case in which the fuzzy rule base is a set of parallel rules, each of which is a granule. This is the case when the fuzzy rule base is a granular approximation of a function. For instance, Figure 11.23 shows an example of a rule

11.4 Basic Functional Modules: Rule Base, Database, and Inference Scheme 301

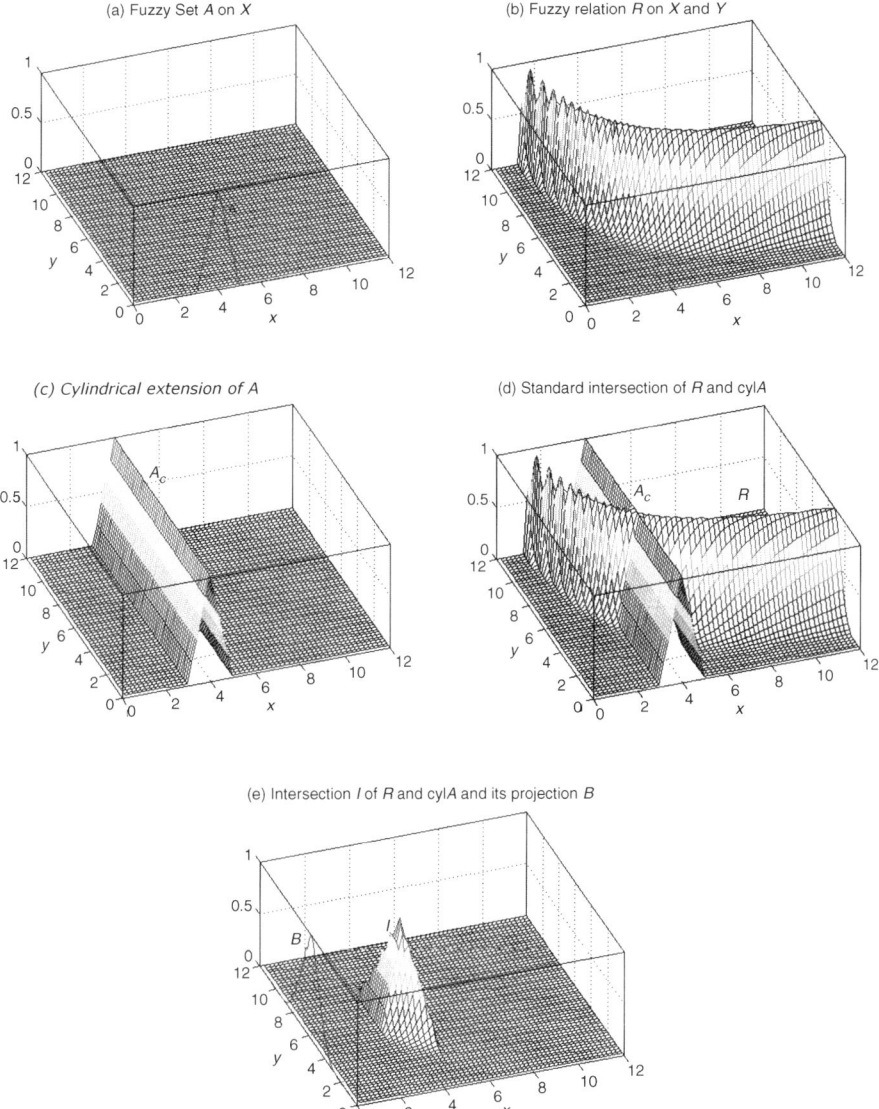

Figure 11.22 Main phases of fuzzy compositional inference.

base composed by eight fuzzy if–then rules viewed as the Cartesian product of triangular fuzzy sets whose contour is the same as in Figure 11.14(b). The fuzzy graph of Figure 11.14(b) is a granular approximation of the function depicted in Figure 11.14(a), which is the same as in (11.17), except that $k = 12$ and $c = 1$. Notice that, looking at the modal value of the input fuzzy set $A(x, 3, 4, 5)$, the corresponding

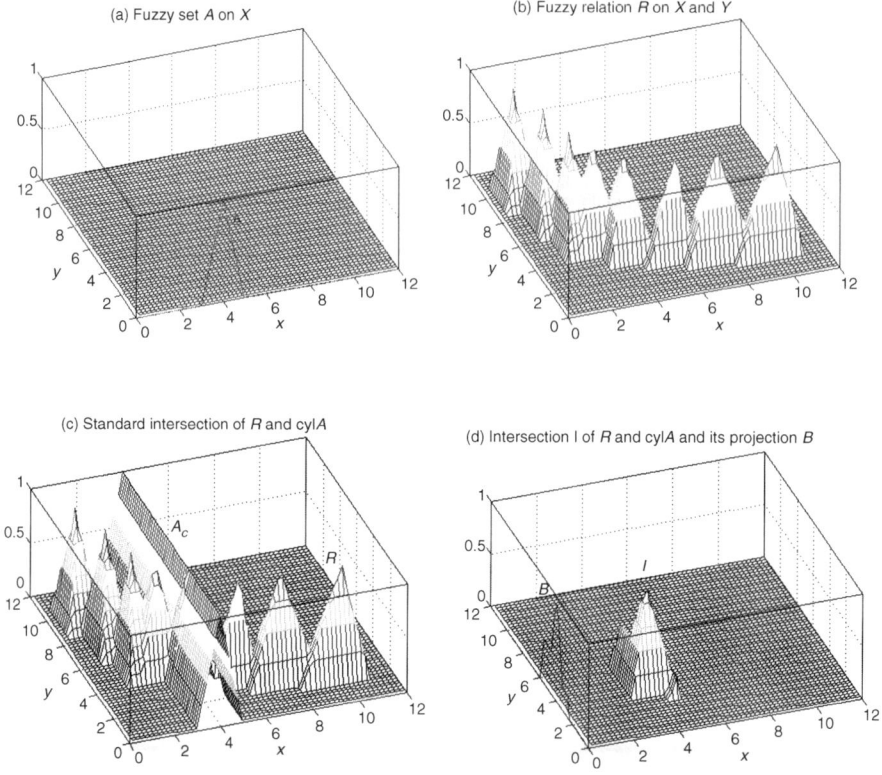

Figure 11.23 Fuzzy inference with a rule base as a fuzzy graph.

inferred fuzzy set B achieves its highest membership value at $y \approx 4$. Observe that A_c intersects with R_3, R_4, and R_5 (from up down in Fig. 11.23(b); also see Fig. 11.14) only and, because of this fact, we say that rules R_3, R_4, and R_5 are active and that the remaining rules are inactive. Here active and inactive means which rules of the rule base contribute to produce an output for a given input.

In addition to the compositional realization of fuzzy inference, there are more abstract, yet formal and more general treatment of fuzzy if–then rules through logical manipulation, namely, multiple-valued first-order logic, perception-based deduction, and logical structures derived from fuzzy-type theory (Novak and Lehmke, 2006).

11.5 TYPES OF RULE-BASED SYSTEMS AND ARCHITECTURES

Fuzzy rule-based computing is prompted by rule semantics and the compositional rule of inference. The previous section provided the general framework to design and develop fuzzy rule-based models. However, computations can be greatly simplified if certain assumptions concerning rule base form, rule semantics, rules aggregation operator, and the t-norm operator of the compositional rule of inference are made.

From the system modeling standpoint, we shall emphasize parallel rules once they produce systems that are highly modular forms of granular models; an expansion or contraction of the model usually requires addition or subtraction of rules while the remaining rules are left intact or adapted to the context, with no need to modify fuzzy inference. This modular character also has high impact on evolving and adaptive fuzzy models and systems.

Looking at the conceptual point of view, we highlight granular, fuzzy graph representation of fuzzy rule-based models because they intuitively provide granular approximations of functions. Most fuzzy system design issues concern the development of appropriate decision functions. Models of physical systems are functions, and fuzzy control strategies are functions, as are many forms of classifiers and diagnostic systems.

From the practical viewpoint, we identify three major classes of fuzzy rule-based system architectures depending on if the fuzzy rules involve fuzzy sets in both antecedent and consequent, but with a rule combination strategy that involves either the union or the sum of the contributions of the active rules. Rules may also have fuzzy sets in the antecedent, but functions of the antecedent base variables in the consequent instead of fuzzy sets. Rules may also involve gradual relationships among antecedent and consequent variables. Thus, we identify three main classes of models: linguistic, functional, and gradual models.

11.5.1 Linguistic Fuzzy Models

A fuzzy rule-based system is a system whose rule base is made up of a set of fuzzy rules of the form

$$R_i : \text{If } X_1 \text{ is } A_i \text{ and } Y \text{ is } B_i \text{ then } Z \text{ is } C_i \quad i = 1, 2, \ldots, N$$

where X, Y, and Z are linguistic variables with base variables x, y, and z, and A_i, B_i, and C_i are fuzzy sets on **X**, **Y**, and **Z**. This is the most prevalent form of linguistic fuzzy model because it linguistically describes, conceptually speaking, behavior, and formally speaking, approximation of functions. However, the operation of the model depends on design choices, namely, the t-norm to aggregate the atomic expressions of rule antecedents X is A_i and Y is B_i, the choice of the fuzzy conjunction to express the meaning of the "If–then" relation between the aggregated antecedent and the consequent, and the s-norm used to aggregate the rules into a global rule base. There are several options, but often rule antecedent aggregation is done via minimum or algebraic product t-norms, "If–then rules" defined as Cartesian products using either the minimum or the algebraic product t-norms, and the maximum s-norm to perform aggregation of rules. The most frequent compositions are the sup–min and sup–product composition. In what follows, we develop one of the most important linguistic models, the one in which rule antecedent aggregation and meaning is taken as the min t-norm and inference via sup–min composition. The general inference pattern is

P:	X is A and Y is Ba	input expression
R_1:	if X is A_1 and Y is B_1 then Z is C_1	
R_2:	if X is A_2 and Y is B_2 then Z is C_2	
R_i:	if X is A_i and Y is B_i then Z is C_i	rule base
R_N:	if X is A_N and Y is B_N then Z is C_N	
Z:	Z is C	inferred expression

The purpose is to determine the fuzzy set C of the inferred expression "Z is C," given an input expression and the fuzzy rule base. Therefore, the task is to determine the fuzzy set C, namely, the membership function of C, given the input and fuzzy rule base.

11.5.1.1 Min-Max Models

This type of fuzzy rule-based models assumes the following pattern:

$$P: \quad P(x,y) = \min\{A(x), B(y)\}$$

$$R_i: \quad R_i(x,y,z) = \min\{A_i(x), B_i(y), C_i(z)\}, \quad i = 1, \ldots, N$$

$$Z: \quad C = P \circ \bigcup_{i=1}^{N} R_i \tag{11.17}$$

In terms of the corresponding membership function, the fuzzy set C is, using standard union

$$C(z) = \sup_{x,y}\{\min[P(x,y), \max(R_i(x,y,z), i = 1, .., N]\} \tag{11.18}$$

Let us adopt, for short, the alternative notation $\vee$ for max and $\wedge$ for min operation. Thus, because the max and min operations are mutually distributive, (11.17) can be rewritten as

$$\begin{aligned}
C(z) &= \sup_{x,y}\{(P(x,y) \wedge [R_1(x,y,z) \vee \cdots \vee R_N(x,y,z)]\} \\
&= \sup_{x,y}\{[P(x,y) \wedge R_1(x,y,z)] \vee \cdots \vee [P(x,y) \wedge R_N(x,y,z)]\} \\
&= \max\{\sup_{x,y}[P(x,y) \wedge R_1(x,y,z)], \cdots, \sup_{x,y}[P(x,y) \wedge R_N(x,y,z)]\}
\end{aligned} \tag{11.19}$$

If we assume

$$C'_i(z) = \sup_{x,y}[P(x,y) \wedge R_i(x,y,z)] \tag{11.20}$$

that is, $C'_i = P \circ R_i$, then (11.19) can be expressed symbolically by

$$C(z) = \bigcup_{i=1}^{N}(P \circ R_i) = \bigcup_{i=1}^{N} C'_i \tag{11.21}$$

and, from (11.17) and (11.21) we have

$$C(z) = P \circ \bigcup_{i=1}^{N} R_i = \bigcup_{i=1}^{N} (P \circ R_i) = \bigcup_{i=1}^{N} C_i' \tag{11.22}$$

Equality (11.22) shows that the sup–min composition and standard union commute. Computationally this means that instead of combining all the rules using the max aggregation operator and next produce a corresponding relation R and proceed with fuzzy inference via sup–min composition, we can first infer the individual fuzzy set due to each rule using sup–min composition and finally produce the desired result combining the individual fuzzy sets via standard union, the max aggregation. This shows that rules can be processed in parallel and if done in hardware, the processing time does not depend upon the number of rules being used in the rule-based model.

It is important to stress that (11.22) holds not only for sup–min composition and max aggregation but also for sup–product composition and, in general case, for any sup-t composition with a continuous t-norm as long as the max aggregation is used.

In the case of sup–min composition, because

$$\sup_{x,y}[P(x,y) \wedge R_i(x,y,z)] = \sup_{x,y}[A(x) \wedge B(y) \wedge A_i(x) \wedge B_i(y) \wedge C_i(z)] \tag{11.23}$$

we have

$$\sup_{x,y}[P(x,y) \wedge R_i(x,y,z)] = \sup_{x,y}[(A(x) \wedge A_i(x)) \wedge (B(y) \wedge B_i(y)) \wedge C_i(z)]$$

and because the supremum is taken with respect to x and y, we can write

$$\sup_{x,y}[P(x,y) \wedge R_i(x,y,z)] = \sup_{x}[A(x) \wedge A_i(x)] \wedge \sup_{y}[(B(y) \wedge B_i(y)] \wedge C_i(z)$$

which means that

$$\sup_{x}[A(x) \wedge A_i(x)] \wedge \sup_{y}[(B(y) \wedge B_i(y)] \wedge C_i(z) = \text{Poss}(A, A_i) \wedge \text{Poss}(B, B_i) \wedge C_i(z)$$

If we assume that $\text{Poss}(A, A_i) = m_i$ and $\text{Poss}(B, B_i) = n_i$, then (11.23) becomes

$$\sup_{x,y}[A(x) \wedge B(y) \wedge R_i(x,y,z)] = m_i \wedge n_i \wedge C_i(z) \tag{11.24}$$

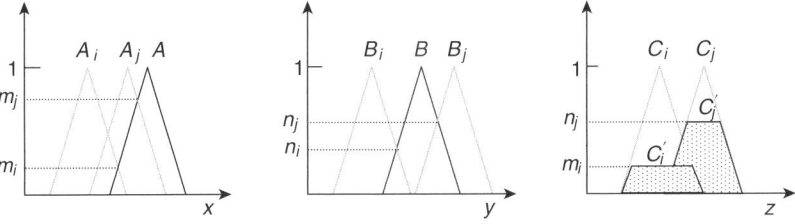

Figure 11.24 Min–max fuzzy model processing.

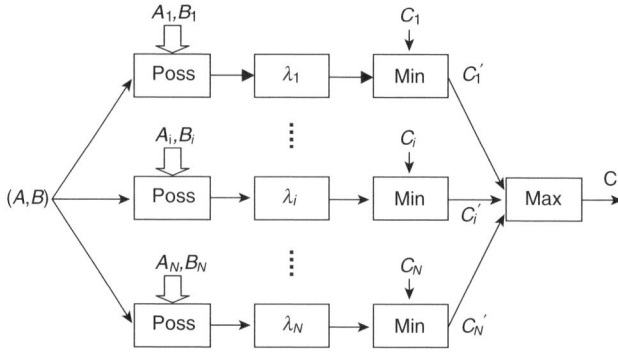

Figure 11.25 Min–max fuzzy linguistic model.

and replacing the corresponding terms in (11.19) by (11.24) we get

$$C(z) = \max\{[m_1 \wedge n_1 \wedge C_1(z)], [m_2 \wedge n_2 \wedge C_2(z)], \cdots, [m_N \wedge n_N \wedge C_N(z)]\} \quad (11.25)$$

The value $\lambda_i = (m_i \wedge n_i)$ is called the degree of activation or firing level of rule R_i. As (11.25) indicates, if $\lambda_i = 0$, then rule R_i does not contribute to infer fuzzy set C because its corresponding portion $\lambda_i \wedge C_i = 0$. Figure 11.24 illustrates (11.25) considering two rules, R_i and R_j.

The developments made so far suggest that linguistic fuzzy models whose rule base are composed by parallel rules are intrinsically modular and can be implemented by a four-layer architecture depicted in Figure 11.25. At the first level, we perform pattern matching computing the possibility measure of the input A and B determined with respect to A_i and B_i. In the sequel, the degree of activation is computed and next the membership functions of the consequent of the active rules are clipped. In the last step, the partial results of inference are aggregated by the standard union operation to produce the fuzzy set C.

For finite universes, the systematization of the processing steps is as follows:

procedure MIN–MAX-MODEL (A,B) **returns** a fuzzy set
local: fuzzy sets: A_i, B_i, C_i, $i = 1, ..., N$
    activation degrees: λ_i
Initialization $C = \emptyset$
for $i = 1 : N$ **do**
  $m_i = \max\,(\min(A, A_i))$
  $n_i = \max(\min\,(B, B_i))$
  $\lambda_i = \min(m_i, n_i)$
  if $\lambda_i \neq 0$ **then** $C_i^{'} = \min(\lambda_i, C_i)$ and $C = \max(C, C_i^{'})$
return C

Let us analyze some essential properties of processing of the inference scheme (11.18)–(11.25).

(a) Boundary conditions: If $A = B = \emptyset$, then the result of inference is also an empty fuzzy set, $C = \emptyset$. If $A = \mathbf{X}$ and $B = \mathbf{Y}$ (which models a situation of *unknown*—We know *nothing* about the input), then the result is a fuzzy set being a union of all C_is, $C = \bigcup_{i=1}^{N} C_i$. This reflects the fact that we "activate" all rules and the result could be any C_i, and in this sense their union is quite reflective of this effect of nonspecificity of the input datum.

(b) Recall properties: Consider that $A = A_i$ and $B = B_i$. In virtue of the max–min composition (11.25) we rewrite this expression as

$$C(z) = \max_{i=1,2,\ldots,N}[\min(\lambda_i, C_i(z))] = \max\{\max_{j \neq i}(\min(\lambda_j, C_j(u)), C_i(z)\}$$
$$= \max(\mu_i(u), C_i(z)) \geq C_i(z)$$

by distinguishing between the membership function which comes from the ith rule and the membership function which comes from all other fuzzy sets. Because there is an overlap between fuzzy sets of conditions in the individual rules, the possibility values assume nonzero values and this contributes to an effect of *crosstalk* in which the results of inference subsumes the original fuzzy set C_i, $C \supset C_i$.

When inputs $A = x_o$ and $B = y_o$ are points, or equivalently, sets with membership functions

$$A(x) = \begin{cases} 1, & \text{if } x = x_o \\ 0, & \text{otherwise} \end{cases} \quad \text{and} \quad B(x) = \begin{cases} 1, & \text{if } y = y_o \\ 0, & \text{otherwise} \end{cases}$$

then the computation of the possibility measures in the first step become much simpler because in this case $m_i = A_i(x_o)$ and $n_i = B_i(x_o)$. Systems with numeric inputs usually require numeric outputs. In this case the output interface must provide a point of the universe representative of the fuzzy set inferred. Linguistic fuzzy models with numeric input and output defuzzification produce in general, nonlinear input–output mappings.

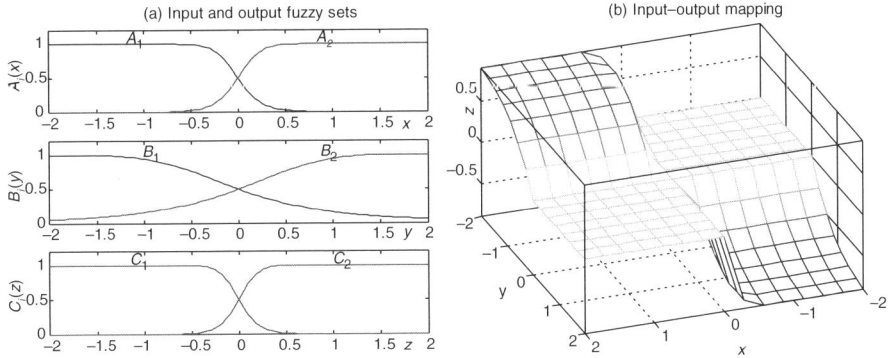

Figure 11.26 Input–output mapping of a fuzzy min–max model.

308 Chapter 11 Rule-Based Fuzzy Models

Table 11.2 Example of Fuzzy Rule Base.

Y X	A_1	A_2
B_1	C_1	
B_2		C_2

A way to find the representative point is to use centroid defuzzification

$$z = \frac{\int_Z zC(z)\mathrm{d}z}{\int_Z C(z)\mathrm{d}z} \qquad (11.26)$$

An alternative and somewhat simplified way is to take a weighted average of the modal values of the fuzzy sets of the rule consequents. Let v_i be the modal value of C_i. The numeric value z produced by an input x_0 is determined as

$$z = \frac{\sum_{i=1}^{N}(m_i \wedge n_i)v_i}{\sum_{i=1}^{N}(m_i \wedge n_i)} \qquad (11.27)$$

where $m_i = A_i(x_0)$ and $n_i = B_i(x_0)$. Figure 11.26 shows a simple two-input and single-output example. The membership functions and respective universes of the input and output variables are also depicted. It may be convenient to express the rule base in a table form to quickly visualize the meaning associated with each rule and to summarize the rule base itself. The table format also helps communication and design verification of fuzzy rule-based models. In the example the rule base is shown in Table 11.2

The main steps of the min-max are as follows:

Step 1: Antecedent matching: For each rule, compute the degree of matching between each atomic expression of the rule antecedent and the corresponding atomic expression of the input.

Step 2: Antecedent aggregation: For each rule, compute the rule activation degree by conjunctively or disjunctively operating on the corresponding degrees of matching depending on whether the atomic expressions of the rule antecedent are conjunctively or disjunctively related, respectively.

Step 3: Rule result derivation: For each rule, compute the corresponding inferred value based on its antecedent aggregation and the rule semantics chosen.

Step 4: Rule aggregation: Compute the inferred value from the complete set of rules by aggregating the result of the inferred values derived from individual rules.

As we shall see next, the general steps summarized above suggest different forms of fuzzy rule-based models, depending on the choices made when selecting triangular norms and rule semantics. However, we must note that, although compositional rule

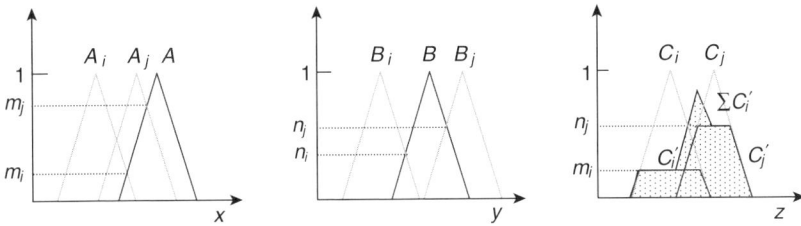

Figure 11.27 Fuzzy additive model processing.

of inference with maximum rule aggregation clearly is an instance of the general approach, many fuzzy rule-based models constructed upon the four steps do not necessarily derive from the compositional rule of inference and, because of this, they are named scaled inference (Driankov et al., 1993) to differentiate from compositional inference. Formally, scaled inference models can be viewed as a generalization of the concept of production systems as modular knowledge processing structures (Russell and Norvig, 2003) because their semantics essentially is procedural in nature (Lucas and Gaag, 1991).

11.5.1.2 Min–Sum Models

Min–sum fuzzy models infer contribution of each fuzzy rule in the same way as (11.20), repeated below

$$C'_i(z) = \sup_{x,y}[P(x,y) \wedge R_i(x,y,z)] = \sup_{x,y}[A(x) \wedge B(y) \wedge A_i(x) \wedge B_i(y) \wedge C_i(z)]$$

However, different from the usual requirement of using fuzzy aggregation operation such as the max s-norm, rules are aggregated using the sum of the inferred membership functions and, strictly speaking, they do not follow the composition rule of inference. In min–sum models, the consequent fuzzy sets that inputs activate are added

$$C(z) = \sum_{i=1}^{N} C'_i \qquad (11.28)$$

as shown in Figure 11.27. These type of models are called fuzzy additive systems (Kosko, 1992). Because the sum in (11.28) may not add up to 1 we can, in principle, normalize the sum of consequent fuzzy sets without affecting the structure of the additive model while keeping the model closer to composition. Strictly speaking, fuzzy additive rule-based systems use scaled inference. When input is numeric and output is also required to be numeric, normalization is not necessary if the centroid defuzzification (11.26) is used because the information in the inferred membership function C resides largely in the relative values of the membership degrees.

In general, additive fuzzy models encode and process parallel rules and the inferred fuzzy set C can be the weighted sum of the contribution of each individual rule $C'_i, i = 1, .., N$, that is,

$$C(z) = \sum_{i=1}^{N} w_i C'_i \qquad (11.29)$$

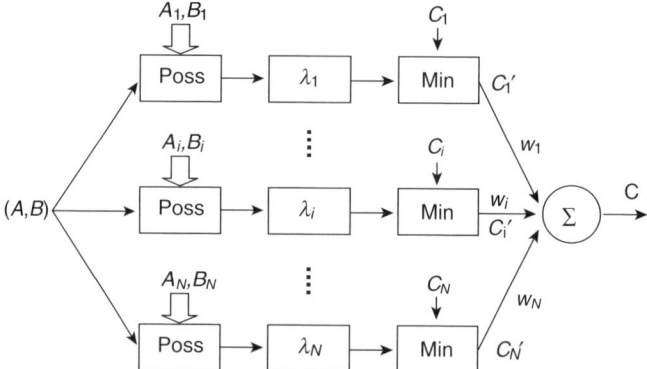

Figure 11.28 Architecture of additive fuzzy models.

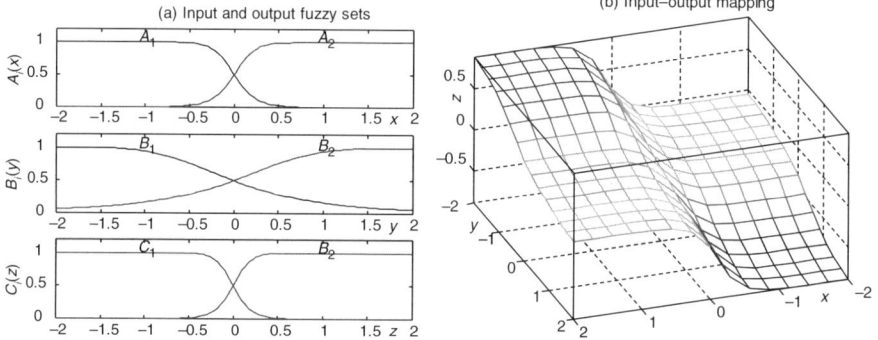

Figure 11.29 Input–output mapping of an additive fuzzy model.

where weights w_i reflect the relevance of the rule R_i. In additive fuzzy models we may change the values of weights w_i, and this ability contributes to the flexibility of these models. An illustration of the underlying architecture of the model is shown in Figure 11.28.

If the inputs are numeric, that is, $A = \{x_0\}$ and $B = \{y_0\}$, then $m_i = A_i(x_0)$ and $n_i = B_i(x_0)$ and (11.29) reads as follows:

$$C(z) = \sum_{i=1}^{N} w_i(\lambda_i \wedge C_i)$$

where $\lambda_i = \min(m_i, n_i)$. Figure 11.29 depicts the input–output function of the same fuzzy rule base in Table 11.1 with centroid defuzzification. We notice in Figure 11.29(b) that additive models produce smoother input–output characteristics (surfaces) than max–min models.

Notice that the systematization of the processing steps of additive models follows the same scheme as min–max models, except the aggregation step in which the sum replaces the max operation that turns additive fuzzy models into a form of scaled fuzzy inference.

11.5.1.3 Product–Sum Models

As discussed above, a broad family of fuzzy rule-based models can be developed if we use scaled inference with different t-norms to aggregate the fuzzy sets of rule antecedents, different rule semantics, and algebraic sum or distinct s-norms to aggregate the fuzzy sets inferred by each fuzzy rule. In what follows, we give examples of fuzzy rule-based models originating from scaled inference. We emphasize models that have been shown to be useful in practice (Mizumoto, 1994). For illustration purposes, we consider numeric inputs and centroid output defuzzification (11.26).

1. Product-probabilistic sum

$$C'_i(z) = m_i n_i C_i(z)$$
$$C(z) = \underset{i=1}{\overset{N}{S_p}} C'_i(z)$$

where the s-norm is the probabilistic sum, namely, $a s_p b = a + b - ab, a, b \in [0, 1]$.

2. Bounded product-bounded sum

$$C'_i(z) = m_i \otimes n_i \otimes C_i(z)$$
$$C(z) = \underset{i=1}{\overset{N}{\oplus}} C'_i(z)$$

where $\otimes$ is a t-norm, and $\oplus$ is its dual s-norm, such as the bounded product (Lukasiewicz t-norm) $a \otimes b = \max\{0, a + b - 1\}$ and the bounded sum (Lukasiewicz s-norm) $a \oplus b = \min\{1, x + y\}, a, b \in [0, 1]$.

3. Product- sum

$$C'_i(z) = m_i n_i C_i(z)$$
$$C(z) = \sum_{i=1}^{N} C'_i(z)$$

Figure 11.30 shows the input–output behavior of the fuzzy rule-based models addressed above.

Except for the bounded product-bounded sum model, the product-probabilistic sum and product-sum show smooth and similar behavior as fuzzy additive models, but both are distinct from the min–max model.

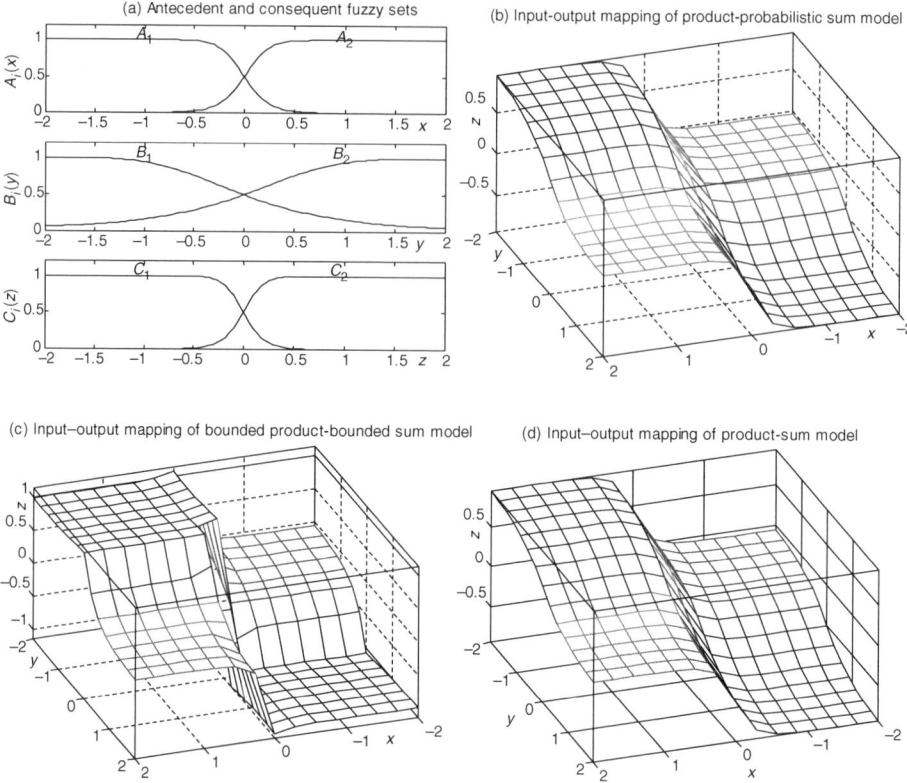

Figure 11.30 Input–output function of product sum fuzzy models.

11.5.2 Functional (Local) Fuzzy Models

Fuzzy functional models have rule bases composed by fuzzy rules whose consequent are functions of the antecedent variables (Takagi and Sugeno, 1985). The rules have the form

$$R_i : \quad \text{if } X \text{ is } A_i \text{ and } Y \text{ is } B_i \text{ then } z = f_i(x, y), \quad i = 1, 2, \ldots, N$$

where X and Y are linguistic variables with values A_i and B_i, and fuzzy sets on $\mathbf{X}$ and $\mathbf{Y}$ with base variables x and y, respectively. Function $f_i(x, y)$ is any function of the antecedent variables that appropriately describes the model output in the region specified by the fuzzy Cartesian product of the antecedent fuzzy sets. Typical examples are polynomials and in this case, the order of the polynomial names the model. For example, when the consequent is a real number, the consequent function is a zero-order polynomial and the functional model is a zero-order functional model. In this particular case, the functional model becomes the same as a linguistic model because the consequent is a real number. If the consequent function is a linear or quadratic polynomial, then we have first- or second-order functional models,

11.5 Types of Rule-Based Systems and Architectures

respectively. Because inputs and outputs of functional models are numeric, inference proceeds computing the activation degree of each rule and aggregating the fuzzy rule outputs using a weighted sum approach.

More specifically, assuming numeric inputs x and y, the degree of activation of rule R_i is computed as follows:

$$\lambda_i(x, y) = A_i(x) t B_i(y)$$

Next, the output z of the model is found as

$$z = \frac{\sum_{i=1}^{N} \lambda_i(x, y) f_i(x, y)}{\sum_{i=1}^{N} \lambda_i(x, y)} \quad (11.30)$$

Introducing an abbreviated notation

$$w_i(x, y) = \frac{\lambda_i(x, y)}{\sum_{i=1}^{N} \lambda_i(x, y)} \quad (11.31)$$

we rewrite (11.30) in the form of a linear combination of functions of the local models $f_i, i = 1, .., N$

$$z = \sum_{i=1}^{N} w_i(x, y) f_i(x, y) \quad (11.32)$$

For example, consider a single input–single output, first-order functional model whose rules are, assuming $\mathbf{X} = [0, 3]$ and $\mathbf{Z} = [0, 3]$,

$$R_1: \text{If } X \text{ is } A_1 \text{ then } z = x$$
$$R_2: \text{If } X \text{ is } A_2 \text{ then } z = -x + 3$$

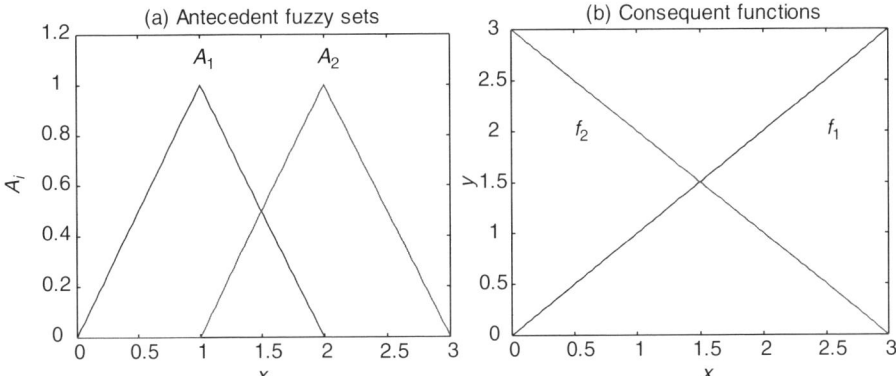

Figure 11.31 Rules of the functional fuzzy model.

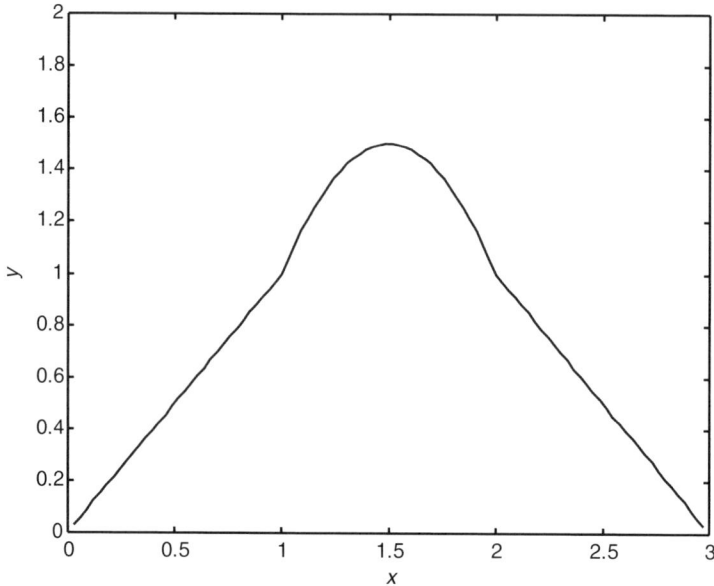

Figure 11.32 Output of the functional fuzzy model.

and triangular membership functions as Figure 11.31 shows. In this case, $\lambda_1(x) = A_1(x)$ and $\lambda_2(x) = 0$ for $x \in (0, 1]$ while $\lambda_2(x) = A_2(x)$ and $\lambda_1(x) = 0$ for $x \in [2, 3)$. Thus, the corresponding output is (note that $A_1(x) + A_2(x) = 1, \forall x \in [0, 3]$)

$$z = \begin{cases} x, & \text{if } x \in (0, 1] \\ A_1(x)x + A_2(x)(-x+3), & \text{if } x \in [1, 2] \\ -x+3, & \text{if } x \in [2, 3) \end{cases}$$

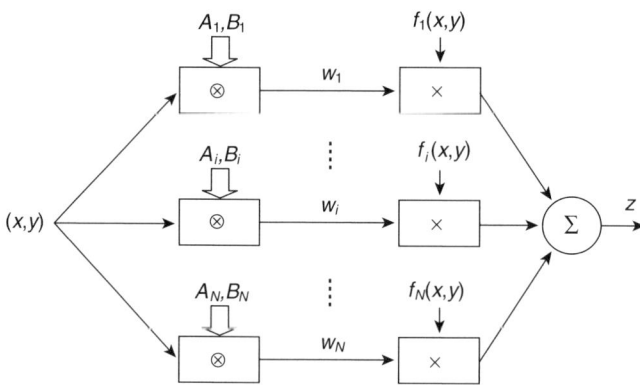

Figure 11.33 Architecture of functional fuzzy models.

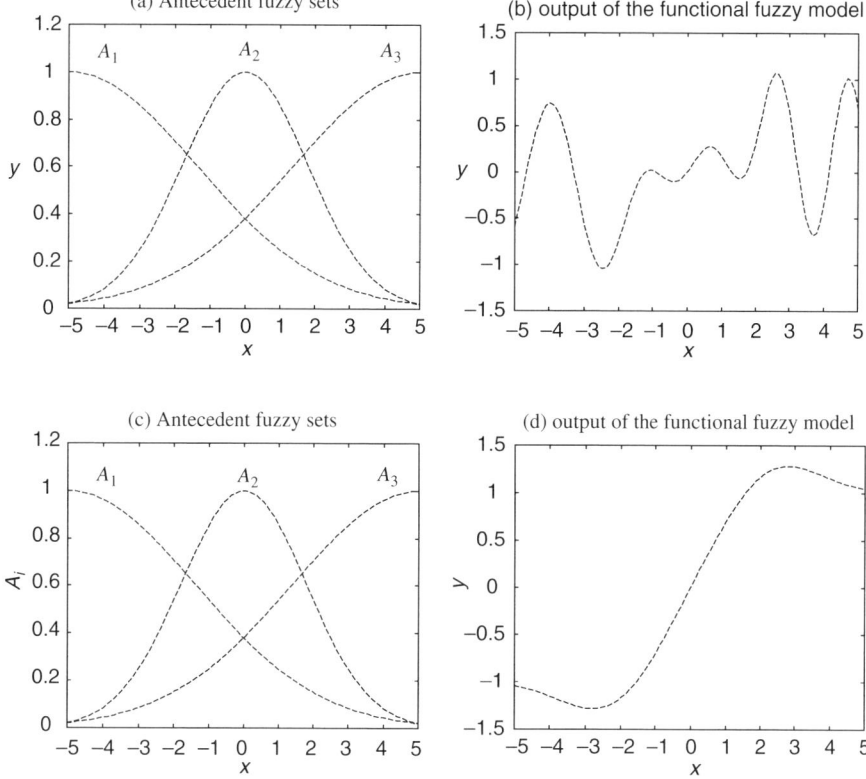

Figure 11.34 Examples of functional fuzzy models.

Looking at Figure 11.31, we note that the output is the same as f_1 for $x \in (0, 1]$ and the same as f_2 for $x \in [2, 3)$. In the middle interval $x \in [1, 2]$, the output is a sum of the product linear functions, which results in a quadratic function, as Figure 11.32 shows.

The overall architecture of functional fuzzy models is given in Figure 11.33. There is an input part that transforms inputs (x, y) into normalized activation levels w_i via t-norms and division operations with the membership degrees $A_i(x)$ and $B_i(y)$ using (11.31), which in turn weight the regression functions to form the consequents of the rules. Finally, the results are combined adding the weighted consequents using (11.32). Fuzzy functional models do not derive from the compositional rule of inference, but can be regarded as an instance of scaled inference.

The input–output characteristics of functional fuzzy rule-based models are affected by the form of the membership functions and of the form of the local models f_i located in the consequent part of the rules. Generally speaking, the input–output mapping results from a nonlinear combination of functions. An exception is the case of single input–single output zero-order models with properly chosen values and with membership functions forming linear complementary partition of the antecedent universe, that is, membership degrees adding up to unity for any value of the input variable.

Figure 11.34 illustrates several choices of antecedent membership functions and consequent functions. In Figure 11.34(a) membership functions are Gaussians and the corresponding consequent functions are $f_1 = -\sin(2x)$, $f_2 = 0.5x$, and $f_3 = \sin(3x)$.

In Figure 11.34(c), membership functions are the same as in (a) but the corresponding consequent functions are $f_1 = -1$, $f_2 = x$, $f_3 = 1$, and $f_3 = \sin(3x)$.

The rule-based systems can be generalized in many ways by admitting different forms of the conclusion part. It could form a family of differential equations:

$$R_i: \quad \text{If } X \text{ is } A_i \text{ and } Y \text{ is } B_i \text{ then } \dot{x} = f_i(x, y), \quad i = 1, 2, \ldots, N$$

The conclusion parts could be realized in the form of local neural networks, $NN_i(x, y, \mathbf{w}_i)$

$$R_i: \quad \text{If } X \text{ is } A_i \text{ and } Y \text{ is } B_i \text{ then } z = NN_i(x, y, \mathbf{w}_i), \quad i = 1, 2, \ldots, N$$

11.5.3 Gradual Fuzzy Models

Gradual fuzzy models assume rule bases composed by parallel gradual rules, which are parallel fuzzy rules of the form "If X is A_i then Z is C_i" understood as

$$R_i: \quad \text{The } more \ X \text{ is } A_i \text{ the } more \ Z \text{ is } C_i, \quad i = 1, 2, \ldots, N$$

In this section we emphasize the simplest instance of pointwise inputs. In this case, inference with gradual rule bases works as follows. Let $\alpha_1 = A_1(x)$ and $\alpha_2 = A_2(x)$ be the compatibility degrees of input x with antecedent fuzzy sets A_1 and A_2, respectively. The semantics of the rule, formally given, for instance, by the Gaines implication

$$R_i(x, z) = \begin{cases} 1, & \text{if } C_i(z) \geq A_i(x) \\ 0, & \text{otherwise} \end{cases}$$

means that the inferred fuzzy set should lie in the intersection of the corresponding consequent α-cuts

$$C = \bigcap_{i=1}^{N} (C'_i)_{\alpha_i} = \bigcap_{i=1}^{N} C_{\alpha_i}$$

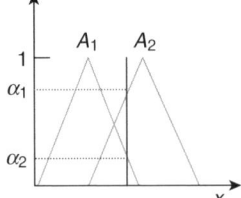 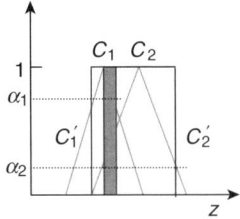

Figure 11.35 Gradual fuzzy model processing.

11.5 Types of Rule-Based Systems and Architectures

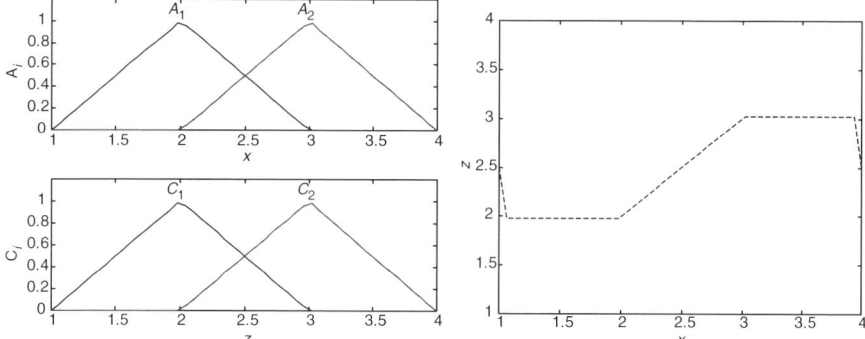

Figure 11.36 Output of the gradual fuzzy model.

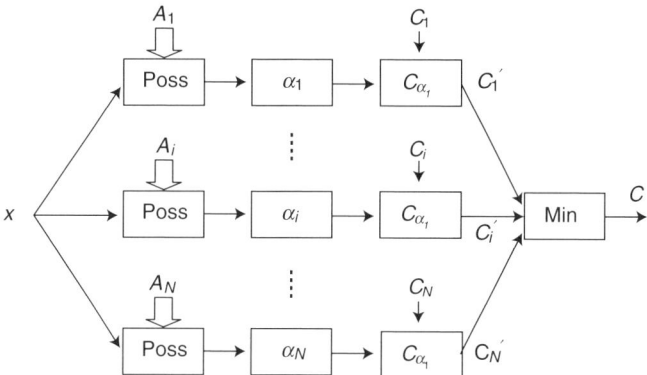

Figure 11.37 Architecture of gradual fuzzy models for pointwise input and bains implication.

where $C_{\alpha_i} = (C'_i)_{\alpha_i}$. The inferred set C is between the core of fuzzy sets C_j and C_k activated by the input x as Figure 11.35 illustrates when $N = 2$. Details of more general cases are covered in Dubois and Prade (1992).

The input–output characteristic of gradual fuzzy models is depicted in Figure 11.36 choosing triangular fuzzy antecedent and consequent fuzzy sets on $\mathbf{X} = \mathbf{Z} = [0, 4]$ and point inputs. We notice from Figures 11.35 and 11.36 that, because the individual inferred sets lie between the cores of the consequent fuzzy sets of active rules, an interpolation effect is obtained up to a bracketing error, which can, theoretically, be made very small by sufficiently increasing the number of rules (Dubois et al., 1994).

The architecture of gradual fuzzy models using bains implication is shown in Figure 11.37. The input part transforms inputs x into compatibility levels α_i via possibility measures and the corresponding α_i cuts of the consequent fuzzy sets C_{α_i} are found. Recall that, because C_{α_i} are α-cuts, they are sets. Finally, the results are combined using intersection as the min operator. Different choices of fuzzy implications induce different architectures.

11.6 APPROXIMATION PROPERTIES OF FUZZY RULE-BASED MODELS

One of the most important properties regarding fuzzy rule-based models concerns their approximation capabilities. Fuzzy rule-based systems can uniformly approximate continuous functions to any degree of accuracy on closed and bounded sets.

Fuzzy systems in which rules are viewed as the product fuzzy conjunction, antecedent expressions with Gaussian membership functions aggregated with the product t-norm, Gaussian consequent membership functions, processed with sup–min composition, and rules aggregation via ordinary sum, pointwise inputs and centroid defuzzification are universal approximators (Wang and Mendel, 1992). Fuzzy additive systems also have the same approximation capabilities (Kosko, 1992). Fuzzy rule-based models with arbitrary t-norms as conjunctions and triangular or trapezoidal membership functions, R-implications or conjunctions with centroid defuzzification are universal approximators as well (Castro, 1995; Castro and Delgado, 1996). Fuzzy functions represented by fuzzy relations have also been shown to be approximate data (Perfilieva, 2004).

Using suitable granulation of one-dimensional antecedent and consequent universes, gradual models provide the same approximation mechanism as do functional fuzzy models. More generally, any single input–single output system can be approximated by a set of gradual rules when using appropriate granulation. (Dubois et al., 1994).

Another important issue concerns rule-based interpolation especially when sparse rule bases developed from data emerge (Koczy and Hirota, 1993). A sparse rule base is a rule base composed by only a subset of all possible rules. For instance, for a system with two input variables, the first with three and the second with four fuzzy values, a total of twelve rules could assemble a full rule base. A rule base with fewer rules, say six rules, would be a sparse rule base. Techniques have been developed for multidimensional spaces (Wong et al., 2005) as a form of interpolative reasoning. Methods of interpolative reasoning are particularly useful to reduce complexity of fuzzy models and to make inference in sparse systems possible. An approach involving scale and move transformations is addressed in Huang and Shen (2006). This variation handles interpolation of multiple antecedents with simple computations and guarantees uniqueness, normality, and convexity of the interpolating fuzzy sets.

11.7 DEVELOPMENT OF RULE-BASED SYSTEMS

By their virtue, fuzzy rule-based systems rely on the computing of information granules. Information granules are in the center of the development of the individual rules. There are two fundamental ways of constructing rule-based models, namely, (a) expert-based and (b) data-driven. Furthermore, there are a number of hybrid approaches that could be positioned somewhere in-between. Each of them exhibits several advantages but is not free of shortcomings. Next we address several representative cases.

11.7.1 Expert-Based Development

The idea is that knowledge can be provided by domain experts. They are individuals who can quantify knowledge about the basic concepts and variables essential to the problem and link them in the form of some rules. Knowledge-based approach has some advantages: knowledge becomes readily available and fuzzy sets are helpful in the quantification process. Rules reflect existing knowledge and thus could be readily quantified. We may enumerate a number of examples in which knowledge comes in the form of a collection of rules. The development time of the rules could be quite short, assuming that we are concerned with a small size of the problem requiring a handful of conditional statements.

This type of handcrafted style development of the rule-based system has been used from the beginning of fuzzy modeling. At that time, we dealt with small rule bases involving small number of rules and having one–three input variables. The typical example in this category comes in the form of fuzzy controllers. A fuzzy controller is a rule-based system that captures the essence of the control strategy as being described by a human operator controlling the process. For instance, one could achieve an effective control by monitoring the values of error and change error and making adjustments to control actions on the basis of these findings. The error is expressed by comparing the current output of the system under control with the given reference. Interestingly, the same type of knowledge is utilized in the classic and commonly used control architecture of the so-called PD controllers. The control rules of the fuzzy controller assume the format

R_i: If Error is A_i and Change of Error is B_i then Control is C_i

where all fuzzy sets standing there are defined in the corresponding universes of discourse (i.e., change of error, and control being regarded as the output of the controller). An example of the control rules is shown in Table 11.3. These rules are highly intuitive and one could easily articulate them after giving some thought to the way in which we affect the process so that we reach the required reference.

Table 11.3 An Example Rule Base of a Fuzzy Controller: NB—negative big, NM–negative medium, NS—negative small, Z—zero, PS—positive small, PM—positive medium, PB—positive big.

Change of error \ Error	NM	NS	ZE	PS	PM
NB	PM	NB	NB	NB	NM
NM	PM	NB	NS	NM	NM
NS	PM	NS	Z	NS	NM
Z	PM	NS	Z	NS	NM
PS	PM	PS	Z	NS	NM
PM	PM	PM	PS	PM	NM
PB	PM	PM	PM	PM	NM

There are several shortcomings that come with this type of the development of the model. Some of them could be detrimental and might manifest quite evidently once we move toward larger rule bases. Let us highlight some of the difficulties and point at possible ways on how to alleviate them:

(a) The rules are highly prescriptive because they reflect the existing domain knowledge. This means that under some circumstances, the construct has to be calibrated to address the specificity of the problem. For instance, the rules could be quite general, yet the fuzzy sets standing in the rules might need calibration so that their semantics fully captures the specificity of the problem. In Chapter 4 we identified some methods that are helpful in addressing this problem. Simply, the generic fuzzy sets are adjusted through a nonlinear mapping of the original spaces.

(b) The rules are more difficult to develop when dealing with high-dimensionality problems. First, the number of rules grows very fast (recall that a complete rule base has $N = p^n$ rules, assuming rules with n input variables each of them with p linguistic values). Second, the quality of the rules could deteriorate quite quickly. In order to alleviate these difficulties, one has to establish some well-developed mechanisms of quality assurance. Measures of completeness and consistency of the rules (see Section 11.9) could be very helpful in this regard as they quantify the quality of each rule and also identify the components of the rule base that are of the lowest quality (say, a rule of the lowest consistency index). One could easily identify some "gaps" in the rule base that might be detrimental when using the system. Unfortunately, in the generic version of the fuzzy rule-based system the curse of dimensionality is present and definitely contributes to the development difficulties and furthers implementation issues. The scalability of the construct is thus highly questionable. There have been a number of alternatives exercised along this line; the one most commonly encountered concerns the so-called hierarchical structures of rule-based systems.

11.7.2 Data-Driven Development

This approach is guided by the use of data. The resulting design captures the main structure existing in the data themselves. As the information granules are predominant components of all rule-based systems, fuzzy clustering plays an essential role here. The results of transforming numeric data into fuzzy sets are in direct usage in the buildup of the rule-based system.

Let us discuss the pertinent details. Consider the data set in the form of the finite set of input–output pairs $(\mathbf{x}_k, y_k), k = 1, 2, \ldots, M$, where $\mathbf{x}_k = [x_{1k}, x_{2k}, \ldots, x_{nk}]$. We combine them into an $(n + 1)$-dimensional vector $\mathbf{z}_k = [\mathbf{x}_k, y_k]$ and cluster them with the use of the FCM (or any other fuzzy clustering technique), which leads to the collection of N clusters, that is, cluster centers, or prototypes, $\mathbf{v}_1, \mathbf{v}_2, \ldots, \mathbf{v}_N$, and the partition matrix $U = [u_{ik}], i = 1, 2, \ldots, N$ in the product space $\mathbf{X} \times \mathbf{Y}$. Figure 11.38

11.7 Development of Rule-Based Systems

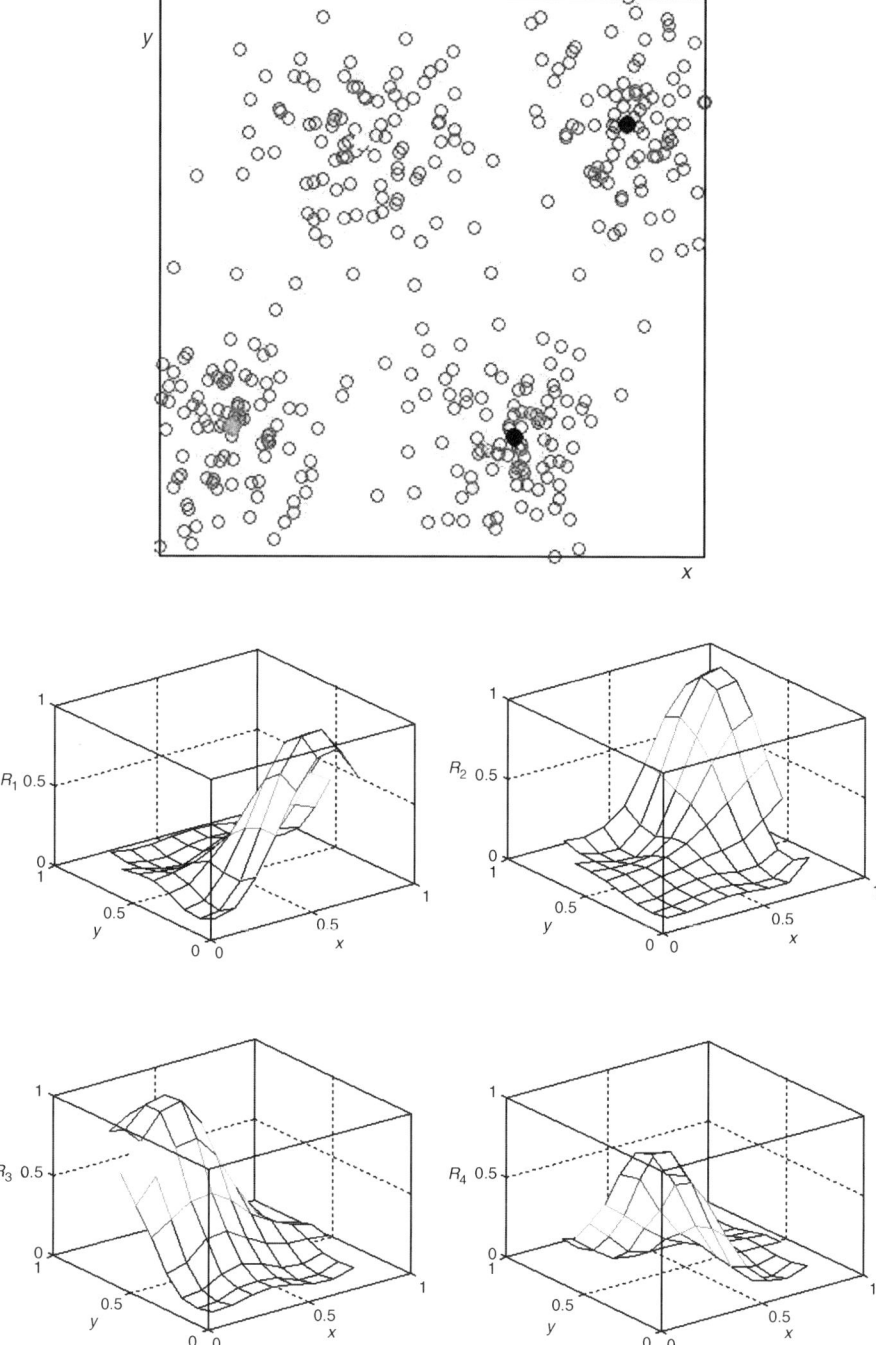

Figure 11.38 Clustering data in the product space with the use of the FCM.

shows an example where we identify four clusters in the data set. Therefore, we have four rules. The membership functions of the fuzzy clusters obtained by the FCM algorithm are also shown. They correspond to the membership functions of the fuzzy relations associated with each fuzzy rule $R_i, i = 1,..,4$ induced by the clusters.

We may also project the prototypes on the output space **Y** by considering their last coordinates. Denote them by $v_1[y], v_2[y], \ldots, v_N[y]$. They give rise to the membership functions that are associated with the way in which we run the FCM. Denote them by $C_1, C_2, \ldots, C_N$. Similarly, project the prototypes on the input space **X** and denote them by $\mathbf{v}_1[\mathbf{x}], \mathbf{v}_2[\mathbf{x}], \ldots, \mathbf{v}_N[\mathbf{x}]$. Building membership functions around them is easy. Denote the resulting fuzzy sets by $A_1, A_2, \ldots, A_N$. In this way we arrive at the collection of rules of the form

$$R_i: \quad \text{If } X \text{ is } A_i \text{ then } Y \text{ is } C_i, \; i = 1, 2, .., N$$

with the fuzzy sets constructed through fuzzy clustering. Because variables can be correlated, projections usually cause loss of information. However, fuzzy clusters are ultimately reflective of the existing numeric data and thus contribute to the increasing rapport of the rule-based system with the available experimental evidence. Furthermore, as they process all variables at the same time, they help reduce the number of rules. It should be emphasized that the condition parts of the rules are defined over the Cartesian product of all input variables.

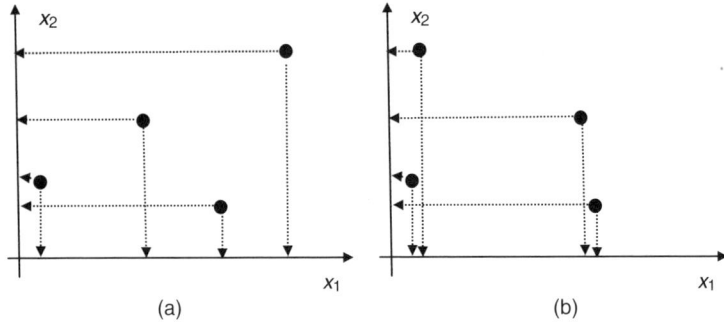

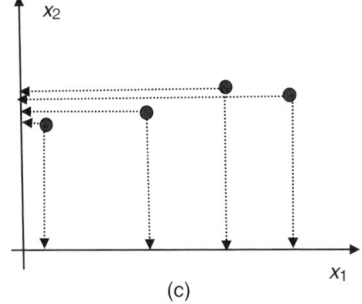

Figure 11.39 Projection of prototypes on the corresponding coordinates of the input space.

11.7 Development of Rule-Based Systems

One may further expand the rules by explicitly expressing the fuzzy sets in the Cartesian product of the input spaces. We project the coordinates of the prototypes on the respective universe of discourse. Figure 11.39 gives an example with $n = 2$ and $N = 4$. Shown are various distributions of the projections: (a) well-delineated clusters in each space, (b) substantial overlap between some prototypes observed for the projection on $\mathbf{X}_1$, and (c) all prototypes projected on $\mathbf{X}_2$ are almost indistinguishable. We next form the corresponding fuzzy sets in each of them.

In this way we arrive at the same format of the rules as discussed earlier. The projection scheme brings another insight into the nature of the data (and resulting rules). As illustrated in Figure 11.39, we can envision a situation where the projected prototypes are visibly well distributed. There might be cases, as illustrated in Figure 11.39(b), in which some projections are almost the same. This suggests that we may consider collapsing of the fuzzy sets as they are semantically the same. Figure 11.39(c) shows another example where with the collapse of all prototypes we may consider dropping the corresponding variable from the rules.

Through this postprocessing of the clustering results in which some fuzzy sets could be combined, the resulting rules may involve different number of fuzzy sets. Similarly, we may expect a situation of dimensionality reduction in which some variables could be dropped (eliminated) from the rules.

In virtue of the fuzzy clustering completed in the way described above, there is no distinction made between the input and output variables. There is an advantage associated with that as we clearly see associations between the fuzzy sets standing in the condition and conclusion parts of the rules. There could be a certain drawback given the fact that the contribution of the output variable could be easily eliminated or even removed from the cluster. Putting it simply, when dealing with n variables, the computations of the distance in the $(n + 1)$-dimensional space is almost exclusively driven by the values of the distance produced by the first n coordinates of it. The role of the output variables in the clustering process has been practically marginalized. To avoid this effect, we may consider clustering carried out independently in the output and input spaces. In other words, we apply the clustering to the data sets $\{y_k\}$ and $\{\mathbf{x}_k\}$. Then the clusters independently formed in these spaces are used to build the rules, yet to do so one has to establish a correspondence between the clusters built in the input and output spaces. We may also envision more specialized clustering such as context-based clustering, C-FCM (these generalizations will be studied in the framework of the so-called granular models). Clustering can also be combined with gradient-based parameter tuning to derive self-generation and simplification of rule bases (Chen and Likens, 2004).

Summarizing the expert-based and data-driven development of rule-based systems we observe that these two approaches are very much complementary. Their advantages and shortcomings complement each other. In practice, there are a number of ways in which both approaches are used in some hybrid manner. Typically, one might develop a skeleton of the system based on the available domain knowledge being a part of the human expertise that is, in the sequel, refined and augmented by the data-driven techniques. In this way we retain some transparency of the construct and add flexibility provided by the use of the data-driven development techniques.

11.8 PARAMETER ESTIMATION PROCEDURE FOR FUNCTIONAL RULE-BASED SYSTEMS

In this section we assume that each local model of fuzzy functional rules is a linear function $\mathbf{a}_i^T \mathbf{x}, i = 1, 2, \ldots, N$, where $\mathbf{a}_i = [a_1, a_2, \ldots, a_n]^T$ and $\mathbf{x}_k = [x_{1k}, x_{2k}, \ldots, x_{nk}]^T$. For estimation purposes given is a finite set of input–output data $(\mathbf{x}_1, y_1), (\mathbf{x}_2, y_2), \ldots, (\mathbf{x}_M, y_M)$. Let us recall that the output $\hat{y}$ of the functional fuzzy model for any $\mathbf{x}_k$ is given as a weighted sum

$$\hat{y}_k = \sum_{i=1}^{N} w_{ik} f_i(\mathbf{x}_k, \mathbf{a}_i) \tag{11.33}$$

with the weights w_{ik} expressed in the form $w_{ik} = \lambda_i(\mathbf{x}_k) / [\sum_{i=1}^{N} \lambda_i(\mathbf{x}_k)]$. Note that we have introduced a new index to underline the fact that the weights depend on the kth input $\mathbf{x}_k$. The output of the model is

$$\hat{y}_k = \sum_{i=1}^{N} \mathbf{z}_{ik}^T \mathbf{a}_i \tag{11.34}$$

Let us use the following vector notation to collect all parameters of the models

$$\mathbf{a} = \begin{bmatrix} \mathbf{a}_1 \\ \mathbf{a}_2 \\ \ldots \\ \mathbf{a}_N \end{bmatrix}$$

which leads to the expression for the model being expressed in the form of some scalar product

$$\hat{y}_k = [\mathbf{z}_{1k}^T \mathbf{z}_{2k}^T \ldots \mathbf{z}_{Nk}^T] \begin{bmatrix} \mathbf{a}_1 \\ \mathbf{a}_2 \\ \ldots \\ \mathbf{a}_N \end{bmatrix}$$

where $\mathbf{z}_{ik} = w_{ik} \mathbf{x}_k$. The collection of M input–output data is organized in the following matrix format:

$$\mathbf{y} = \begin{bmatrix} y_1 \\ y_2 \\ \ldots \\ y_M \end{bmatrix} \quad Z = \begin{bmatrix} \mathbf{z}_{11}^T & \mathbf{z}_{21}^T & \ldots & \mathbf{z}_{N1}^T \\ \mathbf{z}_{12}^T & \mathbf{z}_{22}^T & \ldots & \mathbf{z}_{N2}^T \\ .. & \ldots & \ldots & \ldots \\ \mathbf{z}_{1M}^T & \mathbf{z}_{2M}^T & \ldots & \mathbf{z}_{NM} \end{bmatrix}$$

Ideally, we would expect that the output of the model should follow the experimental data, that is, we require that $\mathbf{y} = Z\mathbf{a}$ where $\mathbf{a}$ should result as a solution to the system of M linear equations. Given that M is typically higher than the number of the parameters of the rule-based system, there is no unique solution to the above system of linear equations and we have to reformulate the problem as a certain optimization task. Instead of solving the equations, we can minimize the distance $\|.\|$ between $\mathbf{y}$ and $Z\mathbf{a}$, that is, solve the optimization problem

$$\operatorname{Min}_{\mathbf{a}} J_G(\mathbf{a}) = \| \mathbf{y} - Z\mathbf{a} \|^2 \tag{11.35}$$

Assuming the Euclidean norm as the distance between $\mathbf{y}$ and $Z\mathbf{a}$, that is,

$$\| \mathbf{y} - Z\mathbf{a} \|^2 = (\mathbf{y} - Z\mathbf{a})^T(\mathbf{y} - Z\mathbf{a}) \tag{11.36}$$

the solution to the above problem is well-known and is expressed with the use of the pseudoinverse of Z,

$$\mathbf{a}_{\text{opt}} = Z^{\#}\mathbf{y} \tag{11.37}$$

where $Z^{\#} = (Z^T Z)^{-1} Z^T$

In case of the polynomial local models, the same estimation procedure applies; however, the input space has to be expanded given the fact that we encounter a number of high-order components of the original variables. The model is linear in terms of its parameters.

A computationally more effective way to compute $\mathbf{a}_{\text{opt}}$ is to put the matrix Z into its singular value decomposition form (Golub and Van Loan, 1989), that is,

$$Z = P \sum Q^T \tag{11.38}$$

where P and Q are orthogonal matrices, $P = [\mathbf{p}_1, \mathbf{p}_2, \ldots, \mathbf{p}_M]$ is an $(M \times M)$ and $Q = [\mathbf{q}_1, \mathbf{q}_2, \ldots, \mathbf{q}_{nN}]$ is an $(nN \times nN)$ matrix, and $\sum$ is the $(M \times nN)$ matrix whose diagonal elements $\sigma_1 \geq \sigma_2 \geq \cdots \geq \sigma_{nN}$ are the singular values of Z. Substituting (11.38) in (11.37) we get

$$\mathbf{a}_{\text{opt}} = \sum_{i=1}^{s} \frac{\mathbf{p}_i^T \mathbf{y}}{\sigma_i} \mathbf{q}_i \quad \text{global estimation} \tag{11.39}$$

where s is the number of nonzero singular values of $\sum$. Expression (11.35) defines a global estimation scheme in the sense that the parameters of the model are estimated using the entire data set.

Alternatively, local estimation computes the parameters of the model through the solution of the following minimization problem:

$$\operatorname{Min}_a J_L(a) = \sum_{i=1}^{N} \| \mathbf{y} - Z_i \mathbf{a}_i \|^2 = \sum_{i=1}^{N} (\mathbf{y} - Z_i \mathbf{a}_i)^T (\mathbf{y} - Z_i \mathbf{a}_i) \tag{11.40}$$

where Z_i is the ith $(M \times n)$ submatrix of Z of the form

$$Z_i = \begin{bmatrix} \mathbf{z}_{i1}^T \\ \mathbf{z}_{i2}^T \\ \vdots \\ \mathbf{z}_{iM}^T \end{bmatrix} \tag{11.41}$$

Similar to the previous case, we find the singular value decomposition of X_i

$$Z_i = P_i \sum_i Q_i^T$$

and compute the smallest Euclidian norm solution as

$$\mathbf{a}_{i,\mathrm{opt}} = \sum_{l=1}^{s_i} \frac{\mathbf{p}_l^T \mathbf{y}}{\sigma_l} \mathbf{q}_l, i = 1, 2, \cdots, N \qquad \text{local estimation} \qquad (11.42)$$

where s_i is the number of nonzero singular values of $\sum_i$ and $\mathbf{p}_l$ and $\mathbf{q}_l$ are the l-th column of P_i and Q_i, respectively. Although global estimation of functional fuzzy models provides approximation with arbitrary accuracy, it does not always guarantee the model to behave well locally and often results in lower interpretability. On the contrary, local models help interpretability and keep models behaving locally well. In terms of computational demand, local estimation is superior to global estimating and this is an important issue when developing complex models. In terms of approximation error, global estimation is favorable with respect to bias error (Nelles, 2001), and whenever good quality data is available, global estimation may be preferable. An alternative is to combine both approaches to achieve an acceptable trade-off between computational demand and model accuracy and interpretability. One mechanism is to formulate the objective function of the optimization problem as a convex combination of the functions J_G (35) and J_L (40) as suggested in Yen et al. (1998). In this way the optimization problem reads as follows:

$$\mathrm{Min}_\mathbf{a} \alpha J_L(\mathbf{a}) + (1-\alpha) J_G(\mathbf{a})$$

where $\alpha \in [0,1]$. Notice that $\alpha = 0$ favors global estimation while $\alpha = 1$ favors local estimation. By choosing intermediate values, the designer can adapt the characteristics of the model to the application requirements.

Precision and interpretability are discussed in Setnes et al. (1998) and Casillas et al. (2003).

11.9 DESIGN ISSUES OF RULE-BASED SYSTEMS—CONSISTENCY, COMPLETENESS, AND THE CURSE OF DIMENSIONALITY

The rule-based systems exhibit a clear logic structure. The quality of the rules can be expressed by taking into account the two main characteristics of completeness and consistency of the rules. We elaborate on them in detail. The evaluation of the rules can be completed on the basis of the available experimental data, refer to Figure 11.40. These data set come as a collection of input–output pairs, say $(\mathbf{x}_k, y_k), k = 1, 2, \ldots, N$, whose corresponding components are defined in the input and the output spaces.

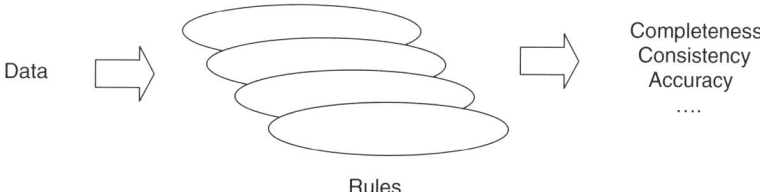

Figure 11.40 Evaluation of the quality of rules with the use of experimental numeric data.

11.9.1 Completeness of Rules

The completeness of the rules with one antecedent is expressed by the following condition:

$$\max_{i=1,2,\ldots,n} A_i(\mathbf{x}_k) > 0 \text{ for all } k = 1, 2, \ldots, N \qquad (11.43)$$

In other words, we require that all data points $\mathbf{x}_k$ are represented through some of the fuzzy sets standing in the conditions of the rules. More descriptively, this requirement states that the input space has to be completely "covered" by the fuzzy sets being used in the conditions of the rules. Putting it more succinctly: We anticipate that for each point of the input space, at least one rule becomes invoked (fired) to a nonzero degree. In this way, the rule-based model does not exhibit any "holes" and one could infer conclusion for any input. Sometimes the requirement of completeness could be made stricter by requesting that

$$\max_{i=1,2,\ldots,n} A_i(\mathbf{x}) \geq \delta \qquad (11.44)$$

where δ is a certain predefined threshold level, say $\delta = 0.2$. This requirement could guard us against situations where there are very small positive values of membership grades, thus (11.43) is satisfied, yet such membership grades are practically meaningless, so there is no underlying semantics. This situation could easily occur when dealing with Gaussian membership functions with infinite support.

11.9.2 Consistency of Rules

The consistency of the rules concerns some rules that could be in conflict, meaning that very *similar* (or the same conditions) could result in completely *different* conclusions. Such a qualitative and intuitive observation can be rephrased by investigating the following four cases that we may encounter when analyzing two rules:

Conditions and conclusions	*Similar* conclusions	*Different* conclusions
Similar conditions	Rules are redundant	Rules are in conflict
Different conditions	Different rules; could be eventually merge	Different rules

328 Chapter 11 Rule-Based Fuzzy Models

Only in one of these cases, there is an evident conflict and hence it should be signaled. A certain consistency index can be developed to describe conflict. To capture the nature of the consistency index that should be reflective of the findings presented in the table above, we consider here an implication operation $g \to h$. This operation exhibits the desired properties as it assumes low values only if the truth values of "g" are high and the values of h are low. In all other cases the implication returns high truth values. Given this, the formal definition of the consistency index of two rules,

It X is A_i then Y is B_i
It X is A_j then Y is B_j

denoted by cons(i,j), given experimental data is defined as follows:

$$\text{cons}(i,j) = \sum_{k=1}^{N} \{|B_i(y_k) - B_j(y_k)| \Rightarrow |A_i(x_k) - A_j(x_k)|\} \quad (11.45)$$

where $\Rightarrow$ is an implication induced by some t-norm. Alternatively, we can define the consistency as follows

$$\text{cons}(i,j) = \sum_{k=1}^{N} \{\text{Poss}(A_i(x_k), A_j(x_k)) \Rightarrow \text{Poss}(B_i(y_k), B_j(y_k))\}$$

Recall that formally we have $a \Rightarrow b = \sup\{c \in [0,1] | atc \leq b\}$ where $a, b \in [0,1]$. Computing an average of the consistency of the ith rule with respect to all other rules ($j = 1, 2, \ldots, i-1, i+1, \ldots, N$), we can talk about an overall consistency of this rule to be equal to

$$\text{cons}(i) = \frac{1}{N} \sum_{j=1}^{N} \text{cons}(i,j) \quad (11.46)$$

Through the computations of the consistency index of the consecutive rules of the model, we can order them in a linear fashion as illustrated in Figure 11.41.

In Figure 11.41, note the emerging sets of rules of different consistency: (a) a uniform spread of the rules with respect to their level of consistency and (b) a subset of rules of low consistency (high conflict rules).

This linear arrangement of the rules with respect to their consistency is essential for several reasons. We gain an important insight into the quality of each rule and isolate the rules that require further attention. For instance, when there are only a few

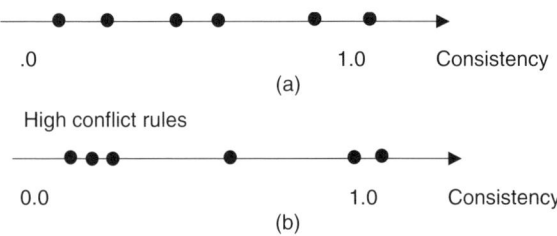

Figure 11.41 A linear order imposed on the rules trough consistency analysis.

rules whose consistency is low (as illustrated in Fig. 11.41(b)), one can revisit their structure as far as the fuzzy sets of condition and conclusion are concerned and suggest some improvements. Let us discuss some alternatives that are worth considering in this setting. By a direct inspection we collect the "weakest" rules, namely, those of the lowest consistency level. As we encounter very different conclusions while have identical or very similar condition parts of these rules, it could well be possible that some input variable was omitted. This omission could be a result of error in the process of knowledge acquisition. For instance, it could well be possible that some variable (which seemed so evident that no attention was paid to its inclusion) has been overlooked. In this case the problem of low consistency can be alleviated by expanding the space of input variables by adding the missing variable along with some of its fuzzy sets. Referring to the consistency index (11.45), augment the rule by some additional condition so that it reads as

$$R_i: \quad \text{If } X \text{ is } A_i \text{ and } Z \text{ is } C_i \text{ then } Y \text{ is } B_i$$

where C_i is a certain fuzzy set defined over the auxiliary input space $\mathbf{Z}$. Now consider the rules with the expanded condition part

$$\begin{aligned} R_i: &\quad \text{If } X \text{ is } A_i \text{ and } Z \text{ is } C_i \text{ then } Y \text{ is } B_i \\ R_j: &\quad \text{If } X \text{ is } A_j \text{ and } Z \text{ is } C_j \text{ then } Y \text{ is } B_j \end{aligned} \quad (11.47)$$

If C_i and C_j are different enough, that is, the expression $|A_i(\mathbf{x}_k)tC_i(z) - A_j(\mathbf{x}_k)tC_j(z)|$ assumes lower values than the one originally computed, $|A_i(\mathbf{x}_k) - A_j(\mathbf{x}_k)|$, then the implication computed as

$$|B_i(y_k) - B_j(y_k)| \Rightarrow |A_i(\mathbf{x}_k)tC_i(z) - A_j(\mathbf{x}_k)tC_j(z)| \quad (11.48)$$

may produce higher values than the one for the case studied before. Note that we do not have experimental data for the additional space ($\mathbf{Z}$), hence we have not used any specific entry by simply indicating some value in $\mathbf{Z}$, that is, "z." To be more formal and avoid this nonuniqueness, the consistency measure is redefined in this case by integrating over the additional input space. We arrive at the expression

$$\{|B_i(y_k) - B_j(y_k)| \Rightarrow \sum_{k=1}^{N} \int_{\mathbf{Z}} |A_i(\mathbf{x}_k)tC_i(z) - A_j(\mathbf{x}_k)tC_j(z)|\} \quad (11.49)$$

It becomes obvious that if C_i and C_j are completely disjoint (which makes the extended conditions "A_i and C_i" and "A_j and C_j" fully disjoint), then the consistency of the rules expanded in this way becomes equal to 1 as the distance between the condition parts of the rules is equal to zero. As we might have a number of rules with some low consistency, it may not be feasible to introduce highly disjoint individual fuzzy sets in a new space. Any augmentation of the rules requires a careful investigation of the existing ones. The proposed approach does help quantify a conflict and assess the quality of the rules resulting from the resolution of the conflict but cannot define fuzzy sets (membership functions) that need to be added to alleviate the existing shortcoming. Similarly, the proposed approach cannot offer any advice as to the formation of the augmented space. It could involve a single variable (**Z**) with a number of pertinent fuzzy sets or a Cartesian product of several new variables (**Z** × **W** × **U**) with a lower number of fuzzy sets defined in each of them.

Further discussions and an algorithm to check coherence and redundancy of fuzzy rule bases with implication and rule combination via intersection can be found in Dubois et al. (1997).

11.10 THE CURSE OF DIMENSIONALITY IN RULE-BASED SYSTEMS

The rules capture domain knowledge in an explicit manner. They are transparent structures conveying knowledge about the system they describe. Unfortunately, this transparency does come with a highly detrimental effect known as a curse of dimensionality. When the number of input variables increases, the number of rules increases exponentially. Consider that for each variable we encounter p fuzzy sets. For n input variables, the number of different rules is p^n. Obviously, some rules might not be relevant (as not being descriptive of the problem at hand), yet potentially we are faced with an exponential growth of the rule base. The scalability of the rule base becomes a serious issue if one moves beyond four–five variables each being quantified in terms of a few fuzzy sets. Considering the generic format of the rules, the problem cannot be avoided. One mechanism to avoid the use of large number of rules is to use dynamic rule bases (Chen and Saif, 2005). Instead of being fixed, dynamic rule bases allow rules to vary with inputs in a piecewise manner.

11.11 DEVELOPMENT SCHEME OF FUZZY RULE-BASED MODELS

We can establish a general view at the development of rule-based systems in the form of a spiral scheme as illustrated in Figure 11.42. As discussed in this chapter, there

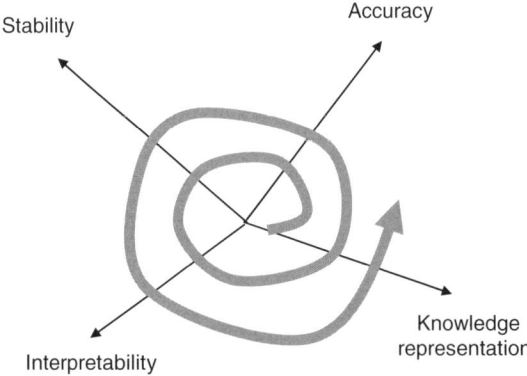

Figure 11.42 A spiral view at the development of rule-based models as shown in a multidimensional space of fundamental characteristics of the models.

are several fundamental qualities that are inherent to rule-based constructs; they form coordinates of the multidimensional space in which assessment is being made. Over time, we witness an ongoing evolution of the rule-based architecture that is illustrated by an unfolding spiral that illustrates the progress supported by new mechanisms of knowledge representation, new learning (development) capabilities, and quality assessment of the resulting rules.

11.12 CONCLUSIONS

In this chapter, the notions of fuzzy expressions, fuzzy rules, and fuzzy rule-based systems have been introduced as a vehicle for knowledge representation and processing. Several classes of fuzzy rule-based models were emphasized together with the associated inference procedures and architectures. Fuzzy rule-based models design issues such as rule-based development, consistency, transparency, and approximation capabilities have been summarized because they provide the foundations upon which fuzzy models are build up and they provide a sound approach for granular models development and applications.

EXERCISES AND PROBLEMS

1. Consider the function $y = f(x) = 2x^2 + 2$ for $x \in [0,4]$. Develop a rule-based model to approximate $f(x)$ in this interval. Assume that $f(x)$ is to be approximated through fuzzy graphs. Experiment with different t-norms.

2. Assume a fuzzy rule of the form "If X is A then Y is B" and suppose that inputs are points $x \in \mathbf{X} = [0,4]$ whereas A and B are triangular fuzzy sets $A(x, 1, 2, 3)$ and $B(y, 2, 3, 4)$. Using the compositional rule of inference, determine the inferred fuzzy set B' assuming that the rule is interpreted as the implication $R(x, y) = 1 - \min[A(x), 1 - B(y)]$. Plot the inferred fuzzy set of conclusion.

3. Repeat the previous problem when interpreting the rule as an implication of the form $R(x, y) = \max[1 - A(x), B(y)]$.

4. Repeat Problem 3 and determine the inferred fuzzy set B' considering the Zadeh implication, that is, $R(x, y) = \min[1, 1 - A(x) + B(y)]$.

5. Assume now a fuzzy rule of the form "If X is A then Y is B" and suppose that inputs are also points $x \in \mathbf{X} = [0,4]$ and A and B triangular fuzzy sets $A(x, 1, 2, 3)$ and $B(y, 2, 3, 4)$. Using the compositional rule of inference, determine the inferred fuzzy set B' assuming that the rule is interpreted first as the min conjunction $R(x, y) = \min[A(x), B(y)]$ and next as the algebraic product $R(x, y) = A(x) B(y)$, illustrating the results graphically.

6. Compare the inferred fuzzy sets of conclusion produced in the previous problems in terms of the resulting fuzzy set of conclusion (say, shape, support, convexity, and normality). Discuss the boundary cases in which the input is empty set and an entire universe of discourse. Offer a comparative analysis of the results. Are they intuitively appealing? Justify your answer.

7. Show that fuzzy min-max linguistic models whose consequent fuzzy sets are degenerate sets (singletons) are equivalent to zero-order functional fuzzy models.

8. Assume you are given the function below. Develop the simplest fuzzy linguistic rule-based system that can exactly approximate the function f (i.e., with no approximation error).

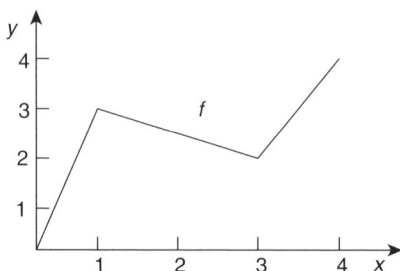

9. Suppose that in a gradual rule of the form

$$\text{the } \textit{more } X \text{ is } A, \text{ the } \textit{less } Y \text{ is } B$$

where the consequent the *less Y* is *B* is understood as the standard complement of *B*. In this case, what is the constraint that the membership functions of *A* and *B* should satisfy to retain semantics compatible with the linguistic description of the rule?

10. Consider the problem of approximating the function $y = f(x) = x^2$ using a functional rule-based model. Assume that the membership functions shown below were chosen to granulate the input universe.

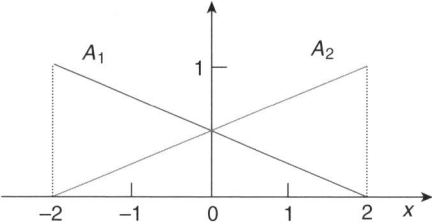

Given the following two rules

$$R_1: \quad \text{if } X \text{ is } A_1 \text{ then } y = ax + b$$
$$R_2: \quad \text{if } X \text{ is } A_2 \text{ then } y = cx + d$$

determine the values for parameters a, b, c, and d such that the rule-based model approximates the function f exactly for $x \in [-2, 2]$.

HISTORICAL NOTES

In 1943 the mathematician Emil Post (see Fischler and Firscein, 1987) proposed the "If–then" rule-based system that demonstrates how strings of symbols could be converted into other symbols. In the 1960s, Newell and Simon embedded the generality of the "If–then" rule-based representation in data-driven control structures of production systems used in Artificial Intelligence (Newell, 1973).

The basic idea of linguistic fuzzy rule-based models was suggested by Zadeh (1973) and implemented for the first time by Mamdani and Assilian (1975) in the form of so-called fuzzy controllers. Procyk and Mamdani (1979) constructed the first fuzzy rule-based learning method and demonstrated its use in fuzzy adaptive process control.

Fuzzy interpolation was introduced by Koczy and Hirota (1993) in the context of linguistic fuzzy models. Later, Bouchon-Meunier et al. (2000) developed an interpolative method using the concept of graduality.

Rules come with some level of readability where their interpretability comes with strong motivation and experimental evidence. Since the 1960s there has been research aimed at understanding what type of rules tend to be easily apprehended by humans. The most interesting finding was that conjunctive types of rules are more easily comprehended than disjunctive rules (Bruner et al., 1960; Feldman, 2006)

REFERENCES

Bouchon-Meunier, B., Marsala, C., Rifqi, M. Interpolative reasoning based on graduality, *Proc. Int.* IEEE Conf. on Fuzzy Systems, 2000, 483–487.

Bruner, J., Goodnow, J., Austin, G. *A Study of Thinking*, John Wiley & Sons, Inc., New York, NY, 1960.

Casillas, J., Cordón, O., Herrera, F., Magdalena, L. *Interpretability Issues in Fuzzy Modeling*, Springer, Heidelberg, 2003.

Castro, J. Fuzzy logic controllers are universal approximators, *IEEE Trans. Systems, Man, and Cybernetics*, 25(4), 1995, 629–635.

Castro, J., Delgado, M. Fuzzy systems with defuzzification are universal approximators, *IEEE Trans. Systems, Man, and Cybernetics*, 26(1), 1996, 149–152.

Chen, M., Linkens, D. Rule-base self-generation and simplification for data drive fuzzy models, *Fuzzy Set Syst.* **142**, 2004, 1243–265.

Chen, W., Saif, M. A novel fuzzy system with dynamic rule base, *IEEE Trams. Fuzzy Syst.* **13**(5), 2005, 569–582.

Driankov, D., Hellendorn, H., Reinfrank M. An Introduction to Fuzzy Control, Springer, Berlin, 1993.

Dubois, D., Grabisch M., Prade, H. Gradual rules and the approximation of control laws, in *Theoretical Aspects of Fuzzy Control*, H. T. Nguyen, M. Sugeno, R. Tong, R. Yager, Eds., Wiley, New York, 1994, 147–181.

Dubois, D., Prade, H. Gradual inference rules in approximate reasoning, *Inform. Sciences*, **61**, 1992, 103–122.

Dubois, D., Prade, H. What are fuzzy rules and how to use them, *Fuzzy Set Syst.* **84**, 1996, 169–185.

Dubois, D., Prade, H., Ughetto, L. Checking the coherence and redundancy of fuzzy knowledge bases, *IEEE Trans. Fuzzy Syst.* **5**(3), 1997, 398–417.

Feldman, J. An algebra of human concept learning, *J. Math. Psychol.* **50**, 2006, 339–368.

Gaweda, A., Zurada, J. Data-driven linguistic modeling using relational fuzzy rules, *IEEE Trans. Fuzzy Syst.* **11**(1), 2003, 121–134.

Golub, G., Van Loan, C. *Matrix Computations*, 2nd ed, John Hopkins University Press, Baltimore, MD, 1989.

Huang, Z., Shen, Q. Fuzzy interpolative reasoning via scale and move transformations, *IEEE Trans. Fuzzy Systems*, 13(2), 2006, 340–359.

Klir, G., Yuan, B. Fuzzy Sets and Fuzzy Logic: Theory and Applications, Prentice Hall, Upper Saddle River, NJ, 1995.

Koczy, L., Hirota, K. Approximate reasoning by linear rule interpolation and general approximation, *Int. J. Approx. Reason.* **9**, 1993, 197–225.

Kosko, B. *Neural Networks and Fuzzy Systems: A Dynamical Systems Approach to Machine Intelligence*, Prentice-Hall, Englewood Cliffs, NJ, 1992.

Lucas, P., Gaag, L. *Principles of Expert Systems*, Addison-Wesley, London, 1991.

Mamdani, E., Assilian, S. An experiment in linguistic synthesis with a fuzzy logic controller, *Int. J. Man Mach. Stud.* **7**, 1975, 1–12.

Mitra, S., Hayashi, Y. Neuro-fuzzy rule generation: survey in soft computing framework, *IEEE Trans. Neural Networ.* **11**(3), 2000, 748–768.

Mizumoto, M. Fuzzy controls under product sum-gravity methods and new fuzzy control methods, in: A. Kandel, G. Langholz (eds.), *Fuzzy Control Systems*, CRC Press, Boca Raton, FL, 1994, 275–294.

Nauck, D., Kruse, R. Obtaining interpretable fuzzy classification rules from medical data, *Artif. Intell. Med.* **16**, 1999, 149–169.

Nelles, O. *Nonlinear System Identification*, Springer, Heidelberg, 2001.

Newell, A. Production systems: Models of control structures, in: W. Chase (ed.), *Visual Information Processing*, Academic Press, NY, 1973.

Pedrycz, W., Gomide, F. An Introduction to Fuzzy Sets: Analysis and Design, MIT Press, Cambridge, MA, 1998.

Perfilieva, I. Fuzzy function as an approximate solution to a system of relation equations, *Fuzzy Sets and Systems*, 147, 2004, 363–383.

Procyk, T., Mamdani, E. A linguistic self-organizing process controller, *Automatica* **15**, 1979, 15–30.

Russell, S., Norvig, P. *Artificial Intelligence: A Modern Approach*, Prentice Hall, Upper Saddle River, NJ, 2003.

Setiono, D. Generating concise and accurate classification rules for breast cancer diagnosis, *Artif. Intell. Med.* **18**, 1999, 205–219.

Setnes, M. Supervised clustering for rule extraction, *IEEE Trans. Fuzzy Syst.* **8**(4), 1998, 416–424.

Setnes, M., Babuska, R., Verbruggen, H. Rule-base modeling: Precision and transparency, *IEEE Trans. Syst., Man Cybern. C: Appl. Rev.* **28**(1), 1998, 165–169.

Takagi, T., Sugeno, M. Fuzzy identification of systems and its applications to modeling and control, *IEEE Trans. on Systems*, Man, and Cybernetics, 15(1), 1985, 116–132.

Wong, K., Tikk, D., Gedeon, T., Koczy, L. Fuzzy rule interpolation for multidimensional input spaces with applications: A case study, *IEEE Trans. Fuzzy Syst.* **13**(6), 2005, 809–819.

Yager, R. Approximate reasoning as a basis for rule-based fuzzy expert systems, *IEEE Trans. Syst. Man Cyb.* **14**(1), 1984, 636–643.

Yen, J., Wang, L., Gillespie, C. Improving the interpretability of TSK fuzzy models by combining global learning and local learning, *IEEE T. Fuzzy Syst.* **6**(4), 1998, 530–537.

Zadeh, L. A. Outline of a w approach to the analysis of complex systems and decision processes, *IEEE Trans. Syst. Man Cyb.* **3**(3), 1973, 28–44.

Zadeh, L. A. The concept of linguistic variable and its application to approximate reasoning (Parts 1 and 2), *Inform. Sciences* **8**, 1975, 199–249, 301–357.

Zadeh, L. A. Fuzzy logic, *IEEE Comput.* **21**(4), 1988, 83–93.

Zadeh, L. A. Knowledge representation in fuzzy logic, *IEEE Trans. Knowl. Data Eng.* **11**(1), 1989, 89–100.

Zadeh, L. A. The calculus of if then rules, *AI Expert* **7**(3), 1992, 23–27.

Zadeh, L. A., Soft computing and fuzzy logic, IEEE Software 11(6), 1994, 48–56.

Chapter 12

From Logic Expressions to Fuzzy Logic Networks

Neural networks and neurocomputing constitute one of the fundamental pillars of Computational Intelligence. They bring a wealth of learning abilities to intelligent systems that can be realized both in supervised and unsupervised modes. These modalities are critical to the functioning of any intelligent system whose plasticity and adaptability to ongoing changes are crucial to any successful implementation. Fuzzy sets, as presented so far, come with a variety of schemes of knowledge representation and processing of information granules being formalized as fuzzy sets. They bring a variety of computing structures that are both transparent and semantically meaningful. Fuzzy sets and neurocomputing are highly complementary when it comes to their strengths. In this sense, they nicely supplement each other.

Having this in mind, our objective is to develop a logic-based mode of neurocomputing in which learning comes hand in hand with the transparency of the resulting structure. The logic facet of the introduced types of neurons is essential for the delivery of the significant interpretation abilities. Once developed, the "content" of the network can be downloaded in the form of some logic expressions.

12.1 INTRODUCTION

Neural networks (Golden, 1996; Jang et al., 1997; Kosko, 1991; Mitra and Pal, 1994, 1995; Pal and Mitra, 1999) are regarded to be a synonym of nonlinear and highly plastic (adaptive) systems equipped with significant learning capabilities. The universal approximation theorem coming with neural networks is highly appealing at least from the theoretical point of view. This means that any continuous functions can be approximated to any desired accuracy by a certain neural network assuming that we are given enough neurons organized in a certain multiplayer topology. The basic processing unit (neuron) realizes a certain nonlinear processing. The multitude of

---

Fuzzy Systems Engineering: Toward Human-Centric Computing, by Witold Pedrycz and Fernando Gomide
Copyright © 2007 John Wiley & Sons, Inc.

learning paradigms is impressive. We have a number of fundamental learning schemes of supervised learning including such mechanisms as perceptron learning and backpropagation. In unsupervised learning mode we often refer to self-organizing maps (Kohonen maps) as a typical neural architecture. Learning itself may pertain to the optimization of the parameters of the network or it can deal with the structural optimization of the network where its topology (configuration) becomes affected. Parametric learning engages various gradient-based techniques. Structural optimization (for which we cannot compute any gradient) requires other optimization tools, and here we usually confine ourselves to evolutionary optimization (Michalewicz, 1996) with this category including genetic algorithms, genetic programming, evolutionary programming, and alike. Some other biologically inspired methods including ant colonies and particle swarm optimization provide another optimization platform.

Owing to the highly distributed character of processing realized by neural networks and a lack of underlying semantics of processing carried out at the level of each individual neuron, we end up with a "black-box" character of computing. In essence, once the network has been designed (trained), we do not have any mechanism using which we can examine the character of the produced mapping and investigate it vis-à-vis the data in hand. This may hamper its future usage because of the lack of comprehension of the structure resulting through the optimization (learning) process. The black-box character of the network does not increase our confidence in the generalization abilities of the network.

This lack of interpretability imposes an important quest as to the future developments of the networks. In this regard, it would be highly desirable to design *transparent* neural networks. There are several evident benefits behind them. First, we can easily interpret the result of learning and produce the corresponding highly compact description of data. Second, the learning of such networks could be facilitated to a significant extent. In many cases this becomes a necessary condition of acceptance of the model. Usually, in solving any problem we usually have some prior domain knowledge. One can take advantage of it by "downloading" such knowledge hints onto the given structure of the network. This could set up a highly promising starting point for further weight adjustments carried out through some well-known schemes of supervised or unsupervised learning. In order to take advantage of this preliminary knowledge one has to be at a position to do this downloading in an efficient manner. This, however, requires the transparency of the network itself so that we know how to affect its structure or set up initial values of the connections. It is worth to emphasize in this context that when using the standard learning schemes we usually assume random values of the connections and start from this configuration (which might be quite inefficient resulting in slow and inefficient learning). The transparency of the network available in this case becomes a genuine asset.

The category of fuzzy neurons (or fuzzy logic neurons) discussed in this chapter addresses these burning issues of transparency of neural networks. We build a network with the aid of conceptually simple and logically appealing nodes (neurons) that complete generic *and* and *or* logic operations. By equipping the neurons with a set of connections, we furnish them with the badly required plasticity; the values of the connections could be easily adjusted by some standard gradient-based learning schemes. Likewise the resulting network could be transformed into a collection of conditional logic statements (rules), thus resulting in the certain rule-based system.

We start with the introduction of the main categories of the fuzzy neurons, elaborate on their main properties, move on to the architectures of networks composed of such neurons, and discuss various facets of interpretation of the networks.

12.2 MAIN CATEGORIES OF FUZZY NEURONS

The logical aspect of neurocomputing we intend to realize requires that the processing elements be endowed with the clearly delineated logic structure. We discuss several types of aggregative and referential neurons. Each of them comes with a clearly defined semantics of its underlying logic expression and is equipped with significant parametric flexibility necessary to facilitate substantial learning abilities.

12.2.1 Aggregative Neurons

Formally, these neurons realize a logic mapping from $[0,1]^n$ to $[0,1]$. Two main classes of the processing units exist in this category (Pedrycz, 1991a, 1991b; Pedrycz and Rocha, 1993; Pedrycz et al., 1995; Hirota and Pedrycz, 1994; Hirota and Pedrycz, 1999 Ciaramella et al., 2005; 2006; Nobuhara et al., 2005; 2006)

or neuron: realizes an *and* logic aggregation of inputs $\mathbf{x} = [x_1, x_2, \ldots, x_n]$ with the corresponding connections (weights) $\mathbf{w} = [w_1 \ w_2 \ldots w_n]$ and then summarizes the partial results in an *or*-wise manner (hence the name of the neuron). The concise notation underlines this flow of computing, $y = \text{OR}(\mathbf{x}; \mathbf{w})$ while the realization of the logic operations gives rise to the expression (commonly referring to it as an s-t combination or s-t aggregation)

$$y = \underset{i=1}{\overset{n}{S}} (x_i t w_i) \tag{12.1}$$

Bearing in mind the interpretation of the logic connectives (t-norms and t-conorms), the OR neuron realizes the following logic expression being viewed as an underlying logic description of the processing of the input signals.

$$(x_1 \text{ and } w_1) \text{ or } (x_2 \text{ and } w_2) \text{ or } \ldots \text{ or } (x_n \text{ and } w_n) \tag{12.2}$$

Apparently the inputs are logically "weighted" by the values of the connections before producing the final result. In other words we can treat y as a truth value of the above statement where the truth values of the inputs are affected by the corresponding weights. Noticeably, lower values of w_i discount the impact of the corresponding inputs; higher values of the connections (especially those being positioned close to 1) do not affect the original truth values of the inputs resulting in the logic formula. In limit, if all connections w_i, $i = 1, 2, \ldots, n$ are set to 1 then the neuron produces a plain *or*-combination of the inputs, $y = x_1 \text{ or } x_2 \text{ or } \ldots \text{ or } x_n$. The values of the connections set to zero eliminate the corresponding inputs. Computationally, the *or*

338   Chapter 12   From Logic Expressions to Fuzzy Logic Networks

neuron exhibits nonlinear characteristics (which are inherently implied by the use of the t- and t-conorms (which are evidently nonlinear mappings). The plots of the characteristics of the OR neuron shown in Figure 12.1 shows this effect (note that the characteristics are affected by the use of some triangular norms). The connections of the neuron contribute to its adaptive character; the changes in their values form the crux of the parametric learning.

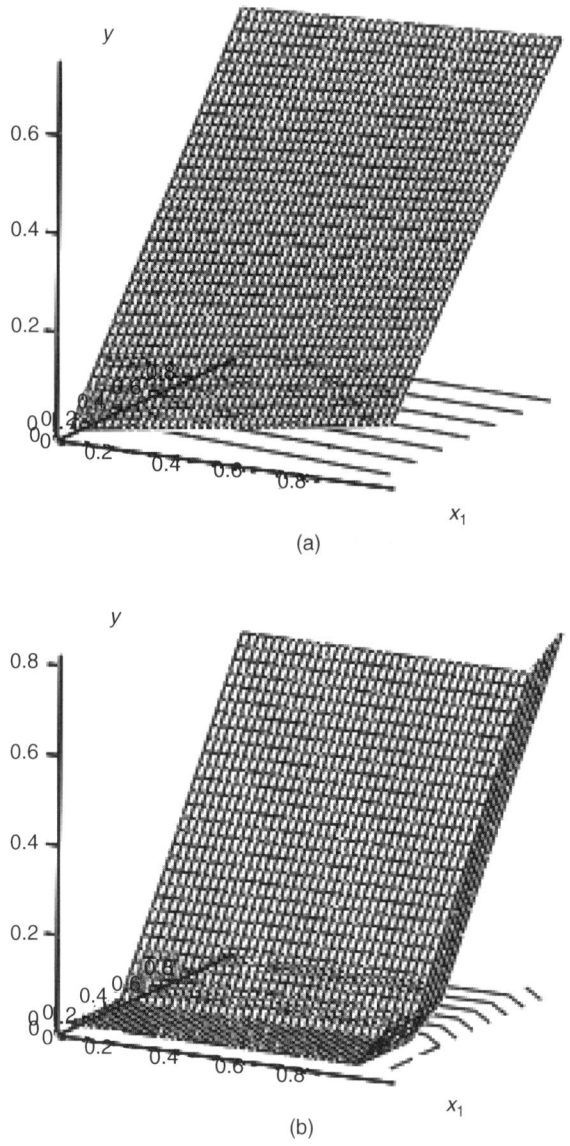

Figure 12.1   Characteristics of the OR neuron for selected pairs of t- and t-conorms. In all cases the corresponding connections are set to 0.1 and 0.7 with an intent to visualize their effects on the input–output characteristics of the neuron: (a) product and probabilistic sum, (b) Lukasiewicz *and* and *or* connectives (b).

12.2 Main Categories of Fuzzy Neurons

and neuron: the neurons in the category, denoted by $y = \text{AND}(\mathbf{x}; \mathbf{w})$ with $\mathbf{x}$ and $\mathbf{w}$ being defined as in case of the OR neuron, are governed by the expression

$$y = \mathop{\text{T}}_{i=1}^{n} (x_i s w_i) \tag{12.3}$$

Here the *or* and *and* connectives are used in a reversed order: first the inputs are combined with the use of the t-conorm and the partial results produced in this way are aggregated *and*-wise. Higher values of the connections reduce impact of the corresponding inputs. In limit $w_i = 1$ eliminates the relevance of x_i. With all w_i set to 0, the output of the AND neuron is just an *and* aggregation of the inputs

$$y = x_1 \text{ and } x_2 \text{ and} \ldots \text{and } x_n \tag{12.4}$$

The characteristics of the AND neuron are shown in Figure 12.2; note the influence of the connections and the specific realization of the triangular norms on the mapping completed by the neuron.

Let us conclude that the neurons are highly nonlinear processing units whose nonlinear mapping depends upon the specific realizations of the logic connectives. They also come with potential plasticity whose usage becomes critical when learning the networks including such neurons.

At this point, it is worth contrasting these two categories of logic neurons with "standard" neurons we encounter in neurocomputing. The typical construct there comes in the form of the weighted sum of the inputs $x_1, x_2, \ldots, x_n$ with the corresponding connections (weights) $w_1, w_2, \ldots, w_n$ being followed by a nonlinear (usually monotonically increasing) function that reads as follows:

$$y = g(\mathbf{w}^T \mathbf{x} + \tau) = g\left(\sum_{i=1}^{n} w_i x_i + \tau\right) \tag{12.5}$$

where $\mathbf{w}$ is a vector of connections, τ is a constant term (bias), and g denotes some monotonically nondecreasing nonlinear mapping. The other less commonly encountered neuron is the so-called π-neuron. Although there could be some variations as to the parametric details of this construct, we can envision the following realization of the neuron

$$y = g\left(\prod |x_i - t_i|^{w_i}\right) \tag{12.6}$$

where $\mathbf{t} = [t_1, t_2 \ldots t_n]$ denotes a vector of translations whereas $\mathbf{w}(> 0)$ denotes a vector of all connections. The plots of the sample characteristics of the two types of the neurons (12.5)–(12.6) are included in Figure 12.3.

As before, the nonlinear function is denoted by g. Although some superficial and quite loose analogy between these processing units and logic neurons could be derived, one has to be cognizant of the fact that these neurons do not come with any underlying logic fabric and hence cannot be easily and immediately interpreted.

Let us make two observations about the architectural and functional facets of the logic neurons we have introduced so far.

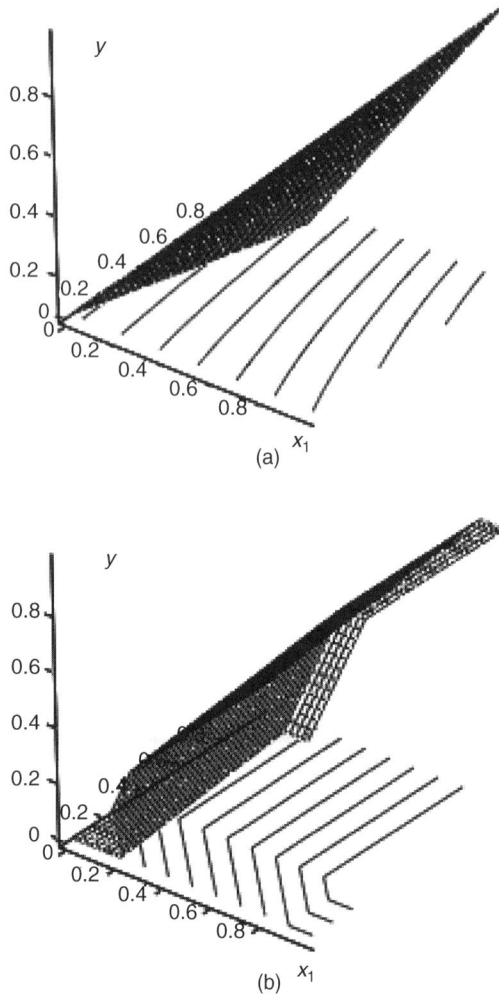

Figure 12.2 Characteristics of AND neurons for selected pairs of t- and t-conorms. In all cases the connections are set to 0.1 and 0.7 with an intent to visualize their effects on the characteristics of the neuron: (a) product and probabilistic sum (b) Lukasiewicz logic connectives.

12.2.1.1 Incorporation of the Bias Term (Bias) in the Fuzzy Logic Neurons

In an analogy to the standard constructs of a generic neuron as presented above, we could also consider a bias term, denoted by $w_0 \in [0, 1]$ that enters the processing formula of the fuzzy neuron in the following manner:
for the *or* neuron

$$y = \mathop{S}_{i=1}^{n} (x_i t w_i) s w_0 \tag{12.7}$$

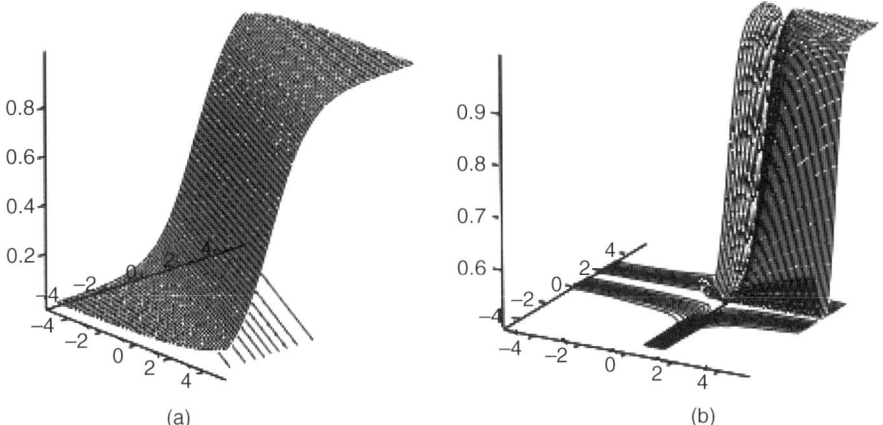

Figure 12.3 Characteristics of the neurons: (a) additive with $\tau = 0.2$, $w_1 = 1.0$, $w_2 = 2.0$ and (b) multiplicative where $w_1 = 0.5$, $w_2 = 2.0$, $t_1 = 1.0$, $t_2 = 0.7$. In both cases the nonlinear function is a sigmoid function, $g(u) = 1/(1 + \exp(-u))$.

for the *and* neuron

$$y = \mathop{T}_{i=1}^{n} (x_i s w_i) t w_0 \qquad (12.8)$$

We can offer some useful interpretation of the bias by treating it as some nonzero initial truth value associated with the logic expression of the neuron. For the *or* neuron it means that the output does not reach values lower than the assumed threshold. For the *and* neuron equipped with some bias, we conclude that its output cannot exceed the value assumed by the bias. The question whether the bias is essential in the construct of the logic neurons cannot be fully answered in advance. Instead, we may include it into the structure of the neuron and carry out learning. Once its value has been obtained, its relevance could be established considering the specific value it has been produced during the learning. It may well be that the optimized value of the bias is close to zero for the OR neuron or close to one in the case of the *and* neuron, which indicates that it could be eliminated without exhibiting any substantial impact on the performance of the neuron.

12.2.1.2 Dealing with Inhibitory Character of Input Information

Owing to the monotonicity of the t-norms and t-conorms, the computing realized by the neurons exhibits an excitatory character. This means that higher values of the inputs (x_i) contribute to the increase in the values of the output of the neuron. The inhibitory nature of computing realized by "standard" neurons by using negative values of the connections or the inputs is not available here as the truth values (membership grades) in fuzzy sets are confined to the unit interval. The inhibitory nature of processing can be accomplished by considering the complement of the original input, $= 1 - x_i$. Hence when the values of x_i increase, the associated values of the complement decrease and subsequently in this configuration we could effectively treat such an input as having an inhibitory nature.

12.2.2 Referential (Reference) Neurons

The essence of referential computing deals with processing logic predicates. The two-argument (or generally multivariable) predicates such as *similar, included in*, and *dominates* (Pedrycz and Rocha, 1993) are essential components of any logic description of a system. In general, the truth value of the predicate is a degree of satisfaction of the expression $P(x, a)$ where a is a certain reference value (reference point). Depending upon the meaning of the predicate (P), the expression $P(x, a)$ reads as "x is similar to a", "x is included in a", "x dominates a", and so on. In case of many variables, the compound predicate comes in the form $P(x_1, x_2, \ldots, x_n, a_1, a_2, \ldots, a_n)$ or more concisely $P(\mathbf{x}; \mathbf{a})$ where $\mathbf{x}$ and $\mathbf{a}$ are vectors in the n-dimensional unit hypercube. We envision the following realization of $P(\mathbf{x}; \mathbf{a})$:

$$P(\mathbf{x}; \mathbf{a}) = P(x_1, a_1) \text{ and } P(x_2, a_2) \text{ and} \ldots \text{and } P(x_n, a_n) \qquad (12.9)$$

meaning that the satisfaction of the multivariable predicate relies on the satisfaction realized for each variable separately. As the variables could come with different levels of relevance as to the overall satisfaction of the predicates, we represent this effect by some weights (connections) $w_1, w_2, \ldots, w_n$ so that (12.9) can be expressed in the following form:

$$P(\mathbf{x}; \mathbf{a}, \mathbf{w}) = [P(x_1, a_1) \text{ or } w_1] \text{ and } [P(x_2, a_2) \text{ or } w_2] \text{ and } \ldots \text{ and} \\ [P(x_n, a_n) \text{ or } w_n] \qquad (12.10)$$

Taking another look at the above expression and using a notation $z_i = P(x_i, a_i)$, it corresponds to a certain AND neuron $y = \text{AND}(\mathbf{z}; \mathbf{w})$ with the vector of inputs $\mathbf{z}$ being the result of the referential computations done for the logic predicate. Then the general notation to be used reads as $\text{REF}(\mathbf{x}; \mathbf{w}, \mathbf{a})$. In the notation below, we explicitly articulate the role of the connections

$$y = \mathop{\text{T}}_{i=1}^{n} (\text{REF}(x_i, a_i) s w_i) \qquad (12.11)$$

In essence, as visualized in Figure 12.4, we may conclude that the reference neuron is realized as a two-stage construct where first we determine the truth values of the predicate (with $\mathbf{a}$ being treated as a reference point) and then treat these results as the inputs to the AND neuron.

So far we have used the general term of predicate-based computing not confining ourselves to any specific nature of the predicate itself. Among a number of available possibilities of such predicates, we discuss the three of them, which tend to occupy an important place in logic processing. Those are inclusion, dominance and match (similarity) predicates. As the names stipulate, the predicates return truth values of satisfaction of the relationship of inclusion, dominance and similarity of a certain argument x with respect to the given reference a. The essence of all these calculations

12.2 Main Categories of Fuzzy Neurons

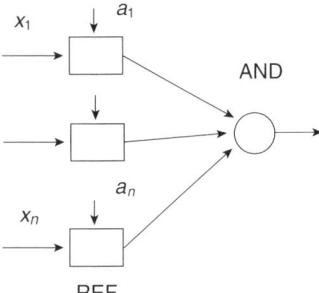

Figure 12.4 A schematic view of computing realized by a reference neuron involving two processing phases (referential computing and aggregation).

is in the determination of the given truth values, and this is done in the carefully developed logic framework so that the operations retain their semantics and interpretability. What makes our discussion coherent is the fact that the proposed operations originate from triangular norms. The inclusion operation, as discussed earlier, denoted by $\subset$ is modeled by an implication $\Rightarrow$, which is induced by a certain left continuous t-norm (Pedrycz and Gomide; 1998)

$$a \Rightarrow b = \sup\{c \in [0,1] | atc \leq b\}, a, b \in [0,1] \qquad (12.12)$$

For instance, for the product the inclusion takes on the form $a \Rightarrow b = \min(1, b/a)$. The intuitive form of this predicate is self-evident: the statement "x is included in a" and modeled as $\text{INCL}(x, a) = x \Rightarrow a$ comes with the truth value equal to 1 if x is less or equal to a (which in other words means that x is included in a) and produces lower truth values once x starts exceeding the truth values of a. Higher values of x (those above the values of the reference point a) start generating lower truth values of the predicate. The dominance predicate acts in a dual manner when compared with the predicate of inclusion. It returns 1 once x dominates a (so that its values exceeds a) and values below 1 for x lower than the given threshold. The formal model can be realized as $\text{DOM}(x, a) = a \Rightarrow x$. With regard to the reference neuron, the notation is equivalent to the one being used in the previous case, which is $\text{DOM}(\mathbf{x}; \mathbf{w}, \mathbf{a})$ with the same meaning of $\mathbf{a}$ and $\mathbf{w}$.

The similarity (match) operation is an aggregate of these two, $\text{SIM}(x, a) = \text{INCL}(x, a) \, t \, \text{DOM}(x, a)$, which is appealing from the intuitive standpoint: we say that x is similar to a if x is included in a and x dominates a. Noticeably, if $x = a$ the predicate returns 1; if x moves apart from "a" the truth value of the predicate becomes reduced. The resulting similarity neuron is denoted by $\text{SIM}(\mathbf{x}; \mathbf{w}, \mathbf{a})$ and reads as

$$y = \underset{i=1}{\overset{n}{T}} \left(\text{SIM}(x_i, a_i) s w_i \right) \qquad (12.13)$$

The reference operations form an interesting generalization of the threshold operations. Consider that we are viewing x as a temporal signal (which changes over time) whose behavior needs to be monitored with respect to some bounds (α and β).

344   Chapter 12   From Logic Expressions to Fuzzy Logic Networks

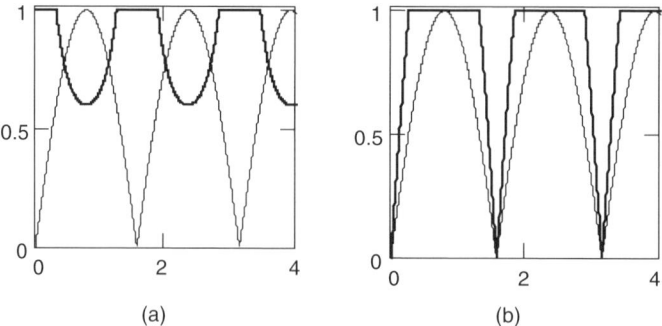

Figure 12.5 Temporal signal $x(t)$ and its acceptance signals (levels of the signals—thick lines) formed with respect to its lower and upper threshold (a) and (b). The complements of the acceptance values are then treated as warning signals.

If the signal does not exceed some predefined threshold α then the acceptance signal should go off. Likewise we require another acceptance mechanism that indicates a situation where the signal does not go below another threshold value of β. In the case of fuzzy predicates, the level of acceptance assumes values in the unit interval rather than being a Boolean variable. Furthermore the strength of acceptance reflects how much the signal adheres to the assumed thresholds. An example illustrating this behavior is shown in Figure 12.5. In this particular case, the values of α and β are set up to 0.6 and 0.5, respectively.

The plots of the referential neurons with two input variables are shown in Figures 12.6 and 12.7 (axis labels missing in most of them); here we have included two realizations of the t-norms to illustrate their effects on the nonlinear characteristics of the processing units.

It is worth noting that by moving the reference point to the origin or the **1**-vertex of the unit hypercube (with all its coordinates being set up to 1), the referential neuron starts resembling the aggregative neuron. In particular, we have

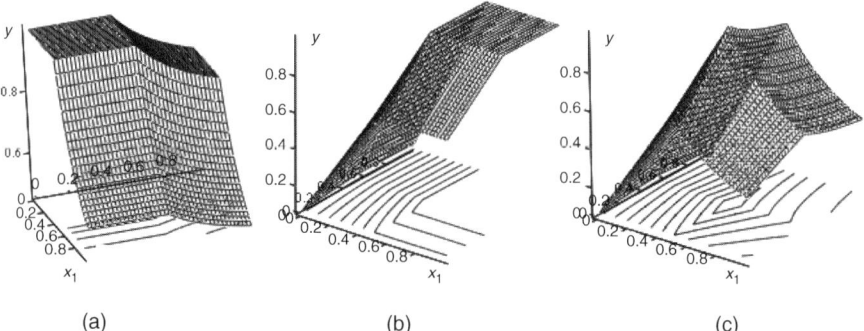

Figure 12.6 Characteristics of the reference neurons for the product (t-norm) and probabilistic sum (t-conorm). In all cases the connections are set to 0.1 and 0.7 with an intent to visualize the effect of the weights on the relationships produced by the neuron. The point of reference is set to (0.5, 0.5): inclusion neuron (a), dominance neuron (b), and similarity neuron (c).

12.3 Uninorm-based Fuzzy Neurons

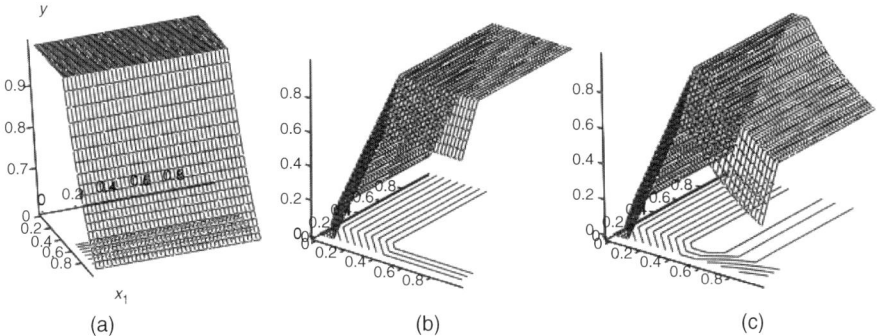

Figure 12.7 Characteristics of the reference neurons for the Lukasiewicz t-norm and t-conorm (which are $a\ t\ b = \max(0, a + b - 1)$ and $a\ s\ b = \min(1, a + b)$). In all cases the connections are set to 0.1 and 0.7 with an intent to visualize the effect of the weights. The point of reference is set to $(0.5, 0.5)$: inclusion neuron (a), and dominance neuron (b), similarity neuron (c).

for $\mathbf{a} = \mathbf{1} = [\ 1\ 1\ 1\ \ldots\ 1\ ]$ the inclusion neuron reduces to the AND neuron

for $\mathbf{a} = \mathbf{0} = [\ 0\ 0\ 0\ \ldots\ 0\ ]$ the dominance neuron reduces to the AND neuron

One can draw a loose analogy between some types of the referential neurons and the two categories of processing units encountered in neurocomputing. The analogy is based upon the *local* versus *global* character of processing realized therein. Perceptrons come with the global character of processing. Radial basis functions realize a local character of processing as focused on receptive fields. In the same vein, the inclusion and dominance neurons are after the global nature of processing whereas the similarity neuron carries more confined and local processing.

12.3 UNINORM-BASED FUZZY NEURONS

The neurons proposed so far dwell on the use of triangular norms and conorms. Some other logic operators such as uninorms could be investigated. Those when applied to the constructs of fuzzy neurons give rise to the construct we can refer to as unineurons. More specifically, we exploit the use of uninorms by combining individual input x_i with some connection w_i giving rise to the expression $u(x_i; w_i, g_i)$ with g_i being the neutral point of this uninorm.

12.3.1 Main Classes of Unineurons

Let us introduce two fundamental categories of logic neurons that will be referred to as *and* unineurons and *or* unineurons, *and_U* and *or_U*, for short. In case it does not produce any confusion, we also refer to them as *and* and *or* neurons. The underlying logical character of processing is schematically captured in Figure 12.8. There are x and g in the y axis and w and g in x axis).

Let us consider two important alternatives.

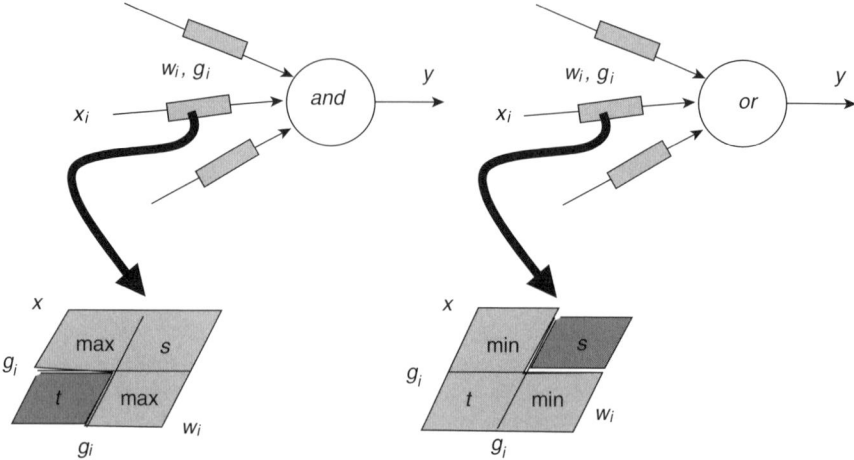

Figure 12.8 A schematic view of the logic processing completed by unineurons: (a) *and* unineuron (neuron) and (b) *or* unineuron (neuron). Note the corresponding realizations of the uninorms.

12.3.1.1 and Unineurons

The *and* unineuron considers a collection of inputs $\mathbf{x} = [x_1, x_2, \ldots, x_n]$ and processes them in the following way:

$$y = AND\_U(\mathbf{x}; \mathbf{w}, \mathbf{g}) \qquad (12.14)$$

where

$$y = \mathop{T}_{i=1}^{n} \left(u(x_i; w_i, g_i) \right) \qquad (12.15)$$

The name of this class of unineurons, *AND_U*, is implied by the *and* type of aggregation of the individual inputs.

12.3.1.2 or Unineurons

The processing realized by this processing unit

$$y = OR\_U(\mathbf{x}; \mathbf{w}, \mathbf{g}) \qquad (12.16)$$

concerns an *or*-type of aggregation of the partial results produced by the uninorm combination of the corresponding inputs. Proceeding with the individual inputs of the neuron, we rewrite (12.16) as follows:

$$y = \mathop{S}_{i=1}^{n} \left(u(x_i; w_i, g_i) \right) \qquad (12.17)$$

Each unineuron is endowed with a collection of the parameters ($\mathbf{w}$ and $\mathbf{g}$) whose role is important for learning the networks involving these processing units.

The "standard" *or* and *and* neurons that we have introduced in the earlier sections, are subsumed by the unineurons in the sense that for some selected values of the neutral points (elements) we end up with the previous constructs. Indeed, by choosing (12.16) all entries of **g** to be equal to **1**, **g** = [1 1 ...1] we obtain the following expression $y = S_{i=1}^{n}(u(x_i; w_i, 1)) = S_{i=1}^{n} t(x_i; w_i)$. The same holds when using the zero vector of the neutral points **g** in the *and_U* neuron, **g** = **0**. In view of the original expression (12.15), we now obtain $y = T_{i=1}^{n}(u(x_i; w_i, 0)) = T_{i=1}^{n} s(x_i; w_i)$ that becomes a "standard" *and* neuron.

12.3.2 Properties and Characteristics of the Unineurons

Given the functionality of the underlying logic operators used in the development of the logic neurons, they come with interesting functional properties that could be beneficial when designing networks formed on their basis. We highlight the two important aspects of the unineurons that become instrumental in the development of the network architectures.

12.3.2.1 Input–output Characteristics of the Unineurons

The characteristics of the individual neurons are inherently nonlinear where the form of the nonlinearity depends upon specific t-norms and co-norms involved. To illustrate this, we show some pertinent plots in the case of $n = 2$; see Figures 12.9 and 12.10.

The nonlinear character of input–output dependencies shows up very clearly. The connections and neutral points impact the resulting characteristics of the neurons in a direct manner. Any changes of their values affect the relationships in a visible way. The plasticity that comes with the adjustable values of **w** and *g* are essential to the learning capabilities of the logic neurons.

In the interpretation of the neurons, we start with the realization of the uninorm that guides the processing of the level of the individual inputs. For the *and* neuron, let us consider the expression $u(x; w, g)$. Once the learning has been completed, we end up with the optimized and fixed values of the connection (w) and the neutral point (g) (see notation Chapter 5). We distinguish between the two cases; refer also to Figure 12.11.

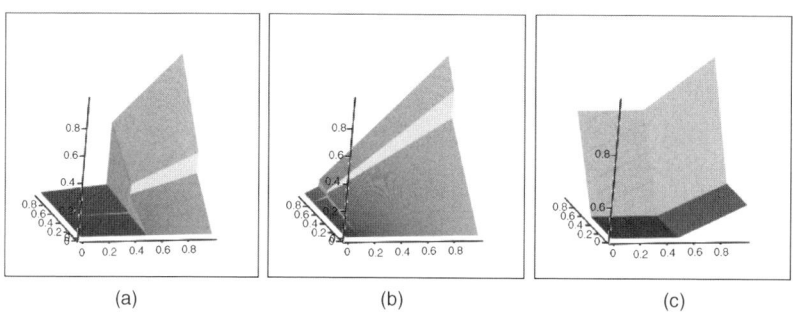

Figure 12.9 Input–output characteristics of *and* unineurons: t-norm: product, t-conorm: probabilistic sum; (a) **w** = [0.05 0.30] **g** = [0.50 0.45] (b) **w** = [0.05 0.50] **g** = [0.1 0.8] (c) **w** = [0.8 0.60] **g** = [0.5 0.5].

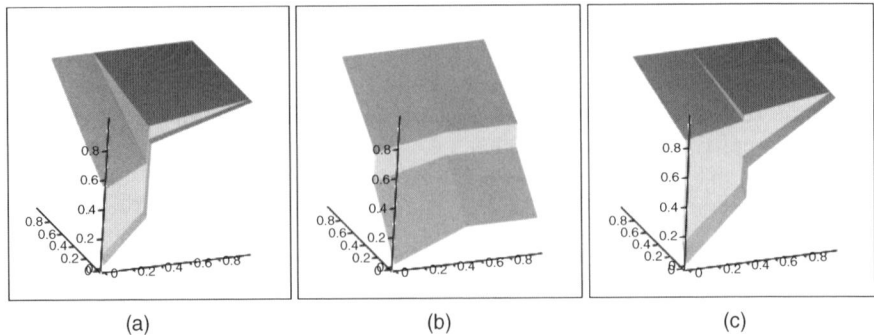

Figure 12.10 Input–output characteristics of OR unineurons: t-norm and t-conorm: Lukasiewicz connectives, (a) **w** = [0.8 0.5] **g** = [0.30 0.05] (b) **w** = [0.2 0.6] **g** = [0.5 0.4] (c) **w** = [0.6 0.8] **g** = [0.4 0.1].

(1) $w \leq g$. This corresponds to the situation where smaller values of x (those lower than g) are suppressed by being combined by some t-norm with the connection, $xtw \leq x$. The values that are higher than the neutral point are combined by the max operation $\max(x, w)$ that returns the original value of the input.

(2) For the values of $w > g$, we encounter a different situation. When $x < g$ the max operation becomes activated, and this yields $\max(x, w) = w$. For the values of x exceeding the neutral point, computing involves the expression $s(x, w) \geq x$.

The same effect of a fairly different processing occurs when dealing with the *or* neuron (see Fig. 12.12). If $w \leq g$ then the suppression of the input x takes place: if $x < g$ we obtain the expression $t(x, w) < x$. On the contrary, $\min(x, w) = w < x$ for $x \geq g$.

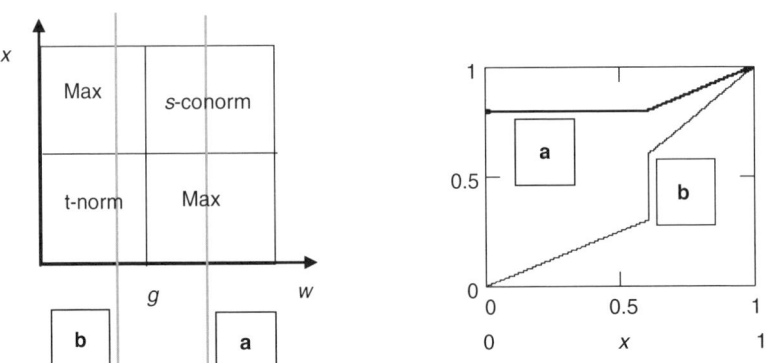

Figure 12.11 Processing at the level of the individual inputs of the *and* neuron; note two cases of the values of w considered with respect to the neutral point. t-norm and t-conorm: product and probabilistic sum.

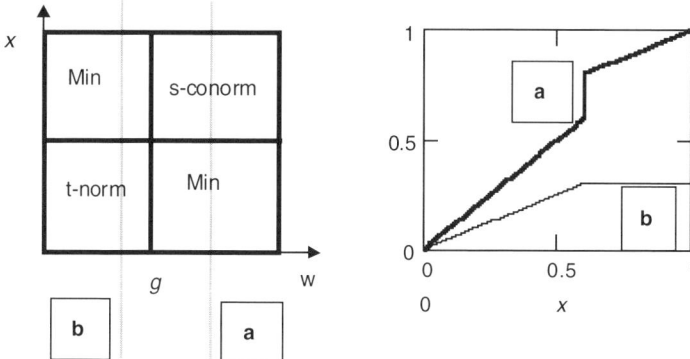

Figure 12.12 Processing at the level of the individual inputs of the *and* neuron; note two cases of the values of *w* considered with respect to the neutral point. t-norm and t-conorm: minimum and maximum.

It is also worth noting that when *g* tends to 0, the computing at the level of the synapse (input) reduces to the model we know in the "standard" logic neurons.

12.3.2.2 The Coverage Property of the Unineuron-based Mappings

When looking at the operation realized at the level of the individual inputs, it is interesting to note that the expression $u(x_i, w_i, g_i)$ produces result that could be either lower or higher than x_i depending upon the values of the parameters. This stands in a sharp contrast with the construct used in previous *and* and *or* neurons where we can only assure one of the following relationships: $x_i t w_i < x_i$ and $x_i s w_i > x_i$ regardless of the values of the connection w_i. In this sense, the uninorm offers a genuine advantage over the previously used *and* and *or* neurons (which are the t- norms and conorms). We will take full advantage of this when learning about the networks built with the use of *and* unineurons and *or* unineurons. To emphasize the fact that the mapping in of the "onto" character, we plot the expression by treating x_i to be fixed and partitioning the regions of (w_i, g_i) into two subregions depending upon the satisfaction of the inequality $u(x_i, w_i, g_i) > x_i$. We note that there are some regions of the values assumed by the two parameters that make this relationship to be satisfactory; refer to Figures 12.13 and 12.14.

Subsequently, we could easily determine the regions of satisfaction of the predicate by solving the corresponding inequalities.

12.4 ARCHITECTURES OF LOGIC NETWORKS

The logic neurons (aggregative and referential) can serve as building blocks of more comprehensive and functionally appealing architectures. The diversity of the topologies one can construct with the aid of the proposed neurons is surprisingly high. This architectural multiplicity is important from the application point of view

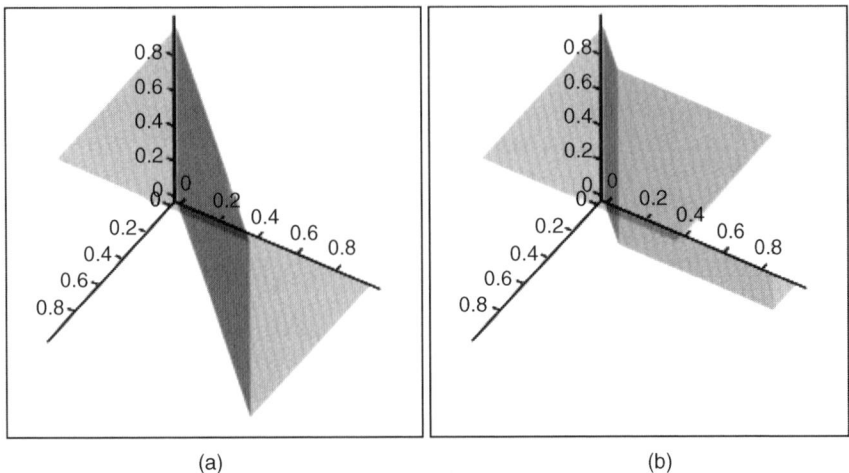

Figure 12.13 Plots of the Boolean predicate $P(w_i, g_i) = \{1, \text{ if } (w_i, gi) > x_i \text{ and } 0, \text{ otherwise}\}$ uninorm described by (1) $x_i = 0.2$ where (a) t-norm and conorm: product and probabilistic sum, (b) minimum and maximum.

as we can fully reflect the nature of the problem in a flexible manner. It is essential to capture the problem in a logic format and then set up the logic skeleton—conceptual blueprint (by forming the and finally refine it parametrically through a thorough optimization of the connections). Throughout the development process we are positioned quite comfortably by monitoring the optimization of the network as well as interpreting its meaning (the issue that will be discussed later on).

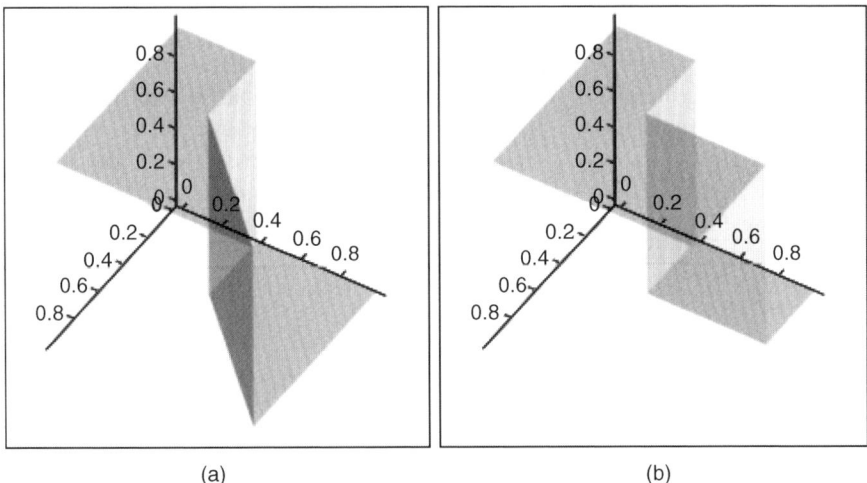

Figure 12.14 Plot of the Boolean predicate $P(w_i, g_i) = \{1, \text{ if}(w_i, g_i) > x_i \text{ and } 0, \text{ otherwise}\}$ uninorm described by (2) and $x_i = 0.4$ where (a) t-norm and conorm: product and probabilistic sum, (b) minimum and maximum.

12.4.1 Logic Processor in the Processing of Fuzzy Logic Functions: A Canonical Realization

The typical logic network that is at the center of logic processing originates from the two-valued logic and comes in the form of the famous Shannon theorem of decomposition of Boolean functions. Let us recall that any Boolean function $\{0,1\}^n \to \{0,1\}$ can be represented as a logic sum of its corresponding minterms or a logic product of maxterms. By a minterm of n logic variables $x_1, x_2, \ldots, x_n$ we mean a logic product involving all these variables either in direct or complemented form. Having "n" variables we end up with 2^n minterms starting from the one involving all complemented variables and ending up at the logic product with all direct variables. Likewise by a maxterm we mean a logic sum of all variables or their complements. Now in virtue of the decomposition theorem, we note that the first representation scheme involves a two-layer network where the first layer consists of *and* gates whose outputs are combined in a single *or* gate. The converse topology occurs for the second decomposition mode: there is a single layer of *or* gates followed by a single *and* gate aggregating *or*-wise all partial results.

The proposed network (referred here as a logic processor) generalizes this concept as shown in Figure 12.15. The *or–and* mode of the logic processor comes with the two types of aggregative neurons being swapped between the layers. Here the first (hidden) layer is composed of the *or* neuron and is followed by the output realized by means of the *and* neuron.

The logic neurons generalize digital gates. The design of the network (viz. any fuzzy function) is realized through learning. If we confine ourselves to $\{0,1\}$ values, the network's learning becomes an alternative to a standard digital design, especially a minimization of logic functions. The logic processor translates into a compound logic statement (we skip the connections of the neurons to underline the underlying logic content of the statement)

if (input$_1$ *and* ... *and* input$_j$) *or* (input$_d$ *and* ... *and* input$_f$) then *output*

The logic processor's topology (and underlying interpretation) is standard. Two LPs can vary in terms of the number of *and* neurons and their connections, but the format of the resulting logic expression is quite uniform (as a sum of generalized minterms).

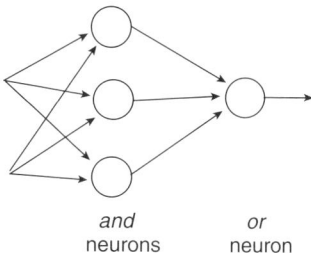

Figure 12.15 A topology of the logic processor in its *and–or* mode.

EXAMPLE 12.1

Consider a simple fuzzy neural network network in which the hidden layer includes two *and* neurons whose outputs are combined through a single *or* neuron located in the output layer. The connections of the first *and* neuron are equal to 0.3 and 0.7. For the second *and* neuron we have the values of the connections equal to 0.8 and 0.2. The connections of the *or* neuron are equal to 0.5 and 0.7, respectively. The input–output characteristics of the network are illustrated in Figure 12.16; to demonstrate the flexibility of the architecture, we included several combinations of the connections as well as used alternative realizations of the triangular norms and conorms.

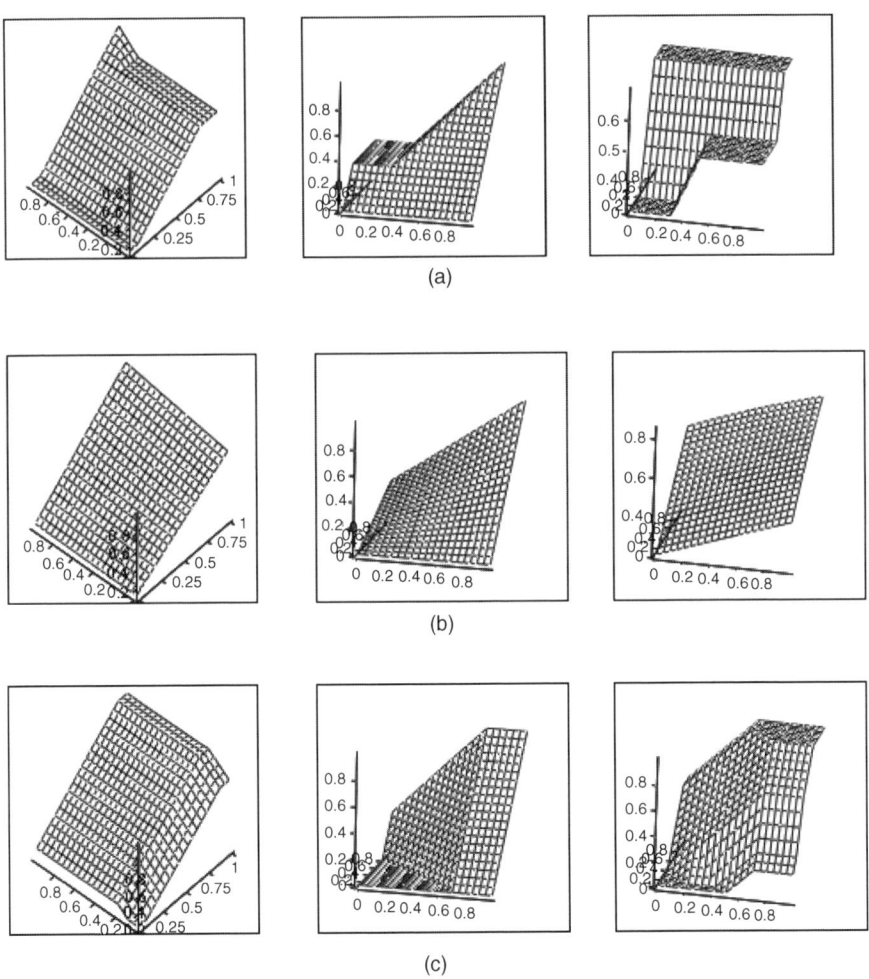

Figure 12.16 Plots of the characteristics of the fuzzy neural network: output of the two *and* neurons and the output of the network (from left to right) for different realizations of the logic operators: (a) min and max, (b) product and probabilistic sum, and (c) Lukasiewicz logic operators.

The resulting networks exhibit a significant diversity in terms of the resulting nonlinear dependencies. More importantly, we note that by choosing certain logic connectives (triangular norms) and adjusting the values of the connections, we could substantially affect the behavior (input-output characteristics) of the corresponding network. This plasticity becomes an important feature that plays a paramount role in the overall learning process.

12.4.2 Fuzzy Neural Networks with Feedback Loops

The architectures of fuzzy neural networks discussed so far are concerned with some static mappings between unit hypercubes, $[0, 1]^n \to [0, 1]^m$. The dynamics of systems modeled in a logic manner can be captured by introducing some feedback loops into the structures. Some examples of the networks with the feedback loops are illustrated in Figure 17. They are based on the logic processor we have already studied. In the first case, the feedback loop is built between the output layer and the inputs of the *and* neurons, Figure 17(a). The form of the feedback (which could be either positive—excitatory or negative—inhibitory) is realized by taking the original signal (thus forming an excitatory loop—higher values of input "excite" the corresponding neuron) or its complement (in which case we end up with the inhibitory effect—higher values of the signal suppress the output of the corresponding neuron). The strength of the feedback loop is modeled by the numeric of the connection. To visualize the effect of inhibition or excitation, we use the dot notation as illustrated in Figure 17(b). The small dot corresponds to the complement of the original variable. Another example of the feedback loops shown in Figure 17(c) embraces both layers of the neurons of the network.

The semantics of such networks is straightforward. Any prior knowledge is essential in forming a structure of the network itself. The problem can be directly mapped onto the network. As an illustration, let us consider a concise example.

EXAMPLE 12.2

We are concerned with two variables (factors) x and y assuming values in the unit interval. The variables are associated with the two nodes of the network. The nodes interact and because of this, they change the values of the variables (x and y) associated with them. The value of x in the current time instant (k) depends upon its previous value, that is, $x(k-1)$, *and* the value $y(k)$ present in the second node. The strength of these relationships is expressed by the values of the corresponding connections. The ensuing logic expression of this node reads as

$$x(k) = and(\mathbf{w}, [x(k-1)y(k-1)]) \qquad (12.18)$$

where **w** stands for the two-argument vector of the connections of the *and* neuron. The functioning of the second node is captured by the following logic statement

> the value of y in the current time instant (k) depends upon its "history" in the two previous steps ($y(k-1)$ and $y(k-2)$) and is inhibited by the value produced by the second node $x(k-1)$.

This expression translates into the *and* neuron governed by the expression

$$y(k) = And\ (\mathbf{v}, [y(k-1)y(k-2)\bar{x}(k-1)]) \qquad (12.19)$$

354 Chapter 12 From Logic Expressions to Fuzzy Logic Networks

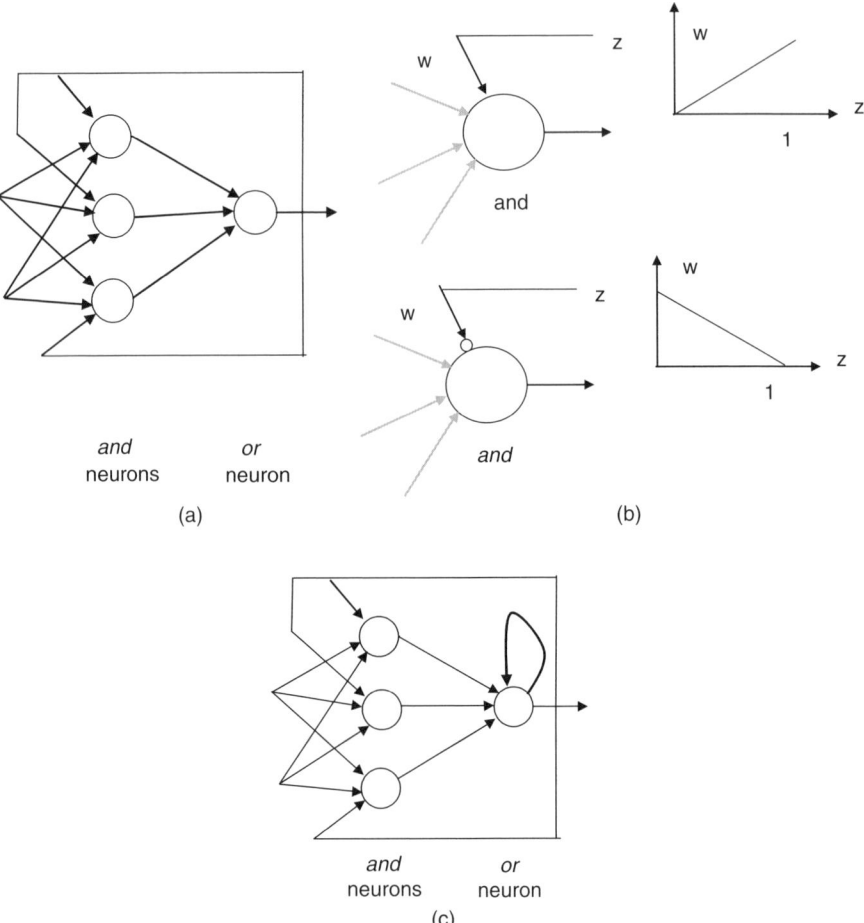

Figure 12.17 Fuzzy neural networks with feedback loops: (a) feedback between output and input layer, (b) notation used in describing excitatory and inhibitory feedback mechanisms along with the underlying computing; t-norm: product, and (c) feedback loop between output layer and the input layer as well as itself.

where $\bar{x}(k-1) = 1 - x(k-1)$. The three-element vector of the connections is denoted by **v**. By putting these two nodes (neurons) together (Fig. 12.18), we form a fuzzy neural network with feedback. The numeric values of the vectors of the connections **w** and **v** are typically adjusted during the learning process assuming that we are provided with some experimental data.

12.5 THE DEVELOPMENT MECHANISMS OF THE FUZZY NEURAL NETWORKS

The learning of fuzzy neural networks is concerned with two fundamental development facets, which are structural learning and parametric learning. Typically, we start

12.5 The Development Mechanisms of the Fuzzy Neural Networks

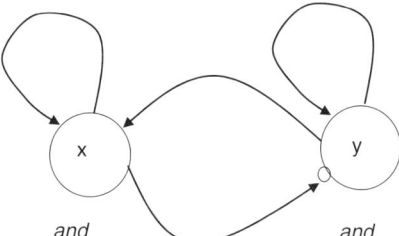

Figure 12.18 A two-node fuzzy neural network with feedback loops.

with the structural mode of learning within which we develop a topology (architecture) of the network and select specific t- norms and t-conorms. For instance, in the network realizing a generalized logic mapping, (Fig. 12.15) the structure of the network is set up by selecting the type of the network (*or–and* or *and–or*) and then choosing a number of *and* (*or*) neurons in the hidden layer. For the networks involving referential neurons, we encounter a far higher level of flexibility, and in a number of cases the use of the prior knowledge would help immensely in setting up the architecture of the network. Once the topology has been fully defined, we concentrate on parametric learning by adjusting the numeric values of the connections of the neurons.

12.5.1 The Key Design Phases

The development of the network is usually carried out in an iterative fashion: We start with a certain topology of the network, optimize it, and assess its performance. If not acceptable, the network has to be revised, for example, by expanding its structure (in which case we add more neurons or include more layers) and adjusting the connections. This iterative process is repeated until the formulated objectives have been met. An overall general view of the development process illustrated in Figure 12.19 highlights its key phases along with the use of data and prior knowledge.

The typical objectives one encounters in the development of fuzzy neural networks deal with their accuracy, interpretability, and stability. When dealing with the accuracy, it is quantified in terms of some performance index expressing a distance between the output of the network and the corresponding experimental data. When it comes to the network interpretability, we are concerned about the size of the network (say, the number of neurons and connections) leading to a concise description of the

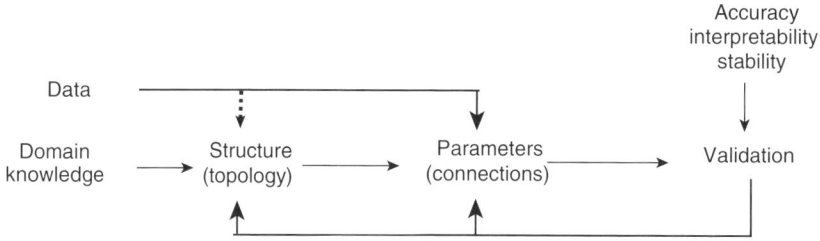

Figure 12.19 An iterative development process of fuzzy neural networks.

data. In this setting, smaller sizes of the networks are preferred. Ultimately, any fuzzy neural network could be easily translated into a collection of rules. Thus more compact form of the network results in the smaller set of rules that are more readable (interpretable). Finally, the stability of the network is associated with the underlying collection of the rules where we would like to see the key relationships revealed by the network to remain stable.

There are several fundamental tools being used in the structural and parametric optimization of the networks. At the structural end, we emphasize the role being played by the techniques of global optimization such as evolutionary methods (genetic algorithms, genetic programming, and evolutionary strategies) and other biologically inspired techniques including particle swarm optimization (PSO), ant colonies, and others. Although these could be also used for parametric optimization, here numerous variants of gradient-based techniques are worth considering.

In the forthcoming discussion, we show that the transparency of the logic neurons plays a pivotal role in the design of the network as we can exploit the two important sources of available knowledge, which is a collection of hints, general observations as to the nature of the underlying logic mapping.

12.5.2 Gradient-based Learning Schemes for the Networks

This mode of supervised learning relies on a collection of some experimental data being available about the process or phenomenon we are interested in modeling with the use of the network. Let us assume, which is a standard learning scenario, that the data are organized into a family of input-output pairs $\{\mathbf{x}(k), target(k)\}$, $k = 1, 2, \ldots, N$, where the successive inputs $x(k)$ and outputs ($target(k)$) are elements of the unit hypercubes $[0, 1]^n$ and $[0, 1]^m$, respectively. The learning is guided by the values of some performance index Q whose values have to be minimized by adjusting the values of the connections. These adjustments are typically completed through some iterative process in which the updates of the connections are governed by the gradient-based scheme that could be schematically described in the form

$$Connection(\text{iter} + 1) = connection(\text{iter}) - \alpha \nabla_{connection(\text{iter})} Q \qquad (12.20)$$

where α is a positive learning rate and $\nabla_{connection(\text{iter})} Q$ stands for the gradient of Q determined with respect to the connections. Successive iterations are denoted here by "iter" and "iter+1." The above notation is very general in the sense that the term *connection* comes as a concatenation of all the connections existing in the networks. It is also meant that the resulting values of the connections are retained in the unit interval that could be easily accomplished by clipping the results to 0 or 1 of the updates if at any iteration they locate themselves outside the required range. We can express this constraint in an explicit way by rewriting (12.20) as follows:

$\langle connection(\text{iter}) - \alpha \nabla_{connection(\text{iter})} Q \rangle$ where $\langle . \rangle$ denotes the truncation operation such as $\langle .a \rangle = 1$, if $a > 1.0$; 0, if $a < 0$; and a, otherwise. The iterations

12.5 The Development Mechanisms of the Fuzzy Neural Networks

governed by (12.18) start with some randomly initialized values of the connections. This helps us avoid any potential bias in the learning as through this type of initialization no particular preference or bias are expressed.

Following this global view of the optimization expressed by (12.18), we next move on with its details. At this step we have to specify the architecture of the network and identify the form of t-norms and t-conorms being used here. Let us develop a detailed learning scheme for the network shown in Figure 12.17 that implements a fuzzy function with "n" inputs and a single output, which is a mapping from $[0, 1]^n$ to $[0, 1]$. The learning is carried out in the supervised mode with the input–output data $\{(x(k), target(k))\}$, $k = 1, 2, \ldots, N$. The performance index Q is expressed as a sum of squared errors

$$Q = \sum_{k=1}^{N} (y(k) - \text{target}(k))^2 \quad (12.21)$$

where $y(k)$ is the output of the network, $y(k) = NN(\mathbf{x}(k))$ corresponding to the input $x(k)$. The detailed computing of the derivatives could be easily traced when adhering to the notation presented in Figure 12.20.

Let us consider an on-line mode of learning meaning that the updates of the connections are made after the presentation of each pair of input–output data. Given this, we could simplify the notation by dropping the index of the data in the training set and simply considering the pair in the form ($\mathbf{x}$, target). For the connections of the *or* neuron in the output layer we have the following expression for the derivative taken with respect to the connections of this neuron.

$$\frac{\partial Q}{\partial w_i} = (y - \text{target}) \frac{\partial y}{\partial w_i} \quad (12.22)$$

$$\frac{\partial y}{\partial w_i} = \frac{\partial}{\partial w_i} \left(\underset{j=1}{\overset{h}{S}} (w_j t z_j) \right) = \frac{\partial}{\partial w_i} \left[A_i s(w_i t z_i) \right] \quad (12.23)$$

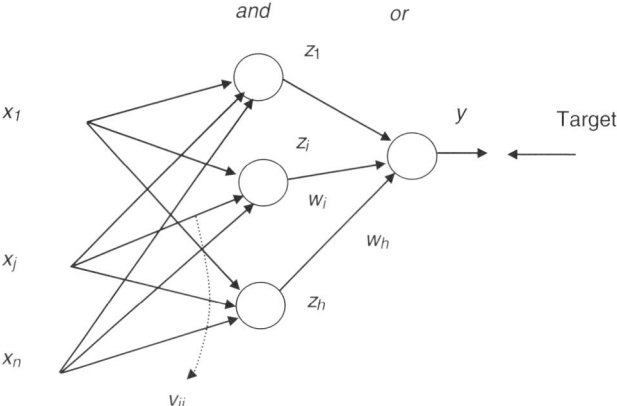

Figure 12.20 Computing the gradient of Q for the connections of the fuzzy neural network- notational details.

where

$$A_i = \underset{\substack{j=1 \\ j \neq i}}{\overset{h}{S}} (w_j t z_j) \tag{12.24}$$

In the sequel, we compute the derivative for the connections of the *and* neurons; see again Figure 12.18 for the pertinent notation. Here we obtain

$$\frac{\partial Q}{\partial v_{ij}} = (y - \text{target}) \frac{\partial y}{\partial v_{ij}} \tag{12.25}$$

$$\frac{\partial y}{\partial v_{ij}} = \frac{\partial y}{\partial z_i} \frac{\partial z_i}{\partial v_{ij}} \tag{12.26}$$

In the sequel one derives

$$\frac{\partial y}{\partial z_i} = \frac{\partial}{\partial z_i}\left(\underset{j=1}{\overset{h}{S}}(w_j t z_j)\right) = \frac{\partial}{\partial w_i}[A_i s(w_i t z_i)] \tag{12.27}$$

$$\frac{\partial z_i}{\partial v_{ij}} = \frac{\partial}{\partial v_{ij}}\left(\underset{l=1}{\overset{n}{T}}(v_{il} s z_l)\right) = \frac{\partial}{\partial v_{ij}}[B_{ij} t(v_{ij} s z_j)] \tag{12.28}$$

with

$$B_{ij} = \underset{\substack{l=1 \\ l \neq i}}{\overset{n}{T}} (v_{il} s z_l) \tag{12.29}$$

At this point, further computing of the gradient becomes possible once we have specified the corresponding t- norm and t-conorm. Thus, as an illustration, let us confine ourselves to the product (t-norm) and probabilistic sum (t-conorm). The above formulas read as follows:

$$\frac{\partial}{\partial w_i}[A_i s(w_i t z_i)] = \frac{\partial}{\partial w_i}(A_i + w_i z_i - A_i w_i z_i) = z_i(1 - A_i) \tag{12.30}$$

$$\frac{\partial z_i}{\partial v_{ij}} = \frac{\partial}{\partial v_{ij}}[B_{ij} t(v_{ij} s z_j)] = \frac{\partial}{\partial v_{ij}}[B_{ij}(v_{ij} + z_j - v_{ij} z_j)] = B_{ij}(1 - z_j) \tag{12.31}$$

An interesting case arises when dealing with the minimum and maximum operators. Their derivatives can be formulated in the form

$$\min(x, w) = \begin{cases} w, \text{if } x \leq w \\ x, \text{if } x > w \end{cases} \tag{12.32}$$

$$\frac{\partial \min(x, w)}{\partial w} = \begin{cases} 1, \text{if } x \leq w \\ 0, \text{if } x > w \end{cases} \tag{12.33}$$

In this way we have addressed a contentious issue of the lack of differentiability of these two functions (formally, the left- and right-hand side derivatives are different). The far more serious problem is completely different and relates to regions of arguments (w) for which the derivative attains zero. These regions could easily lead to a complete stalling of the learning algorithm when the overall gradient

12.5 The Development Mechanisms of the Fuzzy Neural Networks

becomes equal to zero and subsequently no further move in the space of connections is possible. To avoid this effect, we redefine the derivative by admitting intermediate values between 0 and 1 that represent a truth value of the satisfaction of the predicate "included in." Let us consider the following expression:

$$\frac{\partial \min(x, w)}{\partial w} = \| w \subset x \| = w \Rightarrow x \qquad (12.34)$$

where the term $\| w \subset x \|$ denotes the degree of inclusion of w in x, which can be quantified with the use of some multivalued implication; refer to the previous discussion. Similarly, we can generalize the derivative of the maximum operation

$$\frac{\partial \max(x, w)}{\partial w} = \| x \subset w \| = x \Rightarrow w \qquad (12.35)$$

In this sense, we have arrived at the continuous (generalized) version of the derivative of the minimum and maximum by showing how truth values could be efficiently used here.

Considering the Lukasiewicz logic connectives, their derivatives are computed as follows:

$$\frac{\partial \max(0, x + w - 1)}{\partial w} = \begin{cases} 0, & \text{if } w + x - 1 \leq 0 \\ 1, & \text{otherwise} \end{cases} \qquad (12.36)$$

$$\frac{\partial \min(1, x + w)}{\partial w} = \begin{cases} 0, & \text{if } w + x \geq 0 \\ 1, & \text{otherwise} \end{cases} \qquad (12.37)$$

and afterward used in the updates of the values of the connections (12.20) and (12.24).

When carrying out parametric learning, there are two general development modes worth discussing

(a) **Successive expansions**. In this development mode, we start with a small, compact network composed of a minimal number of neurons and successively increase its size (say, by expanding the layer of the *and* neurons when dealing with the architecture illustrated in Fig. 12.15) if the minimized performance index does not meet the predefined value Q^*. If the condition $Q < Q*$ is not met, increase the size of the network.

(b) **Successive reduction**. Here we start with a network of a large size, carry out its learning, and then by inspecting the values of the optimized connections, and if possible we complete its reduction. The essence of this development mode is to admit the use of the excessively large architecture with an intent of its further pruning. The details of the pruning procedure are concerned with the elimination of the "weakest" connections (where the strength of the connection is expressed by means of some well-defined criteria).

The advantage of the successive reduction is that the network after pruning may require further learning, but no further reduction of the architecture may be required afterward. As we are concerned with the (excessively) large size of the network to start with, the learning could be more demanding in terms of the learning time (number of iterations of the learning algorithm). On the contrary, though the learning effort required in successive expansions of the networks is lower (as we are concerned with the far smaller topology), yet we may go through a series of expansions of the networks that are very likely as it is not known in advance how large the network should be.

Note that the networks that are not interpretable cannot take advantage of the prior domain knowledge as we do not have any effective mechanism to "download" these knowledge hints onto the structure of the network. To avoid any bias in the learning process that might have occurred under these circumstances, the learning is started from a random configuration of the values of the connections of the network. This start point of the learning does not offer any possible advantages. Even if the prior domain knowledge were available, it cannot be effectively used in the learning process that has to be started from scratch.

12.6 INTERPRETATION OF THE FUZZY NEURAL NETWORKS

Each logic neuron comes with well-defined semantics that is directly associated with the underlying logic. *or* neurons realize a weighted or combination of their inputs. The higher the values of the connection, the more essential becomes the corresponding inputs. For the *and* neuron the converse relationship holds: lower values of the connections imply higher relevance of the corresponding inputs.

While these two arrangements are taken into consideration, we can generate a series of rules generated from the network. We start with the highest value of the *or* connection and then translate the corresponding *and* neuron into the *and* combination of the inputs. Again the respective inputs are organized according to their relevance proceeding with the lowest value of the corresponding connection.

EXAMPLE 12.3

Let us consider the fuzzy neural network shown in Figure 12.21. The rules can be immediately enumerated from this structure. In addition, we can order them by listing the most dominant rules first. This ordering refers to the relevance of the rules and the importance of each condition standing there. We start from the output (*or*) node and enumerate the inputs in the order implied by the values of the connections of the neuron. We get

if $z_{2|0.9}$

or

$z_{1|04}$

then y

where the subscripts (0.9 and 0.4, respectively) denote the confidence level we associate with the corresponding inputs of the neuron.

Next we expand the terms z_1 and z_2 using the original inputs x_1 and x_2; in this case we report the inputs in the increasing order of the connections starting from the lowest one. Recall that the connections of the *and* neuron realize a masking effect; the higher the value of the

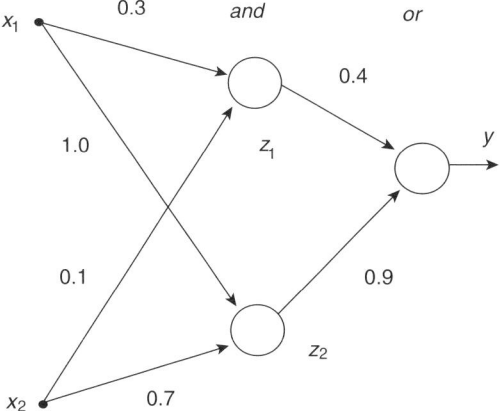

Figure 12.21 An example fuzzy neural network.

connection, the stronger the masking effect. In particular, if the connection is equal to 1, we can drop the associated input as it has been completely masked (eliminated).

Following this interpretation guideline, we translate the network into the following rules:

if $[x_{2|0.7}]_{0.9}$
or
$[x_{2|0.1} \text{ and } x_{1|0.3}]_{0.4}$
then y

Note that the numbers associated with the individual rules and their conditions serve as annotations (quantifications) of the corresponding components.

Although the above guidelines are useful in a systematic enumeration of the rules residing within the network (and owing to the logical underpinnings of its architecture), the interpretation of larger networks still may require more attention and substantially higher effort. It would be beneficial to eliminate the most irrelevant (insignificant) segments of the network before proceeding with the enumeration of the rules. Let us introduce two schemes supporting a methodical reduction of the network. In the first one, we retain the most significant connections. In the second method, we convert the connections of the networks into their Boolean (two-valued) counterparts.

12.6.1 Retention of the Most Significant Connections

In light of the comments about the relevance of the connections, depending on the type of the neuron, we reduce the weakest connections to 0 or 1. This is done by introducing some threshold mappings with the values of thresholds (λ and μ, respectively) coming from the unit interval. For the *or* neuron, we consider the following reduction of the connections $\phi_\lambda : [0, 1] \rightarrow [\lambda, 1] \cup \{0\}$ such that

$$\phi_\lambda(w) = \begin{cases} w, & \text{if } w \geq \lambda \\ 0, & \text{if } w < \lambda \end{cases} \quad (12.38)$$

Hence any connection whose value is lower than some predefined threshold λ becomes reduced to zero (therefore the corresponding input gets fully eliminated) whereas the remaining connections are left intact.

For the *and* neuron, the transformation $\psi_\mu : [0, 1] \to [0, \mu] \cup \{1\}$ reads as

$$\psi_\mu(w) = \begin{cases} 1, & \text{if } w > \mu \\ w, & \text{if } w \leq \mu \end{cases} \quad (12.39)$$

The connections with values greater than *m* are eliminated by making them equal to 1.

12.6.2 Conversion of the Fuzzy Network to the Boolean Version

The connections are converted into the Boolean values. We use some threshold values λ and μ. For the *or* neuron, if the connections whose values are not lower than the threshold are elevated to 1, the remaining ones are reduced to 0. For the *and* neuron, the binarization of the connections is realized as follows: if the connection is higher than the threshold, it is made equal to 1, otherwise we make it equal to 0 meaning that the corresponding input is fully relevant. More formally, the transformations read as follows:

or neuron

$\varphi_\lambda : [0, 1] \to \{0, 1\}$

$$\varphi_\lambda(w) = \begin{cases} 1, & \text{if } w \geq \lambda \\ 0, & \text{if } w < \lambda \end{cases} \quad (12.40)$$

and neuron

$\psi_\mu : [0, 1] \to \{0, 1\}$

$$\psi_\mu(w) = \begin{cases} 1, & \text{if } w > \mu \\ 0, & \text{if } w \leq \mu \end{cases} \quad (12.41)$$

The choice of the threshold values implies a certain number of connections being reduced. Higher values of λ and lower values of μ give rise to more compact form of the resulting network. Although being more readable and interpretable, the modified networks come with the lower approximation capabilities. Consider the data D (denoting a training or testing data set) being originally used in the development of the network. The performance index of the reduced network is typically worse than the originally developed network. The increase in the values of the performance index can be sought as a sound indicator guiding a process of forming a sound balance between the improvement in the transparency (achieved reduction) and accuracy deterioration.

For the referential neurons $y = \text{REF}(\mathbf{x}; \mathbf{w}, \mathbf{a})$ the pruning mechanisms may be applied to the *and* neuron combining the partial results of referential computing as well as the points of reference. Considering that we are concerned with the *and* neurons performing the aggregation, the connections higher than the assumed threshold are practically eliminated from the computing. Apparently we have $(\tilde{w}_i s x_i) t A \approx 1 t = A$, where A denotes the result of computing realized by the neuron for the rest of its inputs. The reference point a_i requires different treatment depending upon the type of the specific referential operation. For the inclusion operation, INCL(x, a_i) we can admit the threshold operation to come in the form

$$\text{INCL}^\sim(x, a_i) = \begin{cases} \text{INCL}(x, a_i), & \text{if } a_i \leq \mu \\ 1 - x, & \text{if } a_i > \mu \end{cases} \quad (12.42)$$

with μ being some fixed threshold value. In other words, we consider that INCL(x, a_i) is approximated by the complement of x (where this approximation is implied by the interpretational feasibility rather than being dictated by any formal optimization problem), INCL$(x, a_i) \approx 1 - x$. For the dominance neuron we have the expression for the respective binary version of DOM, DOM$^\sim$

$$\text{DOM}^\sim(x, a_i) = \begin{cases} \text{DOM}(x, a_i), & \text{if } a_i < \mu \\ x, & \text{if } a_i > \mu \end{cases} \quad (12.43)$$

The connection set up to 1 is deemed essential. If we accept a single threshold level of 0.5 and apply this consistently to all the connections of the network and set up the threshold 0.1 for the inclusion neuron, the statement

$$y = [x_1 \text{ included in } 0.6] \text{ or } 0.2 \text{ and } [x_2 \text{ included in } 0.9] \text{ or } 0.7$$

translates into a concise (yet approximate) version assuming the form of the following logic expression:

$$y = x_1 \quad \text{included in } 0.6$$

The choice of the threshold value could be a subject of a separate optimization phase, but we can also admit some arbitrary values especially if we are focused on the interpretation issues.

EXAMPLE 12.4

Considering the network shown in Figure 12.21, we can analyze and quantify the tradeoff between the accuracy of the network and its pruned version in which a number of the connections have been eliminated. As we are not provided with the original data and the network is very small, we can use the following integral V expressing an absolute difference between the output of the original network (NN) and its reduced version (rNN) that is regarded as a function of λ and μ:

$$V(\lambda, \mu) = \int_0^1 \int_0^1 |\text{NN}(\mathbf{x}) - r\text{NN}(\mathbf{x})| d\mathbf{x} \quad (12.44)$$

The changes of V exhibit a finite number of changes that is understood as we have a finite number of connections so the changes in V occur only when the threshold values impact the modifications of the weights. Based on the reported values of V, we can adjust the values of the thresholds. Another way of presenting the tradeoffs between accuracy and transparency is given in the form of the $\lambda-\mu$ graph where each point in the graph comes with the values of V and the number of the eliminated connections. When counting the number of such connections, we should become aware that the reduction of a certain connection of the *or* neuron triggers the reduction of all connections of the associated *and* neurons.

364 Chapter 12 From Logic Expressions to Fuzzy Logic Networks

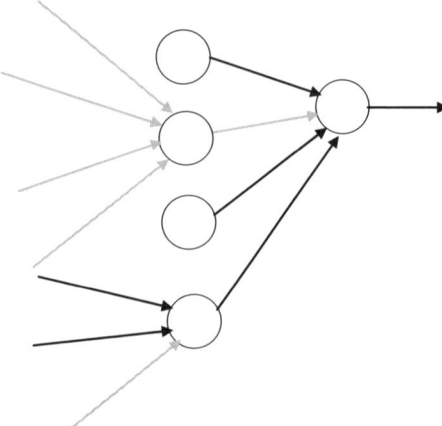

Figure 12.22 Counting an effective reduction in the number of connections of the fuzzy neural network resulting from the elimination of the connections of the neurons and its rippling effect. The eliminated connections are marked in light color.

The computing of the number of connections of the network being eliminated from the architecture requires attention. Observe that once a certain connection of the neuron at the layer close to the output has been eliminated, this implies that effectively the connections of the neurons located in the layers close to the input and therefore affected by the removal of the neuron under discussion. The essence of this way of counting is outlined in Figure 12.22.

We may offer another view of the structural reduction of the network (Fig. 12.23). Let us consider a system of coordinates of the modified (increased) performance index and the number of the eliminated connections. The points shown in the graph are associated with the different configurations of the networks and help visualize the main trend in the interpretability of the network.

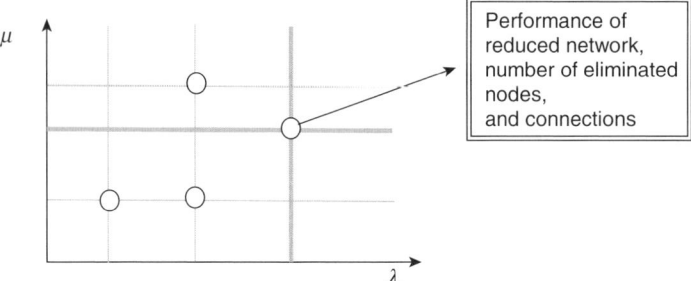

Figure 12.23 A global view of the structural reduction of the network and the associated performance of the reduced network. Each point in the graph corresponds to a certain configuration of the network produced for some specific values of λ and μ.

12.7 FROM FUZZY LOGIC NETWORKS TO BOOLEAN FUNCTIONS AND THEIR MINIMIZATION THROUGH LEARNING

and and *or* neurons generalize (subsume) *and* and *or* logic gates encountered in the realization of digital systems. This observation offers an interesting way of building digital circuits through learning the corresponding fuzzy network and its further refinement. The Boolean input-output pairs $\{(\mathbf{x}(k), \text{target}(k))\}, k = 1, 2, \ldots, N$ where $\mathbf{x}(k) \in \{0, 1\}^n$, $\text{target}(k) \in \{0, 1\}$ are used for training the fuzzy neural network composed of a single layer of *and* neurons with a layer of *or* neurons located in the output layer. As the logic functions also involve the complements of the original variables, we include those as the inputs of the fuzzy neural network (Fig. 12.24). The learning uses all Boolean data being available. The choice of the number of *and* neuron is realized experimentally by accepting one of the development strategies (viz. successive expansions or reduction).

Once the connections have been reduced to their Boolean versions, we end up with the network representing a reduced (simplified) version of the Boolean function. The simplification of the Boolean function (which is typically done with the use of techniques such as Karnaugh maps, K-maps, and algebraic transformation) is completed here through learning, and this offers an attractive alternative to the existing methods.

So far we have discussed a single-output logic mapping from $[0, 1]^n$ to $[0,1]$; these could be easily extended to the multivariable Boolean functions, that is, $[0, 1]^n \rightarrow [0, 1]^m$.

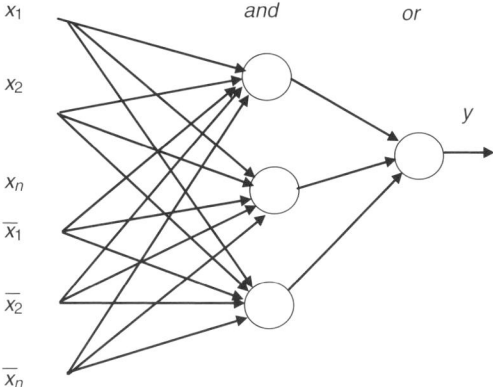

Figure 12.24 The fuzzy logic network used in the learning of the Boolean function; note complemented inputs (denoted by an overbar symbol) entering the network leading to $2n$ inputs.

12.8 INTERFACING THE FUZZY NEURAL NETWORK

Trying to link this logic architecture to the experimental data where the variables are undoubtedly associated with the nature of the problem, we have to consider a certain conceptual interface between the physical system and the logic-driven model of the fuzzy neural network. The main development guidelines we discussed in the context of fuzzy modeling are fully applicable here. For each of the continuous variables in the problem, both input and output ones, we define a finite collection of semantically meaningful information granules—fuzzy sets. The discrete variables assuming a finite number of discrete values are coded using the one-out-n coding scheme. For the continuous output variable we consider using a finite family of fuzzy sets with triangular membership functions. The reason behind their use here comes from their capabilities of lossless reconstruction.

By transforming the original experimental data through the interfaces constructed for the input and output space, we arrive at the vectors of inputs and outputs that can be treated as elements in the unit hypercube and the fuzzy logic network is aimed at the approximation of logic-based relationships. At this stage of the development of the network we are provided with the pairs of the elements located in the input $[0,1]^n$ and output $[0,1]^m$ hypercubes, that is, $\mathbf{x}(k)$ and $\mathbf{y}(k)$ (Fig. 12.25).

Given the topology of the fuzzy neural network, each *and* neuron produces a generalized Cartesian product of the fuzzy sets of the interfaces. In the sequel, these regions in the multidimensional space are combined *or*-wise (with the additional calibration provided by the connections of the *or* neuron). An illustration of this geometry of the rules in the case of two input variables (x and z) is shown in

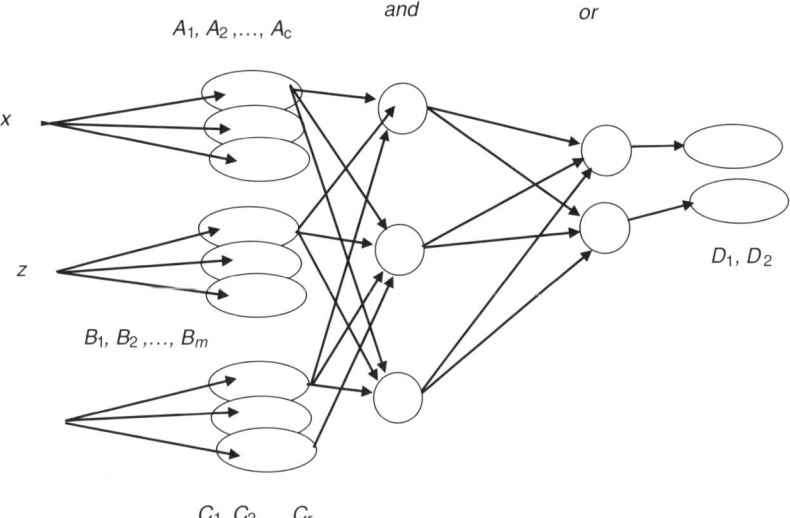

Figure 12.25 An overall architecture visualizing the logic core (fuzzy neural network) and the interfaces composed of families of fuzzy sets formed for the input and output variables.

12.9 Interpretation Aspects—a Refinement of Induced Rule-Based System

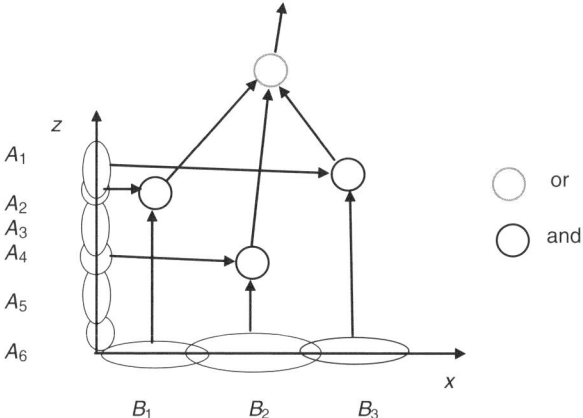

Figure 12.26 The geometry of the rules supported by the fuzzy neural network.

Figure 12.26. Here we have 6 and 3 fuzzy sets defined in the corresponding variables. By a straightforward inspection, we can read the following collection of the rules:

If $(A_4$ and $B_2)$ or $(A_2$ and $B_1)$ or $(A_1$ and $B_3)$ then C

12.9 INTERPRETATION ASPECTS—A REFINEMENT OF INDUCED RULE-BASED SYSTEM

The fuzzy neural network with the inputs and outputs being associated with the collections of fuzzy sets directly translates into a collection of rules. As an illustration, let us consider the following example.

EXAMPLE 12.5

The network portrayed in Figure 12.27 has two inputs (x and z) for which we have defined three and two fuzzy sets (denoted here by A_1, A_2, A_3 and B_1 and B_2, respectively). There is one fuzzy set defined in the output space. Considering the connections of the network, we can incorporate them into the enumeration of the rules

The rules are ordered with respect to their relevance implied by the connections of the OR neuron

If $[A_1|_{0.1}$ and $B_1|_{0.4}]|_{1.0}$
or
$[A_2|_{0.05}$ and $B_2|_{0.5}]|_{0.9}$
or
$[A_3|_{0.1}$ and $B_1|_{0.2}|_{0.6}$
then C

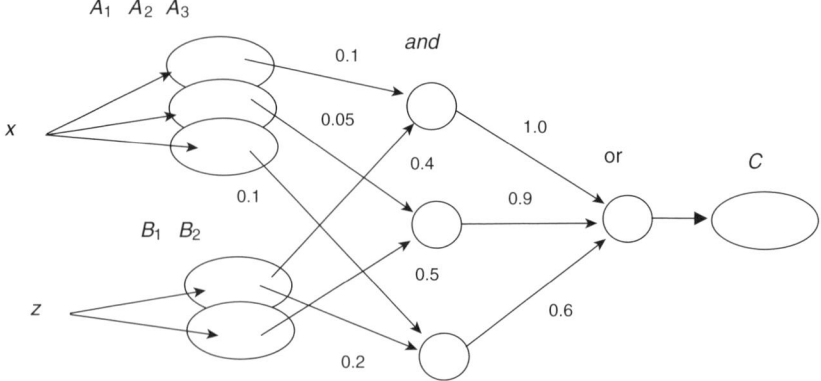

Figure 12.27 An example of fuzzy neural network equipped with the input and output interfaces.

Furthermore the values of the connections of the *and* neuron impact the original fuzzy sets standing in the interface. The expression governing the processing of the *and* neuron, that is, $A(x) s w_i$, can be interpreted by "absorbing" this connection w_i into a new membership function. Denote this modified fuzzy set by $A^\sim$,

$$A^\sim(x) = x_i s w_i = A(x) s w_i \tag{12.45}$$

refer also to Figure 12.28.

Given the nature of the t-conorm, $A^\sim$ is less specific than the original fuzzy set occurring in the rule. In this setting the connections come with an interesting interpretation: their higher values make the membership function of the associated fuzzy set closer to 1. In the limit case when the connection is equal to 1, the associated fuzzy set becomes "masked" (eliminated) and thus the overall rule gains higher generality. Practically, in this way the condition of the rule has been dropped. Hence by carrying out the same absorption process for each condition in the rule (where the intensity of the modification, viz. masking) we arrive at the network whose connections have been eliminated. The plots of the modified fuzzy sets for selected values of the connection are included in Figure 12.29.

In general, given the underlying topology of the network, the rules induced by the fuzzy neural network can be organized in the following way.

$$\text{If } A_i \text{ and } B_j \text{ and } C_l \text{ and } \ldots Z_k \text{ then } W_s$$

or

$\ldots$

with the corresponding quantification of the individual conditions (say, A_i, C_l, etc.) of the rules and their conclusions (W_s).

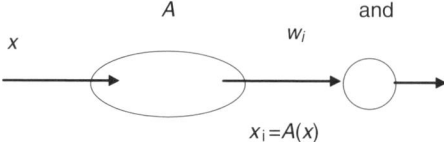

Figure 12.28 Transformation of fuzzy set A of interface through the connection w_i and leading to the modified fuzzy set $A^\sim$.

12.9 Interpretation Aspects—a Refinement of Induced Rule-Based System

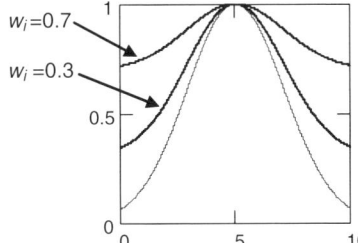
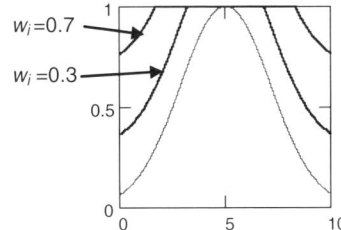

Figure 12.29 An example of the Gaussian fuzzy set (being the original fuzzy set of the interface) and its logic combination (generalization) completed with the aid of the connection w_i with two selected values of 0.3 and 0.7: (a) t-conorm specified as the probabilistic sum, (b) t-conorm specified as the Lukasiewicz *or* connective.

Considering the geometry of the rules implied by the topology of the network, we note that the condition of each rule involves only a single fuzzy set defined in the corresponding input. In some cases it would be convenient to have rules whose conditions are formed by forming a union of some fuzzy sets, say A_i and A_{i+1}, and so on. In the fuzzy neural networks studied so far, there are no particular provisions supporting this format of the rules. To enhance the functionality of the network along this line, we introduce some additional *or* neurons associated with each input variable as illustrated in Figure 12.30.

These additional *or* neurons deliver an option of forming a union of several information granules at the condition level and thus offer an increased generality of the corresponding rule.

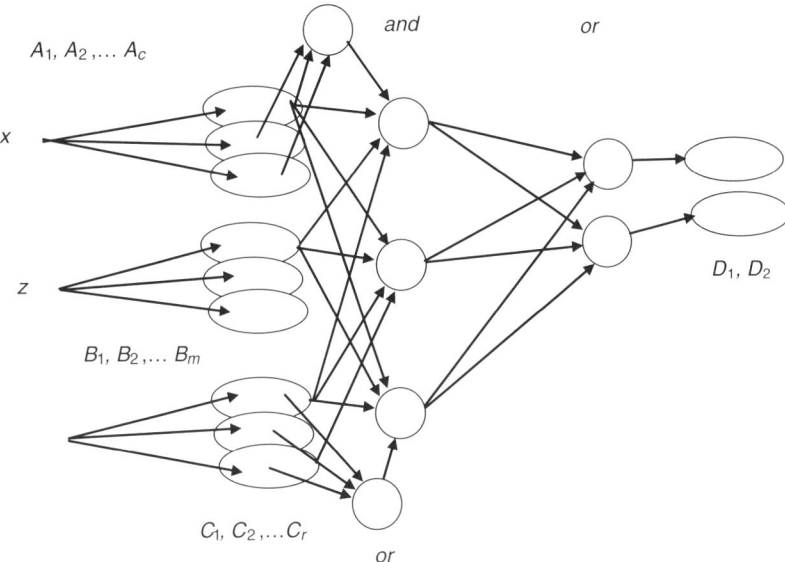

Figure 12.30 An augmented architecture of the fuzzy neural network with additional *or* neurons used to express potential generalization effect present in the input space.

370 Chapter 12 From Logic Expressions to Fuzzy Logic Networks

Subsequently, the geometry of the rules generated by the network fully reflects this phenomenon. In this example (Fig. 12.31), the rules captured by the network read in the form

If $(A_4$ and $(B_1$ or $B_2))$ or $(A_2$ and $B_1)$ or $(A_1$ and $B_3)$ then C

where C denotes the corresponding fuzzy set of conclusion associated with the output of the *or* neuron.

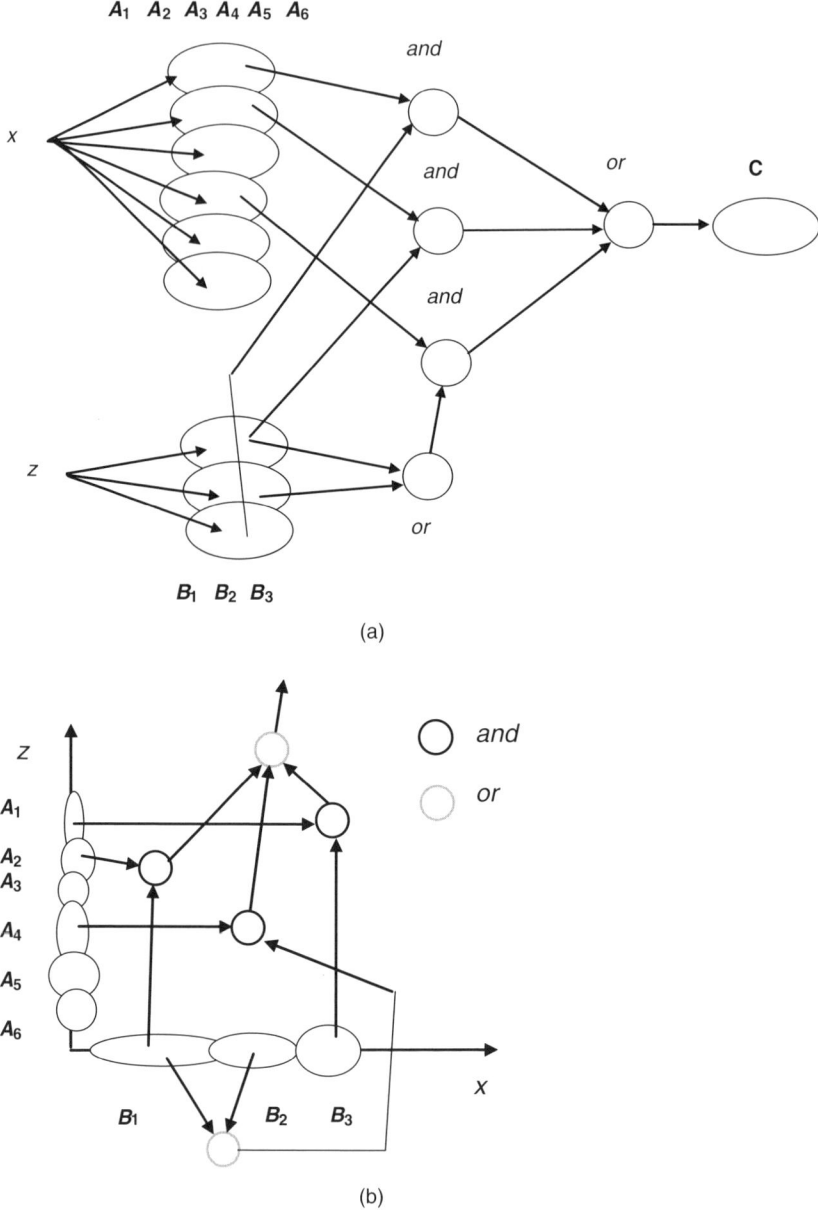

Figure 12.31 An example of the augmented fuzzy neural network (a) and the geometry of the underlying rules supported by it (b).

12.10 RECONCILIATION OF PERCEPTION OF INFORMATION GRANULES AND GRANULAR MAPPINGS

The same information granule—fuzzy set can be perceived in many different ways and cast in fairly different contexts. To model this phenomenon of diversified perception, we may associate with the information granule confidence coefficients (confidence factors). One of the possible ways of modeling the perception effect uses the logic-oriented transformation of the membership function of the original fuzzy set. A mechanism of reconciliation of perception of fuzzy sets can be directly modeled with the use of logic neurons. In this framework, we can capture the logic nature of the problem and benefit from the parametric flexibility residing with the neurons.

When dealing with adaptive environment as presented in Angelov (2004), one may view adjustable membership functions as an integral component of flexible environment in which fuzzy sets can be tuned according to the changing requirements or nonstationary environment. The same concept—fuzzy set can be perceived in many different ways thus leads to the development of several specific membership functions of the same fuzzy set. For instance, the concept of *high* inflation treated as a fuzzy set and coming with its underlying membership function can be perceived by different human observers quite differently and subsequently may produce several and sometimes quite different fuzzy sets. The corresponding membership functions could be then viewed as some modified or distorted versions of the original membership function. Our objective is to reconcile these various perception views by forming some common view of the concept resulting in the form of some fuzzy set. In a nutshell, this leads to the formation of some optimal fuzzy set that takes into consideration the variety of the perceptions. The reconciliation of the perceptions is formed through the formation of a certain optimization problem involving the individual fuzzy sets. The reconciliation of various perceptions could also concern fuzzy mappings. More specifically, some relationships between two spaces can be described by a family of fuzzy relations representing a way in which they are being perceived by various observers. The reconciliation of these relationships produces an optimal fuzzy relation being reflective of some essential commonalities of the existing relationships. This optimal relation becomes then a result of solution to the pertinent optimization problem.

12.10.1 Reconciliation of Perception of Information Granule

Given is a certain fuzzy set A defined in some space (universe) $\mathbf{X}$. It is perceived from different standpoints where each of these perceptions is quantified by c human observers. With this regard, the observers provide some numeric confidence levels $z_1, z_2, \ldots, z_c$ where $z_i \in [0, 1]$. These specific confidence levels are translated into a form of some fuzzy set in which A becomes effectively perceived by these observers. Let us introduce the complements of z_i, $w_i = 1 - z_i$. The conjecture is that our perception transforms A into a less specific construct where a lack of confidence translates into the reduced level of specificity of A. This leads to the disjunctive model

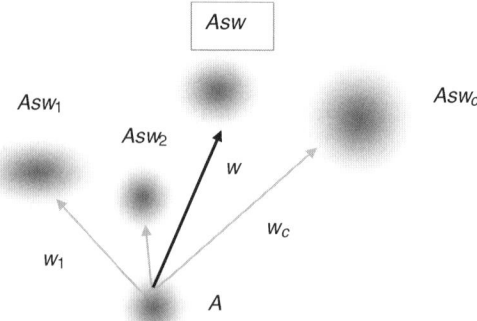

Figure 12.32 Reconciliation of perception of information granule A; note a collection of perceived information granules A_i resulting from A by being affected by the associated levels of confidence $z_i(w_i)$. The result of the reconciliation comes through the optimization of the confidence level z (z, where $z = 1 - w$). The result of this process is expressed as Asw.

of perception in which the perceived information granule A comes with the membership function of the form $A_i(x) = A(x)sw_i$, where s denotes a certain t-conorm (s-norm). This model exhibits several interesting properties. As mentioned, the perceived granule cannot gain in its specificity, but rather depending upon the confidence may exhibit some reduction of detail (lack of specificity). If the confidence about A is high, say $z_i = 1$, then $w_i = 0$ and $A_i(x) = A(x)$, so the original fuzzy set is left unaffected. On the opposite end, if $z_i = 0$ then $w_i = 1$ and subsequently $A_i(x) = 1$. This new membership function demonstrates that the perceived information granule A is modeled as "*unknown*" (being effectively the entire universe of discourse). In other words, the complete lack of confidence transforms A into the information granule that does not bear any specific meaning. The way of perceiving A from different standpoints (quantified by different values of the confidence values) is illustrated in Figure 12.32.

Given a family of fuzzy sets A_i, we are interested in reconciling the variety of the perception standpoints and on this basis construct a certain general (reconciled) viewpoint of A, say $A^\sim$ whose membership function is expressed in the form

$$A^\sim(x) = A(x)sw \qquad (12.46)$$

where the weight $w \in [0, 1]$ is reflective of the reconciliation process. By adjusting the values of w we can effectively capture the contributing components of the process. Graphically, we can envision the overall process described above as illustrated in Figure 12.32.

12.10.2 The Optimization Process

The above reconciliation can be transformed into the following optimization problem in which we are looking for the most suitable realization of perception so that all the individual views are taken into consideration. We introduce the following

12.10 Reconciliation of Perception of Information Granules and Granular Mappings

performance index:

$$Q = \sum_{i=1}^{c} \int_X [A(x)sw_i - A(x)sw]^2 dx \qquad (12.47)$$

The minimization of Q is carried out with respect to the values of "w"; $\mathrm{Min}_w Q$. The necessary condition of the minimum is set as $dQ/dw = 0$. Proceeding with the detailed computing, we obtain

$$\frac{\partial}{\partial w} \sum_{i=1}^{c} \int_X [A(x)sw_i - A(x)sw]^2 dx = -2 \sum_{i=1}^{c} \int_X [A(x)sw_i - A(x)sw] \frac{\partial (A(x)sw)}{\partial w} dx = 0 \qquad (12.48)$$

Denote by $\Phi(x,w)$ the derivative standing in the above expression, that is, $\Phi(x,w) = dA(x)sw/dw$. Then we have

$$\sum_{i=1}^{c} \int_X [A(x)sw_i - A(x)sw]\Phi(x,w) dx = 0 \qquad (12.49)$$

and

$$\sum_{i=1}^{c} \int_X [A(x)sw_i] = \sum_{i=1}^{c} \int_X (A(x)sw)\Phi(x,w) dx \qquad (12.50)$$

Further calculations are possible once we have accepted a certain form of the t-conorm. For instance, let the specific t-conorm under investigation can be realized as a probabilistic sum (that is, $a\ s\ b = a + b - ab$). This implies that the above expression for $\Phi(x,w)$ reads as

$$\Phi(x,w) = \frac{\partial A(x)sw}{\partial w} = \frac{\partial}{\partial w}(A(x) + w - A(x)w) = 1 - A(x) \qquad (12.51)$$

In the sequel we obtain

$$\sum_{i=1}^{c} \int_X [A(x)sw_i] = \sum_{i=1}^{c} \int_X A(x)(1 - A(x)) dx + w \sum_{i=1}^{c} \int_X (1 - A(x))^2 dx \qquad (12.52)$$

Let us rewrite the above expression by rearranging and grouping the terms and setting up its value to zero. We obtain

$$\int_X (1 - A(x))^2 \sum_{i=1}^{c} (w_i - w) dx = 0 \qquad (12.53)$$

As the function under the integral is nonnegative, to make the value of (12.53) equal to zero, we require that the term $\sum_{i=1}^{c} (w_i - w)$ vanishes. Interestingly enough, this happens when w is an average of the individual weights, namely $w = 1/c \sum_{i=1}^{c} w_i$.

12.10.3 An Application of the Perception Mechanism to Fuzzy Rule-based Systems

In a nutshell, fuzzy rule-based systems can be represented as a network of logic relationships (dependencies) between fuzzy sets existing in some input and output spaces. These fuzzy sets form the corresponding conditions and conclusions of the rules. As an illustration, let us consider the rule of the form

If (input$_1$ is A_1 *and* input$_2$ is B_1) *or* (input$_1$ is A_2 *and* input$_2$ is B_2) *or* (input$_1$ is A_3 and input$_2$ is B_3) then conclusion is D (12.54)

Such rule is directly mapped onto a network of *and* and *or* computing nodes. Furthermore the nodes are equipped with some weights (connections) whose role is to calibrate the impact of the individual fuzzy sets standing in the rules and thus affecting the results of the rule-based computing. The network used to realize the above part of the rule-based system is illustrated in Figure 12.33.

Alluding to the realization of the network illustrated in Figure 12.33 composed of a collection of the fuzzy neurons with some specific numeric values of the connections, we can write down the following detailed and numerically quantified "if–then" compound expression:
If

$$\{[(A1 \; or \; 0.7) \; and \; (B1 \; or \; 0.3)] \; and \; 0.9\}$$

or

$$\{[(A2 \; or \; 0.2) \; and \; (B2 \; or \; 0.5)] \; and \; 0.7\}$$

or

$$\{[(A3 \; or \; 0.1) \; and \; (B3 \; or \; 0.2)] \; and \; 1.0\}$$

then D

Each input fuzzy set (A_1, B_1, ...) is "perceived" by the corresponding *and* nodes through their weights (connections). There is an obvious analogy between the problem we formulated above and the formal description of the network formed by the logic neurons.

By repeating the same reconciliation process for all input fuzzy sets, we arrive at the *a* collection of optimized weights $w[1], w[2], w[3]$ (where $w[1]$ comes as a solution of the optimization task in which A_1 is involved, etc.) (Figure 12.34). They serve as a direct mechanism of establishing importance of the input information granule: the one of the lowest value of $w[ii]$ points at the most relevant input (ii).

12.10.4 Reconciliation of Granular Mappings

So far we discussed a way of reconciliation of perception of the same fuzzy set viewed at different standpoints. The problem under consideration is concerned with the reconciliation of the relational mappings. The problem is formulated as follows: given are relational mappings (fuzzy relations) $R_1, R_2, \ldots, R_c$ from space **X** to **Y**.

12.10 Reconciliation of Perception of Information Granules and Granular Mappings

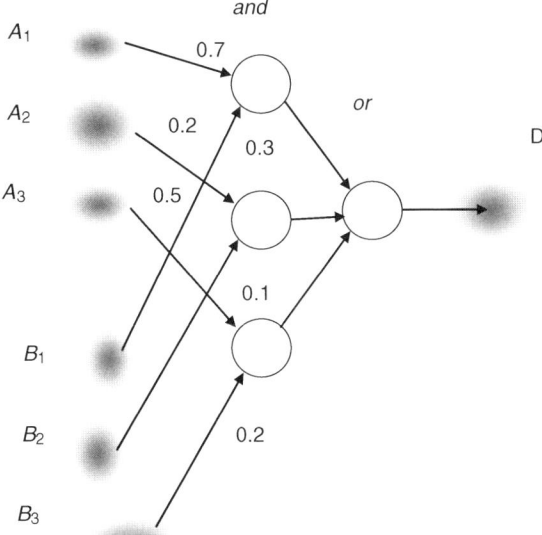

Figure 12.33 A logic network realizing the rule presented by (12.8); note that the *and* and *or* nodes are endowed with some numeric connections.

More specifically, for given A in **X**, the result of mapping comes in the form $A \circ R_1$, $A \circ R_2, \ldots, A \circ R_c$ where $\circ$ is a certain composition operator (in particular, it could be the one such as the sup–min, sup-t, inf-s, inf–min, and alike). Given the finite spaces **X** and **Y**, the expression $A \circ R_1$ can be represented as a collection of *or* neurons with inputs being the corresponding elements of **X**.

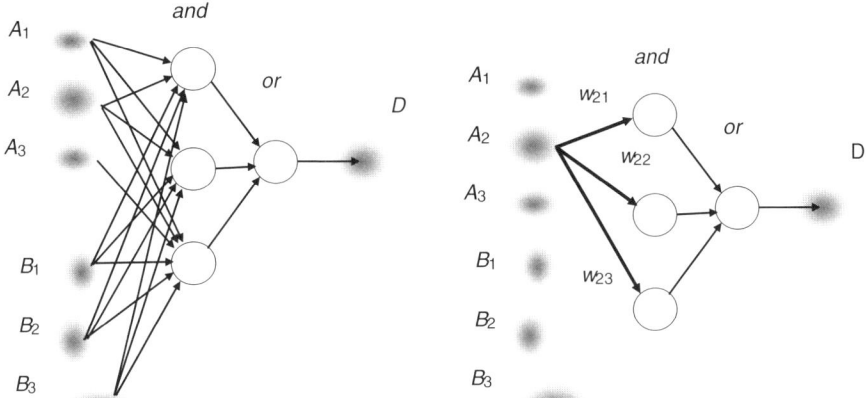

Figure 12.34 Reconciliation of impact of the input on individual *and* nodes through the optimization of the corresponding connections.

Chapter 12 From Logic Expressions to Fuzzy Logic Networks

Determine R such that it forms a reconciliation of the individual fuzzy relations. The reconciliation involves a certain fuzzy set in $\mathbf{X}$, denote it by A, or may deal with a family of fuzzy sets in $\mathbf{X}$, say $A_1, A_2, \ldots, A_N$. Formally speaking, we are looking for R defined in $\mathbf{X} \times \mathbf{Y}$ such that it results from a minimization of the following expression:

$$Q = \sum_{i=1}^{c} \| A \circ R_i - A \circ R \|^2 \tag{12.55}$$

for given A in $\mathbf{X}$ or

$$Q = \sum_{l=1}^{N} \sum_{k=1}^{c} \| A_l \circ R_k - A_l \circ R \|^2 \tag{12.56}$$

for a family of A_l's. $\| \cdot \|$ denotes a distance function (in particular, it could come as the Euclidean one). The essence of the reconciliation process for (12.56) is visualized in Figure 12.35. The optimization problem presented in the form (12.11) can be also regarded as the reconciliation of relational models (being expressed by fuzzy relations $R_1, R_2, \ldots, R_c$) being completed in the context of some granular probes (fuzzy sets)

We may relate the above problem to the issue of designing a unified fuzzy model completed on the basis of a collection of given fuzzy models. The underlying concept is illustrated in Figure 12.36. Given is a finite number of distributed data sets $D_1, D_2, \ldots, D_c$, and each of them is used to construct a certain fuzzy relational model $R_1, R_2, \ldots, R_c$. Each fuzzy relation is formed on the basis of the individual data and because D_i's differ from each other it is very likely that though there are

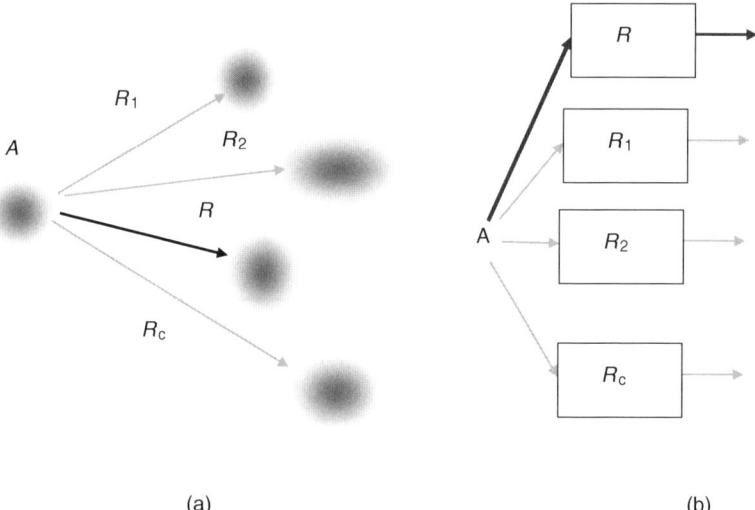

Figure 12.35 Reconciliation of the relational mappings from $\mathbf{X}$ to $\mathbf{Y}$. The result of this process is a certain fuzzy relation R (a). Alternatively one can view this problem as an optimization of the multi-output network composed of fuzzy *or* neurons (b).

12.10 Reconciliation of Perception of Information Granules and Granular Mappings

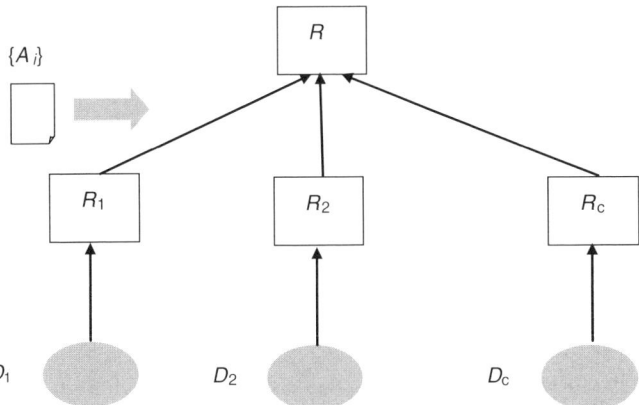

Figure 12.36 Reconciliation of fuzzy models: an origin of the concept and its realization with the use of granular probes $\{A_i\}$.

some commonalities, the resulting models might differ from each other. At the reconciliation phase we are provided with only the fuzzy relations (the data sets could not be shared because of issues of privacy and confidentiality or there could be some technical constraints as well). The role of granular probes $\{A_i\}$ is to invoke the models and then construct the fuzzy relation R on a basis of the responses of the individual models. The minimization of (12.56) produces the fuzzy relation R.

There is an interesting interpretation of this reconciliation process for some class of fuzzy models. Let $\mathbf{X}$ and $\mathbf{Y}$ denote a space of symptoms and diagnosis, respectively. The same fuzzy set of symptoms A leads to different interpretations (diagnoses) depending upon the fuzzy relations $R_1, R_2, \ldots, R_c$. modeling the relational mappings expressed by different experts. The reconciliation is concerned with the development of the fuzzy relation that expresses the relationships between the symptoms and diagnosis.

The minimization of (12.55) and (12.56) can be realized for given once we have agreed upon the form of the composition operator. Similarly the optimization of the fuzzy relation depends upon the fuzzy set(s) available in $\mathbf{X}$.

In what follows, we consider finite spaces $\mathbf{X}$ and $\mathbf{Y}$, $\mathbf{X} = \{x_1, x_2, \ldots, x_n\}$, $\mathbf{Y} = \{y_1, y_2, \ldots, y_m\}$. These cases make sense from a practical standpoint as quite often the spaces of symptoms and diagnoses are of finite dimensionality. Then the fuzzy sets and fuzzy relations can be represented as in vector and matrix form. Considering the general case (12.10), accepting a certain s-t composition of A and R, that is, $S_{i=1}^{n}(A(x_i)tR(x_i, y_j))$ and adopting a Euclidean distance $\|\,.\,\|$, we rewrite the performance index in the following format:

$$Q = \sum_{l=1}^{N} \sum_{k=1}^{c} \| A_l \circ R_k - A_l \circ R \|^2$$

$$= \sum_{l=1}^{N} \sum_{k=1}^{c} \sum_{j=1}^{m} (\overset{n}{\underset{i=1}{S}}(A_l(x_i)tR_k(x_i, y_j)) - \overset{n}{\underset{i=1}{S}}(A_l(x_i)tR(x_i, y_j)))^2 \quad (12.57)$$

The minimization of Q is done through a gradient-based optimization of R, that is, a series of iterations (updates) of the values of R done in successive iterations (i.e., iter and iter+1) that come in the form

$$R(\text{iter} + 1) = R(\text{iter}) - \alpha \nabla_R Q \qquad (12.58)$$

where α is a positive learning rate and denotes a gradient of Q computed with respect to the fuzzy relation. Proceeding with the details, we carry out computations for all elements of the fuzzy relation that leads us to the expression

$$R(x_s, y_t)(\text{iter} + 1) = R(x_s, y_t)(\text{iter}) - \alpha \frac{\partial Q}{\partial R(x_s, y_t)(\text{iter})} \qquad (12.59)$$

$s = 1, 2, \ldots, n; t = 1, 2, \ldots, m.$

The realization of the learning scheme can be completed once the triangular norm and co-norm have been specified. In what follows, we consider the product and probabilistic sum. Then the derivative in (12.59) can be expressed as follows:

$$\frac{\partial Q}{\partial R(x_s, y_t)} = \sum_{l=1}^{N} \sum_{j=1}^{m} \left(\underset{i=1}{\overset{n}{S}} (A_l(x_i) t R_k(x_i, y_j)) - \underset{i=1}{\overset{n}{S}} (A_l(x_i) t R(x_i, y_j)) \right) \qquad (12.60)$$

The derivative in the above expression can be written down in more detailed form

$$\frac{\partial}{\partial R(x_s, y_t)} (B_{l,s,t} + A_l(x_s) R(x_s, y_t) - B_{l,s,t} A_l(x_s) R(x_s, y_t))$$

where

$$B_{l,s,t} = \underset{\substack{i=1 \\ i \neq s}}{\overset{n}{S}} (A_l(x_i) t R(x_i, y_j)) \qquad (12.61)$$

Finally we obtain

$$\frac{\partial}{\partial R(x_s, y_t)} (B_{l,s,t} + A_l(x_s) R(x_s, y_t) - B_{l,s,t} A_l(x_s) R(x_s, y_t)) = A_l(x_s)(1 - B_{l,s,t})$$

$$(12.62)$$

12.11 CONCLUSIONS

Neurocomputing combined with underlying logic fabric builds a unique architecture of fuzzy neurocomputing. We showed that fuzzy neurons with clearly defined semantics give rise to transparent models (Casillas et al., 2003; Dickerson and Lan, 1995; Setnes et al., 1998) whose interpretation results in a certain logical characterization of experimental data. The parametric flexibility coming with the connections of the neurons support all necessary learning abilities. The introduced topologies of the networks lead to the logic–based approximation of data. The unique aggregation of learning and transparency becomes of paramount relevance to the user-centricity of the resulting constructs.

EXERCISES AND PROBLEMS

1. For the fuzzy neural network with feedback shown below.

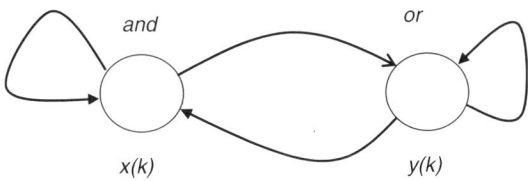

The expressions describing this network are as follows:

$$x(k+1) = and([x(k)y(k)], \mathbf{w})$$
$$y(k+1) = or([y(k)x(k)], \mathbf{v})$$

The values of the connections are $\mathbf{w} = [0.30.1]$ and $\mathbf{v} = [0.90.8]$. Start with $x(0) = 0.3$ and $y(0) = 0.9$. Show the values of $x(k)$ and $y(k)$ in 10 successive discrete time moments.

2. Show that *and* and *or* neurons subsume digital *and* and *or* gates as their special cases.

3. Consider an implication operator induced by Lukasiewicz *and* connective, that is, $\min(1, a + b - 1)$. Show that the similarity measure realized with its use realizes a Hamming distance.

4. Discuss how you could model the tolerance relationship, $tol(x, a, b)$, which returns *1* if x is located in–between a and b, $a < b$ and becomes a monotonically decreasing function for the arguments below "a" and above "b." In your construct use the predicates of inclusion and dominance combined with the use of some t-norm. Next use this predicate in the development of a tolerance neuron.

5. With the use of the tolerance neuron constructed in the previous problem, show how you would model the region of the unit square illustrated in the figure below. The shadowed region comes with the value of 1 whereas the values assumed outside this region are equal to 0.

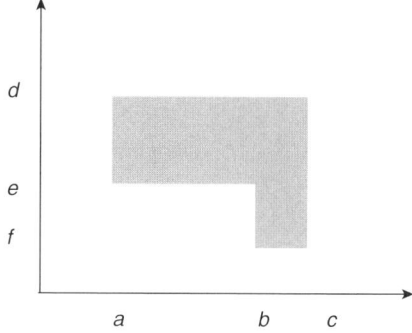

6. Given is the following fuzzy neural network:

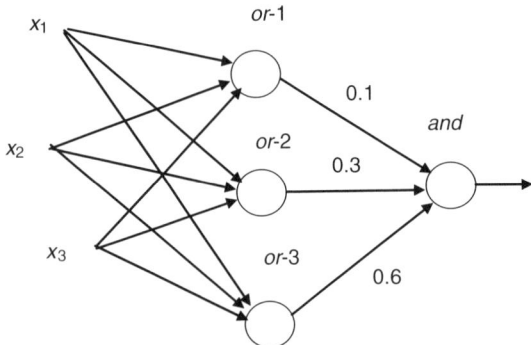

The connections of the *or* neurons (ordered from x_1 to x_3) are the following:

$$\text{or-1}: [0.1\,1.0\,0.7], \quad \text{or-2}: [0.9\,0.7\,0.1], \quad \text{or-3}: [0.8\,0.7\,0.9].$$

Proceed with its reduction by eliminating the weakest connections. Develop a relationship between the number of eliminated connections and the reduced performance. To express performance, consider combinations of input data uniformly distributed in each input at a step of 0.2.

7. Consider a network composed of 4 sensors providing data about the status of a system. The readings of the sensor are located in the unit interval. Propose a warning system that evaluates an overall status of the network (expressed in the unit interval) given the following constraints existing there:

(a) variable x_1 should not exceed threshold "a"
(b) variable x_2 should not fall below threshold "b"
(c) variable x_3 should be similar to the setpoint "c"
(d) variable x_4 should not exceed threshold "d."

The above variables are measured by sensors 1–4. The variables themselves contribute to the overall evaluation of the status of the network to a different extent. The contributions are quantified by running a pairwise comparison leading to the following reciprocal matrix:

$$\begin{bmatrix} 1 & 7 & 5 & 2 \\ 1/7 & 1 & 4 & 3 \\ 1/5 & 1/4 & 1 & 2 \\ 1/2 & 1/3 & 1/2 & 1 \end{bmatrix}$$

The more evident is the violation of the constraints, the higher is the value of the output of the warning system.

8. A two-input fuzzy multiplexer (2–to–1 FMUX) forms a generalization of the well-known functional units used in digital systems is described by the following expression:

$$y = \text{FMUX}(x; c_0, c_1) = (c_0 \, t \, \bar{x})s \, (c_1 \, t \, x)$$

where "x" serves as a switching variable and c_0 and c_1 are two input variables. See also the figure below

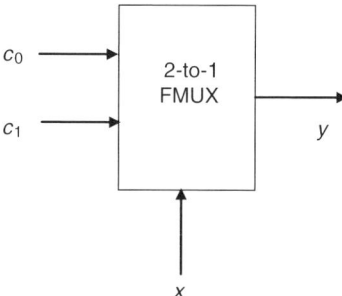

(a) Discuss how FMUX functions; in particular, assume the values of the switching variable to be equal to 0 and 1
(b) Plot the input-output characteristics of the FMUX treating the output (y) as a function of "x" and "c_0"; take $c_1 = 1 - c_0$.
(c) How would you realize switching of four inputs? discuss the use of 2-to-1 FMUXs.

9. Rule-based systems can be represented in the form of a fuzzy neural tree. There are two commonly encountered structures shown below. Interpret them by discussing what type of rules they represent and how the learning in these structures becomes reflected in the calibration of the rules.

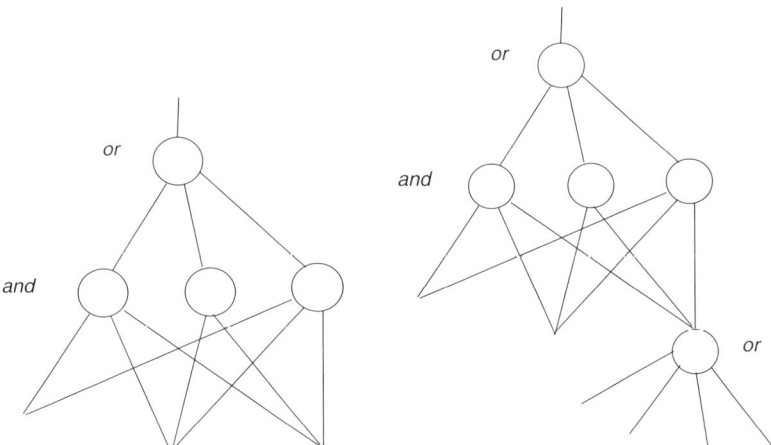

HISTORICAL NOTES

Fuzzy neural networks constitute a general category of logically oriented processing units with the research on this subject originated in Pedrycz (1991a,b, 1993, 2004), Pedrycz and Lam (1995), Pedrycz and Reformat (2005). With regard to the framework of logic processing, one could refer to an interesting note made by J. von Neumann (1958)

> ... we have seen how this leads to a lower level of arithmetical precision but to a higher level of logical reliability: a deterioration in arithmetic has been traded for an improvement in logic...

While the neurofuzzy processing realized in terms of fuzzy neural computing provides a wealth of architectures and learning schemes and helps implement a variety of fuzzy rule-based systems, it could also brings some insight into the development of digital systems through learning. Fuzzy neural networks can be also sought as a direct generalization of the fuzzy relational structures (fuzzy relational equations). In particular, fuzzy relational equations with the s–t composition are examples of *or* neurons whereas the dual fuzzy relational equations (governed by the t-s composition) give rise to *and* neurons.

REFERENCES

Angelov, P. A fuzzy controller with evolving structure, *Inf. Sci.*, **161**(1–2), 2004, 21–35.

Casillas, J. et al. (eds.), *Interpretability Issues in Fuzzy Modeling*, Springer-Verlag, Berlin, 2003.

Ciaramella, A., Tagliaferri, R., Pedrycz, W. The genetic development of ordinal sums, *Fuzzy Set Syst.*, **151**(2), 2005, 303–325.

Ciaramella, A., Tagliaferri, R., Pedrycz, W., Di Nola, A. Fuzzy relational neural network, *Int. J. Approx. Reason.*, **41**(2), 2006, 146–163.

Dickerson, J., Lan, M. Fuzzy rule extraction from numerical data for function approximation, *IEEE T. Syst. Man Cy. B* **26**, 1995, 119–129.

Golden, R. *Mathematical Methods for Neural Network Analysis and Design*, MIT Press, Cambridge, 1996.

Hirota, K., Pedrycz, W. OR/AND neuron in modeling fuzzy set connectives, *IEEE T. Fuzzy Syst.*, **2**, 1994, 151–161.

Hirota, K., Pedrycz, W. Fuzzy relational compression, *IEEE T. Syst., Man Cy. B*, **29**, 1999, 407–415.

Jang, J., Sun, C., Mizutani, E. *Neuro-Fuzzy and Soft Computing*, Prentice Hall, Upper Saddle River, NJ, 1997.

Kosko, B. *Neural Networks and Fuzzy Systems*, Prentice Hall, Englewood Cliffs, NJ, 1991.

Michalewicz, Z. *Genetic Algorithms + Data Structures = Evolution Programs*. Springer-Verlag, Heidelberg, 3rd ed., 1996.

Mitra, S., Pal, S. Logical operation-based fuzzy MLP for classification and rule generation, *Neural Networks* **7**, 1994, 353–373.

Mitra, S., Pal, S. Fuzzy multiplayer perceptron, inferencing and rule generation, *IEEE Trans. on Neural Networks* **6**, 1995, 51–63.

Nobuhara, H., Hirota, K., Sessa, S., Pedrycz, W. Efficient decomposition methods of fuzzy relation and their application to image decomposition, *Appl. Soft Comput.* **5**(4), 2005, 399–408.

Nobuhara, H., Pedrycz, W., Sessa, S., Hirota, K. A motion compression/reconstruction method based on max t-norm composite fuzzy relational equations, *Inf. Sci.*, **176**(17), 2006, 2526–2552.

Pal, S., Mitra, S. *Neuro-Fuzzy Pattern Recognition*, John Wiley & Sons, Inc., New York, 1999.

Pedrycz, W. Processing in relational structures: fuzzy relational equations, *Fuzzy Set Syst.*, **40**, 1991a, 77–106.

Pedrycz, W. Neurocomputations in relational systems, *IEEE T. Pattern anal.* **13**, 1991b, 289–297.

Pedrycz, W. Fuzzy neural networks and neurocomputations, *Fuzzy Set Syst.*, 56, 1993, 1–28.

Pedrycz, W., Rocha, A. Knowledge-based neural networks, *IEEE T. Fuzzy Syst.*, **1**, 1993, 254–266.

W. Pedrycz, Lam, P., Rocha, A. F. Distributed fuzzy modelling, *IEEE T. Syst. Man Cyb. B*, **5**, 1995, 769–780.

Pedrycz, W., Gomide, F. *An Introduction to Fuzzy Sets: Analysis and Design*, MIT Press, Cambridge, MA, 1998.

Pedrycz, W. Heterogeneous fuzzy logic networks: fundamentals and development studies, *IEEE T. Neural Networ.*, **15**, 2004, 1466–1481.

Pedrycz, W., Reformat, M. Genetically optimized logic models, *Fuzzy Set Syst.*, **150**(2), 2005, 351–371.

Setnes, M., Babuska, R., Vebruggen, H. Rule-based modeling: precision and transparency, *IEEE T. Syst. Man Cyb. C* **28**, 1998, 165–169.

Chapter 13

Fuzzy Systems and Computational Intelligence

Fuzzy systems are encountered in numerous areas of applications. They have a long history of development with a diversity of conceptual and algorithmic enhancements. Fuzzy rules, for instance, viewed as a generic mechanism of granular knowledge representation, are positioned in the center of knowledge-based systems. Knowledge representation (including aspects of dealing with granular information) and knowledge processing in general constitute the agenda of fuzzy systems. Neural networks and neurocomputing support quite a different paradigm of information processing that stresses aspects of effective supervised and unsupervised learning, distributed processing, and inherent plasticity of the underlying architectures of neural networks. The agenda of evolutionary computing is aimed at the effective utilization of biologically inspired optimization. The massive parallelism and population-based stochastic search techniques are at the heart of evolutionary systems. Even such a brief characterization of fuzzy systems, neurocomputing, and evolutionary systems helps us not only acknowledge the fundamental differences between these technologies but also appreciate possibilities associated with the development of hybrid systems that take advantage of the unified use of all of them. In this chapter, we discuss the main features of generic models of neurocomputing and evolutionary, biologically inspired systems. Subsequently, we elaborate on various design schemes of their hybrid architectures that give rise to a broad category of neural fuzzy networks and genetic fuzzy systems. Neural fuzzy networks offer an environment in which we seamlessly combine significant learning capabilities of neural networks with the mechanisms of approximate reasoning and logic, and transparent processing inherent to fuzzy systems. Genetic fuzzy systems deliver a powerful tool to develop and design fuzzy and neural fuzzy systems using some prior knowledge and experimental data. Various architectures, learning algorithms, and examples are discussed along with their applications in function approximation, classification, and dynamic system modeling.

Fuzzy Systems Engineering: Toward Human-Centric Computing, by Witold Pedrycz and Fernando Gomide
Copyright © 2007 John Wiley & Sons, Inc.

13.1 COMPUTATIONAL INTELLIGENCE

Computational intelligence (CI) is a field of intelligent information processing related with different branches of computer science and engineering. The core of CI embraces fuzzy systems, neural networks, and evolutionary computation. The ultimate agenda of CI deals with various ways in which these technologies are brought together to constitute some form of hybrid architectures, refer to Figure 13.1. Appendices B and C provide a concise overview of the fundamentals of neural networks and biologically inspired computing.

Growing as a stand-alone field in itself, CI nowadays contains evolving systems (Angelov, 2002) and swarm intelligence (Eberhart et al., 2001; Dorigo and Stutzle, 2004), immune systems (Castro and Timmis, 2002), and other forms of natural (viz. biologically inclined) computation. A key issue in CI is adaptation of behavior as a strategy to handle changing environments and deal with unforeseen situations. CI exhibits interesting links with machine intelligence (Mitchell, 1997), statistical learning (Tibshirani, et al., 2001) and intelligent data analysis and data mining (Berthold and Hand, 2006 Dunham, 2003), pattern recognition (Bishop, 2006) and classification (Duda et al., 2001), control systems (Dorf and Bishop, 2004) and operations research (Hillier and Lieberman, 2005). The name Computational Intelligence has been around in the literature since the 1990s.

At that time, the term Computational Intelligence was coined by Bezdek (1992, 1994). In his view, CI concerns data processing systems that come with capabilities of pattern recognition and adaptive properties, are fault tolerance, and whose performance approximates human performance in human time processing scale with no use of knowledge in the sense as being considered in the realm of Artificial Intelligence (Bezdek, 1994). Later on, CI was seen as a comprehensive framework to

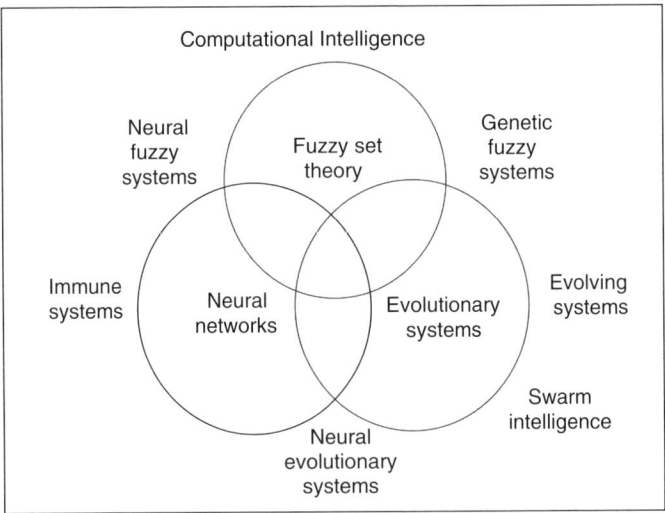

Figure 13.1 The paradigm of CI.

design and analyze intelligent systems with a focus on all fundamentals of autonomy, learning, and reasoning (Duch, 2007). The idea is to consider computing systems that are able to learn and deal with new situations using reasoning, generalization, association, abstraction, and discovery capabilities (Eberhart et al., 1996). In dealing with this grand challenge, CI underlines the need for a full *synergy* between neurocomputing, fuzzy sets (or granular computing, in general), and biologically inspired optimization in the development of intelligent systems (Pedrycz, 1997). Thus, it is not surprising to see a wealth of hybrid systems constructed within the realm of CI. The names neurofuzzy systems, genetic fuzzy systems, and evolutionary neurocomputing are commonly encountered constructs of CI.

In this quest for the construction of intelligent systems, CI shows some conjecture points with Artificial Intelligence agent systems (Russell and Norvig, 2003), control systems, and operations research, as in all these endeavors the constructed computationally intelligent architectures have to deal with effective processes of prediction and decision making. CI is largely human-centered because it relies on humans to be built and humans benefit from computationally intelligent systems to safely achieve goals and solutions. In long range, CI and other forms of artificial and synthetic intelligence aim to develop intelligent systems with comparable human performance from the point of view of collaboration between humans and machines in problem solving and information processing. Affronimation to human performance should be achieved by intelligent systems when performing tasks in which machines replace humans, for example, in hazardous environments. To pave the path toward intelligent systems, CI may benefit from contributions coming from systems science, cognitive sciences, and computational semiotics to improve performance when processing signals, perceptions, and meaning. Figure 13.2 offers some summarization with this regard.

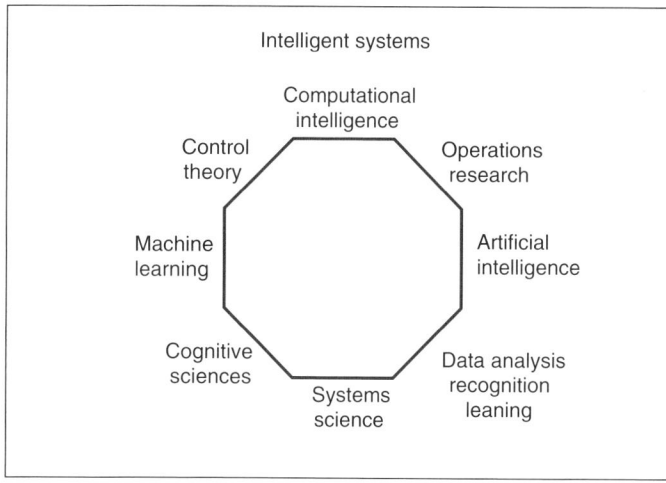

Figure 13.2 Intelligent systems as a collaborative framework of human- and machine-oriented areas.

13.2 RECURRENT NEUROFUZZY SYSTEMS

A class of neural network architectures that has been receiving a great deal of attention during the past years concerns recurrent neural networks. Such networks can be classified as globally or partially recurrent, depending on the feedback connections existing in their architectures. Feedback connections allow to endow the network with memory capabilities, which are essential in processing spatiotemporal information (Haykin, 1998; Santos and Von Zuben, 1999).

Recurrent neural networks perform well in a wide range of applications such as dynamic system modeling, adaptive control, processing of temporal signals, forecasting, and speech recognition. However, a thorough analysis and synthesis of such neural networks becomes a complex task. In particular, in their supervised learning, the process of weight (connection) adjustment is far more demanding when compared with learning algorithms encountered in static (viz. feedforward) networks. Usually gradient learning for recurrent network algorithms is complex and slow (Lee and Teng, 2000; Mandic and Chambers, 2001).

Neural fuzzy networks have emerged as hybrid constructs bringing together the advantages of fuzzy sets, stressing their evident transparency and abilities of processing granular information and benefits of neurocomputing reflected in the plasticity and learning capabilities of the resulting constructs. A major limitation of most fuzzy neural networks comes with their restricted applicability to modeling dynamic mappings. As indicated, this drawback could be attributed to either prevalent feedforward topologies of such networks or a lack of efficient learning procedures for developing feedback connectivity. Recurrent neural fuzzy systems attempt to alleviate these problems (Nürnberger et al., 2001).

In what follows, we discuss recurrent hybrid neural fuzzy networks and present their development with the use of supervised learning. The network structure has a recurrent layer that implicitly performs fuzzy inference followed by an aggregation neural network. The required dynamics of the underlying system—temporal relations—are captured by forming a global feedback in the hidden layer nodes. This modification equips the network with a memory mechanism and hence expands the ability of the neural fuzzy network to store temporal representations. The neural fuzzy network comes as a generalization of the recurrent neural fuzzy structure presented by Ballini et al. (2001) and derived from its feedforward counterpart (Caminhas et al., 1999). The recurrent neural fuzzy network uses aggregative logic neurons of the type *and* and classical neural network with nonlinear neurons as an adaptive aggregation layer.

13.2.1 Recurrent Neural Fuzzy Network Model

The neural network exhibits a multilayer architecture as illustrated in Figure 13.3. There is a feedback loop realized within the *and* neurons located at the second layer of the architecture (Ballini et al., 2001; Hell et al., 2007).

13.2 Recurrent Neurofuzzy Systems

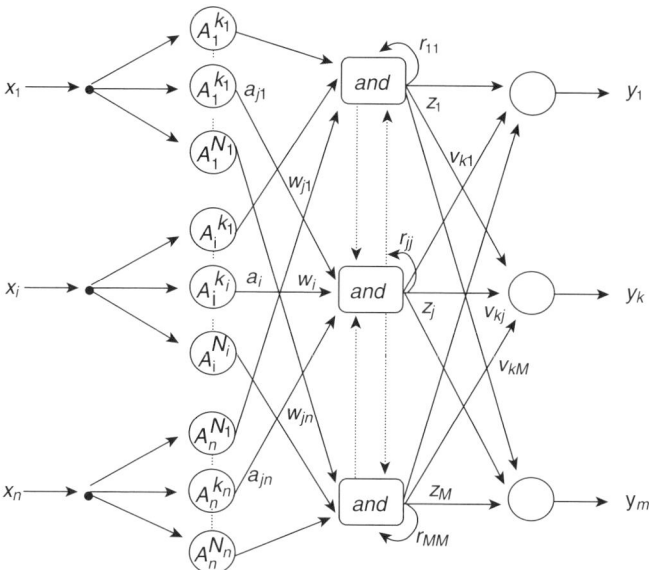

Figure 13.3 An architecture of the recurrent neural fuzzy network.

Let us now discuss the functionality of the fuzzy inference system realized by the input and hidden layer. The input layer consists of neurons whose activation functions are membership functions of fuzzy sets that form the input space partition. For each component $x_i(t)$ of an n-dimensional input vector $\mathbf{x}(t) = [x_1(t), x_2(t), \ldots, x_n(t)]$ there are N_i fuzzy sets $A_i^{k_i}$, $k_i = 1, \ldots, N_i$, whose membership functions are the activation functions of the corresponding input layer neurons. t denotes discrete time instances, that is, $t = 1, 2, \ldots$ The outputs of the first layer are the membership degrees of the associated input variables, that is, $a_{ji} = A_i^{k_i}(x_i)$, $i = 1, 2, \ldots, n$, and $j = 1, 2, \ldots, M$, the number of neurons located in the second layer of the network. The second layer is constructed with the use of fuzzy neuron of *and* type, where the output of the neuron $z : [0, 1]^n \rightarrow [0, 1]$ is a generalized *and* logic expression of its inputs, with each input calibrated by means of the corresponding connection (weight). In contrast to the standard realization of the *and* neuron, the one here, see Figure 13.4, comes with a feedback loop.

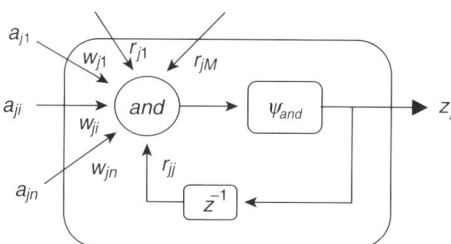

Figure 13.4 Recurrent *and* fuzzy neuron.

In general, in the structure shown in Figure 13.3, we assume the most general case in which there is a global feedback loop between all fuzzy neurons of the second layer. Neural networks with partial feedback and with no recurrent connections are particular instances. The *and* neurons operate on inputs a_{ji} weighted by w_{ji}, and on feedback connections weighted by $r_{j1,j} = 1, \ldots, M$. The activation function ψ_{and} is, in general, a nonlinear mapping, but here we assume it to be the identity function $\psi_{and}(u) = u$. If we assume $a_{jn+l} = z^{-1}z_j$ where z^{-1} is the delay operator and $w_{jn+l} = r_{jl,j} = 1, \ldots, M$ then the net structure encodes represent a set of "If–then" rules $R = \{R_j, j = 1, \ldots, M\}$ of the following form:

R_j: If x_1 is $A_1^{k_1}$ or $w_{j1} \ldots$ and x_i is $A_i^{k_i}$ or $w_{ji} \ldots$ and x_n is $A_1^{k_n}$ or w_{jn} then z is z_j with certainty v_{kj}

where $z_j = w_{j1} s a_{j1} \ldots w_{ji} s a_{ji} s a_{jp} = \underset{i=1}{\overset{p}{T}}(w_{ji}\, s a_{ji})$, $p = M+n$, and s denote some t-conorm.

The output layer contains m nonlinear neurons shown in Figure 13.5.

Here, the output y_k of the k-th neuron, $k = 1, \ldots, m$, is a nonlinear function of its inputs z_j weighted by $v_{kj}, j = 1, \ldots, m$. We assume that the nonlinear activation function ψ of neurons is monotonically increasing. The third-layer neurons perform a form of nonlinear aggregation function with weights found through learning. The neural fuzzy network dynamics works as follows:

1. N_i is the number of fuzzy sets that granulate the ith input;
2. The j indexes the *and* neurons; from the net structure, j is determined from indices k_i as follows:

$$j = k_n + \sum_{i=2}^{M}(k_{(n-i+1)} - 1)\left(\prod_{r=1}^{i-1} N_{(n+1-r)}\right)$$

3. $x_1, \ldots x_i, \ldots, x_n$ are the inputs of the network; $\mathbf{x} = [x_1, \ldots, x_i, \ldots, x_n]$;
4. $a_{ji} = A_i^{k_i}(x_i)$ is the membership degree of x_i in $A_i^{k_i}$, whose output is an input of the jth neuron in the second layer;
5. w_{ji} is the connection weight between the ith input and the jth *and* neuron;
6. z_j is the output of the jth *and* neuron determined by

$$z_j = \underset{i=1}{\overset{n+M}{T}}(w_{ji}\, s a_{ji})$$

7. v_{kj} is the weight for the jth input of the kth output neuron, and r_{jl} is the feedback connection of the lth input of the jth neuron;

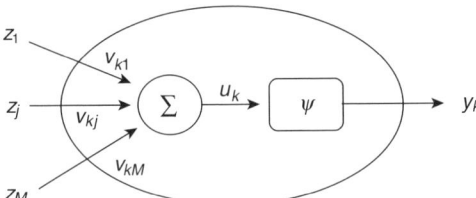

Figure 13.5 Neurons of the output layer.

8. y_k is the output of the kth neuron of the output layer:

$$y_k = \psi(u_k) = \psi\left(\sum_{j=1}^{M} v_{kj} z_j\right)$$

where $\psi: \mathbf{R} \to [0,1]$ is a nonlinear monotonically increasing function such as $\psi(u) = 1/(1 + \exp(-u))_n$

13.2.2 Learning Algorithm

The learning of the neural fuzzy network involves three phases. The first phase uses a clustering procedure based on a modification of the vector quantization approach to granulate the input universes. The second phase constructs the network connections and sets initial values for the weights randomly within [0,1]. In the third phase, we uses gradient descent and associative reinforcement learning to update weights using learning data. Overall, the learning scheme of the recurrent neural fuzzy network is supervised and can be summarized as follows:

procedure NET-LEARNING (**x**, **y**) **returns** a network
input : data **x, y**
local : fuzzy sets
        t,s: triangular norms
        ϵ: threshold

        GENERATE-MEMBERSHIP-FUNCTIONS
        INITIALIZE-NETWORK-WEIGHTS
until   stop criteria $\leq \epsilon$ **do**
        choose an input output pair x and y of the data set
        ACTIVE-AND-NEURONS
        ENCODING
        UPDATE-WEIGHTS
return a network

Typically, the procedure stops when either a specified performance level is reached or a maximum number of iterations has been exceeded. The steps of the learning procedure are shown in detail.

    1. Generation of membership functions

The simplest way to construct membership functions is to assume triangular functions with universes $[x_{i\min}, x_{i\max}]$, $i = 1, \ldots, n$, which overlap with the neighboring fuzzy sets at the level of 0.5. However, uniform partitions may not be adequate if data are concentrated in certain regions of the feature space. In these cases non-uniform partitions, see Figure 13.6, are more suitable.

    One of the possible mechanisms of partitioning the input space is to cluster the training data and treat the resulting centers of the clusters as the modal values of the

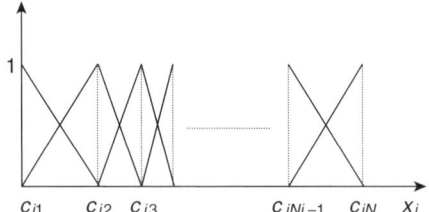

Figure 13.6 Nonuniform partition with triangular membership functions.

membership functions. Here we adopt a certain variation of vector quantization suggested in Caminhas et al. (1999). The essence of this modification is as follows. Clustering is done using a neural network structure whose weights are cluster centers c_{ir}. Initially, the number of output units of the clustering neural network N_i^0 gives an estimation of the number of fuzzy sets that assembles the partition of universe of the ith coordinate (variable). In general, N_i^0 is overestimated and updated via learning. Neurons that rarely win, as quantified by a performance index, are removed. The algorithm is as follows:

procedure GENERATE-MEMBERSHIP-FUNCTIONS (**x**) **returns** membership functions
input : learning data **x**
local : learning rate: α
thresholds: $\epsilon, \delta > 0$

initialization: weights c_{ir}:

$$c_{i1} = x_{i\,min}$$
$$c_{ir} = c_{i(r-1)} + \Delta_i, \text{ for } r = 2, 3, \ldots, N_i^0$$
$$\Delta_i = \frac{x_{i\,max} - x_{i\,min}}{N_i^0 - 1}$$

set performance index $id_i(r) = 0$, for $r = 1, 2, \ldots, N_i^0$.
set membership functions to form a uniform partition

until $|c_{ij}(k+1) - c_{ij}(k)| \leq \varepsilon, \forall j$ **do**
input x_i and update the winning neuron weights as follows:

$$c_{iL}(k+1) = c_{iL}(k) + \alpha(k)[x_i - c_{iL}(k)]$$

L is the winner neuron index $L = \arg\{\min_r |x_i - c_{ir}|\}$
Decrease the step size $\alpha(k)$
Update the performance index of the winner neuron $Id_i(L) = Id_i(L) + 1$

prune all neurons for which $Id_i \leq \delta$.

Let Nne_i be the number of neurons removed for the ith dimension.
Set the number of fuzzy sets of the ith dimension partition as
$N_i = N_i^0 - Nne_i$

Set
$$A_i^r(x_i) = \begin{cases} \alpha_{eir}(x_i - c_{ir}) + 1, & \text{if } c_{ir-1} \leq x_i \leq c_{ir} \\ \alpha_{dir}(x_i - c_{ir}) + 1, & \text{if } a_{ir} < x_{pi} \leq a_{ir+1} \\ 0, & \text{otherwise} \end{cases}$$

$$\alpha_{eir} = \frac{1}{c_{ir} - c_{ir-1}}, \alpha_{dir} = \frac{1}{c_{ir} - c_{ir+1}}, i = 1, \ldots, n \text{ and } r = 1, \ldots, N_i.$$

return c_{ir}

2. Neural fuzzy network weights initialization

The first and second layer of the network are fully connected. Weights w_{ji}, v_{kj} and r_{jl}, for $i = 1, \ldots, n$; $j, l = 1, \ldots, M$ and $k = 1 \ldots, m$ are initialized to some random values in [0,1]. The detailes are presented below.

procedure INITIALIZE-NETWORKS-WEIGHTS **returns** weights.
input: none
local: weights
    t, s: triangular norms

for $i = 1:n$ **do**
  for $j = 1:M$ **do**
    $l \leftarrow j$
    for $k = 1:m$ **do**
      $w_{jk} \leftarrow \text{random}([0, 1])$
      $v_{kj} \leftarrow \text{random}([0, 1])$
      $r_{jl} \leftarrow \text{random}([0, 1])$

return w_{jk}, v_{jk}, r_{jl}

3. Determination of active *and* neurons

By construction, for each input there are at most two nonzero membership degrees for each of its dimension, see Figure 13.6. The corresponding membership functions of the input space partition define the active *and* neurons. They are identified as follows:

Given the input $\mathbf{x} = [x_1, \ldots, x_i, \ldots, x_n]$, let $\mathbf{K}^1 = (k_1^1, \ldots, k_i^1, \ldots, k_n^1)$ be a vector whose components are the indices of the first membership function of each dimension i for which the membership degree is different from zero. Let $\mathbf{K}^2 = (k_1^2, \ldots, k_i^2 \ldots, k_n^2)$ be a vector whose entries are defined as follows:

$$k_i^2 = \begin{cases} k_i^1 + 1, & \text{if } A_i^{k_i^1} \neq 1 \\ k_i^1, & \text{otherwise} \end{cases}$$

The number of active *and* neurons is given as N_a, where $N_a = 2^{P_a} \leq 2^n$, and P_a is the number of elements such that $k_i^1 \neq k_i^2, i = 1, \ldots, n$. Notice that among the M *and* neurons of the second layer, only $2^{P_a} \leq M, P_a \leq n$ are active. Also, because only N_a *and* neurons are active each time an input is presented, the network processing time becomes independent of the number of fuzzy sets in the input space partition. The process of identifying active neurons is summarized as follows:

procedure ACTIVE-AND-NEURONS (input partition, **x**) **returns** active neurons
input: input data **x**
    input partition

  for each dimension of input partition **do**
    compute $\mathbf{K}^1 = (k_1^1, \ldots, k_i^1, \ldots, k_n^1)$
    compute $\mathbf{K}^2 = (k_1^2, \ldots, k_i^2, \ldots, k_n^2)$
  return indices of active neurons

4. Encoding (fuzzification)

This step is straightforward once we note that only membership degrees of active fuzzy sets for which indices in $\mathbf{K}^1$ are computed. If $k_i^1 \neq k_i^2$, then $A_i^{k_i^2}(x_i)$ is immediately found, given that $A_i^{k_i^2}(x_i) = 1 - A_i^{k_i^1}(x_i)$. Therefore, only $2n$ membership degrees are computed.

5. Weight Updating

Weight updating is done in supervised mode and is based on gradient optimization and associative reinforcement learning (Barto and Jordan, 1987). Gradient learning is used to adjust the weights of the output-layer neurons and associative reinforcement learning to adjust the weights of the neurons in the second layer. The first step is to compute the network output for an input **x**. This corresponds to the encoding of the input and the successive forward computation of the outputs of the remaining neuron layers. Next, a supervised learning process is used to minimize an error measure between the network output and the desired output over the set of the learning input–output pairs, that is, to minimize

$$e = \frac{1}{2}\sum_{k=1}^{m}(y_k - \hat{y}_k)^2$$

where $\hat{y}_k$ is the value of the output unit k and y_k is the desired value for the corresponding input vector **x**. Therefore, using a gradient-descendent method, if v_{kj} is the weight connecting the second-layer neuron j to the output-layer neuron k, then

$$\Delta v_{kj} = \eta_k(y_k - \hat{y}_k)\psi'(u_k)z_j$$

If $\psi(u) = 1/(1 + \exp(-u))$ then $\psi'(u_k) = \psi(u_k)(1 - \psi(u_k))$ is the derivative of the activation functions evaluated at $u_k = \sum_{j=0}^{M}(v_{kj}z_j)$ and η_k is the learning rate. If $\psi(u_k) = u_k$, then $\psi'u_k = 1$. The weights of the *and* neurons are updated using a reinforcement signal sent to all *and* units. The updates of the connections of the neuron depends on the values of the reinforcement signal.

Here, we adopt a reinforcement signal $\delta = 1 - e$, which is similar to the one suggested in Barto and Jordan (1987). Large values of δ correspond to the better match occurring between the values of the current output of the network and the desired output. To update the weights of the *and* logic neurons, an update rule based on a reward and punishment scheme (Ballini et al., 2001) is utilized:

13.2 Recurrent Neurofuzzy Systems

$$\Delta w_{ji} = \delta \alpha_1 [1 - w_{ji}] - (1 - \delta) \alpha_2 w_{ji}$$
$$\Delta r_{jj} = \delta \alpha_3 [1 - r_{jj}] - (1 - \delta) \alpha_4 r_{jj}$$
$$0 < \alpha_1 \ll \alpha_2 < 1 \text{ and } 0 < \alpha_3 \ll \alpha_4 < 1$$

where α_1, α_2, α_3 and α_4 are learning rates, $j = 1, \ldots, M$ and $i = 1, \ldots, n$.

procedure UPDATE-WEIGHTS (**x,y**) **returns** weights
input : input / output data (**x,y**)
local : learning rates: $\eta_k, \alpha_1, \alpha_2, \alpha_3, \alpha_4$
       threshold: ε

    until $e \leq \varepsilon, \forall_j$ **do**
        input **x** and compute network outputs
        compute error e
        update output layer weights

$$v_{kj} \leftarrow v_{kj} + \eta_k (y_k - \hat{y}_k) \psi'(u_k) z_j$$

update T-neuron weights

$$w_{ji} \leftarrow w_{ji} + \delta \alpha_1 [1 - w_{ji}] - (1 - \delta) \alpha_2 w_{ji}$$
$$r_{jj} \leftarrow r_{jj} + \delta \alpha_3 [1 - r_{jj}] - (1 - \delta) \alpha_4 r_{jj}$$

    return w_{ij}, r_{jj}

EXAMPLE 13.1

We show the performance of the recurrent neurofuzzy network (NFN) for a set of data generated by an NH_3 laser, a chaotic time series whose behavior has characteristics similar to the integration of Lorenz equation, see Figure 13.7. NH_3 time series data have been used to demonstrate the performance of several forecasting models (Weigend and Gershenfeld, 1992). To test the predication capabilities of the model, the first 1000 samples are used as a testing set. Our task is to predict the next 100 steps.

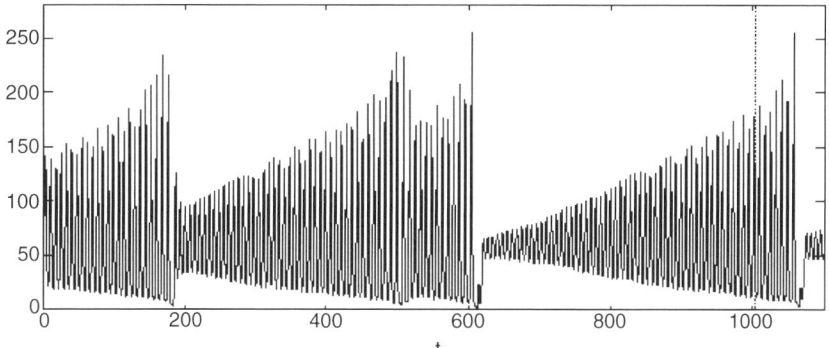

Figure 13.7 NH_3 laser time series data.

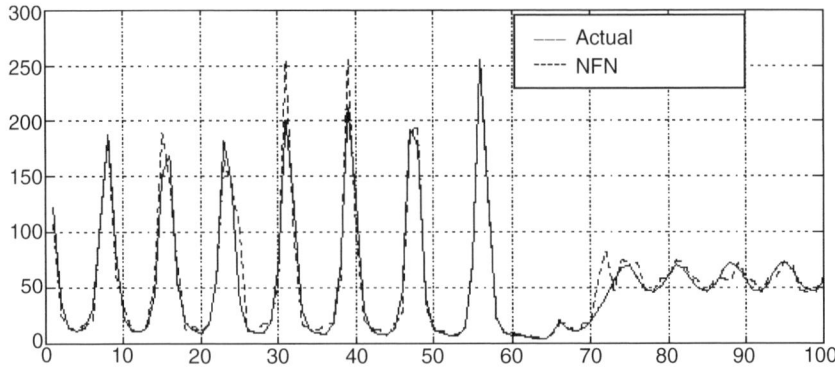

Figure 13.8 NH$_3$ laser time series with 100 step ahead prediction.

Table 13.1 NSE for the NH$_3$ Laser Time Series.

Model	One step ahead	100 steps ahead
FIR	0.0230	0.0551
NFN	0.0139	0.0306

The neural fuzzy network has two inputs, $x(t)$ and $x(t-1)$, one output $x(t+1)$, and uses five fuzzy sets to partition the input space, that is, $N_1 = N_2 = 5$. Therefore, the network comes with 25 *and* fuzzy neurons located in the second layer, one neuron in the output layer, altogether having at most 675 connections. The t-norm and t-conorm are the algebraic product and the probabilistic sum, respectively. The learning rates were set up to be $\alpha_1 = \alpha_3 = 0.01$ and $\alpha_2 = \alpha_4 = 0.001$. The learning process took 5000 iterations to converge. The prediction results are shown in Figure 13.8. To provide some comparative view of the performance of this network, it was contrasted with a feedforward neural network (FIR) with two hidden layers and tapped delay lines that

according to Weigend and Gershenfeld (1992), the one that performed best. The neural network uses 12 neurons in each layer and tapped delay synapses of the order 25, 5, and 5. A backpropagation-like learning procedure adjusts 1080 connections. The normalized squared error (NSE) criterion is adopted for comparison, namely,

$$\text{NSE} = \frac{1}{\sigma^2 N} \sum_{i=1}^{N} (y_i - \hat{y}_i)^2$$

where y_i is the ith actual value, $\hat{y}_i$ is the forecasted value, and σ^2 is the variance computed for the prediction steps $N = 100$. The obtained results are presented in Table 13.1 and Figure 13.8.

13.3 GENETIC FUZZY SYSTEMS

Now we consider a coevolutionary genetic approach to the development of fuzzy functional models. The approach views species as distinct populations of individuals who represent distinct fuzzy constituents of the fuzzy models organized into four hierarchical levels. Individuals at each hierarchical level encode membership functions, rules, rule bases, and fuzzy systems. Individuals are evaluated using a fitness-sharing mechanism. Constraints are observed and particular targets are defined throughout the hierarchical levels, with the purpose of promoting the occurrence of valid individuals and inducing rule compactness, rule base consistency, and partition set visibility. The performance of the approach is evaluated via an example of function approximation with noisy data and a nonlinearly separable classification problem.

In many applications, the information required to develop fuzzy knowledge bases may not be easily available and humans may be unable to extract all relevant knowledge from large amount of numerical data. In these circumstances, computational procedures must be used to extract knowledge and encode it in fuzzy rule bases of fuzzy rule-based models. In fuzzy rule-based modeling, it is essential to trade-off accuracy and interpretability (Setmer et al., 1998). Accuracy is important when developing models for dynamic systems. Interpretability is important to assure basic model characteristics such as consistency, compactness, and visibility. Compactness can be characterized by the number of fuzzy rules in the rule base. Consistency requires the absence of conflicting rules such as rules with the same antecedents, but with different consequents. Visibility is closely related with the granulation of universes and is distinguished by partitions with no gaps and full overlapping (Delgado et al., 2003).

Genetic fuzzy systems (GFS) constitute a powerful approach to the design of fuzzy models. In essence, they are fuzzy systems augmented by learning mechanisms based on genetic algorithms (Cordon et al., 2001). Typically, GFS approaches use evolutionary techniques to adjust the most essential components of fuzzy models such as membership functions, fuzzy rules, and a structure of rule bases. Membership functions, rule base structure and type and number of rules, rule antecedent aggregation, rule aggregation and operators using fuzzy inference procedures affect performance of fuzzy models. Algorithms that simultaneously learn structure and parameters of fuzzy models from data become a key component in GFS.

Evaluation of complete solutions in isolation is one of the features of traditional evolutionary approaches. Because interactions between population members are not handled, there is no evolutionary pressure for co-adaptation and this is inadequate to develop complex models (Potter and De Jong, 2000). In these cases, it is more appropriate to coevolve individuals of different populations, hierarchically structured into levels that represent partial solutions to the problem. The hierarchical structure may be such that individuals of different populations keep collaborative relationships and individual fitness depending on the fitness of individuals of the other populations.

Here a coevolutionary approach to design fuzzy functional models is described (Delgado et al., 2004). A parsimonious functional fuzzy model emerges from a hierarchical and collaborative coevolutionary process. Populations of four hierarchical levels are partial fuzzy models. Individuals of level I encode membership functions, individuals of level II encode fuzzy rules, level III individuals encode rule bases, and individuals of level IV encode fuzzy systems. Fitness of an individual at one level depends on the fitness of individuals at higher hierarchical levels. The approach uses a least square method to compute the parameters of the rule consequents. Constraints are included to guarantee that an individual is valid. The purpose here is to evolve compact and consistent rule bases and visible partitions of the universes. Nonlinear parametric functions of the fuzzy rule consequents assure accuracy using a set optimal set of parameters, and a pruning procedure avoid redundancy and overfitting.

Assume a fuzzy functional model composed of m fuzzy rules $R_j, j = 1, \ldots, m$, of the form

$$R_j : \text{If } x_1 \text{ is } A_1^j \text{ and } \cdots \text{ and } x_n \text{ is } A_n^j \text{ then } y = g_j(\mathbf{w}_j, \mathbf{x})$$

where $\mathbf{x} = [x_1, x_2, \ldots, x_n]$ is an n-dimensional input vector and $\mathbf{w}_j = [w_{j1}, w_{j2}, \ldots, w_{jq}]$ is a q-dimensional vector of parameters. Here we adopt nonlinear functions of the form

$$g_j(\mathbf{w}_j, \mathbf{x}) = w_{j0} + w_{j1}x_1 + \cdots + w_{jn}x_n + w_{j(n+1)}x_1x_1 + \cdots + w_{j(2n)}x_1x_n + \cdots + w_{j(q-1)}x_nx_n$$
(13.1)

where $q = [n(n-1)/2 + 2n + 1]$. The task of the coevolutionary algorithm is to evolve rules and rule base from granulation of the universes and membership function parameters, the linguistic terms that compose the fuzzy rule antecedents, the choice of the t-norm to act as the *and* operator. Rule consequent parameters can be computed using either global or local least squares procedure. For the details about the use of global and local least squares see Section 11.8 or Hoffmann and Nelles (2000) and Delgado et al., (2001a).

13.4 COEVOLUTIONARY HIERARCHICAL GENETIC FUZZY SYSTEM

Coevolutionary approaches can be broadly classified as competitive- or collaborative-species models. In competitive approaches, two different species interact, the hosts and parasites. Hosts estimate their fitness and parasites evaluate the performance of the hosts. Competition between species means that the success of hosts implies failure of the parasites (Rosin and Belew, 1997). The motivation behind the host–parasite approach rests on the idea of a *coevolutionary arms race*: It is expected that each specie becomes increasingly efficient at exploiting the weaknesses of the other. Alternative to competition, individuals of the species can be rewarded based on how well they collaborate with representatives from the other species (Potter and De Jong, 2000).

The coevolutionary approach addressed in this section considers four hierarchically organized populations. Populations of different hierarchical levels represent different species. The parameters of fuzzy functional models can be evolved through collaborative relations involving individuals from different species. The basic coevolutionary genetic fuzzy system model is shown in Figure 13.9.

13.4 Coevolutionary Hierarchical Genetic Fuzzy System

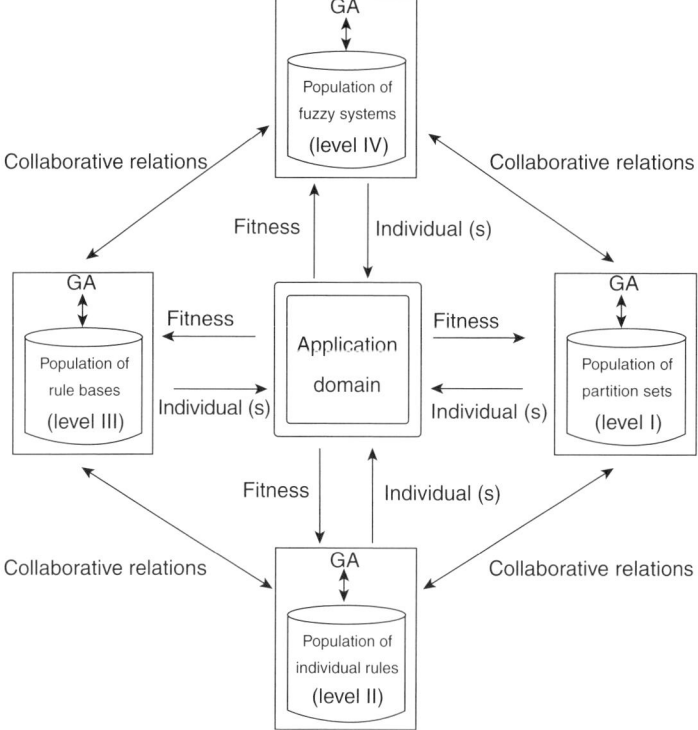

Figure 13.9 Coevolutionary GFS model.

In principle, any evolutionary algorithm can implement the model, but here we adopt genetic algorithm. The fitness of lower level individuals depends on the fitness of higher level individuals. This is because higher level individuals are constructed from lower level individuals. Individuals of one level collaborate by means of structural interdependencies to instantiate individuals at the other levels (Delgado et al., 2002).

Individuals of the four populations represent the four distinct species and encode different parameters. The hierarchical structure considers universe granulation and the corresponding membership functions, called partition sets for short, are at the first hierarchical level. Rules are the individuals of the second level, sets of rules are the individuals of the third level, and fuzzy functional models with their semantics and inference assemble the population of the fourth level. The coding scheme uses real and integer encoding, as shown in Figure 13.10(b). The species coevolve via repeated application of evolutionary operators as summarized in the following procedure:

procedure CO-EVOLVE-GFS (**x**, f) **returns** a functional fuzzy model
input : data **x**
        fitness function f
local : crossover, mutation rates κ, μ
        population: set of individuals
        individual: chromosome

t: triangular norms
max: maximun number of generations

Generation = 1
for level = 1:4
   INITIALIZE (population, level)
if generation ≥ max **return** functional fuzzy model
  else
    for each individual of level IV population
      run LEAST-SQUARES (**x**,**w**) for each rule consequent
    for level = 1:4
      COMPUTE-FITNESS (population, level)
    for level = 1:4
      SELECTION (population, level)
      CROSSOVER (population, level, κ)
      MUTATION (population, level, μ)
      generation ← generation + 1
return best individual

The initialization procedure randomly generates different population for each hierarchical level. The next step involves parameter optimization of the consequents of each fuzzy system of level IV using a least squares procedure such as the ones detailed in Section 11.8 of Chapter 11. Fitness calculation of level IV evaluates the performance of each fuzzy model for the intended application. The details will be given shortly below.

The use of the evolutionary selection, crossover, and mutation operators follows a predefined order, from level IV to level I. Selection is applied first. Selection uses a tournament technique to select 80% of the individuals, and the most distant from the best strategy to chose the remaining 20% to keep diversity. The selection process is elitist. The next evolutionary step uses one-point crossover operator, and the crossover point is randomly chosen. Mutation operates real and integer-encoded parameters. Integer encoding adopts fitness-proportional mutation, that is, alleles with low fitness have higher probabilities of being chosen for mutation. The new value is chosen among all feasible values. Real encoding uses uniform mutation. Computational experiments have indicated that one-point crossover with uniform mutation perform better results when compared with alternative schemes as suggested in the literature (Michalewicz, 1999).

13.5 HIERARCHICAL COLLABORATIVE RELATIONS

Hierarchical collaboration emerges from the interdependencies among individuals of the different species. Figure 13.10(a) overviews the hierarchical collaborative relations and (b) shows the interdependencies involved in the coevolutionary process.

A partition set of level I individual encodes the membership functions that granulate the universes of the variables involved. These individuals are encoded using real-valued chromosomes. Each member of the population of fuzzy rules of

13.5 Hierarchical Collaborative Relations

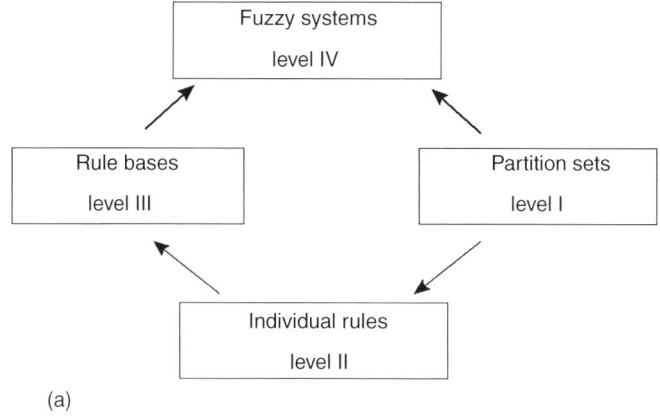

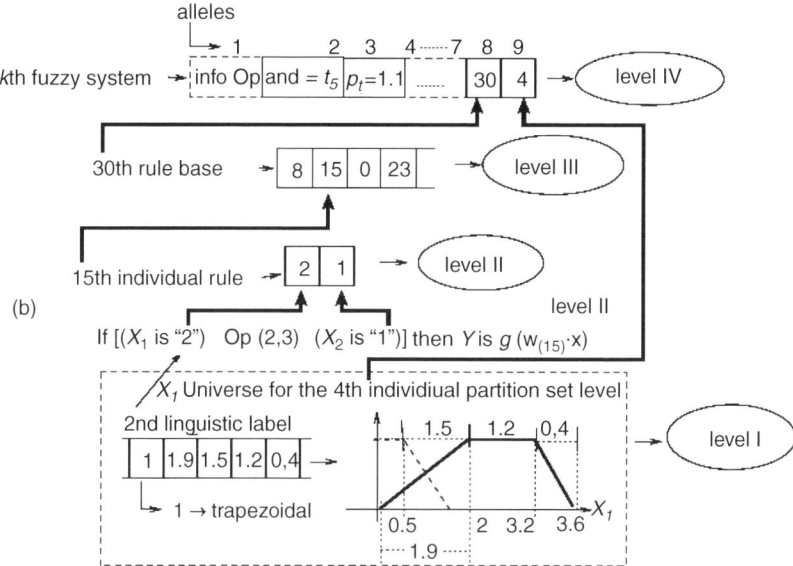

Figure 13.10 Hierarchical collaboration between (a) species and (b) individuals.

level II represents a fuzzy functional rule. This population accepts different combinations of membership functions as identified by their indexes, that is, the order in the partition set. Individuals of the population of set of rules of level III are formed by the indexes that identify the individual rules that assemble the set. The length of the chromosome determines the maximum number of fuzzy rule, and the null index may appear several times to represent the absence of rules. Each individual of level IV represents a fuzzy system. At this level, each chromosome encodes a specific set of rules (allele at 8) and a partition set (allele at 9), with a subset of operators to define rule semantics and processing (alleles at places 1–7). For fuzzy functional models, alleles at places 4–7 one not relevant here, once their purpose is to treat linguistic fuzzy models.

As an example, suppose that, after coevolution, the parameters of specific individuals (at levels IV–I) are as shown in Figure 13.10(b). Thus, the kth fuzzy system shown in level IV is composed by the 30th set of fuzzy rules of level III and the 4th partition set of level I. In the example, each fuzzy rule of the 30th rule base aggregates rule antecedent propositions using the following t-norm:

$$atb = \frac{ab}{p_t + (1 - p_t)(a + b - ab)} \qquad (13.2)$$

with $p_t = 1.1$ (both the t-norm and the value of p obtained by coevolution). To illustrate the hierarchical collaboration, notice that in Figure 13.10 the 15th fuzzy rule, which is part of the kth fuzzy system, is as follows:

If x_1 is "2" $t x_2$ is "1" then $y = g(\mathbf{w}_{15}, \mathbf{x})$

where the linguistic terms labeled here as "2" and "1" are defined (for the kth fuzzy system) by the 4th chromosome of level I population. The nonlinear rule consequent function $g(\mathbf{w}_{15},\mathbf{x})$ is as shown in (13.1) with $q = 2$.

Constraints must be observed to guarantee that only valid individuals are generated for all species. At the partition level, two criteria must be fulfilled by the set of membership functions to guarantee transparency universes partition. The first, γ-completeness, avoid, partition with gaps whereas the second, α-overlapping, limit, overlapping of membership functions of the partition sets. Evolution selects shape and location of membership functions, but the minimum (γ) and the maximum (α) overlapping should always be satisfied. Membership functions are shifted accordingly if they do not satisfy completeness and overlapping criteria. At the second level, crossover and mutation are applied to produce different combinations of linguistic terms in each fuzzy rule. At the rule-base level, crossover and mutation modify the integer indexes associated with the individual rules. New values are chosen in the index set $\{0, 1, \ldots, S_{II}\}$, where S_{II} is the number of individuals of the level II population. Crossover produces combinations of two parent individuals at the fuzzy system level. Mutation replaces the integer allele selected by one among all all possibilities. Uniform mutation modifies alleles at sites encoding the parameter p_t of the t-norm shown in (13.3).

13.5.1 Fitness Evaluation

Figure 13.11 summarizes the fitness evaluation mechanism adopted by the coevolutionary approach. Fitness evaluation of each individual of the four hierarchical population levels is as follows:

1. Fuzzy system—level IV: fit_{FSi} is based on the performance of the fuzzy models;
2. Rule base—level III: $\text{fit}_{RBk} = \max(\text{fit}_{FSb}, \ldots, \text{fit}_{FSd})$ where $b,\ldots,d$ are the fuzzy systems defined by the kth rule base;
3. Individual rule—level II: $\text{fit}_{IRj} = \text{mean}(\text{fit}_{RBm}, \ldots, \text{fit}_{RBp})$ where $m,\ldots,p$ are rule bases containing the jth individual rule.

13.5 Hierarchical Collaborative Relations

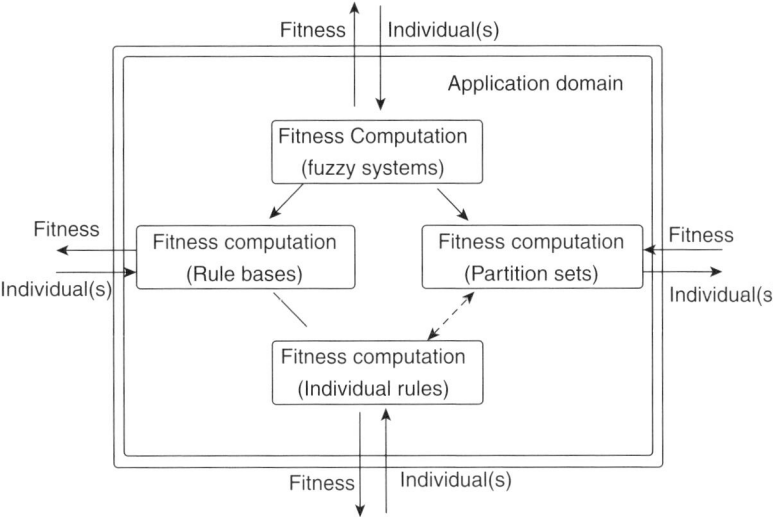

Figure 13.11 Evaluation of fitness in hierarchical collaborative evolution.

4. Partition set—level I: $\text{fit}_{PS_r} = \max(\text{fit}_{IR_x}, \ldots, \text{fit}_{IR_z})$ where $x,\ldots,z$ are fuzzy systems with the rth partition set.

Fitness evaluation uses relationships among hierarchically organized species. At level IV, fitness is evaluated decoding chromosomes to get the corresponding fuzzy models and their performance when solving the target application. Fitness of levels III and I depend on the performance of the fuzzy models in which they appear, and the best fuzzy model defines the fitness value for its rule base and partition set. This is because if a rule base or partition set is part of models with high fitness, their fitness should decrease because of the participation of the same individuals in lower fitness fuzzy models. In the case of individual fuzzy rules, because interaction among them is high, rarely an individual rule influences in isolation the performance of the fuzzy model as a whole. Therefore, the mean of the fitness of the pertinent rule bases makes more sense. A rule is judged measuring how well it cooperates with other rules to compose a fuzzy model at a higher level.

13.5.2 Pruning Algorithm

From Section 11.8 of Chapter 11, the optimal values of parameters w of a fuzzy functional model derived from the least squares optimization is given by

$$\mathbf{w}_{\text{opt}} = Z^{\#}\mathbf{y}$$

where $Z^{\#} = (Z^T Z)^{-1} Z^T$ and $\mathbf{y}$ is a vector corresponding to M training data pairs $(x_p, y_p), p = 1, \ldots, M$. The assumption that matrix Z is of full rank and that

$$M \geq \frac{n(n-1)}{2} + 2n + 1$$

are necessary and sufficient conditions to get w_{opt} in closed form. The full-rank condition can always be satisfied because whenever the rank of Z is not full, we can proceed eliminating the linearly dependent (LD) columns of Z until we get a full-rank matrix. An algorithm to eliminate the LD columns of Z when it is poorly conditioned is as follows (Delgado et al., 2001b). Let the value of the condition number of Z be

1. RCOND(Z) based on the RCOND LINPACK reciprocal condition estimator;
2. an estimate for the reciprocal of the condition of Z in 1-norm;
3. near one if Z is well conditioned, and near zero if Z is poorly conditioned.

The algorithm below improves the reciprocal condition number of matrix Z eliminating the columns of Z that contribute most to reduce RCOND(Z). The main steps are as follows:

procedure MATRIX-COND-LS (Z) **returns** a conditioned matrix
input: data matrix Z
local: threshold

eliminate null columns of Z
set null columns status to 0
set status of the remaining columns to 0.5
let $Z_1 = Z(r)$, where r are the columns of Z with status > 0
if RCOND($Z_1^T Z_1$) $\leq$ threshold **do**
    set $Z_2 = [C_i]$, C_i is the ith column of Z with status $= 0.5$ and biggest Euclidean norm
    $k = i$
    while RCOND($Z_1^T Z_1$) $\geq$ threshold **do**
        set status $[C_k] = 1$,
        let $Z_3 = [Z_2 C_j]$, C_j is the jth column of Z with status $= 0.5$ and highest RCOND($Z_3^T Z_3$),
    $k = j$,
    set $Z_2 = Z_3$,
let $Z_f = Z(r)$, where r are the columns of Z with status $= 1$
return Z_f

Because Z_f is a full-rank matrix, the reduced parameter vector $\mathbf{w}_f^{opt}$ is found using

$$\mathbf{w}_f^{opt} = Z_f^\# \mathbf{y}$$

where $Z_f^\# = (Z_f^T Z_f)^{-1} Z_f^T$

EXAMPLE 13.2

First a function approximation problem with noisy learning data is considered. The functional fuzzy model evolved by the coevoluntionary procedure is compared with an artificial neural fuzzy inference system (ANFIS) trained with least squares and gradient methods (Jang, 1993). In function approximation problem, we emphasize approximation accuracy and degree of

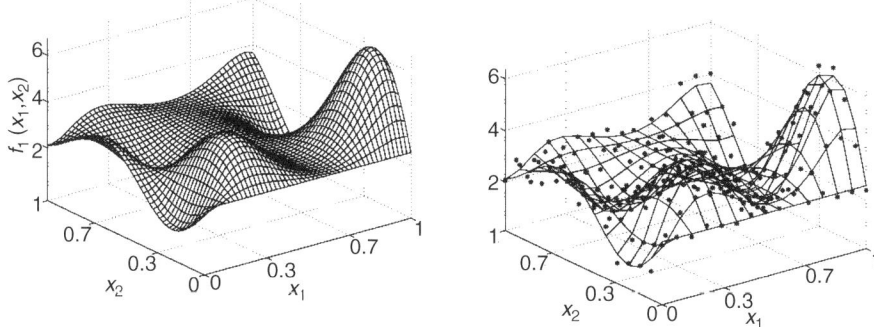

Figure 13.12 Original function and noisy training data.

overfitting. Flexible models with nonlinear consequents are capable of adapting well and they may incorporate peculiarities due to random variations (Berthold and Hand, 2006). The coevolutionary approach deals with overfitting incorporating pruning steps and redundancy control. The task here is to approximate the function

$$F_1 : \Omega \to R, F_1(x_1, x_2) = f_1(x_1, x_2) + N(m, \sigma), \Omega = [0, 1]^2$$
$$f_1(x_1, x_2) = 1.9(1.35 + \exp(x_1) \sin[13(x_1 - 0.6)^2 \exp(-x_2) \sin(7x_2)])$$

where $N(m, \sigma)$ is a Gaussian distribution with mean $m = 0$ and standard deviation $\sigma = 0.3$. Figure 13.12 depicts the original function $f_1(x_1, x_2)$ and training data obtained by equally sampling 255 points from Ω and computing F_1.

Models were evolved after 1000 generations with population sizes of 100 individuals for level IV, 80 individuals for level III, and 20 individuals for level I. Initially, the granularity of the partition set is [3, 3], that is, three fuzzy sets granulate the universes of x_1 and x_2. Each individual, namely, rule base of level III, contains at most nine fuzzy rules. The effective number of fuzzy rules of level III can change during evolution because mutation or crossover may exclude or reintroduce a rule in the rule base. The fitness of the fuzzy models was computed as the inverse of the root mean squared error (RMSE) using learning data set. Selection uses the tournament for reproduction and elitist selection. Crossover and mutation are performed with rates $\kappa = 0.3$ and $\mu = 0.1$ for levels I, III, and IV. In the example, the population of individual rules (level II) contains all possible combinations of the linguistic terms, and no evolution is necessary at this level. Table 13.2 presents the results for the best fuzzy system evolved by the coevolutionary approach (maximun number of generations = 1000). The solution produced by ANFIS after 1000 training epochs is also shown. For comparison purposes, ANFIS uses the same granulation, namely [3, 3].

Experiments have been done using ANFIS with the same membership functions as the ones adopted by the coevolutionary GFS (CoevolGFS) approach, that is, triangular, trapezoidal, and Gaussian. The best performance was achieved using Gaussian membership functions. Both

Table 13.2 RME Errors for Function Approximation.

Approach	Training RME	Test RME	Number of rules
CoevolGFS	0.25	0.13	8
ANFIS	0.32	0.21	9

404 Chapter 13 Fuzzy Systems and Computational Intelligence

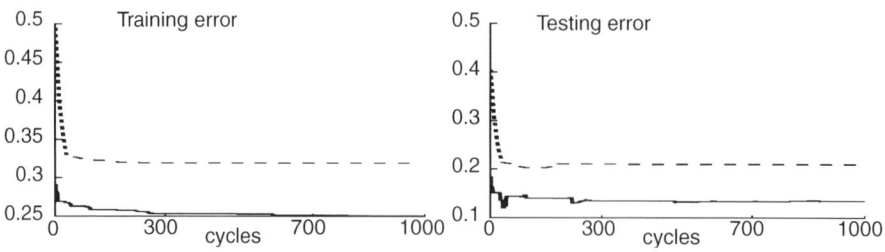

Figure 13.13 Training and testing errors.

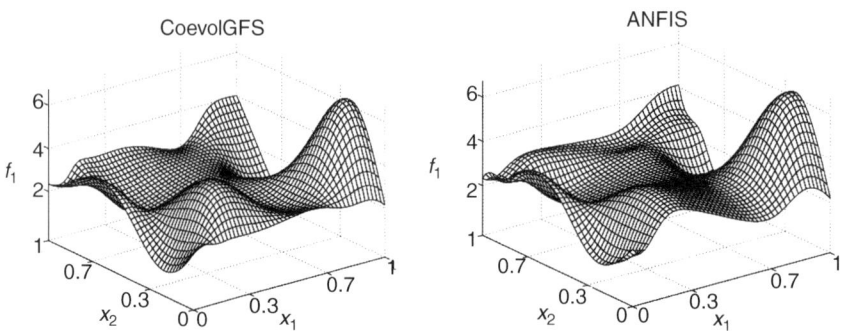

Figure 13.14 Approximation of f_1 by CoevolGFS and ANFIS.

ANFIS and CoevolGFS were tested using 2500 equally spaced samples of Ω. As Table 13.2 indicates, the CoevolGFS approach achieves the best compromise between accuracy, measured by the generalization capability, and compactness, measured by the number of fuzzy rules. Figure 13.13 shows training and testing errors for the best fuzzy model evolved by coevolution (full line) and learned by ANFIS (dotted line). In Figure 13.13 cycles mean generations for the coevolutionary approach and epochs for ANFIS.

Figure 13.13 emphasizes the benefits of coevolution associated with the use of nonlinear consequent functions. As it can be noted, the results of the coevolutionary approach after first generation are better than the ones obtained by ANFIS trained with 1000 epochs. The reason is that, whereas coevolution uses nonlinear adjustable rule consequents, ANFIS uses linear consequent functions. Figure 13.14 illustrates the generalization capabilities of the model evolved by coevolGFS and ANFIS. The coevolutionary process γ-complete and α-overlapped partitions are shown in Figure 13.15a. They are more visible when compared with the overlapping results obtained by ANFIS, Figure 13.15.b.

EXAMPLE 13.3

This second example concerns a classification of intertwined spirals because it is a challenging classification benchmark that has originated from the field of neural networks. (Juillé and

13.5 Hierarchical Collaborative Relations

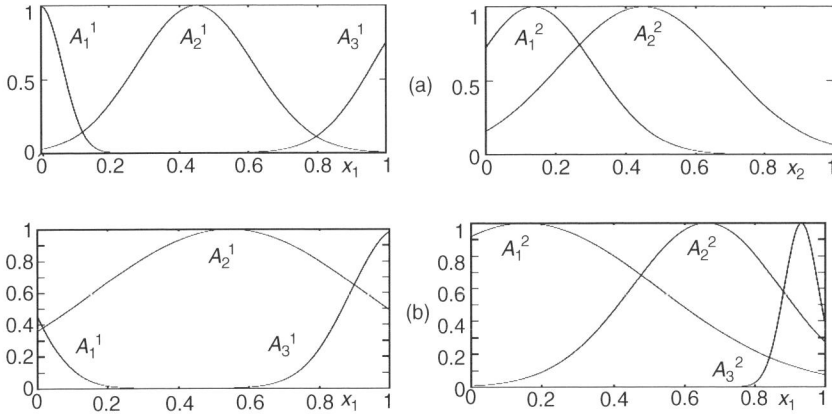

Figure 13.15 Partitions developed by (a) CoevolGFS and (b)ANFIS.

Pollack, 1996). We assume that data are the coordinates of the two spirals. The spirals coil around the origin and around each another as depicted in Figure 13.16. Here, the task is to develop a functional fuzzy rule-based system to associate any (x_1, x_2) coordinate with the proper spiral. The coordinates are labeled to denote which spiral the coordinates refer. Training data is a triple (x_1, x_2, C), where $C = +1$ for one spiral and $C = -1$ for the other. Two sets with 97 points each were generated. Fuzzy models were evolved using a training data composed of 194 points, 97 for each spiral. Denoting fuzzy system output for input x_p by $f(x_p)$, the following assignment can be made:

$$Cl_1 \rightarrow \text{round}(f(x_p)) = +1$$
$$Cl_2 \rightarrow \text{round}(f(x_p)) = -1$$

$$\text{round}(a) = \begin{cases} \text{int}(a) - 1, & \text{if} \quad (a - \text{int}(a) \leq -0.5 \\ \text{int}(a), & \text{if} \quad -0.5 < a - \text{int}(a) < 0.5 \\ \text{int}(a) + 1, & \text{if} \quad a - \text{int}(a) \geq 0.5 \end{cases}$$

As a consequence, if the output is in the range $[-1.5, -0.5]$ then class is Class 1, otherwise, if the output is in the range $[0.5, 1.5]$ then class is Class 2. If the output is out of these two ranges, then the class is unknown.

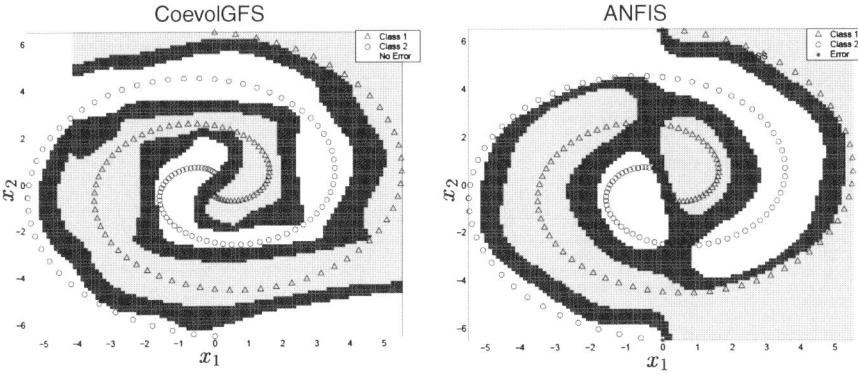

Figure 13.16 Classification performance of CoevolGFS and ANFIS.

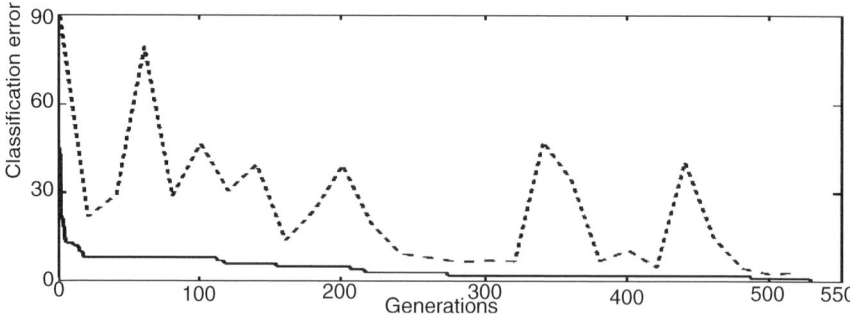

Figure 13.17 Classification error during evolution.

The number of individuals chosen are 20, 80, and 100 for the partition set, rule base, and fuzzy system levels, respectively. Selection for reproduction uses tournament and elitism. Crossover and mutation rates are $\kappa = 0.3$ and $\mu = 0.1$, respectively. Initial partition sets assume at the most three linguistic terms for each variable, equally spaced in the range [6, 6], for x_1, and $[-6.5, 6.5]$ for x_2. The size of each chromosome of the rule-base level is fixed as four, which means four fuzzy rules at the most in each rule base. Figure 13.17 shows the amount of classification errors for best individual (solid line) and for the average of the population (dotted line) during evolution. The best individual misclassified 45 points in the first generation and correctly classified all points after 529 generations.

Table 13.3 summarizes the classification performance of the best individual evolved by the coevolGFS and developed by ANFIS with granularity [3, 3]. Similar to the previous example, different membership functions have been experimented, with the Gaussians being the most efficient.

Figure 13.16 illustrates the classification performance of coevolGFS and ANFIS. Three regions are shown to illustrate the generalization capabilities of the fuzzy models. The regions are constructed using 10,000 uniformly distributed testing data points in $\Omega = [-6,6] \times [-6.5, 6.5]$. Gray areas are data classified as class 1 ($f(x_p) \in (-1.5, -0.5]$), white region corresponds to data classified as class 2 ($f(x_p) \in [0.5, 1.5)$), and black areas correspond to data classified as unknown.

The best fuzzy rule-based system evolved appears in generation 529 and is as follows:

R_1: If x_1 is low *and* x_2 is low then $y = -0.31 + 1.6x_1 - 0.26x_2 + 0.34x_1^2 + 0.17x_2^2 - 0.1x_1x_2$

R_2: If x_1 is medium *and* x_2 is low then $y = 15.3 - 1.3x_1 + 7.7x_2 - 0.05x_1^2 + 0.84x_2^2 - 0.46x_1x_2$

R_3: If x_1 is medium *and* x_2 is high then $y = -17.2 - 2.2x_1 + 7.6x_2 - 0.08x_1^2 - 0.78x_2^2 + 0.45x_1x_2$

R_2: If x_1 is high *and* x_2 is high then $y = 1.14 + 2.0x_1 + 1.24x_2 - 0.25x_1^2 - 0.28x_2^2 - 0.34x_1x_2$

Table 13.3 Classification Performance for Intertwined Spirals.

Approach	Cycles	Misclassification	Number of rules
CoevolGFS	529	0	4
ANFIS	1000	18	9

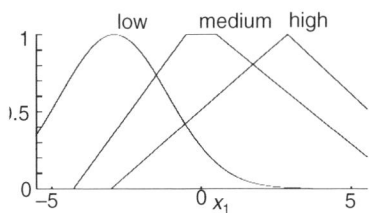 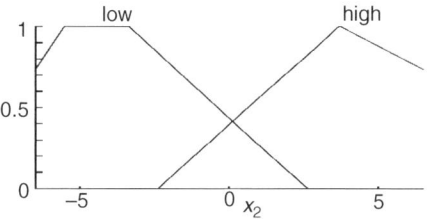

Figure 13.18 Linguistic terms and corresponding membership functions.

Antecedent aggregation (and operator) is a t-norm with $p_t = 9.96$. Linguistic terms low, medium, and high have the membership functions shown in Figure 13.18.

13.6 EVOLVING FUZZY SYSTEMS

As thoroughly discussed in Chapter 11, fuzzy rule-based models are an effective mechanism to combine behaviors that can be individually modeled. To develop fuzzy models means to find a proper structure, namely, the rule base, membership functions, linguistic labels, and parameters. As indicated in the previous section, GFS is a powerful tool to design fuzzy systems when data are available. GFS evolve solutions and help designers to select critical structural elements of fuzzy models. As a design tool, GFS runs off-line and interacts with designers to complement information acquired from data with expert knowledge.

Learning models online requires continuous data acquisition and processing. New data may either reinforce and confirm the current knowledge, or indicate changes and revision of the current model. For instance, in nonstationary environments, operating conditions modify, fault occurs, parameters of a process change, and models must be revised to match the current conditions. In these cases, different evolution methods must be devised to give an answer to a key question: How to update the current model structure using the newest data samples? The evolving mechanism must ensure greater generality of the structural changes to improve the ability of fuzzy rule-based models to describe a number of data samples from the moment of their operation. The mechanism used to modify the rule base must consider replacement of less-informative rules by more informative ones in an online and gradual manner. Evolving fuzzy systems aim at developing more flexible adaptive systems than conventional adaptive system mechanisms mostly based on linearity and fixed structures. Evolving systems target nonstationary processes and embody online learning methods and algorithms that evolve individual fuzzy systems that inherit and gradually change to guarantee life-long learning and self-organization of the system structure (Angelov et al., 2006). The purpose of evolving fuzzy systems is to act online and complement GFS in the sense that models designed offline by GFS can be further adapted to unknown and unpredictable environments using online learning methods and algorithms.

This section presents an evolving participatory learning (ePL) approach to construct evolving fuzzy rule-based models. The approach combines the concept of participatory learning (Yager, 1990) and the evolving fuzzy modeling approach (Angelov, 2002; Angelov and Filev, 2004). Participatory learning (PL) naturally induces unsupervised dynamic fuzzy clustering algorithms (Silva et al., 2005) and provides an effective mechanism to construct evolving algorithms. Here the focus is on functional fuzzy models, but similar approach can be used to develop linguistic fuzzy models as well. In the ePL approach, fuzzy model structure identification means estimation of modal values and dispersion of the Gaussian membership functions associated with the linguistic variables that appear in the antecedents of the fuzzy rules, through participatory learning clustering. The original evolving Takagi–Sugeno (eTS) approach of (Angelov, 2002) uses information potential clustering. After modal values and dispersion (antecedent parameters) are found, both eTS and ePL use the least squares method to find the parameters of the functions that appear in the consequents of the functional fuzzy rules.

13.6.1 Functional Fuzzy Model

Functional fuzzy model identification considers a set of rule-based models with fuzzy antecedents and functional consequents. Assuming linear functions at each rule consequent, the models have the following structure:

$$R_i : \text{If } \mathbf{x} \text{ is } \mathbf{A}_i \text{ then } y_i = a_{i0} + \sum_{j=1}^{n} a_{ij} x_j$$

where R_i is the ith rule, $i = 1, \ldots, c$, and c is the number of fuzzy rules, $\mathbf{A}_i$ is an n-dimensional vector of the antecedent fuzzy sets, a_{ij} are the parameters of the rule consequents, $\mathbf{x} \in [0, 1]^n$, and y_i is the output of the ith rule.

Therefore, the collection of the $i = 1, \ldots, c$ rules assembles a model as a combination of local linear models. The contribution of each local linear model to the overall output is proportional to the degree of activation of each rule. Here we adopt Gaussian membership functions, that is, each element $A_j^i, j = 1, \ldots, n$, of $\mathbf{A}_i$, is

$$A_j^i(x_j) = \exp[-k_{ij}(x_j - v_{ij})^2]$$

where k_{ij} is a positive constant that defines the zone of influence of the ith local model and v_{ij} is the modal value. The dispersion of the Gaussian increases as k_{ij} decreases; too small values lead to averaging and too large values to overfitting. Usually, values in the range [0.3, 0.5] are recommended (Angelov and Filev, 2004). The modal values $v_{ij}, i = 1, \ldots, c, j = 1, \ldots, n$, are the cluster centers (focal points, and prototypes) found using a clustering algorithm, each cluster defining a rule. Clearly, online learning needs online clustering. Clustering can be performed in both input or input–output data space.

The output of the model is determined as the weighted average of individual rule contributions:

$$y = \sum_{i=1}^{c} w_i y_i$$

$$w_i(x) = \frac{\lambda_i(x)}{\sum_{i=1}^{c} \lambda_i(x)}$$

$$\lambda_i = A_1^i(x_1) \; t A_2^i(x_2) \; t \ldots t \; A_n^i(x_n)$$

where λ_i is the degree of activation of rule i and t is a t-norm. In practice, the product t-norm is the one most commonly adopted.

13.6.2 Evolving Participatory Learning Algorithm

Similar to the eTS, the ePL algorithm starts after an initialization step to set initial rule consequent parameters equal to zero, and rule base antecedent parameters whose modal values are usually set at the first data point. Next, the algorithm starts its online operation of reading the next data sample, running the clustering algorithm, modifying or updating the rule base structure, and computing the degree of activation of the rules and the model output. The following procedure summarizes the algorithm.

procedure EVOLVE-PARTICPATORY- LEARNING (x,y) **returns** an output
input : data **x,y**
local: antecedent parameters
       consequent parameters
       INITIALIZE-RULE-PARAMETERS
do forever
       read **x**
       PL-CLUSTERING
       UPDATE-RULE-BASE
       RUN-LEAST-SQUARES(**x,y**)
       COMPUTE-RULE-ACTIVATION
       COMPUTE-OUTPUT
return y

The main step in the ePL algorithm concerns the PL algorithm. Let $\mathbf{v}_i^k \in [0,1]^n$ be the ith, $i = 1, \ldots, c^k$, cluster center at the kth time step. The purpose of the participatory mechanism is to learn the value of $\mathbf{v}_i^k$ from a stream of data $\mathbf{x}^k \in [0,1]^n$. Given an initial cluster structure, that is, a set of initial cluster centers, updates in ePL proceed using a fuzzy compatibility index $\rho_i^k \in [0,1]$ and an arousal index $r_i^k \in [0,1]$. While ρ_i^k measures how much a data point is compatible with the current cluster structure, the arousal index r_i^k acts as a critic to signal when the current cluster centers should be revised in light of new information contained in the data.

The PL clustering procedure may, in each step, create a new cluster or modify the existing ones, depending on the level of the arousal index. If the arousal index is greater than a threshold $\tau \in [0, 1]$, then a new cluster is formed. Otherwise the ith cluster center is updated as follows:

$$\mathbf{v}_i^{k+1} = \mathbf{v}_i^k + G_i^k(\mathbf{x}^k - \mathbf{v}_i^k)$$

where $G_i^k = \alpha \rho_i^k$, $\alpha \in [0, 1]$ is the learning rate, and

$$\rho_i^k = 1 - \frac{\| \mathbf{x}^k - \mathbf{v}_i^k \|}{n}$$

with $\| \cdot \|$ being a distance function. The cluster center to be updated is the one that is the most compatible with $\mathbf{x}^k$. Formally, this means that cluster center i is such that

$$i = \arg\max_p \{\rho_p^k\}$$

Similar to cluster centers, the arousal index r_i^k is updated as follows:

$$r_i^{k+1} = r_i^k + \beta(1 - \rho_i^{k+1} - r_i^k)$$

The value of $\beta \in [0, 1]$ controls the rate of change of arousal: The closer β is to 1, the faster the system can sense compatibility variations. The way in which ePL includes the arousal mechanism is to incorporate the arousal index r_i^k into the effective learning rate G_i^k as follows:

$$G_i^k = \alpha(\rho_i^k)^{1-r_i^k}$$

When $r_i^k = 0$ we have $G_i^k = \alpha \rho_i^k$, which is the PL procedure with no arousal. Notice that if the arousal index increases, the similarity measures have a reduced effect. The arousal index is the complement of the confidence on the truth of the current belief which, in ePL, is the rule base structure. The arousal mechanism checks the performance of the system looking at compatibility of the current model with the observations. Learning is dynamic in the sense that ρ_i^k can be viewed as a belief revision strategy whose effective learning rate $G_i^k = \alpha(\rho_i^k)^{1-r_i^k}$ depends on the compatibility between new data, on the current cluster structure, and on model confidence.

Whenever a cluster center is updated or a new cluster added, the PL fuzzy clustering algorithm verifies if there exist redundant clusters, because updating a cluster center may push the center closer to another one and redundant cluster may be formed. Therefore, a mechanism to exclude redundancy is needed to detect similar outputs due to distinct rules. In PL clustering, a cluster center is declared redundant whenever its similarity with any other center is greater than or equal to a threshold λ. If this is the case, either the original cluster center is maintained or it is replaced by the average between the new data and the cluster center.

Similar to the compatibility index ρ_i^k, the compatibility among cluster centers is computed using

$$\rho_{v_i}^k = 1 - \sum_{p=1}^{n} |v_i^k - v_p^k|$$

Therefore, whenever $\rho_{v_i}^k \geq \lambda$, cluster i is declared redundant.

After clustering, the fuzzy rule based model is constructed using linear consequent functions whose parameters are computed using the local weighted least squares algorithm to obtain locally meaningful models. In off-line applications, model parameters $\mathbf{a}_i = (a_{i0}, a_{i1}, \ldots, a_{in})^T, i = 1, \ldots, c$, are computed to minimize the objective function

$$J_L = \sum_{i=1}^c (\mathbf{y} - X^T \mathbf{a}_i)^T W_i (\mathbf{y} - X^T \mathbf{a}_i) \tag{13.3}$$

where $X \in \mathbf{R}^{d(n+1)}$ is a matrix formed by vectors $(\mathbf{x}_e^k)^T = (1, (\mathbf{x}^k)^T)$, $\mathbf{y}$ is a vector formed by $y^k, k = 1, \ldots, d$, W_i is a diagonal matrix whose elements are $w_i^k = w_i(\mathbf{x}^k)$, and d is the number of training data. The parameters that minimize the objective function (13.3) are given by (Yen et al., 1998):

$$\mathbf{a}_i = (X^T W_i X)^{-1} X^T W_i \mathbf{y}, \quad i = 1, \ldots, c \tag{13.4}$$

In online applications, parameters $\mathbf{a}_i^k$ must be updated, after a new data $\mathbf{x}^k$ is presented at each time step k, using the recursive weighted least squares procedure. In this case, parameters $\mathbf{a}_i^k$ that minimize (13.3) are computed as follows (see, for instance, Angelov and Filev, 2004):

$$\mathbf{a}_i^k = \mathbf{a}_i^{k-1} + q_i^k \mathbf{x}_e^{k-1} w_i^{k-1} [y^k - (\mathbf{a}_i^{k-1})^T \mathbf{x}_e^{k-1}]$$

$$q_i^k = q_i^{k-1} - \frac{w_i^{k-1} q_i^{k-1} \mathbf{x}_e^{k-1} (\mathbf{x}_e^{k-1})^T q_i^{k-1}}{1 + w_i^{k-1} (\mathbf{x}_e^{k-1})^T q_i^{k-1} \mathbf{x}_e^{k-1}}$$

$\mathbf{a}_i^0 = 0$ and $q_i^0 = \theta I$, where θ is a large number and I is the $(n+1) \times (n+1)$ identity matrix, and $i = 1, \ldots, c$. Notice that parameters are computed for each rule. The parameters of a newly added rule are determined as weighted average of the parameters of the remaining rules, that is,

$$\mathbf{a}_{c+1}^k = \sum_{i=1}^c w_i^k \mathbf{a}_i^{k-1}$$

and parameters of rules R_i are left intact, $\mathbf{a}_i^k = \mathbf{a}_i^{k-1}, i = 1, \ldots, c$. Parameter q_{c+1}^k of the new rule is initialized at $q_{c+1}^k = \sigma I$ and that of the remaining rules inherited at $q_i^k = q_i^{k-1}, i = 1, \ldots, c$.

The details of the ePL algorithm steps are given below. The Section 14.8 of Chapter 14 provides a complete description of the PL clustering algorithm.

1. Read new data $\mathbf{x}^k$
2. Compute compatibility ρ_i^k
3. Compute arousal index r_i^k
4. If $r_i^k \geq \tau \; \forall \{1, \ldots, c^k\}$
   $\mathbf{x}^k$ is a new cluster center; set $c^{k+1} = c^k + 1$
   else compute $\mathbf{v}_i^{k+1} = \mathbf{v}_i^k + G_i^k (\mathbf{x}^k - \mathbf{v}_i^k)$

5. Compute $\rho_{v_i}^k$
   if $\rho_{v_i}^k \geq \lambda$ then exclude $\mathbf{v}_i^k$; set $c^{k+1} = c^k - 1$
6. Update rule base structure and rules parameters
7. Compute normalized rule activation degrees w_i^k
8. Compute output using $y = \sum_{i=1}^{c} w_i y_i$

EXAMPLE 13.4

The ePL algorithm is used to forecast average weekly inflows for a large hydroelectric plant. Hydrologic data cover the period of 1931–2000 (Lima et al., 2006). Streamflow forecast is important to plan the operation of water resources systems. One of the major difficulties is the nonstationary nature of streamflow series due to wet and dry periods of the year. The average and standard deviation of each week in the period 1931–2000 are shown in Figure 13.19, where we clearly note seasonality and higher variability of streamflow during wet periods.

The number of model inputs was chosen using the partial autocorrelation function, see Figure 13.20, that suggests three inputs, respectively, $x_{t-3}, x_{t-2}, x_{t-1}$ to forecast $y_t = x_t$. The dotted lines of Figure 13.20 are the two-standard error of the estimated partial autocorrelation. The period from 1991 to 2000 was selected to test the ePL algorithm. This period corresponds to 520 weeks. Data are normalized to fit the unit interval [0, 1] as follows:

$$x^k = \frac{z^k - \min}{\max - \min}$$

where x^k is the normalized data value, and min and max denote the minimum and maximum values of the hydrologic data $z^k, k = 1, \ldots, N$.

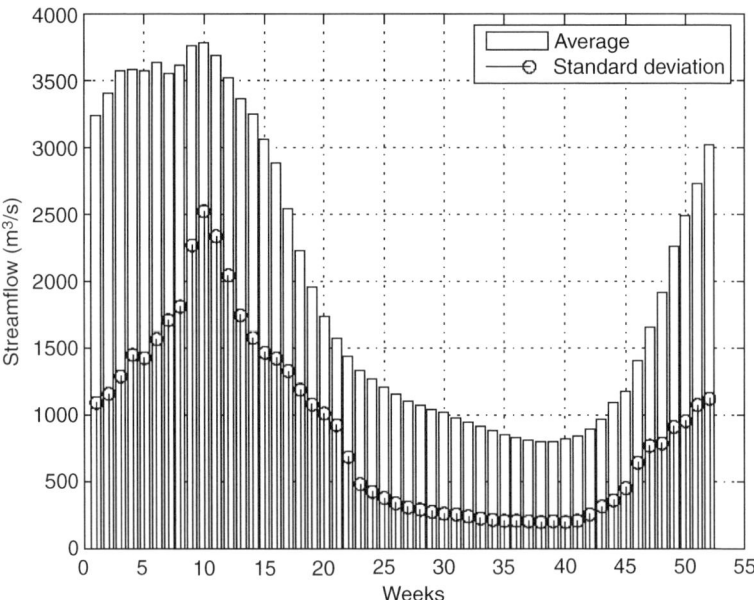

Figure 13.19 Weekly average and standard deviation (m^3/s).

13.6 Evolving Fuzzy Systems

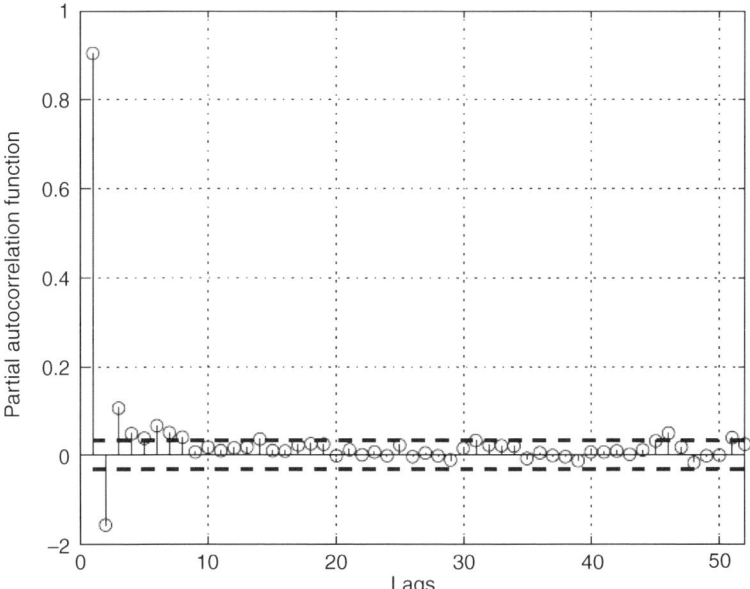

Figure 13.20 Estimated partial autocorrelation function and standard error limits.

The forecasting performance was evaluated using four error criteria: root mean square error (RMSE), mean absolute error (MAD), mean relative error (MRE), and maximum relative error (RE_{max}):

$$RMSE = \sqrt{\frac{1}{P}\sum_{k=1}^{P}(x^k - x_d^k)^2}$$

$$MAD = \frac{1}{P}\sum_{k=1}^{P}|x^k - x_d^k|$$

$$MRE = \frac{100}{P}\sum_{k=1}^{P}\frac{|x^k - x_d^k|}{x_d^k}$$

$$RE_{max} = 100\max\left(\frac{|x^k - x_d^k|}{x_d^k}\right)$$

where $P = 520$ is the number of weeks in the forecasting period, $y^k = x^k$ is the model output, and x_d^k is the actual streamflow value at instant k. Another important evaluation criteria is the correlation coefficient ρ,

$$\rho = \frac{\sum_{k=1}^{P}(x_d^k - \bar{x}_d)(x^k - \bar{x})}{\sqrt{\sum_{k=1}^{P}(x_d^k - \bar{x}_d)^2 \sum_{k=1}^{P}(x^k - \bar{x})^2}}$$

Table 13.4 Forecasting Performance of Evolving Fuzzy Models.

	Models	
Error	ePL	eTS
RMSE (m^3/s)	378.71	545.28
MAD (%)	240.55	356.85
MRE (%)	12.54	18.42
RE$_{max}$ (%)	75.51	111.22
ρ	0.95	0.89
Number of rules	2	2

where $\bar{x}_d$ is the mean valued of the actual streamflow values and $\bar{x}$ is the mean value of the forecasted streamflow values. Correlation coefficient measures how well the forecasts correlate with the actual values. A value of the correlation coefficient closer to 1 indicates higher quality of forecasting.

The ePL algorithm uses the first 28 weeks to initialize the rule base structure. The initialization is done offline. The next phase is done online. During this phase, the number of fuzzy rules and their antecedents and consequents may change depending on the nature of data. The PL clustering procedure starts with two fuzzy rules using the 28 weeks of initialization period.

Table 13.4 summarizes the forecasting performance of the ePL and eTS models for the 520-week testing period and the number of rules evolved.

Figure 13.21 shows the actual inflows and inflows forecasted by ePL and eTS models.

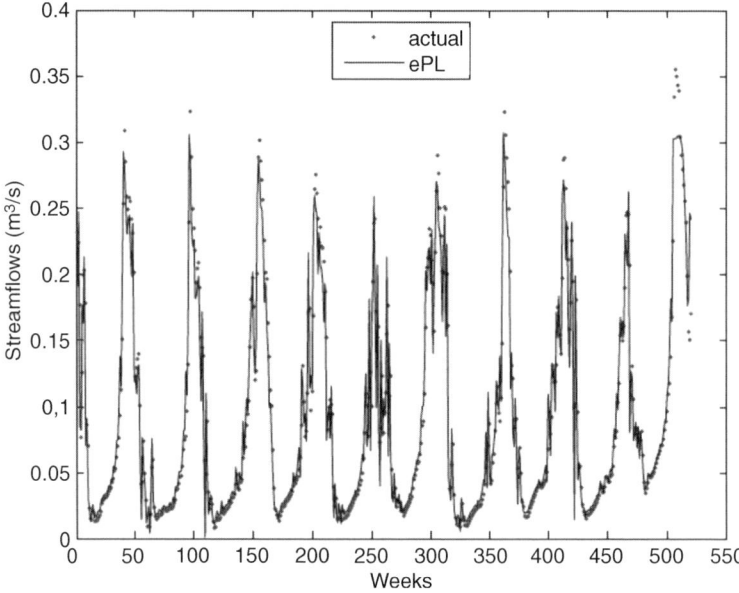

Figure 13.21 Actual (dotted) and forecasted (solid line) streamflow values.

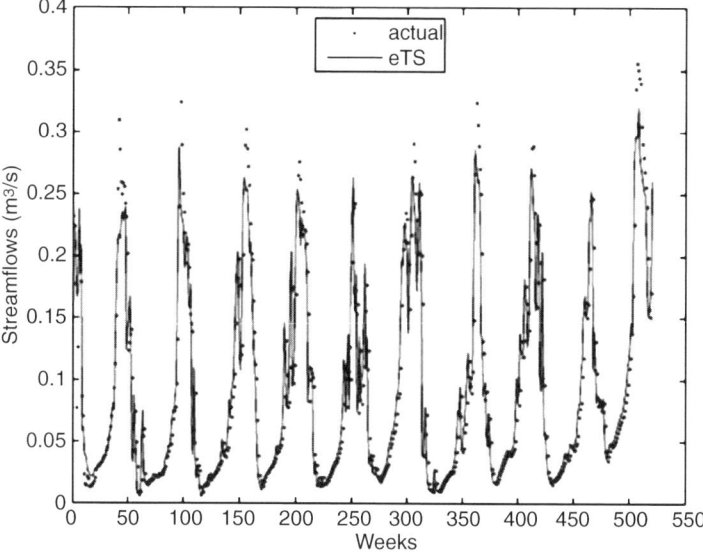

Figure 13.21 (*Continued*).

13.7 CONCLUSIONS

Fuzzy sets and systems are an important constituent of CI because they provide the basis by which imprecision and approximate reasoning can efficiently be handled using computational models. Neural networks are powerful function approximators and clustering devices in which learning procedures provide the key for development. Genetic algorithms are effective stochastic search mechanisms that provide the flexibility required to design complex systems. Because genetic algorithms do not need smoothness to work, they provide an important tool to parameters for which no explicit analytical representation is possible. For instance, it is still a challenge to choose optimal triangular norms using conventional optimization techniques. Hybridizations of fuzzy systems with neural, genetic, and evolving systems explore the benefits and often help to overcome many limitations of each of them. They also augment the area of CI and effectively contribute to solve challenging application problems.

EXERCISES AND PROBLEMS

1. Consider fuzzy *and* and *or* neurons with inputs $x_i \in [0,1]$, $i = 1, \ldots, n$, whose output y_{and} and y_{or} are computed as follows:

$$y_{\text{and}} = (w_1 s x_1) t (w_2 s x_2) t \ldots t (w_n s x_n) = \underset{i=1}{\overset{n}{T}} (w_i s x_i)$$

$$y_{\text{or}} = (w_1 t x_1) s (w_2 t x_2) s \ldots s (w_n t x_n) = \underset{i=1}{\overset{n}{S}} (w_i t x_i)$$

where t and s are t-norm and t-conorm. Assume that $n = 2$ and compute the outputs of the fuzzy neurons for weights $w_i \in \{0, 1\}$ and $t = \min$ and $s = \max$.

2. Consider fuzzy *and* and *or* neurons with inputs $x_i \in [0,1]$ and weights $w_i \in [0,1], i = 1, \ldots, n$, whose outputs y_{and} and y_{or} are given in the previous problem. Plot the input–output function of both types of neurons choosing different t-norms and s-norms.

3. Let us consider a two-input, single-output two-layer fuzzy neural network depicted below. Choose a t-norm, a s-norm, and weights w_{ij} and v_k to solve the exclusive-or classification problems. Use network structures with *and* neurons in the second layer and *or* neuron in the output layer, and vice versa, that is, *or* neurons in the second layer and *and* neuron in the output layer.

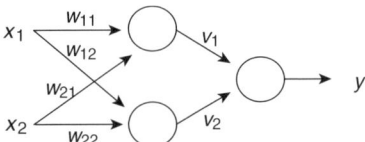

4. Repeat problem 2 for the neural fuzzy network of problem 3.

5. Consider a linguistic fuzzy rule-based system in which membership functions are triangular and restricted to provide strong partitions of the input and output universes. This means that the center of one membership function serves as the left base point of the next, and one membership function is constrained to have its center at the lower boundary of the input universe. In this case, only $(n-1)$ membership function centers need to be specified, where n is the maximum number of fuzzy sets that assemble a partition of a universe. Suggest a binary encoding scheme to evolve a rule-based model in which we must simultaneously determine partitions and fuzzy rules. Assume that rule semantics and remaining parameters of the fuzzy inference are known.

6. The PL clustering algorithm determines cluster centers using an updating equation that looks similar to error correction learning schemes. Assume that the learning rate is fixed and that no arousal mechanism is used. Show that, in this case, the PL updating equation is a form of least squares error-minimization procedure.

7. Suggest alternative mechanisms to update parameters a_i^k and q_i^k in evolving fuzzy systems algorithms for the case in which a new rule is added and the case when existing rules are merged.

HISTORICAL NOTES

Inspiration of natural information processing systems dates back to 1855 when H. Spencer associated neural networks with intelligence through strengths of connections between internal states of neurons (Walker, 1992).

The event marking the birth of the field of CI is often credited to the IEEE World Congress on Computational Intelligence held in Orlando, Florida, 1994 (Craenen and Eiben, 2006). Conferences on fuzzy systems, neural networks, and evolutionary computation were held concurrently during the event.

The use of genetic algorithms as a tool to design and develop fuzzy rule-based systems was suggested by Carr, Pham, Karaboga, Thrift, and Valenzuela-Rendón. A comprehensive review of the area with a wealth of bibliographic references and outlining future developments in the area of GFS was published in the form of the special issue of *Fuzzy Sets and Systems*, in 2004 (vol. 141).

Evolving fuzzy systems in the sense of gradual development of rule-based fuzzy systems structure and their parameters have their roots in the work of Fritzke done in the context of neural networks in 1990. Angelov brought this paradigm to the area of fuzzy systems (Angelov and Zhou, 2006).

REFERENCES

Angelov, P. *Evolving Rule Based Models*, Springer-Verlag, Heidelberg, 2002.

Angelov, P., Filev, D. An approach to Online Identification of Takagi-Sugeno fuzzy model, *IEEE Trans. Syst. Man Cyb. B*, **34**(1), 2004, 484–498.

Angelov, P., Filev, D., Kasabov, N., Cordón, O. Preface, *Proceedings of the 2006 International Symposium on Evolving Fuzzy Systems*, Ambleside, UK 2006.

Angelov, P., Zhou, X. Evolving fuzzy systems from data streams in real-time, *Proceedings of the 2006 International Symposium on Evolving Fuzzy Systems*, Ambleside, UK, 2006, pp. 29–35.

Ballini, R., Soares, S., Gomide, F. A recurrent neurofuzzy network structure and learning algorithm, *IEEE Int. Conf. on Fuzzy Systems*, 3, 2001, 1408–1411.

Barto, A., Jordan, M. Gradient following without back-propagation in layered networks, *Proceedings of the IEEE International Conference on Neural Networks*, San Diego, 1987, pp. 629–636.

Bezdek, J. C. On the relationship between neural networks, pattern recognition and intelligence, *Int. J. Approx. Reason.*, **6**, 1992, 85–107.

Bezdek, J. What is computational intelligence, in: J. Zurada, R. Marks, C. Robinson (eds.), *Computational Intelligence*: Imitating Life, IEEE Press, Pisacataway, NJ, 1994, pp. 1–12.

Berthold, M., Hand, D. *Intelligent Data Analysis*, 2nd ed., Springer-Verlag, Heidelberg, 2006.

Bishop, C. *Pattern Recognition and Machine Learning*, Springer-Verlag, Heidelberg, 2006.

Caminhas, W., Tavares, H., Gomide, F., Pedrycz, W. Fuzzy set based neural networks: Structure, *Learning, and Application*, Journal of Advanced Computational Intelligence, 3(3), 1999, 151–157.

Castro, L., Timmis, J. *Artificial Immune Systems: A New Computational Intelligence Approach*, Springer, Heidelberg, 2002.

Cordón, O., Herrera, F., Hoffmann, F., Magdalena, L. *Genetic Fuzzy Systems: Evolutionary Tuning and Learning of Fuzzy Knowledge Bases*, World Scientific, Singapore, 2001.

Craenen, B., Eiben, A. Computational intelligence, 2006, http://www.cs.vu.nl/~bcraenen.

Delgado, M., Von Zuben, F., Gomide, F. Coevolutionary genetic fuzzy systems: A hierarchical collaborative approach, Fuzzy Sets and Systems, 141, 2004, 89–106.

Delgado, M., Von Zuben, F., Gomide, F. Hierarchical genetic fuzzy systems: Accuracy, interpretability, and design autonomy, in J. Casillas, O. Cordón, F. Herrera, L. Magdalena (Eds.), *Studies in Fuzziness and Soft Computing*, vol. 128, Interpretability Issues in Fuzzy Modeling, Springer, Heidelberg, 2003, 379–405.

Delgado, M. R., Von Zuben, F., Gomide, F. Local and global estimation of Takagi-Sugeno consequent parameters in genetic fuzzy systems, *Proceedings of 9th IFSA World Congress and 20th NAFIPS International Conference*, Vancouver, Canada, 2001a, pp. 1247–1252.

Delgado, M. R., Von Zuben, F., Gomide, F. Hierarchical genetic fuzzy systems, *Inf. Sci.* **136**(1–4), 2001b, 29–52.

Delgado, M. R., Von Zuben, F., Gomide, F. Hierarchical genetic fuzzy systems: accuracy, interpretability and design autonomy, in: J. Casillas, O. Cordon, F. Herrera and L. Magdalena (eds.), *Trade-off Between Accuracy and Interpretability in Fuzzy Rule-Based Modeling*, Physica-Verlag, Heidelberg, 2002.

Dorf, R., Bishop, R. *Modern Control Systems*, 10th ed., Prentice Hall, Upper Saddle River, NJ, 2004.

Dorigo, M., Stutzle, T. *Ant Colony Optimization*, MIT Press, Cambridge, MA, 2004.

Duch, W. What is computational intelligence, in: W. Duch and J. Mandziuk (eds.), *Challenges for Computational Intelligence*, Springer-Verlag, Heidelberg, 2007.

Duda, R., Hart, P., Stork, D. *Pattern Classification*, 2nd ed., John Wiley & Sons, Inc., New York, NY, 2001.

Dunham, M. *Data Mining: Introductory and Advanced Topics*, Prentice Hall, Upper Saddle River, NJ, 2003.

Eberhart, R., Simpson, P., Dobbins, R. *Computational Intelligence PC Tools*, Academic Press, Boston, MA, 1996

Eberhart, R., Shi, Y., Kennedy, J. *Swarm Intelligence*, Morgan Kaufmann, San Francisco, CA, 2001.

Haykin, S. *Neural Networks: A Comprehensive Foundation*, 2nd ed., Prentice Hall, Upper Saddle River, NJ, 1998.

Hell, M., Pires, P., Gomide, F. Recurrent neurofuzzy network in thermal modeling of power transformers, *IEEE Trans. Power Deliver.* 2007, 22(2), 904–910.

Hillier, F., Lieberman, G. *Introduction to Operations Research*, 8th ed., MacGraw-Hill, Boston, MA, 2005.

Hoffmann, F., Nelles, O. Structure identification of TSK-fuzzy systems using genetic programming, *Proceedings of IPMU'99*, Madrid, Spain, 2000, 438–445.

Jang, J. ANFIS: Adaptive-network-based fuzzy inference systems, *IEEE Trans. Syst. Man and Cyb.*, **23**, 1993, 665–685.

Juillé, H., Pollack, J. Co-evolving intertwined spirals, *Proceedings of the 5th Annual Conference on Evolutionary Programming*, San Diego, USA, 1996, 461–468.

Lee, C., Teng, C. Identification and control of dynamic systems using recurrent neural networks, *IEEE Trans. Fuzzy Syst.*, **8**(4), 2000, 349–366.

Lima, E., Gomide, F., Ballini, R. Participatory Evolving Fuzzy Modeling, *Proceedings of the 2006 International Symposium on Evolving Fuzzy Systems*, Ambleside, UK, 2006, pp. 36–41.

Mandic, D. P., Chambers, J. A. *Recurrent Neural Networks for Prediction*, John Wiley & Sons, Ltd, Chichester, 2001.

Michalewicz, Z. Genetic Algorithms + Data Structures = Evolution Programs, Springer, Berlin, 1996.

Mitchell, T. *Machine Learning*, MacGraw-Hill, Boston, MA, 1997.

Nürnberger, A., Radetzky, A., Kruse, R. Using recurrent neuro-fuzzy techniques for identification and simulation of dynamic systems, *Neural Comput.*, **36**, 2001, 123–147.

Pedrycz, W. *Computational Intelligence: An Introduction*, CRC Press, Boca Raton, FL, 1997.

Potter, M., De Jong, K. Cooperative coevolution: an architecture for evolving coadapted subcomponents, *Evol. Comput.*, **8**, 2000, 1–29.

Porter, M., De Jong, K. Coopertive coevolution: An architecture for evolving coadapted subcomponents, Evolutionary Computation, 8, 2000, 1–29.

Rosin, C., Belew, R. New methods for competitive coevolution, *Evol. Comput.*, **5**, 1997, 1–29.

Russell, S., Norvig, P. *Artificial Intelligence: A Modern Approach*, 2nd ed., Prentice Hall, Upper Saddle River, NJ, 2003.

Setnes, M., Babuska, R., Verbruggen, H. Rule-based modeling: precision and transparency, *IEEE Trans. Syst., Man Cyb. C: Appl. Rev.*, **28**(1), 1998, 165–169.

Santos, E., Von Zuben, F. Efficient second-order learning for discrete time recurrent neural networks, in: R. Medsker and L. Jain (eds.), *Recurrent Neural Networks: Design and Applications*, CRC Press, 1999, pp. 47–75.

Silva, L., Gomide, F., Yager, R. Participatory learning in fuzzy Clustering, *Proceedings of the 14nth IEEE International Conference on Fuzzy Systems, Reno*, USA, 2005, pp. 857–861.

Tibshirani, R., Hastie, T., Friedman, J. *The Elements of Statistical Learning*, Springer-Verlag, Heidelberg, 2001.

Walker, S. A brief history of connectionism and its psychological implications, in: A. Clark and R. Lutz (eds.), *The Connectionism in Context*, Springer-Verlag, Berlin,1992, pp. 123–144.

Weigend, A., Gershenfeld, N. *Time Series Prediction*: *Forecasting the Future and Understanding the Past*, Addison Wesley, Santa Fe, NM, 1992.

Yager, R. A model of participatory learning, *IEEE Trans. Syst. Man Cyb.*, **20**(5), 1990, 1229–1234.

Yen, J., Wang, L., and Gillespie, C. Improving the interpretability of TSK fuzzy models by combining global learning and local learning, *IEEE Trans. Fuzzy Syst.*, **6**(4), 1998, 530–537.

Chapter 14

Granular Models and Human-Centric Computing

Human-centric systems and human-centric computing are concerned with a functionality that makes systems highly responsive to the needs of human users. We fully acknowledge a fact that there could be a genuine diversity of requirements and preferences that might well vary from user to user. How could we build systems that are capable of accommodating such needs and offering in this way a high level of user-friendliness? There are numerous interesting scenarios one can envision in which human centricity plays a vital role. For instance, in system modeling, a user may wish to model the reality based on a unique modeling perspective. In this sense the data being available for modeling purposes are to be looked at and used in the construction of the model within a suitable context established by the user. In information retrieval and information organization (no matter whether we are concerned with audio, visual, or hypermedia information), the same collection of objects could be structured and looked at from different standpoints depending upon the preferences of the individual user. In this case, an ultimate functionality of human-centric systems is to achieve an effective realization of relevance feedback provided by the user.

In this chapter, we are concerned with a category of fuzzy modeling that directly explores the underlying ideas of fuzzy clustering and leads to the concept of granular models. The essence of these models is to describe associations between information granules, namely, fuzzy sets formed both in the input and output spaces. The context within which such relationships are being formed is established by the system developer. Information granules are built using specialized, conditional (context-driven) fuzzy clustering. This emphasizes the human-centric character of such models: it is the designer who assumes an active role in the process of forming information granules and casting all modeling pursuits in a right framework. Owing to the straightforward design process, granular models become particularly useful in rapid system prototyping.

Fuzzy Systems Engineering: Toward Human-Centric Computing, by Witold Pedrycz and Fernando Gomide
Copyright © 2007 John Wiley & Sons, Inc.

Fuzzy clustering has established itself as one of the methodological and algorithmic pillars of information granulation. As we have noted in the previous chapters, the algorithms of fuzzy clustering convert masses of numeric data into semantically meaningful information granules. The human centricity brings here a customized view at the structure of data that comes in the form of some knowledge hints offering guidance to the processes of revealing this structure. Here we look thoroughly at the two mechanisms such as clustering with partial supervision and proximity-based fuzzy clustering. We demonstrate how they realize the mechanisms of relevance feedback. Next, we study an interesting mechanism of participatory learning (PL) in fuzzy clustering.

14.1 THE CLUSTER-BASED REPRESENTATION OF THE INPUT–OUTPUT MAPPINGS

Fuzzy clusters (Pedrycz and Vasilakos, 1999; Pedrycz, 2005) establish a sound basis for constructing fuzzy models. By forming fuzzy clusters in input and output spaces, we span the fuzzy model on a collection of prototypes. More descriptively, these prototypes are regarded as a structural *skeleton* or a design *blueprint* of the model. Once the prototypes have been formed, there are several ways of developing the detailed expressions governing the detailed relationships of the model. The one commonly encountered in the literature takes the prototypes formed in the output space, that is, $z_1, z_2, \ldots, z_c \in \mathbf{R}$ and combines them linearly by using the membership grades of the corresponding degrees of membership of the fuzzy clusters in the input space. Consider some given input $\mathbf{x}$. Denote the corresponding grades of membership produced by the prototypes $\mathbf{v}_1, \mathbf{v}_2, \ldots,$ and $\mathbf{v}_c$ located in the input space by $u_1(\mathbf{x}), u_2(\mathbf{x}), \ldots, u_c(\mathbf{x})$. The output reads as follows:

$$y = \sum_{i=1}^{c} z_i u_i(\mathbf{x}) \qquad (14.1)$$

The value of $u_i(\mathbf{x})$ being the degree at which $\mathbf{x}$ is compatible with the ith cluster in the input space is computed in a similar way as encountered in the FCM algorithm, (Bezolek and Pal, 1992) that is,

$$u_i(\mathbf{x}) = \frac{1}{\sum_{j=1}^{c} \left(\frac{\|\mathbf{x} - \mathbf{v}_i\|}{\|\mathbf{x} - \mathbf{v}_j\|} \right)^{2/(m-1)}} \qquad (14.2)$$

where m is a fuzzification factor (fuzzification coefficient). The reader familiar with radial basis function (RBF) neural networks (Karyannis and Mi, 1997; Ridella et al., 1998; Rovetta and Zunino, 2000; Joo et al., 2002; Duda and Hart, 1973; Gath and Geva, 1989; Hecht Nielsen, 1990; Kim and Park, 1997; Park et al., 2002) can easily recognize that a structure of (14.1) closely resembles the architecture of RBF neural networks. There are some striking differences. First, the receptive fields provided by (14.2) are automatically constructed without any need for their further adjustments (which is usually not the case in standard RBFs). Second, the form of the RBF is far more flexible than the commonly encountered RBFs such as, for example, Gaussian functions.

14.1 The Cluster-Based Representation of the Input–Output Mappings

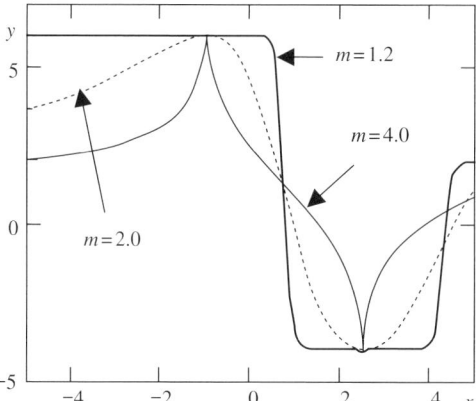

Figure 14.1 Nonlinear input–output characteristics of the cluster-based model. Prototypes are fixed ($v_1 = -1$, $v_2 = 2.5$, $v_3 = 6.1$; $z_1 = 6$, $z_2 = -4$, $z_3 = 2$), whereas the fuzzification coefficient (m) assumes several selected values ranging from 1.2 to 4.0.

It is instructive to visualize the characteristics of the model, which realizes some nonlinear mapping from the input to output space. The character of this nonlinearity depends upon the distribution of the prototypes. It can be easily affected by the values of the fuzzification factor m. Figure 14.1 illustrates input–output characteristics for the fixed values of the prototypes and varying values of the fuzzification factor. The commonly chosen value of m equal to 2.0 is also included. Undoubtedly, this design parameter exhibits a significant impact on the character of nonlinearity being developed. The values of m close to 1 produce a stepwise character of the mapping; we observe significant jumps located at the points where we switch between the individual clusters in input space. In this manner the impact coming from each rule is very clearly delineated. The typical value of the fuzzification coefficient set up to 2.0 yields a gradual transition between the rules, and this shows up through smooth nonlinearities of the input–output relationships of the model. The increase in the values of m, as shown in Figure 14.1, yields quite *spiky* characteristics: we quickly reach some modal values when moving close to the prototypes in the input space whereas in the remaining cases the characteristics switch between them in a relatively abrupt manner positioning close to the averages of the modes.

Figure 14.2 illustrates the characteristics of the model in case of different values of the prototypes. Again it becomes apparent that by moving the prototypes we are able to adjust the nonlinear mapping of the model to the existing experimental data.

It is helpful to contrast this fuzzy model with the RBF network equipped with Gaussian receptive fields that is governed by the expression

$$y = \frac{\sum_{i=1}^{c} z_i G(x; v_i, \sigma)}{\sum_{i=1}^{c} G(x; v_i, \sigma)} \qquad (14.3)$$

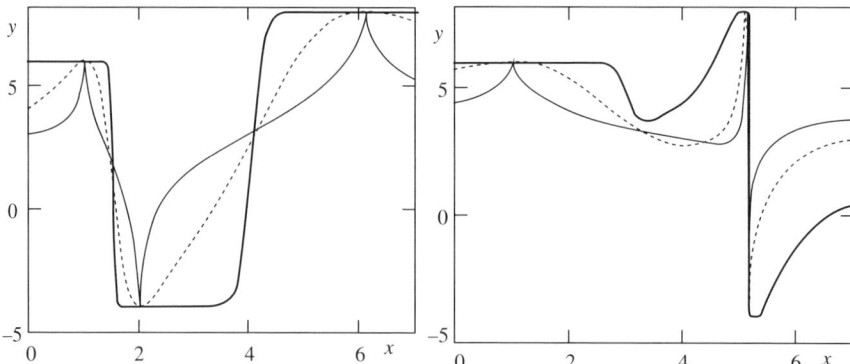

Figure 14.2 Nonlinear input–output characteristics of the cluster-based model. The prototypes in the input space vary; the distribution of the prototypes in the output space is equal to $z_1 = 6$, $z_2 = -4$, $z_3 = 8$; $m = 2$.

where $G(x; m, \sigma)$ denotes a certain Gaussian function (receptive field) characterized by its modal value (m) and spread (σ). Note that in (14.3) we usually require some normalization as the sum of these receptive fields may not always generate the value equal to 1. As shown in Figure 14.3, there is also no guarantee that for the prototypes the model coincides with the prototypes defined in the output space. This effect can be attributed to the smoothing effect of the Gaussian receptive fields. This stands in sharp contrast to the nonlinear relationship formed by the fuzzy partition.

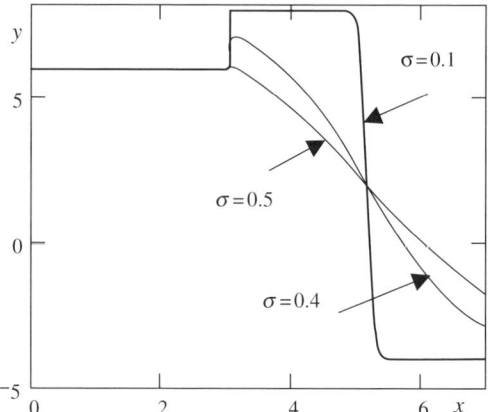

Figure 14.3 Nonlinear input–output characteristics of RBF network; the changes in the spread values of the receptive fields show somewhat a similar effect to the one produced by the fuzzification factor. The remaining parameters of the model are fixed and equal to $z_1 = 6$, $z_2 = -4$, $z_3 = 8$; $v_1 = 1$; $v_2 = 5.2$, $v_3 = 5.1$.

14.2 CONTEXT-BASED CLUSTERING IN THE DEVELOPMENT OF GRANULAR MODELS

Clustering plays a crucial role in granular modeling. First, it helps convert numeric data into information granules without or with some hints of available domain knowledge. The produced information granules form a backbone or a *blueprint* of the model. Although the model could be further refined, it is predominantly aimed at capturing the most essential, numerically dominant features of data by relying on its summarization. The more clusters we intend to capture, the more detailed the resulting blueprint becomes. It is important to note that clustering helps manage dimensionality problem that is usually a critical issue in rule-based modeling. As being naturally based on Cartesian products of input variables rather than individual variables, they offer a substantial level of dimensionality reduction. Let us remind that in any rule-based system, the number of input variables plays a pivotal role when it comes to the dimensionality of the resulting rule base. A complete rule base consists of p^n where p is the number of information granules in each input variable and n denotes the total number of the input variables. Even in case of a fairly modest dimensionality of the problem, (say, $n = 10$) and very few information granules defined for each variable (say, $p = 4$), we end up with a significant number of rules, that is, $4^{10} = 1049 \times 10^6$. By keeping the same number of variables and using eight granular terms (information granules), we observe a tremendous increase in the size of the rule base; here we end up with 2.825×10^8 different rules that amounts to substantial increase. There is no doubt that such rule bases are not practical. The effect of this combinatorial explosion is clearly visible, refer to Figure 14.4.

There are several ways of handling the dimensionality problem. An immediate one is to acknowledge that we do not need a complete rule base because there are various combinations of conditions that never occur in practice and are not supported by any experimental evidence. Although this sounds very straightforward, it is not that easy to gain sufficient confidence as to the nature of such unnecessary rules. The

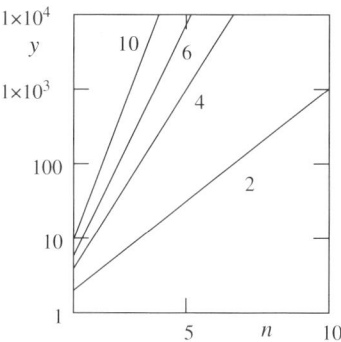

Figure 14.4 Number of rules treated as a function of input variables; the dependency is illustrated for several values of p. Observe a logarithmic scale of the y coordinate.

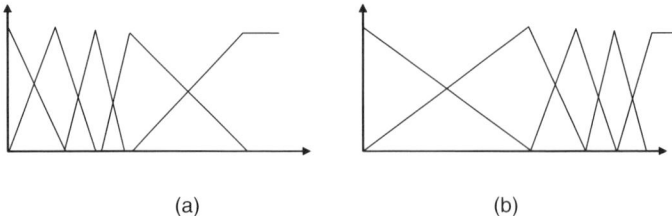

Figure 14.5 Examples of fuzzy sets of context reflecting a certain focus of the intended model: (a) *low*-speed traffic, (b) *high*-speed traffic.

second more feasible approach would be to treat all variables at the same time by applying fuzzy clustering. The number of clusters is far smaller than the number of rules involving individual variables.

In what follows, we discuss various modes of incorporating clustering results into the blueprint of the fuzzy model. It is very important to understand the implications of the use of the clustering technique in forming information granules especially in the setting of the models. The most critical observation concerns a distinction between relational aspects of clustering and directional features of models. By their nature, unless properly endowed, clustering looks at multivariable data as relational constructs, so the final product of cluster analysis results in a collection of clusters as concise descriptors of data where each variable is treated in the same manner irrespectively where it is positioned as a modeling entity. Typical clustering algorithms do not distinguish between input and output variables. This stands in sharp contrast with what we observe in system modeling. The role of the variable is critical as most practical models are directional constructs, namely, they represent a certain mapping from independent to dependent variables. The distinction between these two categories of variables does require some modifications to the clustering algorithm to accommodate this requirement. To focus our attention, let us consider a many input–many output (MIMO) model involving input and output variables **x** and **y**, respectively, that is, **y** = f(**x**). The experimental data come in the format of ordered tuples $\{(\mathbf{x}_k, \mathbf{target}_k)\}$, $k = 1, 2, \ldots, N$. If we are to ignore the directionality aspect, the immediate approach to clustering the data would be to concatenate the vectors in the input and output space so that $\mathbf{z}_k = [\mathbf{x}_k\ \mathbf{target}_k]$ and carry our clustering in such augmented feature space. By doing that we have concentrated on the relational aspects of data and the possible mapping component that is of interest and has been completely ignored. To alleviate the problem, we may like to emphasize a role of the output variables by assigning to them higher values of weights. In essence, we emphasize that the clustering needs to pay more attention to the differences (and similarities) occurring in the output spaces (i.e., vectors **target**$_1$, **target**$_2$, …, **target**$_n$). This issue was raised in the past and resulted in a so-called directional fuzzy clustering, D-fuzzy clustering, for short (Hirota and Pedrycz, 1995). An easy realization of the concept would be to admit that the distance function has to be computed differently depending upon the coordinates of **z** by using a certain

14.2 Context-Based Clustering in the Development of Granular Models

positive weight factor γ. For instance, the Euclidean distance between $\mathbf{z}_k$ and $\mathbf{z}_l$ would read as $\| \mathbf{x}_k - \mathbf{x}_l \|^2 + \gamma \| \mathbf{target}_k - \mathbf{target}_l \|^2$ where $\gamma > 0$. The higher the value of γ, the more the attention focused on the distance between the output variables. As usually the dimensionality of the input space is far higher than the output one, the value of γ needs to be properly selected to reach a suitable balance. Even though the approach might look sound, the choice of the weight factor becomes a matter of intensive experimentation.

Conditional clustering (Pedrycz, 1998) is naturally geared toward dealing with direction-sensitive (direction-aware) clustering. The context variable(s) are those being the output variables used in the modeling problem. Defining contexts over these variables become an independent task. Once the contexts have been formed, the ensuing clustering is induced (or directed) by the provided fuzzy set (relation) of context. Let us link this construct to rule-based modeling. In context-based clustering, the role of the conclusion is assumed by the context fuzzy set. The clusters formed in the input space form the conditions of the rule. Being more specific, the rule is of the form

If $\mathbf{x}$ is A_1 or A_2 or ... or A_c, then y is B

where B is a context fuzzy set and $A_1, A_2, \ldots, A_c$ are fuzzy sets (clusters) formed in the input space.

Given the context, we focus the pursuits of fuzzy clustering on the pertinent portion of data in the input space and reveal a conditional structure there. By changing the context we continue search by focusing on some other parts of the data. In essence, the result produced in this manner becomes a web of information granules developed conditionally upon the assumed collection of the contexts. Hence, the directional aspects of the model we want to develop on the basis of the information granules become evident. The design of contexts is quite intuitive. First, these are fuzzy sets whose semantics is well defined. We may use terms such as *low*, *medium*, and *large* output. Second, we can choose fuzzy sets of context so that they reflect the nature of the problem and our perception of it. For instance, if for some reason we are interested in modeling a phenomenon of a slow traffic on a highway, we would define a number of fuzzy sets of context focused on low values of speed. To model highway traffic with focus on high speed, we would be inclined to locate a number of fuzzy sets at the high end of the output space. The customization of the model by identifying its major focus is thus feasible through setting the clustering activities in a suitable context.

We can move on with some further refinements of context fuzzy sets and, if required, introduce a larger number of granular categories. Their relevance could be assessed with regard to the underlying experiment (Pedrycz, 1998). To assure full coverage of the output space, it is advisable that fuzzy sets of context form a fuzzy partition. Obviously, we can carry out clustering of data in the output space and arrive at some membership functions being generated in a fully automated fashion. This option is particularly attractive in case of many output variables to be treated together where the manual definition of the context fuzzy relations could be too tedious or even impractical.

In what follows, we briefly recall the essence of conditional (context-based) clustering and elaborate on the algorithmic facet of the optimization process. This clustering, which is a variant of the FCM, is realized for individual contexts, W_1, $W_2, \ldots, W_p$. Let us consider a certain fixed context (fuzzy set) W_j described by some membership function (the choice of its membership will be discussed later on). For the sake of convenience, we consider here a single output variable. A certain data point (target$_k$) located in the output space is then associated with the corresponding membership value of W_j, $w_{jk} = W_j(\text{target}_k)$. Let us introduce a family of the partition matrices induced by the jth context and denote it by $U(W_j)$

$$U(W_j) = \left\{ u_{ik} \in [0, 1] \mid \sum_{i=1}^{c} u_{ik} = w_{jk} \ \forall \ k, \text{ and } 0 < \sum_{k=1}^{N} u_{ik} < N \ \forall \ i \right\} \quad (14.4)$$

where w_{jk} denotes a membership value of the kth data point to the jth context (fuzzy set). The objective function guiding clustering is defined as the following weighted sum of distances:

$$Q = \sum_{i=1}^{c} \sum_{k=1}^{N} u_{ik}^m \|\mathbf{x}_k - \mathbf{v}_i\|^2 \quad (14.5)$$

The minimization of Q is realized with respect to the prototypes $\mathbf{v}_1, \mathbf{v}_2, \ldots, \mathbf{v}_c$ and the partition matrix $U \in U(W_j)$. The optimization is realized iteratively by updating the partition matrix and the prototypes in a consecutive manner (Pedrycz, 1996). More specifically, the partition matrix are computed as follows:

$$u_{ik} = \frac{w_{jk}}{\sum_{j=1}^{c} \left(\frac{\|\mathbf{x}_k - \mathbf{v}_i\|}{\|\mathbf{x}_k - \mathbf{v}_j\|} \right)^{\frac{2}{m-1}}} \quad i = 1, 2, \ldots, c, \ k = 1, 2, \ldots, N \quad (14.6)$$

Let us emphasize here that the values of u_{ik} pertain here to the partition matrix induced by the jth context. The prototypes $\mathbf{v}_i$, $i = 1, 2, \ldots, c$ are calculated in the

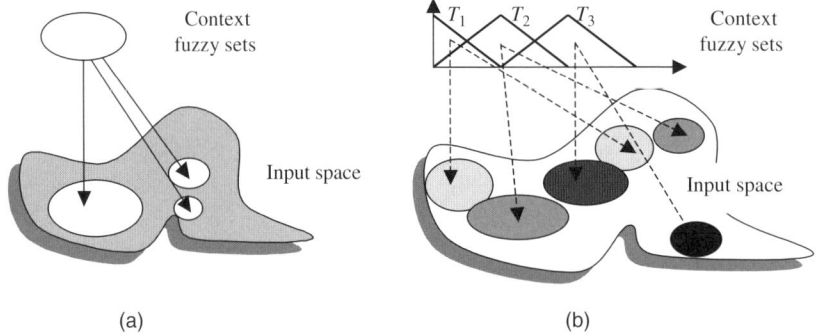

Figure 14.6 (a) A blueprint of a granular model induced by some predefined fuzzy sets or relations of context defined in the input space; and (b) a detailed view at the model in case of three contexts and two clusters per context.

well-known form of the weighted average

$$\mathbf{v}_i = \frac{\sum_{k=1}^{N} u_{ik}^m \mathbf{x}_k}{\sum_{k=1}^{N} u_{ik}^m} \quad (14.7)$$

We iterate (14.6) and (14.7) until some predefined termination criterion has been satisfied.

The blueprint of the model, Figure 14.6, has to be further formalized to explicitly capture the mapping between the information granules. This leads us to a detailed architecture of an inherently granular network whose outputs are information granules. The concept of a granular or granular neuron becomes an interesting construct worth exploring in this setting.

14.3 GRANULAR NEURON AS A GENERIC PROCESSING ELEMENT IN GRANULAR NETWORKS

As the name suggests, by the granular neuron we mean a neuron with granular connection. More precisely, we consider the transformation of many numeric inputs $u_1, u_2, \ldots, u_c$ (confined to the unit interval) of the form

$$Y = N(u_1, u_2, \ldots, u_c, W_1, W_2, \ldots, W_c) = \sum_{\oplus} (W_i \otimes u_i) \quad (14.8)$$

with $W_1, W_2, \ldots, W_c$ denoting granular weights (connections), see Figure 14.7. The symbols of generalized (granular) addition and multiplication (i.e., $\oplus, \otimes$) are used here to emphasize a granular character of the arguments being used in this aggregation. When dealing with interval-valued connections, $W_i = [w_{i-}, w_{i+}]$, the operations of their multiplication by some positive real input u_i produce the results in the form of the following interval:

$$W_i \otimes u_i = [w_{i-} u_i, \ w_{i+} u_i] \quad (14.9)$$

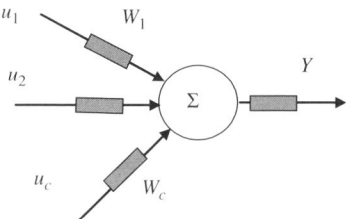

Figure 14.7 Computational model of a granular neuron; note a granular character of the connections and ensuing output Y.

When adding such intervals being produced at the level of each input of the neuron, we obtain

$$Y = \left[\sum_{i=1}^{n} w_{i-} u_i, \sum_{i=1}^{n} w_{i+} u_i \right] \qquad (14.10)$$

For the connections represented as fuzzy sets, the result of their multiplication by a positive scalar u_i is realized through the use of the extension principle

$$(W_i \otimes u_i)(y) = \sup_{w:y=wu_i}[W_i(w)] = W_i(y/u_i) \qquad (14.11)$$

Next, the extension principle is used to complete additions of fuzzy numbers being the partial results of this processing. Denote by Z_i the fuzzy number $Z_i = W_i \otimes u_i$. We obtain

$$Y = Z_1 \oplus Z_2 \oplus \cdots \oplus Z_n \qquad (14.12)$$

that is,

$$Y(y) = \sup\{\min(Z_1(y_1), Z_2(y_2), \ldots, Z_n(y_n))\}$$
$$\text{s.t.} \ y = y_1 + y_2 + \cdots + y_n \qquad (14.13)$$

Depending on a specific realization, these connections can be realized as intervals, fuzzy sets, shadowed sets, rough sets, and so on. In spite of the evident diversity of the formalisms of granular computing under consideration, the output Y is also a granular construct, Figure 14.7.

EXAMPLE 14.1

Consider the granular neuron with two inputs $u_1 = \alpha$ and $u_2 = 1 - \alpha$ as shown in Figure 14.8. Assume the granular weights represented as intervals and being equal to $W_1 = [0.3, 3]$ and

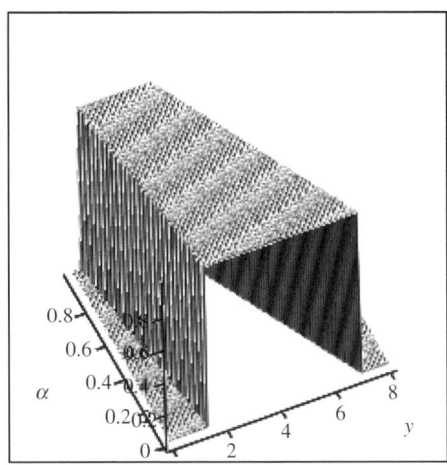

Figure 14.8 The interval output of the granular neuron in case of interval connections.

14.4 Architecture of Granular Models Based on Conditional Fuzzy Clustering

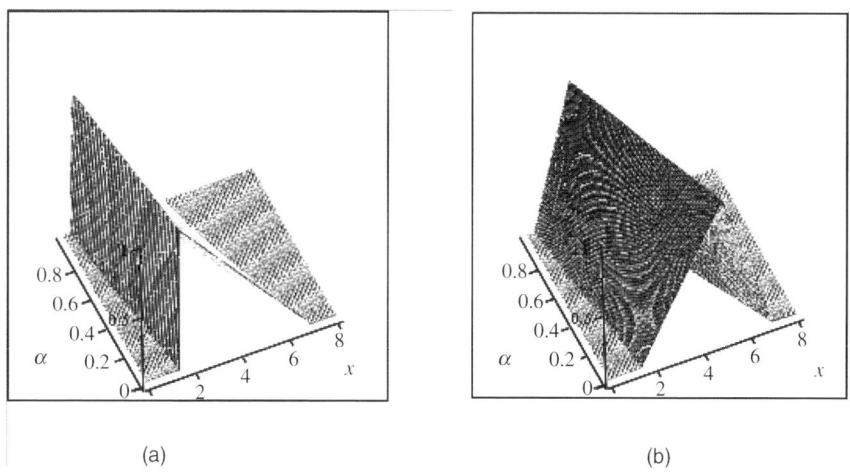

Figure 14.9 The output of the granular neuron in case of connections as triangular fuzzy numbers: (a) $W_1 = \langle 0.3, 0.5, 3.0 \rangle$, $W_2 = \langle 1.4, 1.5, 7.0 \rangle$ and (b) $W_1 = \langle 0.3, 2.0, 3.0 \rangle$, $W_2 = \langle 1.4, 5.0, 7.0 \rangle$.

$W_2 = [1.4, 7]$, respectively. The output of the neuron is an interval; as α changes from 0 to 1, it moves along the y axis.

In case of fuzzy sets used to implement the connections, we end up with more complicated formulas for the resulting output. They simplify quite profoundly if we confine ourselves to triangular fuzzy sets (fuzzy numbers) of the connections. Following the calculus of fuzzy numbers, we note that the multiplication of W_i by a positive constant scales the fuzzy number yet retains the piecewise character of the membership function. Furthermore, the summation operation does not affect the triangular shapes of the membership function, so at the end the final result can be again described in the following format:

$$Y = \langle \sum_{i=1}^{c} a_i u_i, \sum_{i=1}^{c} m_i u_i, \sum_{i=1}^{c} b_i u_i \rangle \qquad (14.14)$$

where each connection W_i is fully characterized by the triple of real numbers (parameters) $W_i = \langle a_i, m_i, b_i \rangle$. Here m_i denotes a modal value of the connection whereas a_i and b_i stand for the lower and upper bound of the triangular number describing this fuzzy set of the connection. The plot of the output of the neuron for u_1 and u_2 defined as above is included in Figure 14.9.

The granular neuron exhibits several interesting properties that generalize the characteristics of (numeric) neurons. Adding a nonlinearity component (g) to the linear aggregation does not change the essence of computing; in case of monotonically increasing relationship ($g(Y)$), we end up with a transformation of the original output interval or fuzzy set (in this case we have to follow the calculations using the well-known extension principle).

14.4 ARCHITECTURE OF GRANULAR MODELS BASED ON CONDITIONAL FUZZY CLUSTERING

The conditional fuzzy clustering has provided us with a backbone of the granular model (Pedrycz and Vasilakos, 1999). Following the principle of conditional

430 Chapter 14 Granular Models and Human-Centric Computing

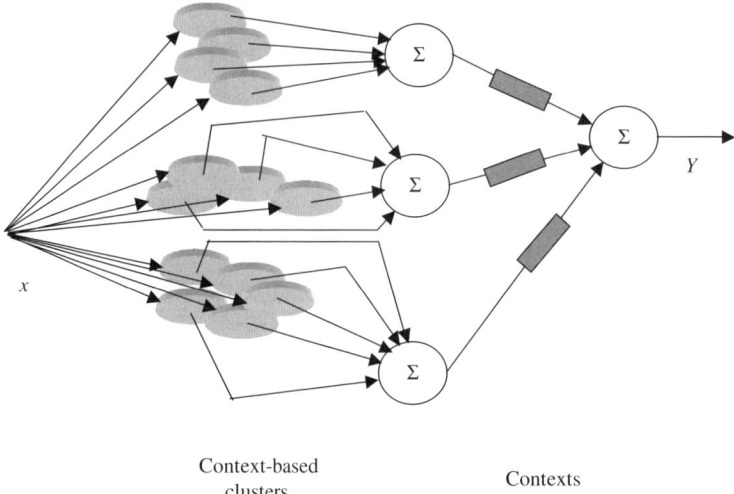

Figure 14.10 The overall architecture of the granular models. Context are shown as rectangular connections of the output V neurons.

clustering, we end up with a general topology of the model shown in Figure 14.10. It reflects a way in which the information granules are put together following the way the information granules have been formed.

The computations of the output fuzzy sets are completed in two successive steps: (a) aggregation of activation levels (membership grades) of all clusters associated with a given context and (b) linear combination of the activation levels with the parameters of the context fuzzy sets. In case of triangular membership functions of the context fuzzy sets, the calculations follow the scheme described by (14.8).

The development of the granular model comprises of two main phases, that is,

1. forming fuzzy sets of context, and
2. conditional clustering for the already available collection of contexts.

These two processing phases are tied together: once contexts have been provided, the clustering uses this information in the directed development of the structure in the input space.

The granular models come with several important features:

- In essence, the granular model is nothing but a web of associations between information granules that have been constructed.
- The model is inherently granular; even for a numeric input, the model returns some information granule, in particular some triangular fuzzy set (fuzzy number).
- The model is built following a design scheme of rapid prototyping. Once the information granules have been defined in the output space and constructed in the input space, no further design effort is required. Simply, we organize them in the topology as presented in Figure 14.10.

14.5 REFINEMENTS OF GRANULAR MODELS

The granular model can be augmented by several features with the intension of improving its functionality and accuracy. The first modification is straightforward and concerns a bias component of the granular neuron. The second one focuses on an iterative scheme of optimization of the fuzzy sets of context.

14.5.1 Bias of Granular Neurons

So far, the web of the connections between the contexts and their induced clusters was very much reflective of how the clustering has been realized. Although the network can be assembled without any further computing effort, it becomes useful to look into its further refinement. In particular, we would look whether the model is not biased by a systematic error and if so, make some modifications to the topology of the granular model to alleviate this shortcoming. A numeric manifestation of the granular model can be viewed in the form of the modal value of the output fuzzy set of the granular model, see Figure 14.11. Denote it by y_k considering that the fuzzy set itself is given as $Y(\mathbf{x}_k)$. If the mean of the error being computed as $\text{target}_k - y_k$ $k = 1, 2, \ldots, N$ is nonzero, we are faced with a systematic error. Its elimination can be realized by involving a bias term as illustrated in Figure 14.11. This bias augments the summation node at the output layer of the network.

The bias is calculated in a straightforward manner.

$$w_0 = -\frac{1}{N} \sum_{k=1}^{N} (\text{target}_k - y_k) \qquad (14.15)$$

In terms of granular computing, the bias is just a numeric singleton that could be written down as a degenerated fuzzy number (singleton) of the form $W_0 = (w_0, w_0, w_0)$.

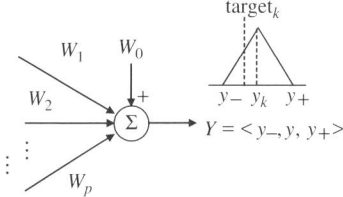

Figure 14.11 Inclusion of the bias term in the granular model.

Subsequently, the resulting granular output Y reads in the form

1. the lower bound $\quad \sum_{t=1}^{p} z_t w_{t-} + w_0$

2. modal value $\quad \sum_{t=1}^{p} z_t w_t + w_0$

3. the upper bound $\quad \sum_{t=1}^{p} z_t w_{t+} + w_0$

14.5.2 Refinement of the Contexts

The conditional FCM has produced the prototypes or, equivalently, clusters in the input space. Using them we generate inputs to the granular neuron. The connections of the neuron are the fuzzy sets of context. In essence, the parameters of the network are downloaded from the phase of fuzzy clustering. This constitutes an essence of rapid prototyping.

There is however some room for improvement if one might involve in further optimization activities. The refinement may be necessary because of the fact that each conditional FCM is realized for some specific context, and these developments tasks are independent. As a consequence of putting all pieces together, the prototypes may need some shifting. Furthermore, the contexts themselves may require refinement and refocus. Note also that the result of the granular model is an information granule (interval, fuzzy set, fuzzy relation, etc.), and this has to be compared with a numeric datum y_k. Again, for illustrative purposes we have to confine ourselves to a single output granular model. Thus, the optimization has to take this into account. As we are concerned with numeric and granular entities, there could be several ways of assessing the quality of the granular model and its further refinement, see Figure 14.12.

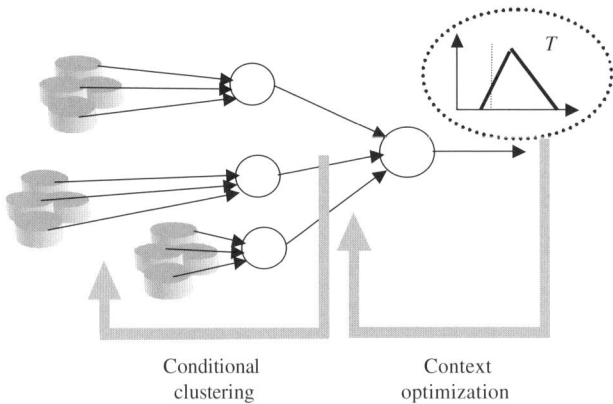

Figure 14.12 An optimization scheme of the granular model.

We put forward two optimization problems in which the minimization (and maximization) is carried out with respect to the parameters (modal values) of the fuzzy sets of context. The maximization of the average agreement of the granular output of the model with available numeric data is a straightforward one. Let us consider that Y is a granular output of the granular model produced for input $\mathbf{x}_k$, $Y(\mathbf{x}_k)$. As it is described by the membership function defined in the output space, we compute the membership degree at the value target$_k$, that is, $Y(\mathbf{x}_k)$ (target$_k$).

$$\max_{\mathbf{P}} \frac{1}{N} \sum_{k=1}^{N} Y(\mathbf{x}_k)(\text{target}_k) \qquad (14.16)$$

As mentioned earlier, the maximization of (14.16) is completed with respect to the parameters of the context fuzzy sets where these parameters are collectively denoted by $\mathbf{P}$.

Alternatively, we can consider a minimization of the average spread of the granular output of the network obtained for the corresponding inputs

$$\min_{\mathbf{P}} \frac{1}{N} \sum_{k=1}^{N} (b_k - a_k) \qquad (14.17)$$

where a_k and b_k are the lower and upper bounds of the triangular fuzzy number produced for $\mathbf{x}_k$. In both cases the optimization can be confined to the portion of the network requiring the refinement of the context fuzzy sets. Furthermore, we can make the optimization more manageable by assuming that the successive contexts overlap at the level of $\frac{1}{2}$. Given this condition, the optimization concentrates on the modal values of the triangular fuzzy sets of context. Once these values have been adjusted, the conditional FCM is repeated and the iteration loop of optimization is repeated.

14.6 INCREMENTAL GRANULAR MODELS

We can take another, less commonly adopted principle of fuzzy modeling whose essence could be succinctly captured as follows:

Adopting a construct of a linear regression as a first-principle global *model, refine it through a series of local fuzzy rules that capture remaining and more localized nonlinearities of the system.*

More schematically, we could articulate the essence of the resulting fuzzy model by stressing the existence of the two essential modeling structures that are combined together using the following symbolic relationship:

fuzzy model = linear regression and local granular models.

By endorsing this principle, we emphasize the tendency that in system modeling we always proceed with the simplest possible model (Occam's principle), assess its performance, and afterward complete a series of necessary refinements. The local

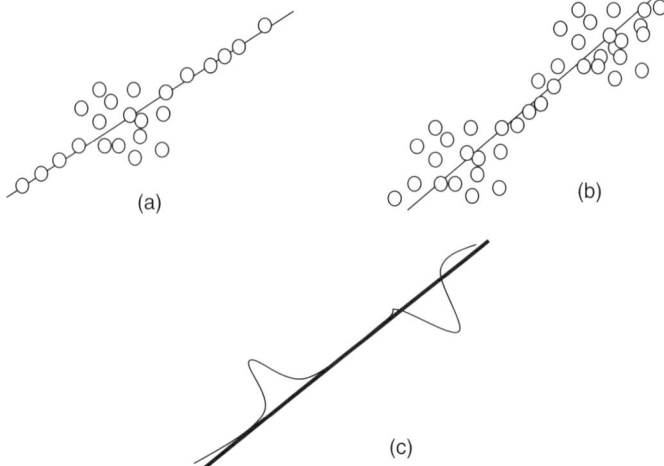

Figure 14.13 Examples of nonlinear relationships and their modeling through a combination of linear models of global character and a collection of local rule-based models.

granular models handling the residual part of the model are conveniently captured through some rules.

Let us proceed with some illustrative examples, shown in Figure 14.13, that help underline and exemplify the above development principle. In the first case, Figure 14.13 (a), the data are predominantly governed by a linear relationship whereas there is only a certain cluster of points that are fairly dispersed within some region. In the second one, (b), the linearity is visible, yet there are two localized clusters of data that contribute to the local nonlinear character of the relationship. In (c) there is a nonlinear function, yet it exhibits quite dominant regions of linearity. This is quite noticeable when completing a linear approximation; the linear regression exhibits a pretty good match with a clear exception of the two very much compact regions. Within such regions, one could accommodate two rules that capture the experimental relationships present there. The nonlinearity and the character of data vary from case to case. In the first two examples, we note that the data are quite dispersed, and the regions of departure from the otherwise linear relationship could be modeled in terms of some rules. In the third one, the data are very concentrated and with no scattering, yet the nonlinear nature of the relationship is predominantly visible.

14.6.1 The Principle of Incremental Fuzzy Model and Its Design and Architecture

The fundamental scheme of the construction of the incremental granular model is covered as illustrated in Figure 14.14. There are two essential phases: the development of the linear regression being followed by the construction of the local granular

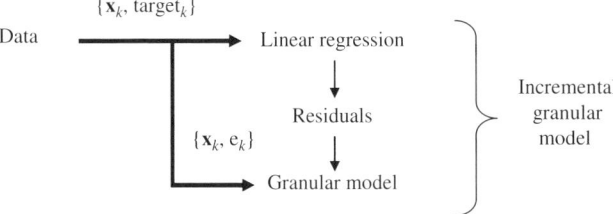

Figure 14.14 A general flow of the development of the incremental granular models.

rule-based constructs that attempt to eliminate errors (residuals) produced by the regression part of the model.

Before proceeding with the architectural details, it is instructive to start with some required notation. The experimental data under discussion are the pairs of the n-dimensional inputs and scalar inputs. They come in the following form $\{\mathbf{x}_k, \text{target}_k\}$ $k = 1, 2, \ldots, N$, where $\mathbf{x}_k \in \mathbf{R}^n$ and $\text{target}_k \in \mathbf{R}$. The first-principle linear regression model comes in the standard form of $z_k = L(\mathbf{x}, \mathbf{b}) = \mathbf{a}^T \mathbf{x}_k + a_0$ where the values of the coefficients of the regression plane, denoted here by a_0 and $\mathbf{a}$, namely $\mathbf{b} = [\mathbf{a}\ a_0]^T$, are determined through the standard least-square error method as encountered in any statistics textbooks. The enhancement of the model at which the granular part comes into the play is based on the transformed data $\{\mathbf{x}_k, e_k\}$, $k = 1, 2, \ldots, N$ where the residual part manifests through the expression $e_k = \text{target}_k - z_k$ that denotes the error of the linear model. In the sequel, those data pairs are used to develop the incremental and granular rule-based part of the model. Given the character of the data, this rule-based augmentation of the model associates input data with the error produced by the linear regression model in the form of the rules if input then error. The rules and the information granules are constructed by means of the context-based clustering.

Let us recall the main design phases of the model; refer also to Figure 14.15 showing how the two functional modules operate.

Initial setup of the modeling environment: Decide upon the granularity of information to be used in the development of the model, namely, the number of contexts and the number of clusters formed for each context. Similarly, decide upon the parameters of the context-based clustering, especially the value of the fuzzification coefficient. The choice of the (weighted) Euclidean distance is a typical choice in the clustering activities.

1. Design of a linear regression in the input–output space, $z = L(\mathbf{x}, \mathbf{b})$ with $\mathbf{b}$ denoting a vector of the regression hyperplane of the linear model, $\mathbf{b} = [\mathbf{a}\ a_0]^T$. On the basis of the original data set a collection of input-error pairs is formed, $(\mathbf{x}_k, e_k)$ where $e_k = \text{target}_k - L(\mathbf{x}_k, \mathbf{a})$.

2. Construction of the collection of contexts fuzzy sets in the space of error of the regression model $E_1, E_2, \ldots, E_p$. The distribution of these fuzzy sets is optimized through the use of fuzzy equalization whereas the fuzzy sets are characterized by triangular membership functions with a 0.5 overlap between

436   Chapter 14   Granular Models and Human-Centric Computing

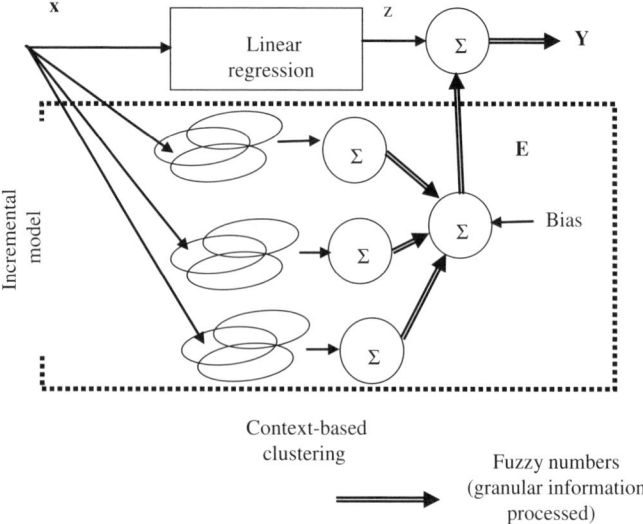

Figure 14.15 The overall flow of processing realized in the design of the incremental granular model. Note a flow of numeric and granular computing.

neighboring fuzzy sets (recall that such arrangement of fuzzy sets leads to a zero decoding error).

3. Context-based FCM completed in the input space and induced by the individual fuzzy sets of context. For p contexts and c clusters for each context, obtained are $c \times p$ clusters.

4. Summation of the activation levels of the clusters is induced by the corresponding contexts and their overall aggregation through weighting by fuzzy sets (triangular fuzzy numbers) of the context leading to the triangular fuzzy number of output, $E = F(\mathbf{x}; E_1, E_2, \ldots, E_p)$ where F denotes the overall transformation realized by the incremental granular model. Furthermore, note that we eliminated eventual systematic shift of the results by adding a numeric bias term. These two functional modules are illustrated in Figure 14.15.

5. The granular result of the incremental model is then combined with the output of the linear part; the result is a shifted triangular number Y, $Y = z \oplus E$.

EXAMPLE 14.2

One-dimensional spiky function, spiky(x), used in this experiment is a linear relationship augmented by two Gaussian functions described by their modal values m and spreads σ,

$$G(x) = \exp\left(\frac{-(x-c)^2}{2\sigma^2}\right) \qquad (14.18)$$

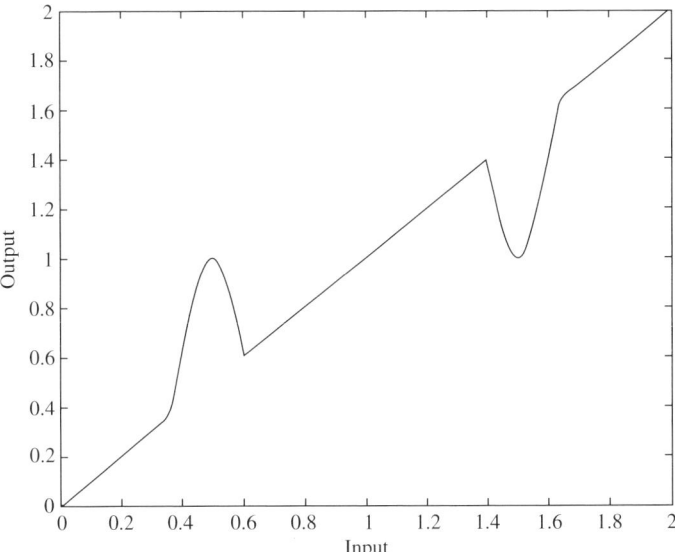

Figure 14.16 Example spiky function used in the experiment.

This leads to the overall expression of the function to be in the form

$$\text{spiky}(x) = \begin{cases} \max(x, G(x)), & \text{if } 0 \leq x \leq 1 \\ \min(x, -G(x)+2), & \text{if } 1 < x \leq 2 \end{cases} \quad (14.19)$$

with $c = 0.5$ and $\sigma = 0.1$, refer to Figure 14.16. Each training and test data set consists of 100 pairs of input–output data.

As could have been anticipated, the linear regression is suitable for some quite large regions of the input variable but becomes quite insufficient in the regions where these two spikes are positioned. As a result of these evident departure from the linear dependency, the linear regression produces a high approximation error of 0.154 ± 0.014 and 0.160 ± 0.008 for the training and testing set, respectively. The augmented granular modification of the model was realized by experimenting with the two essential parameters controlling the granularity of the construct in the input and output space, that is, p and c. The corresponding results are summarized in Tables 14.1 and 14.2. All of them are reported for the optimal values of the

Table 14.1 RMSE Values (Mean and Standard Deviation)—Training Data.

		No. of contexts (p)			
		3	4	5	6
No. of clusters per context (c)	2	0.148 ± 0.013	0.142 ± 0.018	0.136 ± 0.005	0.106 ± 0.006
	3	0.141 ± 0.012	0.131 ± 0.008	0.106 ± 0.008	0.087 ± 0.006
	4	0.143 ± 0.006	0.124 ± 0.007	0.095 ± 0.007	0.078 ± 0.005
	5	0.131 ± 0.012	0.111 ± 0.007	0.077 ± 0.008	0.073 ± 0.006
	6	0.126 ± 0.011	0.105 ± 0.005	0.072 ± 0.007	0.061 ± 0.007

Table 14.2 RMSE Values (Mean and Standard Deviation)—Testing Data.

		No. of contexts (p)			
		3	4	5	6
No. of clusters per context (c)	2	0.142 ± 0.016	0.139 ± 0.028	0.139 ± 0.012	0.114 ± 0.007
	3	0.131 ± 0.007	0.125 ± 0.017	0.115 ± 0.009	0.096 ± 0.009
	4	0.129 ± 0.014	0.126 ± 0.014	0.101 ± 0.009	0.085 ± 0.012
	5	0.123 ± 0.005	0.119 ± 0.016	0.097 ± 0.008	0.082 ± 0.010
	6	0.119 ± 0.016	0.114 ± 0.015	0.082 ± 0.011	0.069 ± 0.007

fuzzification coefficient as listed in Table 14.3, namely, its values for which the root mean squared error (RMSE) attained its minimum,

$$\text{RMSE} = \sqrt{\frac{1}{N}\sum_{k=1}^{N}(y_k - \text{target}_k)^2}$$

where y_k is the modal value of Y produced for input $\mathbf{x}_k$. The numeric range of this coefficient used in the experiments is between 1.5 and 4.0 with the incremental step of 0.1. The increase in the specificity of the granular constructs (either in the output space—via the number of contexts and the input space when increasing the number of the clusters) leads to the reduction of the RMSE values. The number of clusters for a fixed number of contexts exhibits a less significant effect on the reduction of the performance index in comparison to the case when increasing the number of the contexts. Figure 14.17 shows the variation of the RMSE values caused by the values of the fuzzification factor for these four cases. Here the optimal values of the parameters are such for which the testing error becomes minimal. As shown there, the values of the optimal fuzzification coefficient depend on the number of context and cluster, but there is a quite apparent tendency: the increase in the values of "p" and "m" implies lower values of the fuzzification coefficient. This means that the preferred shape of the membership functions becomes more *spiky*.

As Figure 14.17 illustrates, the increase in the number of the contexts and clusters leads to higher optimal values of the fuzzification factor. The optimal results along with the

Table 14.3 Optimal Values of the Fuzzification Coefficient for Selected Number of Contexts and Clusters.

		No. of contexts (p)			
		3	4	5	6
No. of clusters per context (c)	2	3.5	4.0	3.8	3.1
	3	3.2	3.9	3.5	3.1
	4	3.0	2.7	2.6	2.6
	5	3.1	2.8	2.2	2.4
	6	3.0	2.5	2.2	2.0

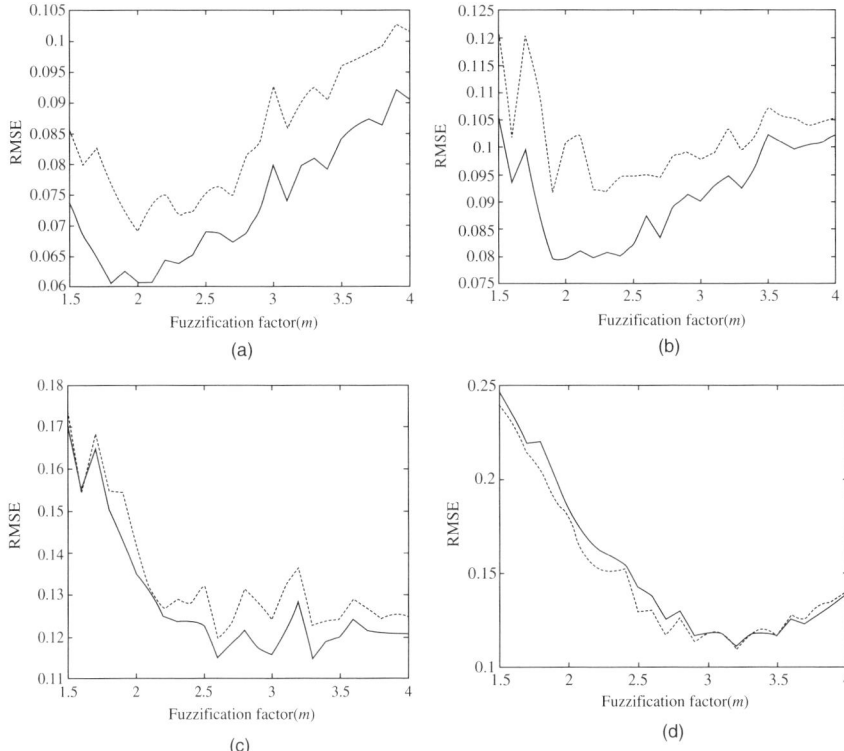

Figure 14.17 Performance index (RMSE) versus values of the fuzzification coefficient for some selected combinations of p and m: (a) $p = c = 6$, (b) $p = c = 5$, (c) $p = c = 4$, and (d) $p = c = 3$, solid line: training data, dotted line: testing data.

visualization of the prototypes when $c = 5$ and $p = 5$ are displayed in Figure 14.18 (this is one of the ten runs). The plot shows the modal values as well as the lower and upper bounds of the resulting fuzzy number produced by the incremental model. Here we have used the optimal value of the fuzzification factor (with the value being equal to 2.2).

14.7 HUMAN-CENTRIC FUZZY CLUSTERING

Typically, fuzzy clustering is carried out on a basis of numeric data. Algorithms such as, for example, FCM produce a local minimum of the given objective function that leads to the formation of a collection of information granules. If there is an input from a human that is taken into consideration as a part of the clustering activities, these pursuits become human-centric. In turn, such clustering produces information granules that are reflective of the human-driven customization. In parallel to the name human-centric fuzzy clustering, we also use the term knowledge-based fuzzy clustering (Pedrycz, 2005). The crux of these clustering activities relies on a seamless incorporation of auxiliary knowledge about the data structure and problem at hand that is taken into consideration when running the clustering algorithm. In this

440 Chapter 14 Granular Models and Human-Centric Computing

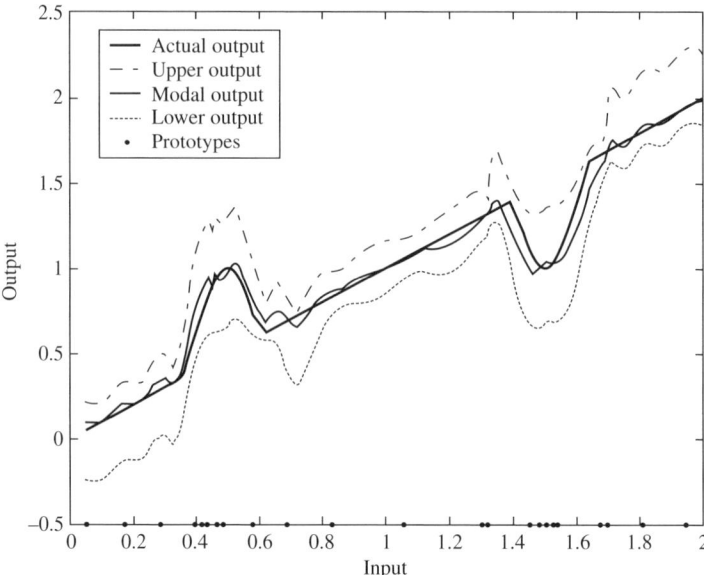

Figure 14.18 Modeling results for $c = 5$ and $p = 5$ ($m = 2.2$); a distribution is also shown of the prototypes in the input space. Note that most of them are located around the spikes that is quite legitimate as they tend to capture the nonlinearities existing in these regions.

manner, we can effectively navigate the clustering activities and impact the geometry of the resulting clusters. In Figure 14.19 we contrast human-centric clustering with the (data-driven) clustering by pointing at the role of knowledge tidbits.

There are several fundamental ways the knowledge tidbits could be incorporated into the generic clustering technique. Here we elaborate on the two well-motivated approaches. The first concerns clustering with partial supervision whereas in the second one we deal with proximity type of knowledge tidbits.

14.7.1 Fuzzy Clustering with Partial Supervision

Partially supervised fuzzy clustering is concerned with a subset of patterns (data) whose labels have been provided. A mixture of labeled and unlabeled patterns could be easily encountered in many practical situations.

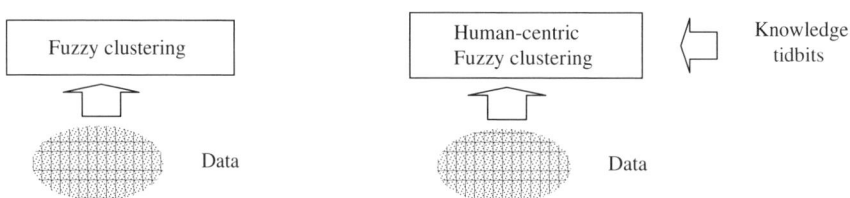

Figure 14.19 The principle of fuzzy clustering versus human-centric fuzzy clustering.

As an example, consider a huge data set of handwritten characters (say, digits extracted from various handwritten postal codes). We want to build a classifier for these characters. The structure of the character sets (groups of digits) revealed through clustering becomes definitely helpful in the design of the classifier. The characters are not labeled; hence, the unsupervised mode of learning is an obvious alternative. Let us now consider that we are provided with some knowledge-based hints about a small subset of labeled digits. In this subset, the characters are labeled by an expert. Practically, with hundreds of thousands of handwritten characters, only a small fraction of them could be effectively labeled. A class assignment process comes with some extra cost and time. Given these constraints, a subset itself could consist of highly "challenging" characters, namely, those that are quite difficult to decipher, so a human intervention in such cases becomes highly desirable. When coming with class labels provided by the expert, such characters can play an important role in enhancing the clustering process. More descriptively, the role of the labeled patterns would be to serve as "anchor" points when launching clustering: we expect that the structure discovered in the data will conform to the class membership of the reference (labeled) patterns. As this example illustrates, we are provided with a mixture of labeled and unlabeled data. It is always worth analyzing how much labeling is useful and really helpful in the ensuing clustering.

Another scenario in which partial supervision could play an important role originates at the conceptual end. Consider that the patterns have been labeled, so on surface we can view that they imply the use of full supervision and call for the standard mechanisms of supervision of classifier design. However, the labeling process could have been very unreliable, and therefore our confidence in the already assigned labels could be relatively low. Then we resort to the clustering mode and accept only a small fraction of patterns that we deem to be labeled quite reliably. The design scenarios similar to those presented above could occur quite often. We need to remember that labeling is a time consuming process, and labeling comes with extra cost. The clustering, on the contrary, could be far more effective. There is a spectrum of learning spread between "pure" models of unsupervised and supervised learning, and this could be schematically visualized in Figure 14.20. Here the underlying criterion discriminating between various cases concerns a mixture of labeled and unlabeled patterns. In the two extreme situations, we end up with 100% of patterns falling in one of the two modes. Note that in general there is no dichotomy of supervised and unsupervised learning.

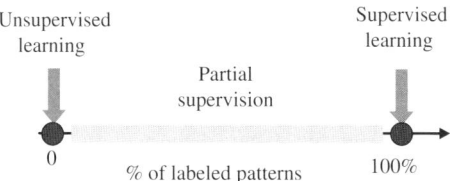

Figure 14.20 A schematic visualization of (infinite) possibilities of partial supervision quantified by the fraction of labeled patterns.

The effect of partial supervision involves a subset of labeled patterns, which come with their class membership. To achieve an efficient use of such knowledge tidbits, we include them as a part of an objective function. During its optimization, we anticipate that the structure to be discovered conforms to the membership grades already conveyed by these selected patterns. We consider an additive form of the objective function (Pedrycz, 1985; Pedrycz and Waletzky, 1997a, b).

$$Q = \sum_{i=1}^{c} \sum_{k=1}^{N} u_{ik}^2 d_{ik}^2 + \alpha \sum_{i=1}^{c} \sum_{k=1}^{N} (u_{ik} - f_{ik} b_k)^2 d_{ik}^2 \qquad (14.20)$$

The first term of (14.20) is aimed at the discovery of the structure in data and is the same as used in the generic version of the FCM algorithm. The second term (weighted by the positive weight factor α) captures an effect of partial supervision. Its interpretation becomes clear once we identify the two essential data structures containing information about the labeled data.

1. The vector of labels, denoted by $\mathbf{b} = (b_1, b_2, \ldots, b_N)^T$: Each pattern $\mathbf{x}_k$ comes with a Boolean indicator function. We assign b_k equal to 1 if the pattern has been already labeled as a part of the available knowledge tidbits. Otherwise we consider the value of b_k equal to zero.

2. The partition matrix $F = [f_{ik}]$, $i = 1, 2, \ldots, c$; $k = 1, 2, \ldots N$ that contains membership grades assigned to the selected patterns (already identified by the nonzero values of b): If $b_k = 1$ then the corresponding column shows the provided membership grades. If $b_k = 0$ then the entries of the corresponding kth column of F do not matter; technically we could set up all of them to zero. Let $b_k = 1$. The optimization of the membership grades u_{ik} is aimed at making them close to u_{ik}.

The nonnegative weight factor (α) helps set up a suitable balance between the supervised and unsupervised mode of learning. Apparently when $\alpha = 0$, then we end up with the standard FCM. Likewise, if there are no labeled patterns ($\mathbf{b} = \mathbf{0}$), then the objective function reads as

$$Q = (1 + \alpha) \sum_{i=1}^{c} \sum_{k=1}^{N} u_{ik}^2 d_{ik}^2 \qquad (14.21)$$

and becomes nothing but a scaled version of the standard objective function guiding the FCM optimization process. If the values of α increase significantly, we start discounting any structural aspect of optimization (where properly developed clusters tend to minimize the objective function) and rely primarily on the information contained in the labels of the patterns. Subsequently, any departure from the values in F would be heavily penalized by significantly increasing the values of the objective function. The choice of a suitable value of the weight factor α could be made by considering a ratio of the data that have been labeled (M) versus all data (N). To achieve a sound balance, we consider the values of α to be proportional to the ratio N/M.

One could also consider a slightly modified version of the objective function

$$Q = \sum_{i=1}^{c} \sum_{k=1}^{N} u_{ik}^2 d_{ik}^2 + \alpha \sum_{i=1}^{c} \sum_{k=1}^{N} (u_{ik} - f_{ik})^2 b_k d_{ik}^2 \qquad (14.22)$$

where the labeling vector **b** shows up in a slightly different format. For $b_k = 1$, we involve the differences between u_{ik} and f_{ik} and they are minimized.

For some variations on the issue of partial supervision, the reader is referred to the work by Bensaid et al. (1989), Kersten (1996), Timm et al. (2002), Abonyi and Szeifert (2003), Coppi and D'Urso (2003), Liu and Huang (2003).

14.7.2 The Development of the Human-Centric Clusters

As usual, the optimization of the objective function (14.22) is completed with respect to the partition matrix and prototypes of the clusters. The first part of the problem is a constraint-based minimization. To minimize it, we consider Langrage multipliers to accommodate the constraints imposed on the membership grades. Hence, the augmented objective function arises in the form

$$V = \sum_{i=1}^{c} u_{ik}^2 d_{ik}^2 + \alpha \sum_{i=1}^{c} (u_{ik} - f_{ik} b_k)^2 d_{ik} - \lambda (\sum_{i=1}^{c} u_{ik} - 1) \qquad (14.23)$$

To compute the gradient of V with respect to the partition matrix U, we note that choosing the value of the fuzzification factor equal to 2 would be quite helpful. By doing that we avoid solving a high-order polynomial equation with respect to the entries of the partition matrix.

By solving the optimization problem, the resulting entries of the partition matrix U assume the form

$$u_{ik} = \frac{1}{1+\alpha} \left[\frac{1 + \alpha \left(1 - b_k \sum_{i=1}^{c} f_{ik}\right)}{\sum_{j=1}^{c} \left(\frac{d_{ik}}{d_{jk}}\right)^2} + \alpha f_{ik} b_k \right] \qquad (14.24)$$

Moving on to the computations of the prototypes, the necessary condition for the minimum of Q with respect to the prototypes v_i comes in the form $\partial Q / \partial v_{st} = 0$, $s = 1, 2, \ldots, c$; $t = 1, 2, \ldots, n$. Calculating the respective partial derivatives one derives

$$\frac{\partial Q}{\partial v_{st}} = \frac{\partial}{\partial v_{st}} \left[\sum_{i=1}^{c} \sum_{k=1}^{N} u_{ik}^2 \sum_{j=1}^{n} (x_{kj} - v_{ij})^2 + \alpha \sum_{i=1}^{c} \sum_{k=1}^{N} (u_{ik} - f_{ik} b_k)^2 \sum_{j=1}^{n} (x_{kj} - v_{ij})^2 \right]$$

$$= \frac{\partial}{\partial v_{st}} \left[\sum_{i=1}^{c} \sum_{k=1}^{N} [u_{ik}^2 + (u_{ik} - f_{ik} b_k)^2] \sum_{j=1}^{n} (x_{kj} - v_{ij})^2 \right] \qquad (14.25)$$

Let us introduce the following shorthand notation:

$$\psi_{ik} = u_{ik}^2 + (u_{ik} - f_{ik}b_k)^2 \qquad (14.26)$$

This leads to the optimality condition of the form

$$\frac{\partial Q}{\partial v_{st}} = 2\sum_{k=1}^{N} \psi_{sk}(x_{kt} - v_{st}) = 0 \qquad (14.27)$$

and finally we derive

$$\mathbf{v}_s = \frac{\sum_{k=1}^{N} \psi_{sk}\mathbf{x}_k}{\sum_{k=1}^{N} \psi_{sk}} \qquad (14.28)$$

EXAMPLE 14.3

For illustrative purposes, we consider a small synthetic two-dimensional data as shown in Figure 14.21.

The partial supervision comes with the classification results of several patterns—their labels are shown in Figure 14.22 as well. Those are two patterns with the membership grades [0.5 0.5] and [0.0 1.0]. These hints suggest to consider two clusters, $c = 2$.

The clustering was completed for several increasing values of α, and this development gives a detailed view of the impact the classification hints exhibit on the revealed structure of the patterns. This is shown in two different ways, which is by visualizing the entries of the partition matrices (Fig. 14.22). We note that by changing α the discovered structure tend to conform to the available classification constraints. For reference, we have shown the results for $\alpha = 0$ so that no supervision effect is taken into consideration.

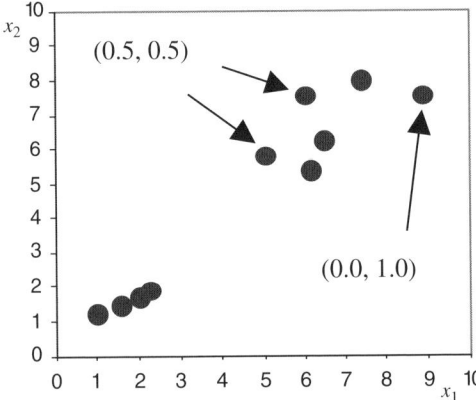

Figure 14.21 A two-dimensional synthetic data set; visualized are the knowledge tidbits (hints)-labeled patterns that are used to guide the clustering process.

14.7 Human-Centric Fuzzy Clustering **445**

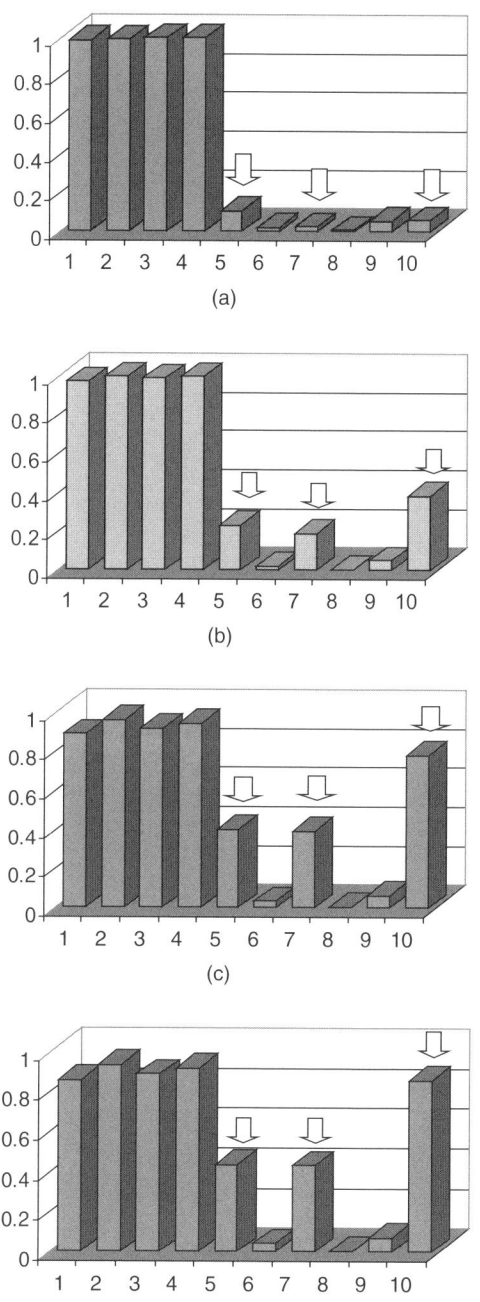

Figure 14.22 Membership grades of patterns for selected values of α: (a) $\alpha = 0.0$ (no supervision), (b) $\alpha = 0.5$, (c) $\alpha = 3.0$, and (d) $\alpha = 5.0$. Small arrows point at the three labeled patterns.

14.7.3 Proximity-Based Fuzzy Clustering

Proximity tidbits provide another interesting mechanism of knowledge-based guidance available in fuzzy clustering. In contrast to the mechanism of partial supervision discussed in the previous section, we are provided with evaluations of proximity between selected pairs of data. If two data points are close to each other, we quantify this by assigning to them a high value of proximity. On the contrary, if the patterns are very different, this becomes reflected by low proximity values being close to zero. This form of knowledge hints is quite intuitive. In contrast to the previous knowledge tidbits, we do not require to specify the number of classes (clusters). For instance, when dealing with a collection of digital photographs, we want to organize; it is quite easy to quantify proximity between some pairs of them. On the contrary, as we do not know the number of classes or clusters in advance, it would be highly impractical to estimate a vector of membership grades as its dimensionality has not been specified.

Let us quantify the concept of proximity between two objects (patterns). Formally, given two patterns a and b, their proximity, denoted by prox(a, b), is a mapping from the pairs of data (patterns) to the unit interval such that it satisfies the following two conditions:

1. prox $(a, b) =$ prox (b, a) symmetry
2. prox $(a, a) = 1$ reflexivity

The notion of proximity is the most generic one that relies on a minimal set of requirements. What we have to impose is straightforward: a exhibits the highest proximity when compared to itself. It is also intuitive that the proximity relation is symmetric. In this sense, we can envision that in any experimental setting, these two properties can be easily realized. Given a collection of patterns, the proximity results obtained for all possible pairs of patterns are usually arranged in a matrix form known as a proximity relation P.

It is worth mentioning that the concept of similarity is more demanding from a formal standpoint as it requests some sort of transitivity (say, max–min transitivity, etc.). In practice, experimental results (that come as a consequence of some comparison between pairs of objects) do not guarantee a satisfaction of transitivity.

Fuzzy partitions produced by the FCM algorithm are directly linked with the proximity relation in the following way. Given the fuzzy partition U, a well-known transformation (Bezdek, 1981; Pedrycz, 2005) to its proximity counterpart is governed by the expression

$$\hat{p}[k_1, k_2] = \sum_{i=1}^{c} \min(u_{ik_1}, u_{ik_2}) \tag{14.27}$$

where k_1 and k_2 indicate the corresponding patterns (data) $\mathbf{x}_{k_1}, \mathbf{x}_{k_2}$. The proximity matrix $\hat{P} = [\hat{p}[k_1, k_2]], k_1, k_2 = 1, 2, \ldots, N$, organizes proximity values for all pairs of data. As this matrix is symmetric, it is enough to compute and save only its upper part located above the main diagonal. Owing to the well-known properties of the partition matrix, we observe that for $k_1 = k_2$ we end up with the value of $\hat{p}[k_1, k_2]$ equal to 1. Evidently, $\hat{p}[k_1, k_1] = \sum_{i=1}^{c} \min(u_{ik_1}, u_{ik_1}) = \sum_{i=1}^{c} u_{ik_1} = 1$. The satisfaction of the symmetry property of $\hat{p}[k_1, k_2]$ becomes self-evident.

The algorithm accepting the proximity hints consists of two main phases that are realized in interleaved manner. The overall computing process is summarized below.

procedure P-FCM-CLUSTERING (**X**) **returns** cluster centers and partition matrix
input: data set $\mathbf{X} = \{\mathbf{x}_k, k = 1, \ldots, N\}$
local: fuzzification coefficient: m
thresholds: δ, ε

    INITIALIZE-PARTITION-MATRIX
    repeat until distance two successive partition matrices $\leq \delta$
      run FCM
        repeat until values of V over successive iterations $\leq \varepsilon$
          minimize V
            compute u_{ik}
            compute $\mathbf{v}_s$
return cluster centers and partition matrix

Owing to the first phase, we consider the data driven optimization realized by the FCM. The second phase concerns an accommodation of the proximity-based hints and involves some gradient-oriented updating of the partition matrix. All optimization activities will be referred to as proximity-based FCM, or P-FCM, for brief.

The first phase of the P-FCM is straightforward and follows the well known scheme of the FCM optimization. The inner part, however, deserves detailed discussion.

Given: Specify number of clusters, fuzzification coefficient, distance function and initiate a partition matrix (generally it is started from a collection of random entries), and termination condition (small positive constant ε).

The accommodation of the proximity requirements (knowledge tidbits) has to be completed with the use of a suitable performance index V. As stated in the problem formulation, we are provided with pairs of patterns and their associated levels of proximity. Denote them by $p[k_1, k_2]$. Furthermore, we introduce an indicator function $b[k_1, k_2]$ that assumes 1 if the value of proximity has been provided for this specific pair of data. Otherwise the value of this indicator function is set up to zero.

Turning on to the optimization problem, we require that the values of proximity induced by the partition matrix U determined by the FCM (the first phase of the algorithm) are made as close as possible to the given values of proximity by adjusting the values of the partition matrix. Bearing this in mind, the performance index can be formulated as the following sum:

$$V = \sum_{k_1=1}^{N} \sum_{k_2=1}^{N} (\hat{p}[k_1, k_2] - p[k_1, k_2])^2 b[k_1, k_2] d[k_1, k_2] \tag{14.28}$$

The notation $\hat{p}[k_1,k_2]$ is used to describe the corresponding proximity level induced by the partition matrix. Introducing these induced proximity values we obtain

$$V = \sum_{k_1=1}^{N} \sum_{k_2=1}^{N} (\sum_{i=1}^{c} (\min(u_{ik_1}, u_{ik_2}) - p[k_1,k_2])^2 b[k_1,k_2] d[k_1,k_2] \quad (14.29)$$

The optimization of V with respect to the partition matrix does not lend itself to a closed-form expression and requires some iterative optimization. We resort ourselves to the gradient-based minimization that leads to the expression

$$u_{st}(\text{iter}+1) = \left[u_{st}(\text{iter}) - \alpha \frac{\partial V}{\partial u_{st}(\text{iter})} \right]_{0,1} \quad (14.30)$$

$s = 1, 2, \ldots, c, t = 1, 2, \ldots, N$ where $[\ ]_{0,1}$ indicates that the results of adjustment of the membership grades are clipped to the unit interval; α stands for a positive learning rate. Successive iterations are denoted as "iter" and "iter + 1."

The detailed computations of the above derivative are straightforward. Taking the derivative of V with respect to u_{st}, $s = 1, 2, \ldots, c, t = 1, 2, \ldots, N$, one obtains

$$\frac{\partial V}{\partial u_{st}(\text{iter})} = \sum_{k_1=1}^{N} \sum_{k_2=1}^{N} \frac{\partial}{\partial u_{st}} (\sum_{i=1}^{c} \min(u_{ik_1}, u_{ik_2}) - p[k_1,k_2])^2$$

$$= 2 \sum_{k_1=1}^{N} \sum_{k_2=1}^{N} \sum_{i=1}^{c} \min(u_{ik_1}, u_{ik_2}) - p[k_1,k_2]) \frac{\partial}{\partial u_{st}} \sum_{i=1}^{c} \min(u_{ik_1}, u_{ik_2})$$

$$(14.31)$$

The inner derivative assumes binary values depending on the satisfaction of the conditions

$$\frac{\partial}{\partial u_{st}} \sum_{i=1}^{c} \min(u_{ik_1}, u_{ik}) = \begin{cases} 1 & \text{if } t = k_1 \text{ and } u_{sk_1} \leq u_{sk_1} \leq u_{sk_2} \\ 1 & \text{if } t = k_2 \text{ and } u_{sk_2} \leq u_{sk_2} \leq u_{sk_2} \\ 0, & \text{otherwise} \end{cases} \quad (14.32)$$

Making this notation more concise, we can treat the above derivative as a binary (Boolean) predicate, denoting it by $\phi[s,t,k_1,k_2]$, and plug it into (14.31). This produces the following expression:

$$\frac{\partial V}{\partial u_{st}(\text{iter})} = 2 \sum_{k_1=1}^{N} \sum_{k_2=1}^{N} \sum_{i=1}^{c} (\min(u_{ik_1}, u_{ik_2}) - p[k_1,k_2] \varphi[s,t,k_1,k_2] \quad (14.33)$$

The organization of the computing of the P-FCM scheme is illustrated in Figure 14.23. The two interleaving phases (FCM and the proximity-based optimization) are clearly identified. The results of the FCM optimization (which is guided by a set of its own parameters) are passed on the gradient-based procedure of the

14.7 Human-Centric Fuzzy Clustering

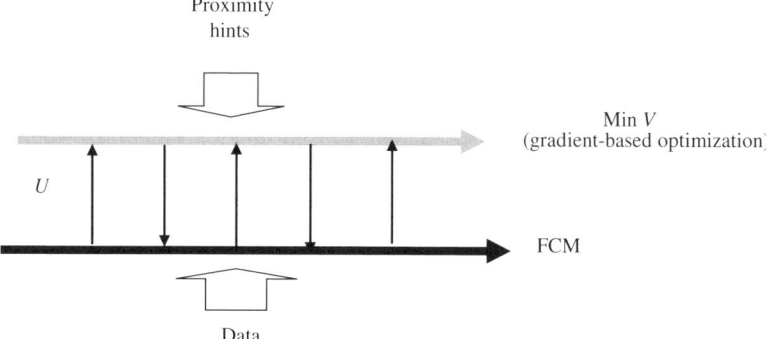

Figure 14.23 P-FCM: a general flow of optimization activities.

proximity-based optimization. The results obtained there are normalized (to meet the requirements of the partition matrix) and then made available to the next phase of the FCM optimization. For each iteration of the FCM, we encounter a series of iterations of the gradient-based optimization.

14.7.4 Interaction Aspects of Sources of Information in the P-FCM

The P-FCM clustering environment brings into picture an important and fundamental issue of collaboration and/or competition between different sources of information and an algorithmic manifestation of such interaction. In the discussed framework, we are inherently faced with two diverse source of data/knowledge. FCM aims at "discovering" the structure in the data by minimizing a certain objective function and in its pursuits relies exclusively on the available data. The gradient-based learning concentrates on the proximity hints (thus relying on human-oriented guidance), and this is the point where the interaction with the data starts to unveil. The strength of this interaction is guided by the intensity of the gradient-based learning. More specifically, for higher values of the learning rate, we put more confidence in the hints and allow them to affect the already developed structure (partition matrix) to a high extent. It may be that the collaboration may convert into competition when the two mechanisms start interacting more vigorously and these two sources of information are fully consistent (which is difficult to quantify in a formal way). The existence of the competition starts manifesting through substantial oscillations during the minimization of V; to avoid them we need to loosen the interaction and lower the value of the learning rate (α).

The P-FCM exhibits some resemblance with the well-known and interesting trend of fuzzy relational clustering, cf. Hathaway et al. (1989, 1996) and Hathaway and Bezdek (1994). Instead of individual patterns, in this mode of clustering, we consider *relational* objects that are new objects describing properties of pairs of the original data. More specifically, these objects are organized in a single matrix $R = [r_{kl}]$, $k, l = 1, 2, \ldots, N$ where r_{kl} denotes a degree of similarity, resemblance, or more generally dependency (or association) between two patterns (k and l). We assume that this type of *referential* information about patterns is available to the clustering purposes, and the algorithms are developed along this line. The importance of the relational character data is motivated by the lack of interpretation of single patterns, whereas their pairwise comparison (leading to the relational patterns) makes perfect sense. Computationally, the number of relational patterns is substantially higher than the original data (N vs. $N(N-1)/2$). The computational advantage can arise with regard to the dimensionality of the space; the original patterns may be located in a highly dimensional space (n), whereas the relational data could often be one-dimensional.

In the P-FCM, we have a number of differences when comparing with the relational clustering:

(a) As already underlined, the relational clustering operates in the homogeneous space of relational objects. The P-FCM augments FCM by adding an extra optimization scheme, so it still operates on patterns (rather than on their relational counterpart).

(b) P-FCM attempts to reconcile two sources of information (structural and domain hints); this mechanism is not available to the relational clustering.

(c) Computationally, P-FCM does not affect the size of the original dataset; we can be provided with a different number of hints (being a certain percentage of the dataset). Relational clustering increases the size of the dataset while operating in a low-dimensional space.

(d) P-FCM dwells on the core part of the FCM optimization scheme by augmenting it by an additional gradient-based optimization phase; in contrast, the relational clustering requires substantial revisions to the generic FCM method (which sometimes leads to optimization pitfalls and is still under further improvements).

14.8 PARTICIPATORY LEARNING IN FUZZY CLUSTERING

Here we consider an augmentation of clustering based on a concept of participatory learning (PL) (Yager, 1990). In essence, PL assumes that learning and beliefs about a system depend on what the learning agent knows about the system itself. The current knowledge about the system is part of the learning process itself and influences in a way in which new observations are used for learning purposes. This feature of learning is very important in learning tasks such as fuzzy clustering. An essential characteristic of PL is data impact in causing revision of a cluster structure depending on its compatibility with the current cluster structure.

Formally, let $\mathbf{v} \in [0, 1]^n$ be a variable that encodes a prototype, a cluster center. The aim is to learn the values of this vector. Without any loss of generality, we assume that knowledge about the values of this vector comes in through a sequence of

14.8 Participatory Learning in Fuzzy Clustering

data $\mathbf{x}_k \in [0,1]^n$. In other words, $\mathbf{x}_k$ is used as a vehicle to learn about $\mathbf{v}$. We say that the learning process is *participatory* if the contribution of each data $\mathbf{x}_k$ to the learning process depends upon its acceptance by the current estimate of $\mathbf{v}$ as being a valid observation. Implicit to this idea is that, to be useful and to contribute to the learning of $\mathbf{v}$, observations $\mathbf{x}_k$ must somehow be compatible with the current estimates of $\mathbf{v}$. Let $\mathbf{v}_k$ be the estimate of $\mathbf{v}$ after k observations of the data stream $\{\mathbf{x}_k\}$. To be appropriate for learning purposes, in PL we assume that $\mathbf{x}_k$ should be close to $\mathbf{v}_k$. Intuitively, PL updates $\mathbf{v}_k$ if information received through data $\mathbf{x}_k$ agrees, in some sense, with $\mathbf{v}_k$. A mechanism to updated $\mathbf{v}$ is a smoothing procedure governed by the following expression

$$\mathbf{v}_{k+1} = \mathbf{v}_k + \alpha \rho_k (\mathbf{x}_k - \mathbf{v}_k) \tag{14.34}$$

where $k = 1, \ldots, N$, and N is the number of observations, whereas $\mathbf{v}_{k+1}$ is the new cluster center (belief); $\mathbf{v}_k \in [0,1]^n$ is the current cluster center; $\mathbf{x}_k \in [0,1]^n$ the current observation or data input; $\alpha \in [0,1]$ the learning rate; and $\rho_k \in [0,1]$ denotes a compatibility degree (compatibility index) determined for $\mathbf{x}_k$ and $\mathbf{v}_k$, given by

$$\rho_k = F(S_{1k}, S_{2k}, \ldots, S_{nk})$$

where in general S_{jk} is a similarity measure,

$$S_{jk} = G_{jk}(v_{jk}, x_{jk})$$

$S_{jk} \in [0,1]$, $j = 1, \ldots, n$, and F an aggregation operator. $S_{jk} = 1$ indicates full similarity and $S_{jk} = 0$ means no similarity. G_{jk} maps pairs (v_{jk}, x_{jk}) into a similarity degree and this frees the values v_{jk} and x_{jk}, $j = 1, \ldots, n$, $k = 1, \ldots, N$, from being in the unit interval. Moreover, G_{jk} allows that two vectors $\mathbf{v}_k$ and $\mathbf{x}_k$ have $S_{jk} = 1$ even if they are not exactly equal. This formulation also allows for different perceptions of similarity for different components of the vectors, that is, for different j's. A possible formulation of the compatibility index could come in the form

$$\rho_k = 1 - \frac{1}{n} \sum_{j=1}^{n} d_{jk} \tag{14.35}$$

where $d_{jk} = |x_{jk} - v_{jk}|$ is the Hamming distance between x_{jk} and the current value v_{jk}. More generally, we can express the compatibility index to be in the following form

$$\rho_k = 1 - d_k \tag{14.36}$$

where d_k is a certain distance function, $d_k = \|\mathbf{x}_k - \mathbf{v}_k\|$.

One important issue ignored so far concerns a situation in which a stream of conflicting data is provided during a certain period of time. In this case, the system perceives a sequence of low values of ρ_k because of the incompatibility reported between belief and data. Although in the short-term low values of ρ_k cause aversion to learning, actually, they should act to turn the system more receptive to learning

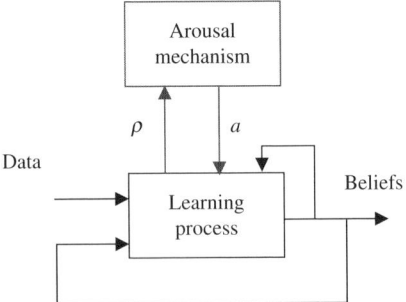

Figure 14.24 PL with arousal.

because it may be the case that the current cluster structure is wrong or is changing. There is a need of a type of *arousal* mechanism to monitor the compatibility of current cluster structure with observations. One alternative, shown in Figure 14.24, is to use an arousal index to influence the learning process. The higher the arousal rate, the less confident is the system with the current cluster structure, and conflicting observations become important to update the cluster structure.

One way to update the arousal index $a_k \in [0, 1]$ can be described in the following form:

$$a_{k+1} = a_k + \beta((1 - \rho_{k+1}) - a_k) \qquad (14.37)$$

The value of $\beta \in [0, 1]$ controls the rate of change of arousal. The closer the values of β to 1, the faster the system senses compatibility variations. The arousal index can be viewed as the complement of the confidence in the cluster structure currently held.

A way of the PL procedure to appropriately consider the arousal mechanism is to incorporate the arousal index in the basic procedure as follows:

$$\mathbf{v}_{k+1} = \mathbf{v}_k + \alpha(\rho_k)^{1-a_k}(\mathbf{x}_k - \mathbf{v}_k) \qquad (14.38)$$

The update formula of the arousal index introduces a self-control mechanism in the PL algorithm. Although ρ_k expresses how much the system changes its credibility in its own beliefs, the arousal index a_k acts as a critic to remind when current belief should be modified in front of new evidences. A detailed analysis of PL principles is found in (Yager, 1990). The details of the PL algorithm for unsupervised fuzzy clustering are given next.

First, notice that the compatibility ρ_{ik} gives the compatibility degree between the cluster center $\mathbf{v}_i$ and data $\mathbf{x}_k$. The same interpretation can be developed for the arousal index a_{ik} as it measures the incompatibility between $\mathbf{v}_i$ and data $\mathbf{x}_k$. More precisely, the arousal index gives an evaluation on how far data $\mathbf{x}_k$, $k = 1, \ldots, N$, are from current cluster centers $\mathbf{v}_i$, $i = 1, \ldots, c$. The set of cluster centers is a manifestation of the current cluster structure, the clustering algorithm found after processing a data stream $\{\mathbf{x}_k, k = 1, \ldots, N\}$ (Silva et al., 2005).

To partition data set X into clusters, the PL fuzzy clustering algorithm requires an extra threshold τ. Its purpose is to advise if a new data should be declared incompatible with the current cluster structure. This happens when a new cluster has been found or when existing clusters merge into one. In these cases, the cluster structure must be revised to accommodate new knowledge. More specifically, in terms of clustering this means that if a new data $\mathbf{x}_k$ is far enough from all cluster centers, then there is enough evidence to form a new cluster. The simplest alternative, the one adopted here, is to choose $\mathbf{x}_k$ as the new cluster center.

There are several ways to compute the values of ρ_{ik} depending upon a selection of a distance. One of the obvious alternatives would be the Hamming distance as used in (14.35). The other commonly encountered alternative is the Euclidean distance. More generally, we may consider the Mahalanobis distance (Gustafson and Kessel, 1992) where

$$d_{ik} = (\mathbf{x}_k - \mathbf{v}_i)^T (\det(F_i)^{1/N} F_i^{-1})(\mathbf{x}_k - \mathbf{v}_i) \qquad (14.39)$$

where F_i is the covariance matrix associated with the ith cluster,

$$F_i = \frac{\sum_{k=1}^{N} u_{ik}^m (\mathbf{x}_k - \mathbf{v}_i)(\mathbf{x}_k - \mathbf{v}_i)^T}{\sum_{k=1}^{N} u_{ik}^m} \qquad (14.40)$$

As the Euclidean distance limits the geometry of clusters to some spherical shapes, the use of Mahalanobis and similar scatter measures is advisable when computing ρ_{ik}. The partition matrix is computed in a standard way, that is,

$$u_{ki} = \frac{1}{\sum_{i=1}^{c} (d_{ik}/d_{jk})^{2/m-1}} \qquad (14.41)$$

The participatory fuzzy clustering algorithm can be summarized as follows. Given values for α, β, and τ, two random points of X are chosen as cluster centers, represented here by V^0, a set whose elements are the initial cluster centers $\mathbf{v}_1^0$ and $\mathbf{v}_2^0$. Next, the compatibility index ρ_{ik} and arousal index a_{ik} are computed. If for all $\mathbf{v}_i$ the arousal index of $\mathbf{x}_k$ is greater than threshold τ, then $\mathbf{x}_k$ forms a center of a new cluster. Otherwise, the cluster center that is the closest to $\mathbf{x}_k$ (viz., the highest compatibility index) is updated.

Whenever a cluster center has been updated or a new cluster has been added, it is necessary to check if redundant clusters are being formed. This is because when a cluster is updated, its center may be moved closer to another cluster center and redundant cluster may be formed. Therefore, we require a mechanism to exclude redundant cluster centers. A simple mechanism is to exclude a center whenever the compatibility degree between cluster centers, computed using the Euclidean distance, is greater than a threshold λ whose value typically is $\lambda = 0.95\tau$. Summing up, cluster center i is excluded whenever its compatibility $\lambda_{\mathbf{v}_i}$ with any another center is less than or equal to λ, that is, whenever the compatibility between two prototypes is high enough. The idea to compute $\lambda_{\mathbf{v}_i}$ is, similar to the one supported by the arousal index, namely, to set

$$\lambda_{v_i} = \beta(1 - \rho_{v_i})$$

Given the finite number of data (N), the algorithm stops when either the maximum number of iterations l_{max} has been reached or there is no significant variation of the cluster centers from an iteration to the next, that is, when $\|\Delta V\| \leq \varepsilon$, where ε is a small positive real number. Since the cluster centers are updated whenever data points are input, as soon as the algorithm stops, the fuzzy partition matrix U must be updated considering the cluster centers produced in the last iteration. The detailed steps of the algorithm are outlined as follows.

procedure OFFLINE-PARTICIPATORY (X) **returns** cluster centers and partition matrix
input: data set: X= $\{\mathbf{x}_k, k = 1,\ldots,N\}$
local: cluster membership parameter: m
        threshold: τ
        learning rates: α, β
        parameters: ε, l_{max}
        V = INITIALIZE-CLUSTER-CENTERS(X)
        $l = 1$
        until stop = TRUE **do**
          for $k = 1 : N$
            CLUSTER-LEARNING($\mathbf{x}_k, V$)
        if $\|\Delta V\| \leq \varepsilon$ and $l \geq l_{max}$ **then** update U, set stop = TRUE
        else $l = l + 1$
        return V

The procedure cluster-learning is the one that needed the online version of the algorithm:

procedure ONLINE-PARTICIPATORY (x) **returns** cluster centers and partition matrix
input: data **x**
local: cluster membership parameter: m
        threshold: τ
        learning rates: α, β
        parameters: ε, l_{max}
        V = INITIALIZE-CLUSTER-CENTERS(**x**)
        do forever
          CLUSTER-LEARNING(**x**, V)

procedure CLUSTER-LEARNING(x) **returns** cluster centers and partition matrix
    input: $\mathbf{x}_k = \mathbf{x}$
    for $i = i : c$
      compute d_{ik}
      compute ρ_{ik}
      compute a_{ik}
        if $a_{ik} \leq \tau$ for all $i = 1, \ldots, c$
          then update $\mathbf{v}_s$, $s = \arg\max_i\{\rho_{ik}\}$, compute U
          else create new cluster center
            for $i = i : c$
              for $j = (i + 1) : c$
                compute ρ_{v_i}
                compute λ_{v_i}
                if $\lambda_{v_i} \leq 0.95\tau$
                  then eliminate $\mathbf{v}_i$ and update U
    return U, V

EXAMPLE 14.4

Several clustering examples are addressed here to show the behavior of offline PL fuzzy clustering alternative fuzzy clustering algorithms such as Gustafson–Kessel, GK, (Gustafson and Kessel, 1992) and the modified fuzzy k-means MFKM (Gath et al., 1997). We recall that the GK is as a generalization of the fuzzy c-means in which the Mahalanobis distance is used in objective function. GK is particularly effective to find spherical, ellipsoidal, and convex clusters. The MFKM uses a data induced metric with the Dijkstra shortest path procedure in a graph-based representation. The outstanding feature of the MFKM is its ability to find nonconvex clusters. The GK and MFKM are supervised clustering algorithms.

(a) Consider the classic Iris data and the results are depicted in Figure 14.25. Both PL and GK algorithms capture the correct cluster structure. PL performs closely to GK because it adopts Mahalanobis distance as GK does, but MFKM fail to find appropriate clusters.

(b) This example was suggested by Gustafson and Kessel (1992). Here clusters are difficult to discern, specially in the regions where they intersect or are close to each other. Despite the complex structure of the clusters, PL and GK successfully group the data whereas MFKM failed to do not.

(c) The intertwined spiral, Figure 14.27, is a classic and difficult problem to be solved by partition-based fuzzy clustering algorithms. In this case both GK and PL fail, but MFKM succeeds in finding the proper cluster.

Overall, PL clustering performs as well as any partition-based clustering algorithm and suffers from similar inconveniences. Computational complexity of the PL clustering depends on the choice of the distance measure to compute the compatibility degree ρ_{ik}.

Contrary to GK and MFKM, the PL algorithm runs in unsupervised mode. The number of cluster it chooses, however, depends on the choice of the value of the threshold τ. In general, when running in offline mode, if the number of clusters remains constant after a few iterations, the algorithm finds a cluster structure that mirrors the one existing in the data set. When there is

456 Chapter 14 Granular Models and Human-Centric Computing

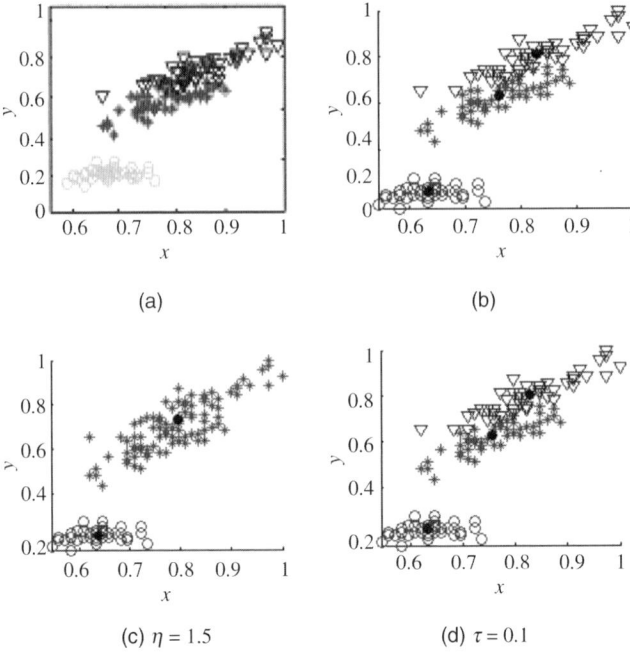

Figure 14.25 Iris data and resulting clusters (a) GK, (b) MFKM, (c) PL, and (d) Cluster centers are marked with "●".

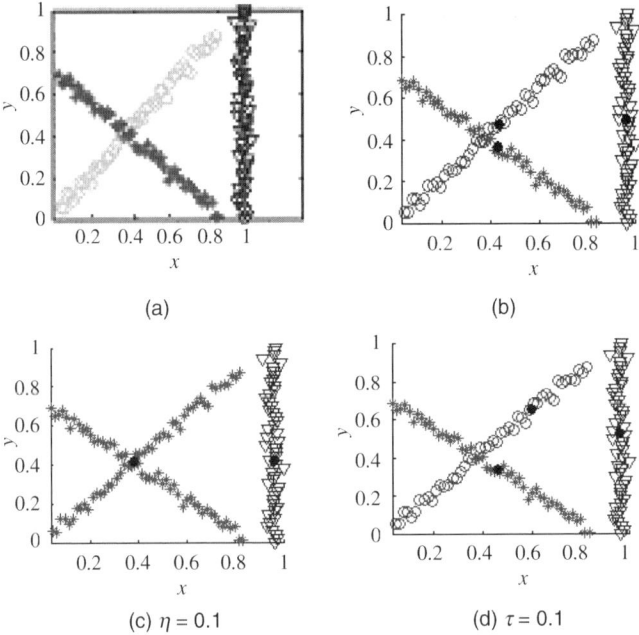

Figure 14.26 Original data and clusters (a) GK, (b) MFKM, (c) PL, and (d) Cluster centers are marked with "●".

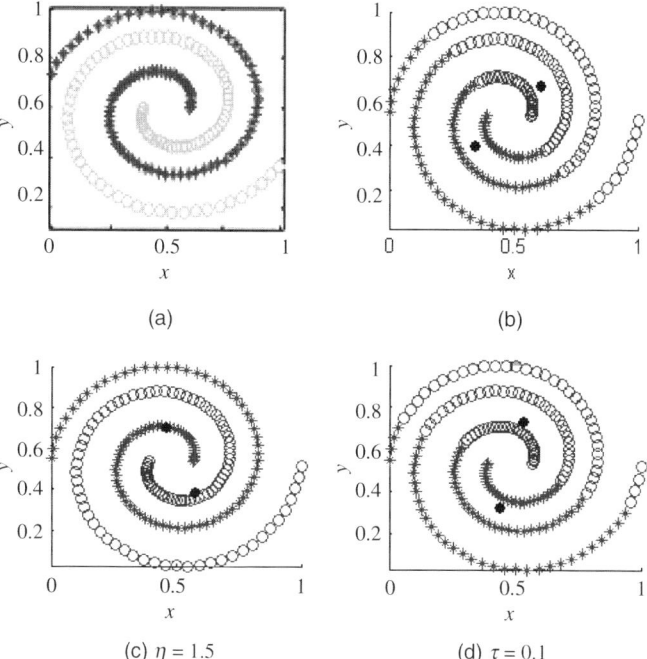

Figure 14.27 Original data and clusters (a) GK, (b) MFKM, (c) PL, and (d) Cluster centers are marked with "●".

oscillation of the number of clusters, experience shows that slight adjustment of the threshold produce cluster structures that closely represent the one found in data set.

14.9 CONCLUSIONS

Fuzzy clustering and granular modeling are rooted in the fundamental concept of information granules regarded as semantically meaningful conceptual entities that are crucial to the overall framework of user-centric modeling. The user is at position to cast modeling activities in a particular way that becomes a direct reflection of the main purpose of the given modeling problem. For instance, in data mining the user is ultimately interested in revealing relationships that could be of potential interest given the problem under consideration.

The algorithmic diversity of fuzzy clustering is particularly well suited to address the key objectives of granular modeling. Fuzzy clusters fully reflect the character of the data. The search for the structure is ultimately affected by some specific well-articulated modeling needs of the user. We have demonstrated that fuzzy sets of context can play an important role in shaping up modeling activities and help handle dimensionality issues decomposing the original problem into a series of subproblems guided by specific contexts. By linking context fuzzy sets and the induced clusters, we directly form the modeling blueprint of the model. It is relevant to note that granular models are expressed at a certain level of granularity. Instead of single numeric results that are typical for numeric models, the user is provided with a fuzzy set of result that can be communicated in some granular format, compatible with the vocabulary of granular terms being originally exploited in the design of the model, and presented visually in terms of membership functions.

EXERCISES AND PROBLEMS

1. Consider three prototypes in the two-dimensional input space, $\mathbf{v}_1 = [1.5, -3.0]^T$, $\mathbf{v}_2 = [0.0, -1.0]^T$, and $\mathbf{v}_3 = [4.2, 0.5]^T$. The corresponding prototypes in the output space are equal to -2, 5, and -3. Plot the input–output characteristics of the model governed by (xx) assuming three values of m equal to 1.05, 2.0, and 3.5.

2. Discuss computational complexity of the conditional FCM and contrast it with the one of the FCM. How does it depend upon the number of contexts being used?

3. How could you decompose a certain fuzzy relation A defined in $\mathbf{R}^n$ into another fuzzy relation B defined in $\mathbf{R}^{n-1}$ and a fuzzy set F defined in $\mathbf{R}$ so that $A = B \times F$. Discuss a general idea and offer algorithmic details.

4. Consider a mixed-mode two-input granular neuron in which the connections are described as an interval $[-3, 4]$ and a triangular fuzzy set with the modal value of 1.5 and lower and upper bounds equal to -1.0 and 2.0. Derive input–output characteristics of this neuron assuming that the two inputs u_1 and u_2 are positive and sum up to 1.

5. In an n-input granular neuron with interval-valued connections $[w_{i-}, w_{i+}]$, the positive inputs satisfy the relationship $\sum_{i=1}^{n} u_i = 1$. Determine the values of inputs under which the granularity of the output quantified as the following sum

$$Q = \sum_{i=1}^{n} (w_{i+} - w_{i-}) u_i^2$$

achieves its maximum. What happens to the solution if all connections are the same?

6. In the P-FCM algorithm, we use an indicator function that assumes binary values. What suggestions could you offer if you were provided with a flexibility of using values in the [0,1] interval.

7. Consider a situation when for the given data set $\mathbf{D}$ there are several collections of proximity tidbits coming from various sources (e.g., experts). How could you modify the objective function to accommodate this scenario?

8. A subset of data set $\mathbf{D}$, that is, $\mathbf{D}'$ was clustered into c clusters.

   How could you use its results in clustering data set $\mathbf{D}$ into the same number of clusters. What if the number of clusters to be determined in $\mathbf{D}$ is going to be different from c?

9. We are about to cluster data set $\mathbf{D}$ and were told that some other data generated by the same phenomenon as for $\mathbf{D}$ have been already clustered. How could you take advantage of these results in clustering $\mathbf{D}$?

10. Suggest some examples in which you could consider proximity-based clustering as a viable alternative. Identify problems in which you can view clustering with partial supervision as a viable model of human-centric unsupervised learning.

11. In conditional FCM developed so far, we have assumed the same number of clusters for each context. Is this a viable assumption? How could you revisit it? Suggest a way in which you could adjust (optimize) the number of clusters for each context fuzzy set.
12. Suggest a mechanism to incorporate context-based information in clustering algorithms based on PL. Discuss its role in both offline and online system modeling.
13. Elaborate on procedures to learn context information in fuzzy rule-based models.
14. Develop a PL procedure to train neural granular networks arranged in a one or two-dimensional array of neurons.

HISTORICAL NOTES

The role of fuzzy sets as a basis to analyze complex, human-centric systems was identified by Zadeh (1973) where he explicitly stressed that conventional quantitative techniques of system analysis could be inherently inappropriate to cope with systems where a human factor becomes integral to its functioning. The incompatibility principle developed by him underlines existence of important trade-offs between accuracy of models and their descriptive power and interpretability (and transparency). In this setting, the evolution of the fundamental concept of information granulation plays a pivotal role (Zadeh, 1973, 1979, 1997).

The notion of granular models promotes rapid prototyping in fuzzy modeling. Here, the specialized version of fuzzy clustering, known as conditional or context-based clustering, brings a collection of building blocks—information granules that are instantaneously put together in a form of a web of connections. This class of fuzzy models promotes the conjecture that when dealing with information granules, the results of modeling should be fully reflective of the granular character of the developed construct (Pedrycz and Vasilakos, 1999; Pedrycz and Kwak, 2006).

REFERENCES

Bezdek, J. C., Pal, S. K. *Fuzzy Models for Pattern Recognition: Methods That Search for Structures in Data*, IEEE Press, Piscataway, NY, 1992.

Duda, R., Hart, P. *Pattern Classification and Scene Analysis*, Wiley, New York, USA, 1973.

Gath, I., Geva, A. Unsupervised optimal fuzzy clustering, *IEEE T. Pattern Anal.* **11**(7), 1989, 773–781.

Gath, I., Iskoz, A., Cutsem, B., Van, M. Data induced metric and fuzzy clustering of non-convex patterns of arbitrary shape, *Pattern Recogn. Lett.* **18**, 1997, 541–553.

Geva, A., Steinberg, Y., Bruckmair, S., Nahum, G. A comparison of cluster validity criteria for a mixture of normal distributed data, *Pattern Recogn. Lett.* **21**, 2000, 511–529.

Gonzalez, J., Rojas, I., Pomares, H., Ortega, J., Prieto, A. A new clustering technique for function approximation, *IEEE T. Neural Network.* **13**, 2002, 132–142.

Gustafson, D., Kessel, W. Fuzzy clustering with a fuzzy covariance matrix, in: J. Bezdek, S. Pal (eds.), *Fuzzy models for Pattern Recognition: Methods That Search for Structures in Data*, IEEE Press, New York, NY, 1992.

Hathaway, R. J., Bezdek, J. C. NERF-c means: non-Euclidean relational fuzzy clustering, *Pattern Recogn.* **27**, 1994, 429–437.

Hathaway, R. J., Bezdek, J. C., Davenport, J. W. On relational data versions of c-means algorithms, *Pattern Recogn. Lett.* **17**, 1996, 607–612.

Hathaway, R. J., Davenport, J. W., Bezdek, J. C. Relational dual of the C-means clustering algorithms, *Pattern Recogn.* **22**(2), 1989, 205–212.

Hecht Nielsen, R. *Neurocomputing*, Addison-Wesley, Menlo Park, CA, 1990.

Hirota K., Pedrycz, W. D-fuzzy clustering, *Pattern Recogn. Lett.* **16**, 1995, 193–200.

Karyannis, N. B., Mi, G. W. Growing radial basis neural networks: merging supervised and unsupervised learning with network growth techniques, *IEEE T. Neural Networ.* **8**, 1997, 1492–1506.

Kim, E., Park, M., Ji, S., Park, M. A new approach to fuzzy modeling, *IEEE T. Fuzzy Syst.* **5**, 1997, 328–337.

Lazzerini, B., Marcelloni, F. Classification based on neural similarity, *Electron. Lett.* **38**(15), 2002, 810–812.

Loia, V., Pedrycz, W., Senatore, S. P-GCM: a proximity based fuzzy clustering for user-centered web applications, *Int. J. Approx. Reason.* **34**, 2003, 121–144.

Park, B. J., Pedrycz, W., Oh, S. K. Fuzzy polynomial neural networks: hybrid architectures of fuzzy modeling, *IEEE T. Fuzzy Syst.* **10**, 2002, 607–621.

Pawlak, Z. Rough sets, *Int. J. Comput. Inf. Sci.* **11**, 1982, 3421–356.

Pedrycz, W. Algorithms of fuzzy clustering with partial supervision, *Pattern Recogn. Lett.* **3**, 1985, 13–20.

Pedrycz, W. Conditional fuzzy C-Means, *Pattern Recog. Lett.* **17**, 1996, 625–632.

Pedrycz, W. Conditional fuzzy clustering in the design of radial basis function neural networks, *IEEE T. Neural Networ.* **9**, 1998, 601–612.

Pedrycz, W. Knowledge-Based Clustering, J. Wiley, Hoboken, NJ, 2005.

Pedrycz, W., Vasilakos, A. Linguistic models and linguistic modeling, *IEEE T. Syst. Man Cyb.* **29**, 1999, 745–757.

Pedrycz, W., Kwak, K. Granular models as a framework of user-centric system modeling, *IEEE T. Syst. Man Cyb. A* 2006, 727–745.

Pedrycz, W., Waletzky, J. Fuzzy clustering with partial supervision, *IEEE T. Syst. Man Cyb.* **5**, 1997a, 787–795.

Pedrycz, W., Waletzky, J. Neural network front-ends in unsupervised learning, *IEEE T. Neural Networ.* **8**, 1997b, 390–401.

Ridella, S., Rovetta, S., Zunino, R. Plastic algorithm for adaptive vector quantization, *Neural Comput. Appl.* **7**, 1998, 37–51.

Roubens, M. Fuzzy clustering algorithms and their cluster validity, *Eur. J. Oper. Res.* **10**, 1992, 294–301.

Rovetta, S., Zunino, R. Vector quantization for license-plate location and image coding, *IEEE T. Ind. Electron.* **47**, 2000, 159–167.

Silva, L., Gomide, F., Yager, R. Participatory learning in fuzzy Clustering, *Proceedings of the 14nth IEEE International Conference on Fuzzy Systems*, Reno, USA, 2005, 857–861.

Sugeno, M., Yasukawa, T. A fuzzy-logic-based approach to qualitative modeling, *IEEE T. Fuzzy Syst.* **1**, 1993, 7–31.

Takagi, T. Sugeno, M. Fuzzy identification of systems and its applications to modeling and control, *IEEE T. Syst. Man Cyb.* **15**, 1985, 116–132.

Yager, R. R. A model of participatory learning, *IEEE T. Syst. Man Cyb.* **20**(5), 1990, 1229–1234.

Zadeh, L. A. Outline of a new approach to the analysis of complex systems and decision processes, *IEEE T. Syst. Man Cyb.* 1973, 28–44.

Zadeh, L. A. Fuzzy sets and information granularity, in: M. Gupta, R. Ragade, R. Yager (eds.) *Advances in Fuzzy Set Theory and Appli*cations, North-Holland, Amsterdam, 1979, 3–18.

Zadeh, L. A. Toward a theory of fuzzy information granulation and its centrality in human reasoning and fuzzy logic, *Fuzzy Set Syst.* **90**, 1997, 111–127.

Chapter 15

Emerging Trends in Fuzzy Systems

Fuzzy sets are essential to the development of applications in which complexity plays an essential role. The role of the conceptual and algorithmic framework of fuzzy sets is helpful in coping with the factor of complexity. In this chapter, we discuss some selected areas in which fuzzy sets have already started playing a vital role or their potential becomes tangible. Information retrieval systems have been a part of the history of successful applications of fuzzy sets, but they still pose considerable challenges especially when involving some mechanisms of semantics that need to be considered to improve the quality of retrieval process. On the opposite side of the spectrum of applications are positioned multiagent systems and distributed information processing applications. Multiagent and distributed systems recognize that software and other form of agent implementation have limited capabilities. Nevertheless, through effective interaction they are able to produce required solutions. Multiagent behavior paraphrases, within larger grain framework, the natural multiagency faculty existing in the human brain.

15.1 RELATIONAL ONTOLOGY IN INFORMATION RETRIEVAL

Information search in large database systems is important for many activities. Several systems have been developed to make information search more effective. Information retrieval systems are predominantly based on keywords. This retrieval may result in a large number of answers, requiring considerable effort from the user to analyze and find relevant information. A great deal of effort has been devoted to improve retrieval performance of information search systems by employing the technology of Computational Intelligence (Pasi, 2002). Fuzzy set theory, in particular, has been successfully employed in indexing, clustering techniques,

Fuzzy Systems Engineering: Toward Human-Centric Computing, by Witold Pedrycz and Fernando Gomide
Copyright © 2007 John Wiley & Sons, Inc.

recommendation systems, data mining, and distributed information retrieval (Herrera-Viedma and Pasi, 2003).

Many fuzzy information systems use knowledge bases encoded in ontologies, thesauri, and conceptual networks (Ogawa et al., 1991; Chen and Wang, 1995; Horng et al., 2001; Takagi and Kawase, 2001; Widyantoro and Yen, 2001). Here, we look at models of information retrieval systems based on fuzzy relational ontologies whose knowledge base is encoded in a form of the fuzzy relational ontology, a fuzzy relation on the space of words and categories (Ricarte and Gomide, 2004; Pereira et al., 2006).

15.1.1 Fuzzy Relational Ontological Model

The fuzzy relational ontological model uses two-layer ontology. The first layer of this architecture contains category names whereas the second layer involves keywords associated with the category names occurring at the first layer. Given the contents of a document collection, a system designer selects category names and keywords and forms their associations. Figure 15.1 presents an illustrative example of a fuzzy relational ontology with two categories, c_1 and c_2, and three keywords, k_1, k_2, and k_3. The degree of association between category c_i and keyword k_j is quantified by $r_{ij} \in [0, 1]$. This figure also shows how documents d_1, d_2, and d_3 are connected with the fuzzy relational ontology.

Overall, the fuzzy relational ontology is characterized by a fuzzy relation R defined in the Cartesian product of keywords and categories, $\mathbf{K} \times \mathbf{C}$, where $\mathbf{K} = \{k_1, k_2, \ldots, k_i, \ldots, k_n\}$ and $\mathbf{C} = \{c_1, c_2, \ldots, c_j, \ldots, c_m\}$

$$R = \begin{array}{c} \\ k_1 \\ k_2 \\ \vdots \\ k_n \end{array} \begin{bmatrix} c_1 & c_2 & \cdots & c_m \\ r_{11} & r_{12} & \cdots & r_{1m} \\ r_{21} & r_{22} & \cdots & r_{2m} \\ \vdots & \vdots & \ddots & \vdots \\ r_{n1} & r_{n2} & \cdots & r_{nm} \end{bmatrix}$$

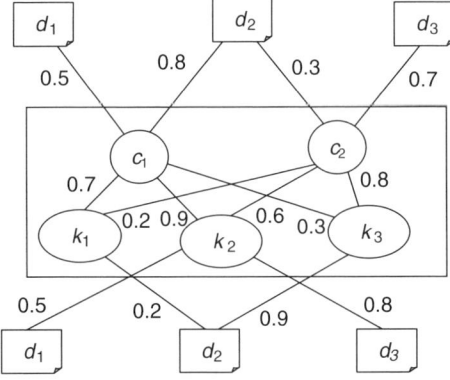

Figure 15.1 An example of a fuzzy relational ontology.

where m is the number of categories present at the first layer, n is the number of keywords at the second layer, and r_{ij} denotes a degree of association between k_i and c_j.

15.1.2 Information Retrieval Model and Structure

Formally, the information retrieval (IR) model is a quadruple $\langle \mathbf{D}, \mathbf{Q}, V, F(q_i, d_{\text{doc}}) \rangle$ where

1. **D** is a set of document representation;
2. **Q** is a set of query representation;
3. *V* is a framework to represent document, queries, and their relationships;
4. $F(q_i, d_{\text{doc}})$ is a function that associates a real number to document–query pairs, the order (ranking) that reflects the document relevance with respect to the user's query (Baeza-Yates, and Ribeiro-Neto, 1999).

An information retrieval system consists of two functional components. The first one is responsible for building a retrieval database from the set of documents. The second component is responsible for providing access to the retrieval database to retrieve relevant documents for the user. In the construction of the retrieval database, we define a suitable model for document representation that involves extracting from each document a set of its representative terms. The second functional component starts with a user query, which is used by the system to access relevant information from the retrieval database. This information is then returned to the user, preferably ranked in the order of relevance.

The structure of an IR system constructed on the basis of the fuzzy relational ontological model is illustrated in Figure 15.2.

Let us now briefly outline the underlying components.

15.1.3 Documents Representation

Let **D** denote a collection of documents, $\mathbf{D} = \{d_1, d_2, \ldots, d_{\text{doc}}, \ldots, d_u\}$; **K** the set of keywords, $\mathbf{K} = \{k_1, k_2, \ldots, k_i, \ldots, k_n\}$; and **C** the set of categories,

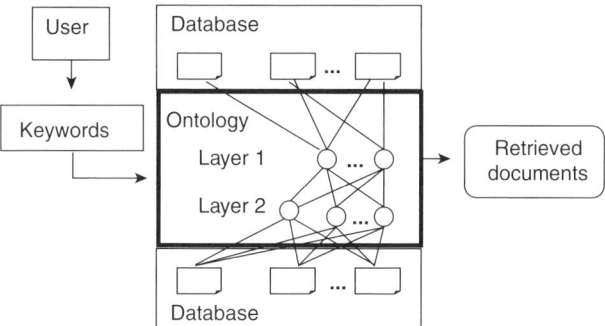

Figure 15.2 Information retrieval system structure.

$\mathbf{C} = \{c_1, c_2, \ldots, c_j, \ldots, c_m\}$. The characterization of documents is represented by two matrices, $\mathbf{T}_k$ and $\mathbf{T}_c$:

$$T_k = \begin{array}{c} \\ d_1 \\ d_2 \\ \vdots \\ d_u \end{array} \begin{array}{cccc} k_1 & k_2 & \cdots & k_n \\ \left[\begin{array}{cccc} \alpha_{11} & \alpha_{12} & \cdots & \alpha_{1n} \\ \alpha_{21} & \alpha_{22} & \cdots & \alpha_{2n} \\ \vdots & \vdots & \ddots & \vdots \\ \alpha_{u1} & \alpha_{u2} & \cdots & \alpha_{un} \end{array}\right] \end{array} \tag{15.1}$$

$$T_c = \begin{array}{c} \\ d_1 \\ d_2 \\ \vdots \\ d_u \end{array} \begin{array}{cccc} c_1 & c_2 & \cdots & c_m \\ \left[\begin{array}{cccc} \beta_{11} & \beta_{12} & \cdots & \beta_{1m} \\ \beta_{21} & \beta_{22} & \cdots & \beta_{2m} \\ \vdots & \vdots & \ddots & \vdots \\ \beta_{u1} & \beta_{u2} & \cdots & \beta_{um} \end{array}\right] \end{array} \tag{15.2}$$

where $\alpha_{\text{doc},i} \in [0,1]$ is the relevance degree between the ith document d_{doc}, and keyword $(k_i, \beta_{\text{doc},j}) \in [0,1]$ is the relevance degree between the document d_{doc} and the category c_j, $1 \leq \text{doc} \leq u$, $1 \leq i \leq n$, $1 \leq j \leq m$.

15.1.4 Query Representation

Let $Q = \{k_i, c_j\}$ be the set of keywords and categories in a query, eventually linked by operators *and* or *or*. A query Q is represented by two vectors, $\mathbf{x} = [x_1, x_2, \ldots, x_i, \ldots, x_n]^T$, $1 \leq i \leq n$ and $\mathbf{y} = [y_1, y_2, \ldots, y_j, \ldots, y_m]^T$, $1 \leq j \leq m$, such that

$$x_i = \begin{cases} 1, & \text{if } k_i \in Q \\ 0, & \text{otherwise} \end{cases}$$

$$y_j = \begin{cases} 1, & \text{if } c_j \in Q \\ 0, & \text{otherwise} \end{cases}$$

15.1.5 Information Retrieval with Relational Ontological Model

Let $\mathbf{x}$ be the vector of query keywords and let R be a fuzzy relational ontology. The composition of $\mathbf{x}$ and R results in a fuzzy set G_c that associates categories with query terms, that is,

$$G_c = \mathbf{x} \circ R \tag{15.3}$$

Similarly, if $\mathbf{y}$ is the vector of query categories, then the composition of R and $\mathbf{y}$ is the fuzzy set G_k representing the association between keywords and query terms, namely,

$$G_k = R \circ \mathbf{y} \tag{15.4}$$

15.1 Relational Ontology in Information Retrieval

where ∘ denotes the max–min composition. In general, we could consider here any max-t composition.

To simplify the exposition of the underlying concept, we assume queries to be in the form of subqueries connected by the *and* (∧) operator whose realization is provided in terms of some t-norm. Each subquery is composed by a set of terms of **K** and **C** connected by the *or* (∨) operator (with the underlying realization offered by some t-norm). In what follows, we assume a specific form of t-norms and t-conorms realized as the minimum and maximum operators. For example, referring to Figure 15.1, we have $\mathbf{C} = \{c_1, c_2\}$ and $K = \{k_1, k_2, k_3\}$ and the query could assume the following form:

$$Q = (k_1 \vee k_2 \vee c_1) \wedge (k_3 \vee c_2)$$

Let $F_c = [f_{c1}, f_{c2}, \ldots, f_{cj}, \ldots, f_{cm}]^T$ be a fuzzy set formed on the basis of G_c (15.3), and $F_k = [f_{k1}, f_{k2}, \ldots, f_{ki}, \ldots, f_{kn}]$ be a fuzzy set formed using G_k (15.4) as follows:

$$f_{ki} = \begin{cases} g_{ki}, & \text{if } g_{ki} > z_1 \\ 0, & \text{otherwise} \end{cases}$$

$$f_{cj} = \begin{cases} g_{cj}, & \text{if } g_{cj} > z_1 \\ 0, & \text{otherwise} \end{cases}$$

where g_{cj} is the jth component of G_c, g_{ki} is the ith component of G_k, and z_1 is a threshold value confined to the unit interval and chosen by the system designer. The composition of F_k with T_k is the fuzzy set V_{DK} that gives the relevance degrees between the collection documents and the keywords k_i, that is,

$$V_{DK} = T_k \circ F_k \qquad (15.5)$$

Similarly, the composition of F_c with T_c produces the fuzzy set V_{DC} that gives the relevance degree between the collection documents and categories c_j,

$$V_{DC} = T_c \circ F_c^T \qquad (15.6)$$

where T stands for a transpose operation of the fuzzy relation.

The documents are retrieved in decreasing order of relevance. This relevance ordering is obtained by arranging according to the increasing values of the membership degrees of

$$V_D = V_{DK} \cup V_{DC} \qquad (15.7)$$

to form the retrieval vector V.

In summary, the retrieval algorithm can be presented in the following format:

procedure INFORMATION-RETRIVAL (Q) **returns** documents
input: query Q
local: thresholds: z_1, z_2
       fuzzy relations: G_c, G_k

  set $Q = \{k_i, c_j\}, 1 \leq i \leq n, 1 \leq j \leq m$
  split $Q : Q_1 = \{k_i\}$ and $Q_2 = \{c_j\}$

construct queries vectors **x** and **y**
compute $G_c = [g_{cj}]$
compute $G_k = [g_{ki}]$
select categories $c_j \in G_c$ with $g_{cj} > z_1$
select keywords $k_i \in G_k$ with $g_{ki} > z_1$
find database documents related with the categories c_j
find database documents related with the keywords k_i
find database documents related with the categories presented in Q_2
find database documents related with the keywords presented in Q_1
if subqueries *and* connected **then** select the common documents
if subqueries *or* connected **then** select all documents
compute V_{DK}
compute V_{DC}
set $V_D = V_{DK} \cup (V_{DC}$
set $V =$ rank ordered V_D according to increasing values
retrieval documents for which V component values are greater than or equal to z_2
return documents

EXAMPLE 15.1

Let us consider the model depicted in Figure 15.1 along with the following fuzzy relations:

$$R = \begin{bmatrix} 0.7 & 0.2 \\ 0.9 & 0.6 \\ 0.3 & 0.8 \end{bmatrix} \quad T_c = \begin{bmatrix} 0.5 & 0 \\ 0.8 & 0.3 \\ 0 & 0.7 \end{bmatrix} \quad T_k = \begin{bmatrix} 0 & 0.5 & 0 \\ 0.2 & 0 & 0.9 \\ 0 & 0.8 & 0 \end{bmatrix}$$

and let $Q = \{k_2 \text{ and } c_1\}$, $z_1 = 0.65$, and $z_2 = 0.4$. Following the steps of the procedure presented above, we obtain

$$Q_1 = \{k_3\} \quad \text{and} \quad Q_2 = \{c_1\}$$

$$G_c = x \circ R = \begin{bmatrix} 0 & 0 & 1 \end{bmatrix} \circ \begin{bmatrix} 0.7 & 0.2 \\ 0.9 & 0.6 \\ 0.3 & 0.8 \end{bmatrix} = \begin{bmatrix} 0.3 & 0.8 \end{bmatrix}$$

The categories c_1 and c_2 are associated with user query at degrees of 0.3 and 0.8, respectively.

$$G_k = R \circ y = \begin{bmatrix} 0.7 & 0.2 \\ 0.9 & 0.6 \\ 0.3 & 0.8 \end{bmatrix} \circ \begin{bmatrix} 1 \\ 0 \end{bmatrix} = \begin{bmatrix} 0.7 \\ 0.9 \\ 0.3 \end{bmatrix}$$

The keywords k_1, k_2, and k_3 are associated with user query at the level of 0.7, 0.9, and 0.3, respectively. The category with $g_{cj} > 0.65$ is c_2 only and keywords with $g_{cj} > 0.65$ are k_1 and k_2. From Figure 15.1 we note that documents related with category c_2 are d_2 and d_3 The documents related with the keywords k_1 and k_2 are d_1, d_2, and d_3. From these documents, the ones related with the category c_1, the category present in Q_2, is d_2 and the documents related with the keyword k_3, the keyword present in Q_1, is d_2 (see Fig. 15.1). Because query terms are connected by the *and* operator, the common document between these sets of documents is d_2.

At the next step we compute the relevance degrees of documents d_1, d_2, and d_3 with respect to category c_2, that is,

$$V_{DC} = T_C \circ F_C = \begin{bmatrix} 0.5 & 0 \\ 0.8 & 0.3 \\ 0 & 0.7 \end{bmatrix} \circ \begin{bmatrix} 0 \\ 0.8 \end{bmatrix} = \begin{bmatrix} 0 \\ 0.3 \\ 0.7 \end{bmatrix}$$

which means that documents d_1, d_2, and d_3 are related with category c_2 with degrees 0.0, 0.3, and 0.8, respectively. Similarly, the relevance degrees of documents d_1, d_2, and d_3 with respect to keywords k_1 and k_2 can be computed as

$$V_{DK} = T_K \circ F_K = \begin{bmatrix} 0 & 0.5 & 0 \\ 0.2 & 0 & 0.9 \\ 0 & 0.8 & 0 \end{bmatrix} \circ \begin{bmatrix} 0.7 \\ 0.9 \\ 0 \end{bmatrix} = \begin{bmatrix} 0.5 \\ 0.2 \\ 0.8 \end{bmatrix}$$

meaning that documents d_1, d_2, and d_3 are related with keywords k_1 and k_2 with degrees 0.5, 0.2, and 0.8, respectively. Now we can compute

$$V_D = V_{DK} \cup V_{DC} = \begin{bmatrix} 0.5 \\ 0.3 \\ 0.8 \end{bmatrix}$$

and increasingly order the components of vector V_D to obtain

$$V = \begin{bmatrix} 0.8 \\ 0.5 \\ 0.3 \end{bmatrix}$$

The retrieved documents are those for which the corresponding components of V satisfy $V \geq 0.4$, namely, documents d_1 and d_3 in the example.

The performance of information retrieval systems is evaluated using recall (R) and precision (P) being defined as follows:

$$R = \frac{\text{Number of relevant documents retrieved}}{\text{Total number of relevant documents in database}}$$

$$P = \frac{\text{Number of relevant documents retrieved}}{\text{Number of documents retrieved}}$$

The measure of recall quantifies how many of the relevant documents have been retrieved. The precision indicates the percentage of retrieved documents that are relevant. To evaluate retrieval algorithms, a number of distinct queries are considered to calculate an average precision at each recall level (Baeza-Yates and Ribeiro-Neto, 1999) as follows:

$$\overline{P}(r) = \sum_{i=1}^{N_q} \frac{P_i(r)}{N_q} \qquad (15.8)$$

where $\overline{P}(r)$ is the average precision at the recall level r, N_q is the number of queries, and $P_i(r)$ is the precision at recall level r for the ith query. In a similar way, we can determine the average

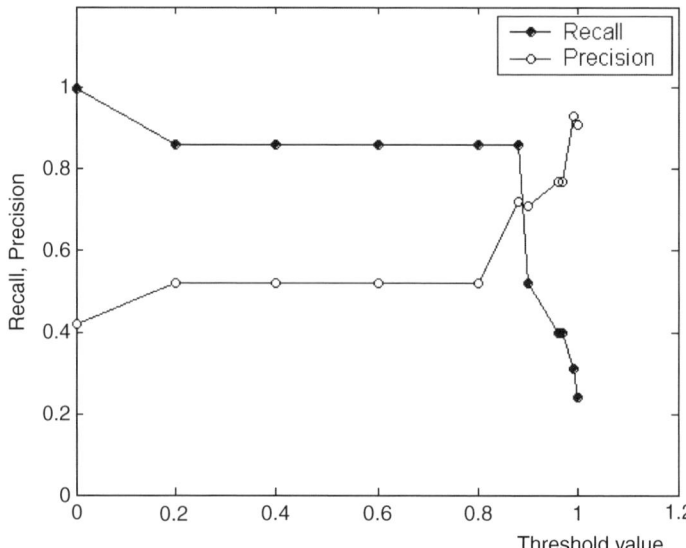

Figure 15.3 Threshold value versus recall and precision.

recall. Detailed information on this evaluation procedure is found in Baeza-Yates and Ribeiro-Neto (1999).

Here, for illustrative purposes, we consider a database containing 100 scientific papers dealing with *computational intelligence*. Using this database, we extracted from the documents 61 words, 6 category names, and 55 keywords. These words form a part of the fuzzy relational ontology. Figure 15.3 shows how the values of recall and precision change versus the values of the relevance threshold value z_2.

Figure 15.4 shows the values of recall and precision for five distinct composite queries namely, *Agent and Information Retrieval, Fuzzy Logic and Information Retrieval, Information Retrieval and Search Engine, Agent and Genetic Algorithm*, and *Genetic Algorithm and Hybrid System* with (a) $z_2 = 0.2$ and (b) $z_2 = 0.75$.

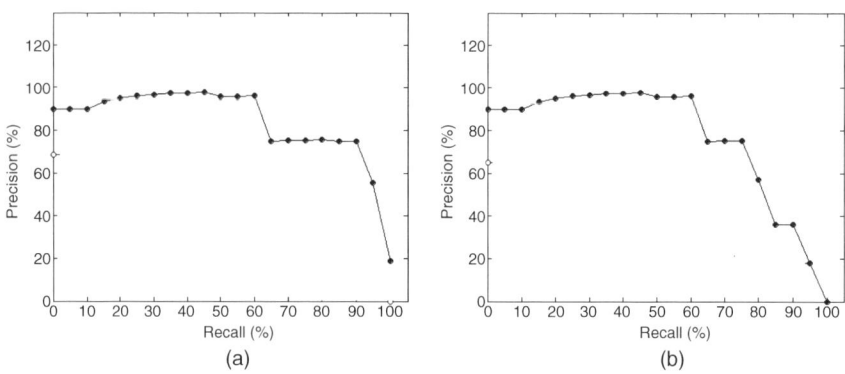

Figure 15.4 Precision versus recall for composite queries.

15.2 MULTIAGENT FUZZY SYSTEMS

This section concentrates on multiagent fuzzy systems and emphasizes the use of genetic fuzzy systems in the design of interaction strategies. Our focus is on auction protocols because they are among the most important ones in many real-world applications such as power markets, transportation systems, electronic markets, and manufacturing control systems. For instance, in power markets, agents must possess profitable bidding strategies to preserve system integrity and improve system goals. A particularly useful way to model bidding strategies is realized through fuzzy rule-based systems whose behavior is formed by using genetic algorithm. Evolution of bidding strategies uncovers unforeseen agent behaviors and allows for a richer analysis of auction mechanisms and their role as a multiagent interaction protocol.

15.2.1 Agents and Multiagents

Software agents are computer programs that perceive their environment, process environment information, and run decision-making procedures to select actions and execute them (Russell and Norvig, 2003). In most cases, the environment contains a number of agents whose actions affect each other, see Figure 15.5. Interdependencies arise because different agents may have different objectives and may operate in a common environment in which resources and capabilities are limited. Therefore, different types of interactions occur between agents. Cooperation is a form of interaction in which agents work together to achieve a common objective. Competition admits self-interested agents to coexist and assumes that they can reach mutually acceptable agreements through negotiation.

An important mechanism for exchanging resource and conducting negotiations among multiple agents is a market. In competitive environments, software agents behave as bounded rational individuals acting in their own self-interest. There are two important issues that need to be addressed when designing multiagent systems (Dash and Jennings, 2003). The first one concerns the specification of the protocols,

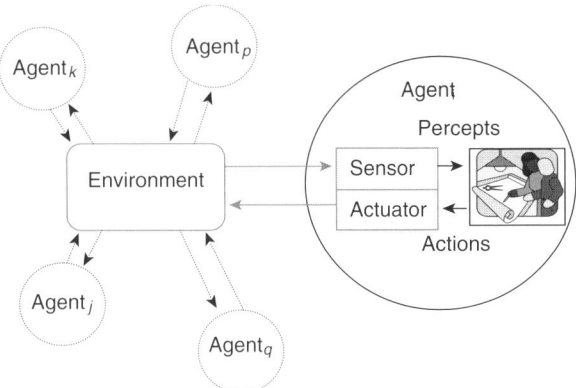

Figure 15.5 An architecture of a multiagent system.

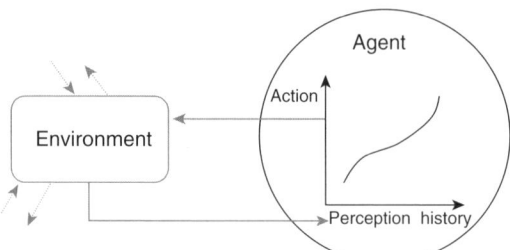

Figure 15.6 Interaction strategy of an agent.

namely, the rules that govern the interactions. These protocols deal with issues such as how the actions of the agents translate into outcomes and set a suite of feasible actions available to the agents, and whether interaction is completed in a single step or it requires multiple steps, or whether we deal with static or dynamic interactions. The second design issue concerns the decision-making policy of the agents, that is, the interaction strategies the agents adopt to play in an environment ruled by the prevailing protocol. These decision-making policies are mappings from perception history into actions, Figure 15.6.

In auction-based markets, auction is a form of resource allocation. The agents submit bids and a center solves a problem to determine the allocation that maximizes a reported demand, given agent bids. Sellers then receive their income based on prices set by a pricing mechanism.

In what follows, we focus on electricity power markets as a vehicle to illustrate the development of interaction strategies for fuzzy agent systems. Many power markets worldwide use a form of auction mechanism to decide on power generation dispatch and energy pricing. In particular, we highlight the development of an economical agent, a power generation plant, to compete on an auction environment, the power market. The objective of the agent is to decide how to bid to offer energy. In other words, given an auction protocol, the agents must decide which bidding strategy will result in their best payoff. We use genetic fuzzy systems to evolve bidding strategies for agents. This approach is especially important in auction-based markets because in practice most auctions deviate from ideal auction models for which theoretical results exist (McAfee and McMillan, 1987).

15.2.2 Electricity Market

Many of the restructured power industries worldwide use auctions as a mechanism of resource allocation and system coordination. A power supplier competing on a power market must decide upon how much energy to offer and at what price. In a perfectly competitive market, risk averse participants have an incentive to offer energy at a price equivalent to their marginal costs (Green, 2000). However, electricity markets are much more an oligopoly than a *laissez-faire*, with low or no demand elasticity on the short term, barriers to entry, and physical constraints.

Thus, a power supplier may have an incentive to offer energy at a price other than its running marginal costs and extract some surplus from such an imperfect market. The behavior of generator bidding other than marginal costs in an effort to exploit market imperfections is called strategic bidding. The most common strategy is to maximize the expected profit, but other strategies may interest a supplier exercising its market power such as competing for being a base-load generator (Visudhipan and Ilic, 1999), increasing *market share*, and increasing profit margin (Monclar and Quatrain, 2001). A detailed survey on strategic bidding is found in David and Wen (2000).

15.2.2.1 *Demand Characteristics*

The load demand is supposed to be publicly known and the auction is an *ex-ante* mechanism to allocate the power to be produced by each plant and to define the hourly energy price. Figure 15.7 shows the hourly load profile for a 3-week period. Load data from the first week will later be used by a genetic fuzzy rule-based system to evolve the bidding agent. The remaining 2 weeks' load data are used for testing and performance analysis. Overall we have 7 days times 24 h producing 168 training examples and 14 days times 24 h producing a total of 336 samples for testing.

15.2.2.2 *Running Cost Function*

A pool of thermal plants is chosen, and actual public plant data is considered. Electrical constraints and geographical location have been neglected to simplify the presentation. The fuel cost of producing power on coal, gas, and oil plants (GJ/h)

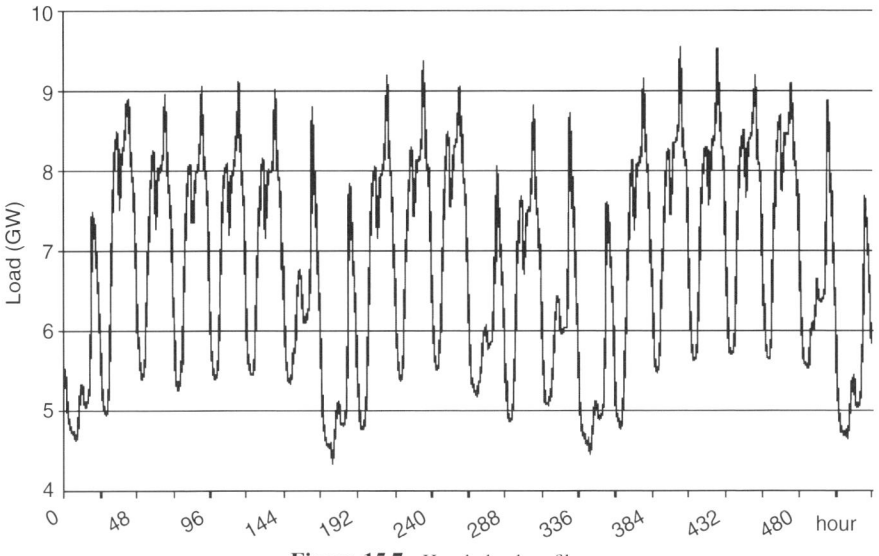

Figure 15.7 Hourly load profile.

are modeled as a quadratic function of the active power generation (g_{jh}) supplied by plant j at hour h, (El-Hawary and Christensen, 1979).

$$F(g_{jh}) = a + bg_{jh} + cg_{jh}^2 \tag{15.9}$$

Nuclear plants assume linear cost functions. The supplier cost function $C_j(.)$ is given by $F(g_{jh})$ multiplied by the fuel cost in \$/GJ. Hence, we arrive at the following quadratic cost function:

$$C_j(g_{jh}) = \alpha + \beta g_{jh} + \gamma g_{jh}^2 \tag{15.10}$$

15.2.2.3 Uniform Price Auction

The pricing mechanism is a uniform price sealed bid auction for a day ahead market. In each trading day the auctioneer performs the following functions:

1. Opens the auction;
2. Publishes the day ahead hourly load forecast;
3. Accepts bids from suppliers;
4. Stops receiving bids;
5. Applies the pricing algorithm (merit order);
6. Publishes the hourly price $\pi_h, h = 1, \ldots, 24$;
7. Informs each supplier about the power to be produced for 24 h;
8. Closes the auction.

The central auctioneer decides the hourly dispatch, choosing the plants to produce and the corresponding price to minimize overall energy cost. Power demand D_h is considered to be inelastic with price. Hence, the auctioneer must assure that for each hour h, we have $\sum_{j=1}^{T_h} g_{jh} = D_h$, where g_{jh} is the power supplied by plant j and T_h is the number of suppliers. Thus, allocation would cost the market an amount of $D_h \pi_h(D_h)$, and suppliers would profit an amount of

$$P_{jh} = \pi_h g_{jh} - C_j(g_{jh}) \tag{15.11}$$

Suppliers must internalize all costs to a simple bid, a pair (q_{jh}, p_{jh}) of the quantity offered (MW) and its price (\$). The quantity q_{jh} is less than or equal to the plant capacity, G_j. Description of more complex bidding mechanisms is given in Contreras et al. (2001).

We assume that conservative agents bid pairs $(G_j, MC_j(G_j))$, where $MC_j(G_j)$ is the marginal cost at capacity G_j

$$MC_j(G_j) = \left.\frac{\partial C_j(g_{jh})}{\partial g_{jh}}\right|_{g_{jh}=G_j} \tag{15.12}$$

Contrary to conservative agents, the intelligent agent is free to choose both bid price and quantity. To simplify, only the load demand is taken into consideration to decide the bid.

Table 15.1 Thermal Plants Characteristics.

Plant	Type	G_j	$MC_j(G_j)$	$C_j(.)$
Angra 1	Nuclear	657	8.5	$8.5g$
Angra 2	Nuclear	1309	8.5	$8.5g$
P.Medici 3-4	Coal	320	32.95	$865.3 + 28.914g + 0.0063g^2$
P.Medici 1-2	Coal	126	33.33	$343.34 + 28.53g + 0.01905g^2$
TermoBahia	Gas	171	34.38	$580.54 + 30.985g + 0.00992g^2$
TermoCeara	Gas	153	34.72	$505.29 + 30.558g + 0.0136g^2$
Canoas	Gas	450	37.54	$1575.22 + 33.869g + 0.00408g^2$
N. Fluminense	Gas	426.6	37.63	$1484.92 + 33.759g + 0.00454g^2$
Araucaria	Gas	441.6	37.70	$1505.65 + 33.782g + 0.00443g^2$
Tres Lagoas	Gas	324	37.76	$1115.29 + 33.595g + 0.00643g^2$
Corumba	Gas	79.2	38.03	$278.97 + 33.256g + 0.03016g^2$
Juiz de Fora	Gas	103	38.73	$323.68 + 33.088g + 0.0274g^2$
Ibirite	Gas	766.5	39.07	$3632.08 + 31.966g + 0.00463g^2$
TermoRio	Gas	824.7	39.11	$3904.05 + 31.912g + 0.00436g^2$
Argentina I	Gas	1018	41.04	$4459.61 + 32.775g + 0.00406g^2$
Argentina II	Gas	1000	41.05	$4379.82 + 32.774g + 0.00414g^2$
J. Lacerda C	Coal	363	52.64	$1547.15 + 45.962g + 0.00919g^2$
J. Lacerda B	Coal	262	63.30	$1407.65 + 56.198g + 0.0136g^2$
J.Lacerda-A1-2	Coal	100	67.10	$549.89 + 57.895g + 0.04605g^2$
J.Lacerda-A3-4	Coal	132	67.35	$728.6 + 57.65g + 0.03674g^2$
Charquendas	Coal	69.1	67.72	$414.59 + 60.037g + 0.05559g^2$
FAFEN	Gas	57.6	74.78	$417.18 + 66.857g + 0.06879g^2$
Uruguaiana	Gas	582	82.77	$4306.82 + 76.729g + 0.00519g^2$

15.2.2.4 Market Configuration

Table 15.1 shows data of the thermal plants, the agents that are part of the power market. Plant name and type are indicated. The plant capacity G_j is given in MW and the marginal cost at full capacity, $MC_j(G_j)$ in \$/MW, corresponds to the conservative bid. The cost functions $C_j(.)$ for each of the thermal plants are given by the last column of Table 15.1, where g_{jh} is denoted by g for short.

Given the capacity of plants and assuming that agents bid their marginal cost at full capacity, the result is the conservative merit order that corresponds to the supply curve shown in Figure 15.8.

15.2.3 Genetic Fuzzy System

A first step to design a genetic fuzzy rule-based system (GFRBS) is to decide which components of the knowledge base will be optimized by the genetic algorithm (Cordón et al., 2001; Cordón et al., 2004). The choice involves a setup of some trade-off between granularity and search efficiency. Among all possible selections, the following have been chosen for evolution: database granularity, that is, number of linguistic terms; membership functions parameters; and rule base, namely, rule

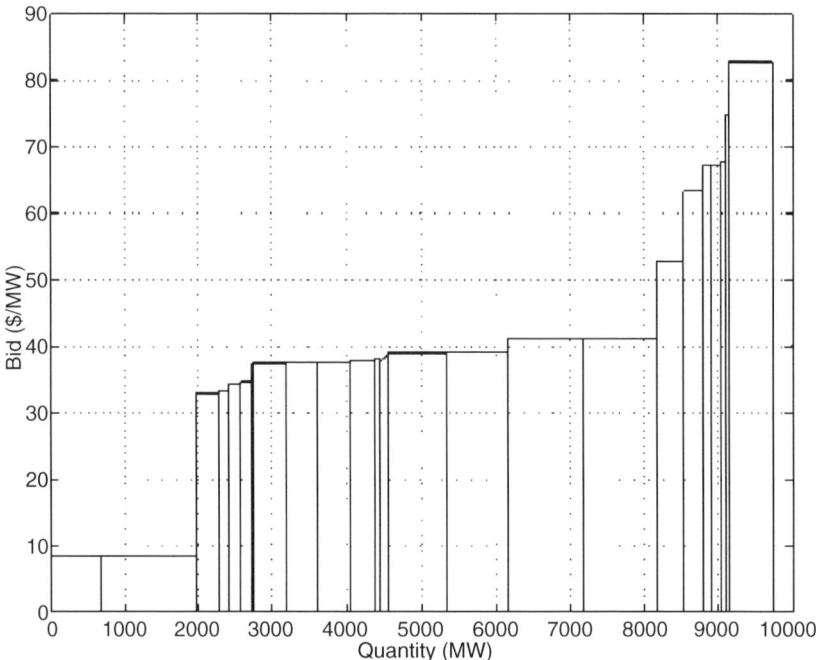

Figure 15.8 Market supply function.

syntax and rule status. Rule status serves to indicate if a rule is active or inactive. An active rule effectively belongs to the rule base and is used during fuzzy inference. An inactive rule remains in the rule base genotype, but does not take part in fuzzy inference. Evolve rule status as a means to obtain the number of rules that composes the rule base. The approach adopted here evolves simultaneously the data and rule bases (Walter and Gomide, 2006). Each individual of a population represents a complete fuzzy rule-based system.

15.2.3.1 Chromosome Representation

Granularity. The granularity of the GFRBS, the number of linguistic terms adopted, defines the component Cr_1 of the chromosome and is encoded by a variable length chain of integers (N, M, n, m), with $n = (n_1, \ldots, n_i, \ldots, n_N)$ and $m = (m_1, \ldots, m_j, \ldots, m_M)$. N is the number of antecedents, M the number of consequents, n_i the number of linguistic terms of the input variable i, and m_j is the number of terms of output variable j, with $i = 1, \ldots, N, j = 1, \ldots, M$.

The total number of linguistic terms is given by equation (15.13) where L_a is the number of linguistic terms of the antecedents and L_c the number of linguistic terms of the consequents.

$$L = L_a + L_c = \sum_{i=1}^{N} n_i + \sum_{j=1}^{M} m_j \qquad (15.13)$$

15.2 Multiagent Fuzzy Systems

Membership Functions Following the approach suggested in Glorennec (1996) and Cordón et al. (2001), one considers the use of fuzzy partitions. The number of fuzzy sets is kept the same and the change of only one parameter results in a new fuzzy partition of the input or output universe. This approach allows global adjustment of membership functions. Moreover, the fuzzy rule-based system evolved is likely to be more transparent because it produces frames of cognition and therefore these frames satisfy the requirements as identified in Chapter 3. Trapezoidal membership functions are used. Therefore, there is no need to evolve shapes of membership functions. Because partitions are used, each linguistic input variable i requires $2(n_i - 1)$ real numbers to define the partition of the corresponding universe. The output variable j requires $2(m_j - 1)$ real numbers to define its universe partition. These real numbers are encoded in the component Cr_2 whose length is given in (15.13). Because no normalization is performed, each value must lie within the corresponding interval $[v^X_{\min}, v^X_{\max}]$.

$$L_{mf} = \sum_{i=1}^{N} 2(n_i - 1) + \sum_{j=1}^{M} 2(m_j - 1) \qquad (15.14)$$

For instance, if $N = 1$ and $M = 2$, then Cr_1 is $(1, 2, (n_1), (m_1, m_2))$. The parameters of the membership functions encoded in Cr_2 corresponding to this example Cr_1 component are as follows:

$$((r_1^{X_1}, 1_2^{X_1}, r_2^{X_1}, 1_2^{X_1}, \ldots, 1_k^{X_1}, \ldots, 1_{n_1}^{X_1}),$$
$$(r_1^{Y_1}, 1_2^{Y_1}, r_2^{Y_1}, 1_2^{Y_1}, \ldots, 1_k^{Y_1}, \ldots, 1_{m_1}^{Y_1}),$$
$$(r_1^{Y_2}, 1_2^{Y_2}, r_2^{Y_2}, 1_2^{Y_2}, \ldots, 1_k^{Y_2}, \ldots, 1_{m_2}^{Y_2}))$$

where $l_k^{X_1}$ denotes the parameter on the left of the top of the kth trapezium associated with the kth fuzzy set of variable X_1, whereas r_k is the parameter on the right of the top. Figure 15.9 shows an example of a fuzzy partition when $n_1 = 4$. Note that in this case $l_1^{X_1} = v^{X_1}_{\min}$ and $r_4^{X_1} = v^{X_1}_{\max}$.

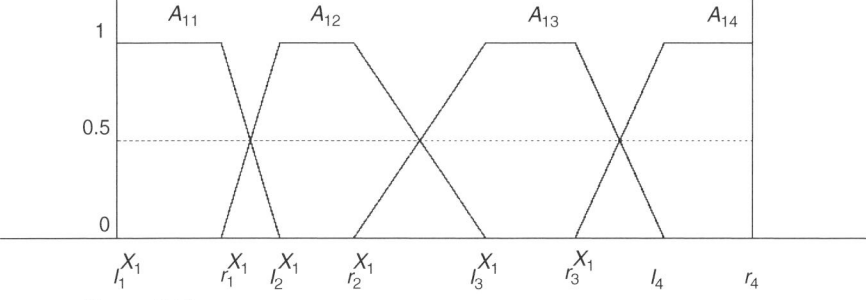

Figure 15.9 Example of a fuzzy partition with trapezoidal membership functions.

Rule Base We assume that the fuzzy bidding strategy is described by the union of a set of linguistic fuzzy rules,

$$R = \bigcup_{l=1}^{k} R_k$$

Each rule is interpreted by a fuzzy conjunction of disjunctions. A rule has the following format:

If X_i is $(A_{io} \text{ or } A_{ip})$ and ... then Y_j is $(C_{jq} \text{ or } C_{jr})$

where A_{io} and A_{ip} are the fuzzy sets associated with the input variable X_i, and C_{jq} and C_{jr} are the fuzzy sets associated with the output variable Y_j, with $o, p \leq n_i$ and $q, r \leq m_j$. Each rule is encoded in a chain of bits of variable length L given by (15.13), with one additional bit used to indicate an (in)active rule. If the antecedent of a rule contains an entry like "X_i is A_{ij}," then the corresponding bit at position $p = j + \sum_{k=1}^{i-1} n_k$ is 1, otherwise it is 0. For instance, if $N = 3, M = 1, n = \{5, 3, 5\}, m = \{7\}$, then $L_a = 13$. Thus the chain representing the rules has length of $(20 + 1)$ entries. A rule such as

If X_1 is $(A_{13} \text{ or } A_{14})$ and X_3 is (A_{31}) ... then Y_1 is $(C_{14} \text{ or } C_{15})$

is encoded by the sequence

(1) 00110 000 10000 $\Rightarrow$ 0001100

where the first bit in parenthesis indicates that the rule is active.

Each rule base contains L_r rules, where L_r is randomly chosen in the interval, that is,

$$\min(L_a, L_c) \leq L_r \leq L_a \, L_c \tag{15.15}$$

The number of rules, rule size, and the rules themselves define chromosome component Cr_3. Number of rules and rule size are encoded for convenience because they could have been calculated from the rule base. Encoding the number of rules make it readily accessible by the algorithm. Thus, component Cr_3 is composed by two integers and L_r chains of length $(L + 1)$, including the first bit indicating (in)active rules.

Evolutionary Algorithm The algorithm to evolve fuzzy rule-based systems is described as follows.

procedure GFRBS-ALGORITHM (**X**,**Y**, f) **returns** a rule base
input: universes **X**,**Y**
    fitness function: f
local: population: set of individuals
    crossover rate, mutation rate
    max: maximum number of generations

    INITIALIZE (population, number individuals)
    repeat
    evaluate each individual using f

select parents in population using relative fitness
apply crossover and mutation on parents
create new population
until number generations $\geq$ max
return rule base

Below we offer details on the realization of the genetic operators such as selection, crossover, and mutation.

15.2.3.2 Genetic Operators

Choosing appropriate genetic operators is important in GFRBS due to the specific chromosome encoding structure that represents the knowledge base. Because there are direct relationships among chromosome components, the genetic operators work synchronously to keep genotype integrity. Selection mechanism uses the roulette wheel with *elitism*, that is, the best individual of a population is always retained and carried over to the next population.

15.2.3.3 Crossover

Two different crossover operators are used depending on whether the selected parents have the same granularity. We say that two individuals with Cr_1 given by

$$(N, M, (n_1, \ldots, n_i, \ldots n_N), (m_1, \ldots, m_j, \ldots, m_M))$$
$$(N, M, (\eta_1, \ldots, \eta_i, \ldots \eta_N), (\mu_1, \ldots, \mu_j \ldots, \mu_M))$$

have the same granularity, if $n_i = h_i, i = 1, \ldots, N$, and $m_j = m_j, j = 1, \ldots, M$.

When the granularity of parents is the same, a promising region in the search space is found and could be exploited (Cordón et al., 2001). In this case, the granularity of the database, the chromosome Cr_1 component, is kept the same for the offspring and the membership functions parameters (component Cr_2) combined using the max-min-arithmetic algorithm of Herrera et al. (1997).

The max-min-arithmetic algorithm produces, for each pair of chromosomes, four offspring via pairwise combination of the minimum, maximum, and two linear combinations of each element. The best two individuals among the offspring can be added to the next generation (Cordón et al., 2001). Here all offsprings are added and the algorithm proceeds with mutation. Population size is limited in the next generation selection.

When the parents have different granularity, a random crossover position $p, 1 \leq p \leq (N + M - 1)$, is chosen. Both granularity (component Cr_1) and the corresponding parameters of the membership functions (component Cr_2) are recombined. Two individuals whose component Cr_1 are

$$((n_1, \ldots, n_i, \ldots, n_N), (m_1, \ldots, m_j, \ldots, m_M))$$
$$((\eta_1, \ldots, \eta_i, \ldots, \eta_N), (\mu_1, \ldots, \mu_j, \ldots, \mu_M))$$

478 Chapter 15 Emerging Trends in Fuzzy Systems

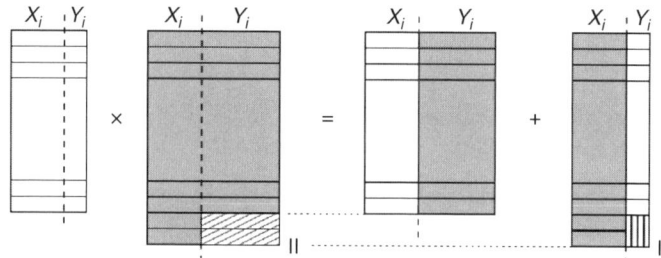

Figure 15.10 Rule base crossover.

are crossed at a position $p, p \leq N$, the result are offspring

$$((n_1,\ldots,n_{p-1},\eta_p,\ldots,\eta_N),(\mu_1,\ldots,\mu_j,\ldots,\mu_M))$$
$$((\eta_1,\ldots,\eta_{p-1},n_p,\ldots,n_N),(m_1,\ldots,m_j,\ldots,m_M))$$

The result is similar for $N < p \leq (N + M - 1)$.

Rule base crossover between individuals of different granularities is shown in Figure 15.10. The length of each rule and the rule base size (number of rules) can be different for the selected rule bases. A specific crossover operator, denoted by × in Figure 15.10 is used. To realize a crossover of the rule bases, the same crossover position p corresponding to the position where the granularities have been crossed is kept. Rule antecedents of a rule are combined with rule consequents of other rule to form new rules. The difference of size, area I of the offspring of Figure 15.10, is completed. The corresponding portion of fuzzy rules from the mating individuals are chosen randomly to complete the remaining part, and area II is discarded.

For example, assume two individuals with granularity (1,1,3,5) and (1,1,4,3) and rule bases shown at the top of Figure 15.11 (A and I indicate an active or inactive rules respectively).

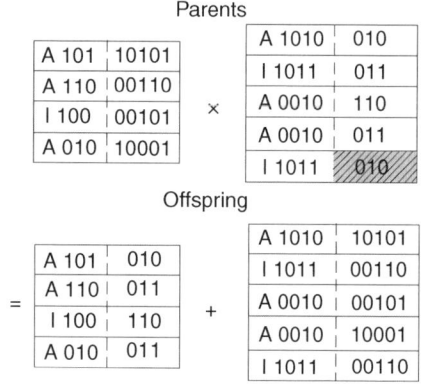

Figure 15.11 Example of crossover of rule bases.

Crossover of these two individuals results in two offspring with granularity (1,1,3,3) and (1,1,4,5) shown in Figure 15.11. The last rule of offspring (1,1,4,5) is a recombination of the last rule I1011 ⇒ 010 of the second parent with the second rule I110 ⇒ 00110 randomly chosen from its mate.

15.2.3.4 Mutation

Depending on the granularity of individuals, different mutation operators are used. Granularity is an integer in the interval [3,9], and mutation produces a local variation adding or subtracting 1 with equal probability. This mutation scheme is suggested in Cordón et al. (2001). When the granularity of individuals increases, a new randomly chosen pair of membership functions parameters is added to Cr_2 and a set of linguistic terms added to the rule base component Cr_3. When the granularity decreases, a pair of membership function parameters at a randomly chosen position of Cr_2 is deleted together with the set of corresponding bits of the linguistic terms at the same position in the rule base.

Mutation of membership function parameters uses non-uniform mutation operator (Michalewicz, 1996) and rule bases mutated using the standard, bitwise reversing operation. This means that if the entry of the chromosome is equal to 1, its value is flipped to 0 and vice versa.

15.2.3.5 Fitness Evaluation

The fitness function is the profit obtained during auctions of the training period. To avoid negative fitness, a constant whose value equals the total number of hours H times fixed costs C_{fixed} is added to the function. When modeling the costs as in expression (15.10), we have $C_{fixed} = a$. For 1-week period $H = 168$ and hence the fitness ϕ_j is computed using the expression

$$\phi_j = \sum_{h=1}^{168} P_{jh} + 168\alpha$$

where P_{jh} is the profit computed using (15.11).

The evaluation of individuals (fitness function) in the population uses the following procedure:

procedure fITNESS-EVALUATION (P, f) **returns** fitness values
input: population P
        fitness function: f

  for each individual in P **do**
      decode individual
      add individual in the market
      run auctions for the training period
      store individual fitness ϕ
      remove individual from the market
  return fitness values

EXAMPLE 15.2

We consider the case in which thermal plant Argentina II plays the role of an intelligent agent whose strategy is evolved using the genetic fuzzy systems (Walter and Gomide, 2006). The remaining power plants bid based on their marginal costs. The agent strategy uses demand as input and bid price and quantity as outputs. Given the price, there is a quantity that maximizes profit. If price is lower than β, then from (15.10) to (15.11) the quantity is 0. For prices greater than $(\beta + 2\gamma G_j)$ the quantity that maximizes profit is the plant capacity G_j. For prices π_h such that $\beta < \pi_h < \beta + 2\gamma G_j$ the quantity that maximizes profit is

$$q* = \frac{\pi_h - \beta}{2\gamma} \qquad (15.16)$$

Figure 15.12 shows the evolution of fitness considering a population of 36 individuals, with probability of crossover (crossover rate) of 0.7 and probability of mutation (mutation rate) of 0.1. The granularity in this example ranges between 3 and 11. Evolution stops after a maximum of 500 generations.

The membership functions and partitions of the input and output membership functions evolved are depicted in Figure 15.13. Decoded values of outputs Y_1 and Y_2 are values that multiply the marginal cost at full capacity and plant capacity to produce the bid price and quantity, respectively. We adopt the interval [0, 2] as the universe of Y_2, and hence whenever the decoded (defuzzified) value of Y_2 is greater than 1, the agent bids its maximum capacity. The rule base evolved corresponds to the one whose active rules are as follows:

10010 ⇒ 01010101 111011
11111 ⇒ 01100000 110011
11011 ⇒ 11100010 010110
00010 ⇒ 01010101 011011
00011 ⇒ 01011010 010111
01010 ⇒ 11110110 011111

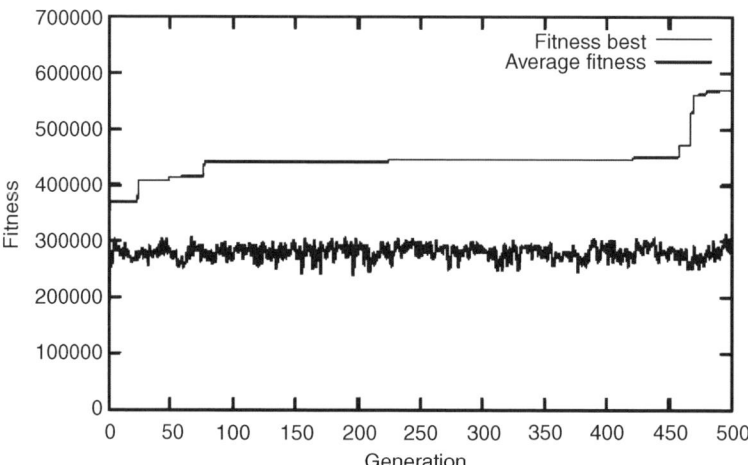

Figure 15.12 Values of the fitness function in successive generation of evolution.

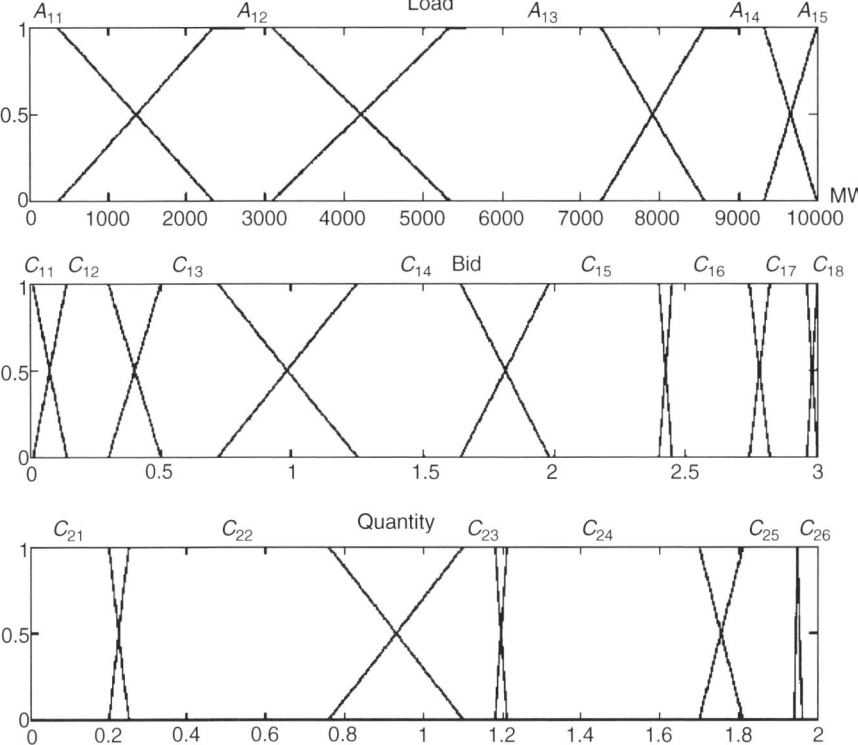

Figure 15.13 Membership functions and partitions of the input and output universes.

For instance, the first rule is decoded as

If X_1 is (A_{11} or A_{14})
then Y_1 is (C_{12} or C_{14} or C_{16} or C_{18})
and Y_2 is (C_{21} or C_{22} or C_{23} or C_{25} or C_{26})

The best bidding strategy playing during the two test-week period results in a profit 36.7% higher and produces 91.3% more energy than the conservative strategy. The plant is dispatched for 306 h over the total of 336. The agent fuzzy rule-based strategy obtains higher payoff combining two policies. First, when demand is low and price is below its marginal cost at full capacity, the agent tries to be dispatched by bidding lower price and quantity that minimizes losses. For example, during the first hour of the test period and a demand of 305.9 MW, the agent bids (774.3, 23.77) in (MW, $/MW). This results in a loss of $2246 because energy price is defined by Juiz de Fora plant as $38.73/MW. With this price, the quantity that minimizes losses is, 719.3 MW from (15.14). Hence the strategy is successful in choosing the best bidding quantity. The conservative strategy for the same 1-h period would not let the plant be dispatched and would incur a loss equivalent to its fixed costs, $4379.82. The overall system energy price would, in this case, be slightly higher, $39.07. Second is to increase price when it has the opportunity to be the marginal generator. For instance, looking at the 19h day of the first testing day as an example, the agent plant produces 299.3 MW at $58.38/MW and the overall system price is defined by its bid. This results in a profit of $2912 against a production of 662.3 MW at $41.05/MW and a loss of $715 for conservative strategy. Minimizing losses for low demand increases the total profit during the test period.

15.3 DISTRIBUTED FUZZY CONTROL

One of the major challenges of control theory is control of complex dynamic systems. Complex systems are high-dimensional and uncertain and information structures are constrained by the number of channels and a topology of the communication networks. Often, the system arises as an interconnection of several subsystems and both the communication structure and the subsystems themselves are not precisely known. Computer controlled systems have distributed architectures and are connected via communication networks. Controllers are local and run asynchronously in the distributed architecture. Conventional centralized control systems often are impractical to control complex systems, and decentralized control approaches become a key requisite to handle system complexity (Palm, 2004).

During the last decade, sophisticated control strategies using fuzzy sets and neural networks were used successfully in practice, usually to manage processes with only a few inputs and output variables. These controllers work in a stand-alone mode and only several decentralized approaches for large-scale systems have been recently developed (Tong et al., 2004)

An approach to decentralized control based on multiagent systems and market-based schemes have emerged recently (Voos, 1999). It uses principles of economy to distribute and allocate resources. Market-based control approaches have the ability to allocate resource efficiently in large-scale systems using simple algorithms that can be decentralized. A distinguishing property of distributed, multiagent control systems comes with an ability to perform intelligently only by interaction in the sense that all agents together show a behavior that no single agent would be able to perform.

In this section, we examine the application of market-based control mechanism as a vehicle to design distributed fuzzy control systems. A simple example considering a set of couple tank system is used to illustrate the approach.

15.3.1 Resource Allocation

Resource allocation problems can be formulated as an optimization problem of the form

$$\max J(\mathbf{u}_1, \mathbf{u}_2, \ldots, \mathbf{u}_n)$$
$$\text{s.t.} \sum_{i=1}^{n} \mathbf{u}_i \leq \mathbf{r} \quad (15.17)$$
$$\mathbf{u}_i \geq 0, i = 1, \ldots, n$$

The type of resource **u** to allocate is limited by a total amount smaller than or equal to **r**.

Economic systems can be viewed as composed by agents, a price system, and a number of commodities. Commodity can be services and goods. Because commodities are limited, each is associated with a price mechanism that forms a pricing system. In general, agents are either producers or consumers. The aim of an agent in an economy is to decide on the amount of his input or output for each commodity. Producers choose their supply and consumers their demand. Producers aim at maximizing profit and consumers to maximize utility. Market equilibrium requires the sum of all demands to be equal to the sum of all supplies.

Let us assume an amount of ℓ types of resources to form a resource vector $\mathbf{u} \in R^\ell$ and a number of n consumer agents and m producer agents. Considering that the consumer agents aim to maximize their utilities $f_i(\mathbf{u}_i^c), \mathbf{u}_i^c \in R^\ell$ and that producer agents aim to maximize their profits $f_j(\mathbf{u}_j^p), \mathbf{u}_j^p \in R^\ell$; under limited amount of resources, optimal allocation requires solution of the following optimization problem:

$$\max_{\mathbf{u}_i^c} f_i(\mathbf{u}_i^c) \quad \text{consumer}$$

$$\max_{\mathbf{u}_j^p} f_j(\mathbf{u}_j^p) \quad \text{producer}$$

$$\text{s.t.} \sum_{i=1}^n \mathbf{u}_i^c = \sum_{j=1}^m \mathbf{u}_j^p \quad (15.18)$$

$$\mathbf{u}_i^c, \mathbf{u}_j^p \geq 0, i = 1, \ldots, n, \ j = 1, \ldots, m$$

Optimization problem (15.18) is a distributed allocation problem and the objective functions have to simultaneously be optimized. The set of all supplies and demands define the state of an economy. An equilibrium with respect to a price system means that no consumer can increase his utility without increasing his expending, and no producer can increase his profit (MasColell et al., 1995).

15.3.2 Control Systems and Economy

Consider a set of n coupled dynamic systems of the form

$$\begin{aligned} \mathbf{x}_i(t+1) &= f_i(\mathbf{x}(t), \mathbf{u}(t), t) \\ \mathbf{y}_i(t) &= g_i(\mathbf{x}(t), t) \end{aligned} \quad (15.19)$$

where $\mathbf{x}_i(0) = \mathbf{x}_{io}, \mathbf{u}_i \in \mathbf{R}^\ell$ is the input variable, $\mathbf{x}_i \in \mathbf{R}^q$ is the state variable, and $\mathbf{y}_i \in R^s$ is the output of the ith subsystem. To apply market mechanism to control systems, the control problem must be translated into an economic optimization problem such as the resource allocation problem (15.18) for which resource variables have to be defined. In control systems, the input variables can be considered as the resource variables to be distributed. Therefore, control variables must be limited. The general formulation of a control problem as a resource allocation problem in the case of distributed systems can therefore be formulated as follows (Voos, 1999):

$$\min_{\mathbf{u}} f_i(|\mathbf{u}_1|, \ldots, |\mathbf{u}_n(t)|, \mathbf{x}_o)$$
$$\text{s.t. } \mathbf{x}_i(t+1) = f_i(\mathbf{x}(t), \mathbf{u}(t), t)$$
$$\mathbf{y}_i(t) = g_i(\mathbf{x}(t), t)$$
$$\sum_{i=1}^{n} \mathbf{u}_i(t) = \mathbf{R} \qquad (15.20)$$
$$|\mathbf{u}_i(t)| \geq 0$$
$$\mathbf{x}_i(0) = \mathbf{x}_{io}$$
$$i = 1, \ldots, n$$

The remaining problem is the transformation of the optimization problem (15.20), a form of resource allocation problem, into a distributed optimization problem (15.18). Therefore consumer and producer agents have to be defined together with the corresponding utility and profit functions. These functions shall be chosen such that their simultaneous optimization produces a solution that also solves (15.20). In complex systems, a possible choice is to assign each subsystem with a consumer and a producer agent. After the translation, the distributed allocation optimization problem is solved using market-based mechanisms such as exchange and pricing to find an equilibrium solution. Solution of optimal control problems using transformations in economy optimization in the form of distributed resource allocation problems is called market-based control (Clearwater, 1996). There are several approaches to develop such principles (Voos, 1999; Voos and Litz, 1999, 2000). A fuzzy set-base approach is introduced next.

15.3.3 Fuzzy Market-Based Control

We address a fuzzy distributed control approach based on market principles using a coupled tank system example as suggested by Voos (1999). The control problem deals with the regulation of water levels in a four tank coupled system shown in Figure 15.14.

Each tank can be filled through a valve i. The tank level is denoted by h_i. The valve is an actuator whose output is a mass flow to tank i proportional to the valve position u_i. A pump with limited power is used to circulate water in the system. Therefore, the mass flow can be viewed as limited resource. The reference operation levels of each tank, the set points, are assembled in a single vector $\mathbf{h}_s$. We assume that the desired operation point $(\mathbf{u}_o, \mathbf{h}_s)$ of the tanks is known. The control task is to regulate the levels at the operation point. Each of them is considered as a subsystem associated with an agent, the controller. The input variable of each agent i is the deviation $\Delta h_i(t)$ of water level and the output variable is the valve position $u_i(t) = u_{io} + \Delta u_i(t)$, the sum of the desired operation level of tank i and the control action $\Delta u_i(t)$. Because tank set points are known, the agent only has to compute

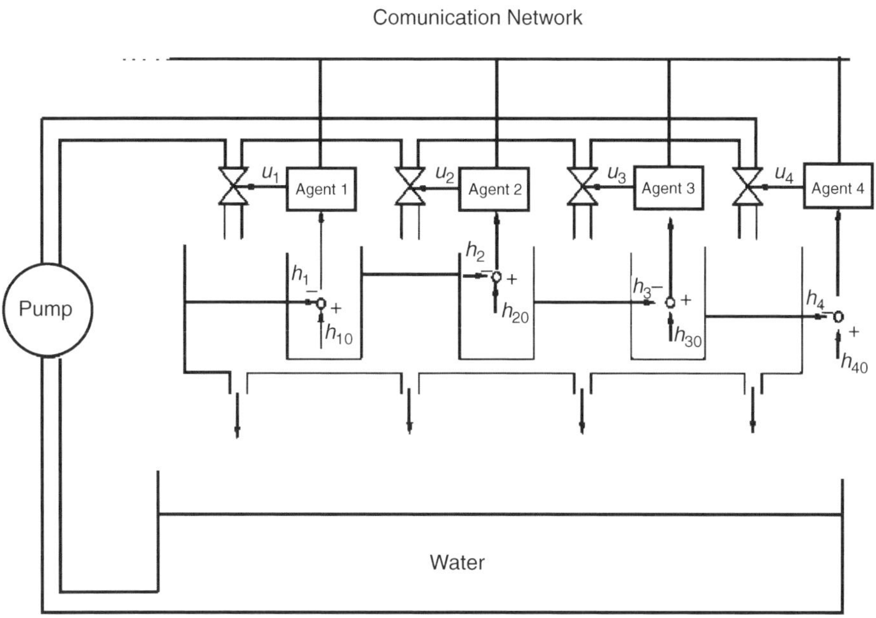

15.3 Distributed Fuzzy Control

Figure 15.14 Coupled tank system.

the control action $\Delta u_i(t)$. All agents are connected through a communication network. The structure of the distributed, market-based control system is also shown in Figure 15.14.

Because the volume of water that flows in the tank system is limited, the absolute values $|\Delta u_i(t)|$ are considered as resource variables. Each agent can, at any time instant, play the role of the consumer of the producer. A positive deviation $\Delta h_i(t)$ means that the tank needs more water to reach the required operation level. In this situation, the associated agent behaves like a consumer that has to acquire a certain quantity of water. This is the demand of consumer i at time instant t. When deviation $\Delta h_i(t)$ is negative, the tank contains too much water and the agent is considered as a producer that wants to furnish water to the consumers. This is the supply of producer i at time instant t.

Assuming n coupled tanks, the actual number of consumers at time instant t is q and the number of producers is m, with $n = m + q$. The market-based mechanism approach for distributed control is summarized in the procedure below.

procedure DISTRIBUTED-FUZZY-AUCTIONER (d, s) returns price
input : demand: d
        supply: s
        for each agent **do**
            if demand agent **then** get demand
            if consumer agent **then** get supply
            compute equilibrium price
            run auctions for the training period

store individual fitness ϕ
remove individual from the market
return price

The procedure assumes that each consumer agent computes his demand while each producer agent computes his supply. All values are sent to an auctioneer agent using the communication network. The auctioneer can be any of the agents. Next, the auctioneer computes an equilibrium price using the constraint that the sum of all supplies has to be equal to the sum of all demands, that is,

$$r_{pp}(t) + \sum_{j=1}^{m} r_j^p(t) = r_{cp}(t) + \sum_{i=1}^{q} r_i^c(t) \tag{15.21}$$

where $r_{pp}(t)$ and $r_{cp}(t)$ are the supply and demand of the pump. The pump is modeled as a permanent producer and consumer. The terms $r_j^p(t)$ and $r_i^c(t)$ are the supply and demand functions of the agents.

Agents compute their supply and demand functions using market knowledge (Yager, 1998; Silva, 2004) in the form of fuzzy rules as follows. Let price be a linguistic variable with values *low* (A_1) and *high* (A_2), and deviation be a linguistic variable representing the absolute value of $\Delta h_i(t)$ with values *small* (B_1) and *large* (B_2).

1. Consumer agent

   If price is *low* and deviation is *small* then demand is g_{do}.

   If price is *low* and deviation is *large* then demand is $g_{d\max}$. (15.22)

   If price is *high* then demand is g_{do}.

   where *low* and *high* defined by fuzzy sets A_1 and A_2 whose membership functions are

   $$A_1(p(t)) = \begin{cases} -\dfrac{p(t)}{P_{\max}} + 1, & \text{if } 0 \leq p(t) \leq P_{\max} \\ 0, & \text{otherwise} \end{cases}$$

   $$A_2(p(t)) = \begin{cases} \dfrac{p(t)}{P_{\max}} + 1, & \text{if } 0 \leq p(t) \leq P_{\max} \\ 0, & \text{otherwise} \end{cases}$$

as shown in Figure 15.15.

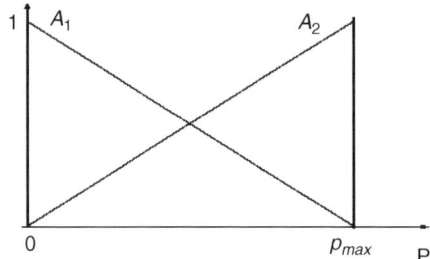

Figure 15.15 Membership functions of A_1 (*low*) and A_2 (*high*).

15.3 Distributed Fuzzy Control

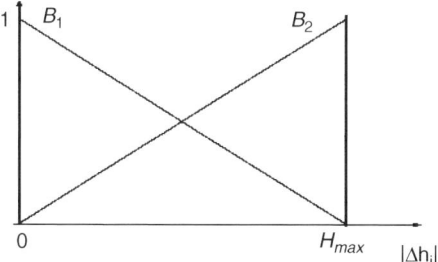

Figure 15.16 Membership functions of B_1 (*small*) and B_2 (*large*).

Similarly, the membership functions of fuzzy sets B_1 and B_2 for *small* and *large* are

$$B_1(\Delta h_i(t)) = \begin{cases} -\dfrac{\Delta h_i(t)}{H_{\max}} + 1, & \text{if } 0 \le \Delta h_i(t) \le H_{\max} \\ 0, & \text{otherwise} \end{cases}$$

$$B_2(\Delta h_i(t)) = \begin{cases} \dfrac{\Delta h_i(t)}{H_{\max}} + 1, & \text{if } 0 \le \Delta h_i(t) \le H_{\max} \\ 0, & \text{otherwise} \end{cases}$$

as depicted in Figure 15.16.
The rules consequents are as follows:

$$g_{do}(p(t), \Delta h_i(t)) = 0, \quad \forall p(t), \Delta h_i(t)$$
$$g_{d\max}(p(t), \Delta h_i(t)) = r_{d\max}, \quad \forall p(t), \Delta h_i(t)$$

Assuming the product t-norm in the rule consequent and noticing that rules in (15.22) are functional fuzzy rules we obtain

$$\begin{aligned} r_i^c(t) &= (k_{d_1} - k_{d_2} p(t)) \Delta h_i(t) \\ k_{d_1} &= (r_{d\max}/H_{\max}) \text{ and } k_{d_2} = (k_{d_1}/P_{\max}) \end{aligned} \quad (15.23)$$

Figure 15.17 shows the demand function assuming normalized universes.

2. Producer agent
   If price is *low* then supply is g_{so}.
   If price is *high* and deviation is *small* then supply is g_{so}. (15.24)
   If price is *high* and deviation is *large* then $g_{s\max}$.
   The consequents of the fuzzy rules are

$$g_{so}(p(t), \Delta h_i(t)) = 0, \quad \forall p(t), \Delta h_i(t)$$
$$g_{s\max}(p(t), \Delta h_i(t)) = r_{s\max}, \quad \forall p(t), \Delta h_i(t)$$

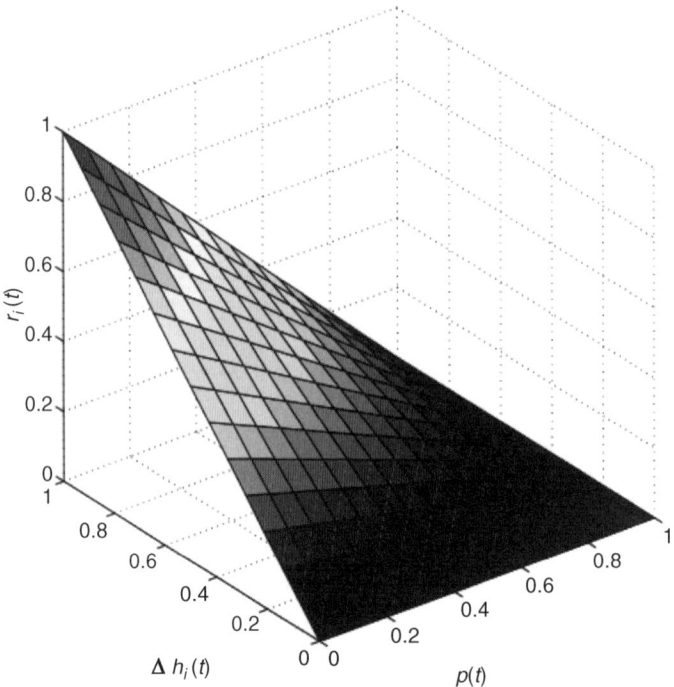

Figure 15.17 Demand function of an agent as a consumer.

and from the functional fuzzy rules (15.24), for normalized universes, we get the supply function

$$r_j^p(t) = k_s p(t)|\Delta h_j(t)|$$
$$k_s = (r_{s\max}/(P_{\max}\Delta H_{\max})) \tag{15.25}$$

as depicted in Figure 15.18.
Assuming the pump as a permanent producer provides

$$r_{pp}(t) = \delta p(t) \tag{15.26}$$

where δ is a constant parameter. Likewise, when the pump behaves as a permanent consumer it demands

$$r_{cp}(t) = \alpha_1 \left(1 - \frac{p(t)}{\alpha_2}\right) \tag{15.27}$$

where α_1 and α_2 are constants.
From expressions (15.21), (15.23), and (15.25–15.27) we obtain the equilibrium prices as follows:

$$p(t) = \frac{\alpha_1 + k_{d_1}\sum_{i=1}^{q}\Delta h_i(t)}{\delta + (\alpha_1/\alpha_2) + k_s\sum_{j=1}^{m}|\Delta h_j(t)| + k_{d_2}\sum_{i=1}^{q}\Delta h_i(t)}$$

15.3 Distributed Fuzzy Control **489**

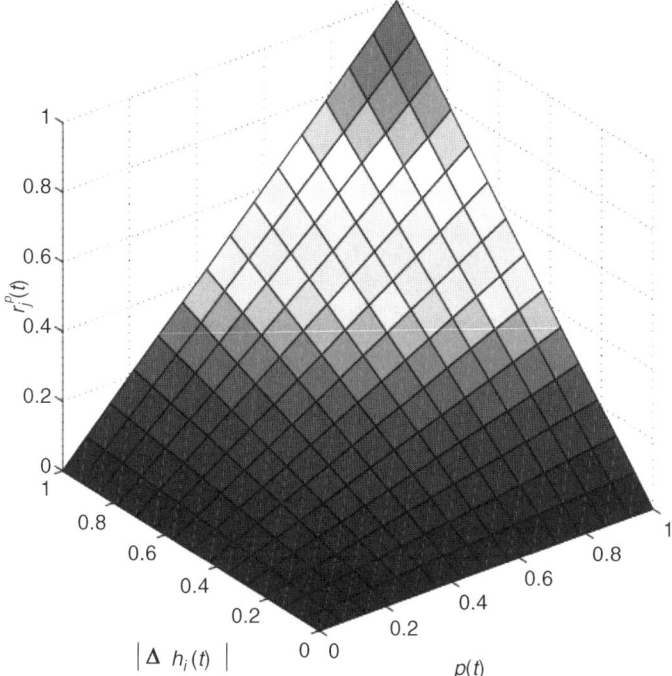

Figure 15.18 Supply function of an agent as a producer.

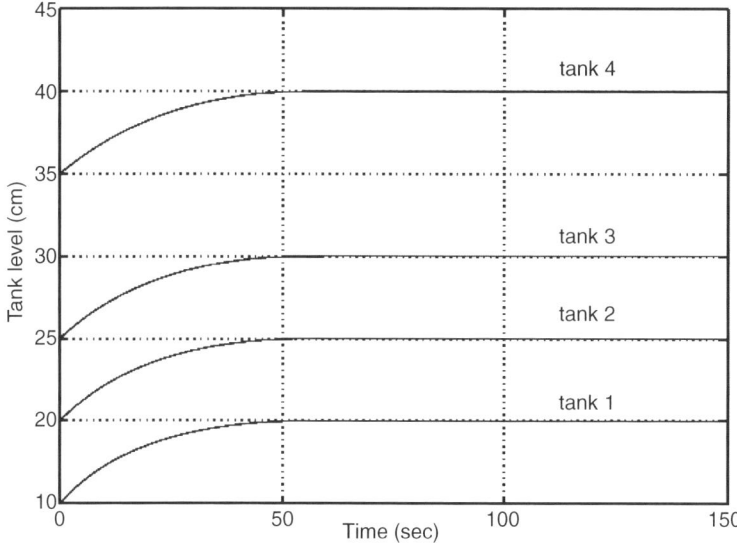

Figure 15.19 Tank level trajectories.

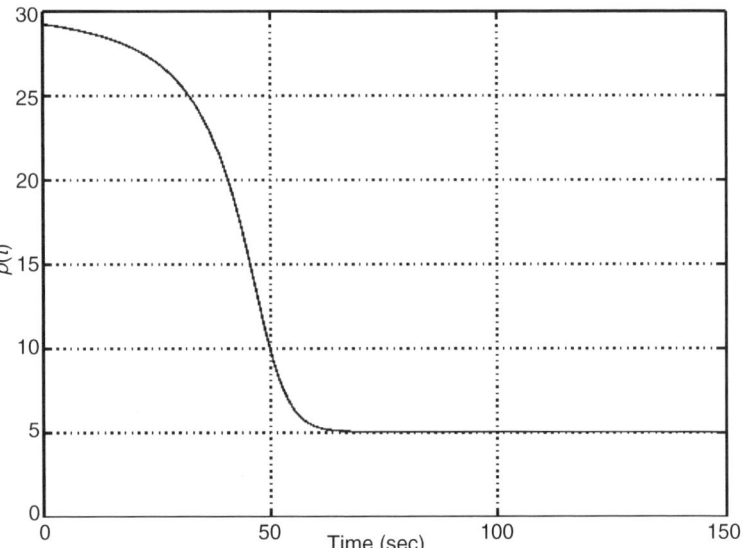

Figure 15.20 Evolution of resource prices.

and from the respective demand and supply functions, the consumer and producer agents compute their outputs as follows:

$$\Delta u_i(t) = (k_{d_1} - k_{d_2} p(t)) \Delta h_i(t)$$
$$\Delta u_j(t) = k_s p(t) \Delta h_j(t)$$

Figure 15.19 shows the trajectories of the tank levels assuming $h(0) = (15, 20, 25, 35)$ and choosing $\alpha_1 = 10, \alpha_2 = 10, \delta = 1, r_{smax} = 1, r_{dmax} = 500, P_{max} = 30$, and $H_{max} = 5$.

The evolution of the resource prices is shown in Figure 15.20. Figures 15.19 and 15.20 show that the prices act as a coordination factor among multiagents to induce equilibrium, namely, to achieve the respective tanks set points.

It is worth to note the modularity and simplicity of the control algorithms obtained via market-based control principles, especially when economic equilibrium notions are expressed using fuzzy rules.

15.4 CONCLUSIONS

Fuzzy sets and systems are important to develop complex systems due to their ability to handle imprecision and linguistic knowledge and to add intelligence to agents and applications. Conventional tools to analyze market models and equilibrium pose substantial challenges to system modelers and developers when imprecise and partially known systems must be handled. This is the case with many complex systems in diverse areas such as energy, transportation, control, information processing, and logistic systems. Together with the notion of agent and multiagents as both software engineering and decision-making systems, fuzzy set theory has a enormous potential to proceed to its next leap to develop large-scale intelligent systems.tpb

EXERCISES AND PROBLEMS

1. Sketch a multiagent system architecture and its main interface and decision modules for a distributed traffic control system.
2. Suggest an alternative to the crossover operator illustrated in Figure 15.10.
3. Using your knowledge of inference with functional fuzzy models, develop detailed steps to derive expression (15.23).
4. Similar to Problem 3, develop detailed steps to arrive at expression (15.25).
5. Develop a mechanism to obtain granular models of demand and supply versus price, and suggest a procedure to find the equilibrium point.

HISTORICAL NOTES

Multiagent systems have their origins in distributed artificial intelligence started in the 1970s. The current notion of agent assumes that they operate and exist in some environment that typically is both computational and physical (viz. software and hardware in general). Software agents are specifically referred to as computer programs (Weiss, 1999).

The use of economy principles in computer systems was suggested by Wellman (1995) as a mechanism for resource allocation and distributed problem solving.

REFERENCES

Baeza-Yates, R., Ribeiro-Neto, B. *Modern Information Retrieval*, ACM Press/Addison Wesley, Menlo Park 1999.

Chen, S., Wang, J. Document retrieval using knowledge-based fuzzy information retrieval techniques, *IEEE Trans. Syst. Man Cyb.* **25**(5), 1995, 1–6.

Clearwater, S. (ed.), *Market-Based Control: A Paradigm for Distributed Resource Allocation*, World Scientific, Singapore, 1996.

Contreras, J., Candiles, O., de la Fuente, J. I., Gómez, T. Auction design in day-ahead electricity markets, *IEEE T. Pow. Syst.* **16**(3), 2001, 409–417.

Cordón, O., Herrera, F., Hoffman, F., Magdalena, L. Genetic Fuzzy Systems: evolutionary tuning and learning of fuzzy knowledge bases, *Advances in Fuzzy Systems: Applications and Theory*, vol. 19, World Scientific, Singapore, 2001.

Cordón, O, Herrera, F., Magdalena, L., Villar, P. A genetic learning process for scaling the factors, granularity and contexts of the fuzzy rule-based system data base, *Inf. Sci.* **136**, 2001, 85–107.

Cordón, O., Herrera, F., Villar, P. Generating the knowledge base of a fuzzy rule-based system by the genetic learning of the data base, *IEEE Trans. Fuzzy Syst.* **9**(4), 2001, 667–674.

Cordon, O., Herrera, F., Gomide, F., Hoffmann, F., Magdalena, L. Ten years of genetic fuzzy systems: current framework and new trends, *Fuzzy Set. Syst.* **141**, 2004, 5–32.

Dash, R., Jennings, N. Computational mechanism design: a call to arms, *IEEE Intell. Syst.* **18**(6), 2003, 40–47.

David, A., Wen, F. Strategic bidding in competitive electricity markets: a literature survey, in: *Proceedings of the IEEE PES 2000 Summer Power Meeting*, IEEE Power Engineering Society, IEEE, 2000, pp. 2168–2173.

El-Hawary, M. E., Christensen, G. S. *Optimal Economic Operation of Electric Power Systems, Mathematics in Science and Engineering*, vol. 142 Academic Press, New York, 1979.

Glorennec, P. Constrained optimization of FIS using an evolutionary method, in: F. Herrera and J. L. Verdegay (eds.), *Genetic Algorithms and Soft Computing, Studies in Fuzziness and Soft Computing*, Physica-Verlag, Wurzburg, 1996, pp. 349–368.

Green, R. Competition in generation: the economic foundations, *Proc. IEEE* **88**(2), 2000, 128–139.

Herrera, F., Lozano, M., Verdegay, J. Fuzzy connectives based crossover operators to model genetic algorithms population diversity, *Fuzzy Set. Syst.* **92**(1), 1997, 21–30.

Herrera-Viedma, E., Pasi, G. Fuzzy approaches to access information on the web: recent developments and research trends, *Proceedings of the Third Conference of the European Society for Fuzzy Logic and Technologies*, 2003, 25–31.

Horng, Y., Chen, S., Lee, C. Automatically constructing multi-relationship fuzzy concept in fuzzy information retrieval systems, *Proceedings of the 10th IEEE International Fuzzy Systems Conference*, 2001, 606–609.

MasColell, A., Whinston, M., Green, J. *Microeconomic Theory*, Oxford University Press, Oxford, 1995.

McAfee, R., McMillan, J. Auctions and bidding, *J. Econ. Lit.* **25**, 1987, 699–738.

Michalewicz, Z. *Genetic Algorithms + Data Structures = Evolution Programs*, Springer-Verlag, Berlin, 1996.

Monclar, F., Quatrain, R. Simulation of electricity markets: a multi-agent approach, *International Conference on Intelligent System Application to Power Systems*, Budapest, Hungary, 2001, 207–212.

Ogawa, Y., Morita, T., Kobayashi, K. A fuzzy document retrieval system using the keyword connection matrix and a learning method, *Fuzzy Set. Syst.* **39**, 1991, 163–179.

Palm, R. Synchronization of decentralized multiple-model systems market-based optimization, *IEEE Trans. Syst. Man Cyb.B* **34**(1), 2004, 665–671.

Pasi, G. Flexible information retrieval: some research trends, *Math. Soft Comput.* **9**, 2002, 107–121.

Pereira, R., Ricarte, I., Gomide, F. Fuzzy relational ontological model in information search and retrieval systems, in: E. Sanchez (ed.), *Fuzzy Logic and the Semantic Web*, Elsevier, Amsterdam, 2006, pp. 395–412

Ricarte, I., Gomide, F. A reference model for intelligent information search, in: Nikravesh, M., Zadeh, L., Azvine, B., R. Yager (eds.), *Enhancing the Power of Internet*, Springer-Verlag, Berlin, 2004, pp. 327–346.

Russell, S., Norvig, P. *Artificial Intelligence: A Modern Approach*, Prentice Hall, Upper Saddle River, NJ, 2003.

Silva, M. Market-based distributed fuzzy control, *Internal Report*, Unicamp-FEEC-DCA, Campinas, São Paulo, December 2004.

Takagi, T., Kawase, K. A trial for data retrieval using conceptual fuzzy sets, *IEEE Trans. Fuzzy Syst.* **9**(4), 2001, 497–505.

Tong, S., Li, H., Chen, G. Adaptive fuzzy decentralized control for a class of large-scale nonlinear systems, *IEEE Trans. Syst. Man Cyb. B* **34**(1), 2004, 770–774.

Visudhipan, P., Ilic, M. Dynamic games-based modeling of electricity markets, *Proceedings of the 1999 IEEE Power Engineering Society Winter Meeting*, New York City, NY, 1999, pp. 274–281.

Voos, H. Market-based control of complex dynamic systems, *Proceedings of the IEEE International Symposium on Intelligent Control ISIC'99*, Cambridge, MA, 1999, pp. 284–289.

Voos, H., Litz, L. A new approach to optimal control using market-based algorithms, *Proceedings of the European Control Conference ECC'99*, Karlsruhe, 1999, DA-2,4.

Voos, H., Litz, L. Market-based optimal control: a general introduction, *Proceedings of the American Control Conference*, Chicago, Illinois, June 2000.

Walter, I., Gomide, F. Design of coordination strategies in multiagent systems via genetic algorithms, *Soft Co.* **10**(10), 2006, 903–915.

Wllman, M. The economic approah to artifical intelligience ACM Computing Surveys, 27(3), 1995, 369–362.

Weiss, G. *Multiagent Systems: An Approach to Distributed Artificial Intelligence*, MIT Press, Cambridge, MA, 1999.

Widyantoro, D., Yen, J. A fuzzy ontology-based abstract search engine and its user studies, *Proceedings of the 10nth IEEE International Conference on Fuzzy Systems*, 2001, pp. 1291–1294.

Yager, R. On the Solution of simultaneous fuzzy models, *Proceedings of the IEEE World Congress on Computational Intelligence*, 1998, pp. 956–960.

Appendix A

MATHEMATICAL PREREQUISITES

Here we offer a very concise and focused review of the notation, basic notions, and pertinent facts related to linear algebra (vectors and matrices), analysis, and optimization that relate directly to the computing with fuzzy sets. For an extended treatment of the subject matter, the reader is referred to a wealth of specialized texts, some of which are included in the list of references.

Vectors and Matrices

An n-dimensional vector $\mathbf{x}$ organizes a collection of numbers $x_1, x_2, \ldots, x_n$ in the following form:

$$\mathbf{x} = \begin{bmatrix} x_1 \\ x_2 \\ \ldots \\ x_n \end{bmatrix}$$

The transpose of $\mathbf{x}$, denoted by $\mathbf{x}^T$, is a row of elements, $[x_1, x_2, \ldots, x_n]^T$. Vectors are denoted by boldface letters such as $\mathbf{a}$, $\mathbf{b}$, $\mathbf{x}$, $\mathbf{y}$, and so on. A collection of all n-dimensional vectors forms an n-dimensional Euclidean space denoted by $\mathbf{R}^n$.

The sum of two vectors, $\mathbf{x}$ and $\mathbf{y}$, is another vector, denoted by $\mathbf{x} + \mathbf{y}$, whose ith component is equal to $x_i + y_i$. The product of a scalar α by a vector $\mathbf{x}$ denoted by $\alpha\mathbf{x}$ is obtained by multiplying each component of $\mathbf{x}$ by α, $\alpha\mathbf{x} = [\alpha x_1, \quad \alpha x_2, \ldots, \alpha x_n]$.

We say a collection of vectors $\mathbf{x}_1, \mathbf{x}_2, \ldots, \mathbf{x}_n$ in $\mathbf{R}^n$ is linearly independent if $\sum_{i=1}^{n} \alpha_i \mathbf{x}^i = 0$ implying that $\alpha_i = 0$ for $i = 1, \ldots, n$. A vector $\mathbf{y}$ in $\mathbf{R}^n$ is a linear combination of the vectors $\mathbf{x}_1, \mathbf{x}_2, \ldots, \mathbf{x}_k$ in $\mathbf{R}^n$ if it can be expressed in the form $\mathbf{y} = \sum_{i=1}^{k} \alpha_i \mathbf{x}^i$ for some scalars α_i, $i = 1, \ldots, k$.

A collection of vectors $\mathbf{x}_1, \mathbf{x}_2, \ldots, \mathbf{x}_n$ in $\mathbf{R}^n$ form a basis of $\mathbf{R}^n$ if they are linearly independent. In this case, we say that the basis spans $\mathbf{R}^n$.

Fuzzy Systems Engineering: Toward Human-Centric Computing, by Witold Pedrycz and Fernando Gomide
Copyright © 2007 John Wiley & Sons, Inc.

The inner (scalar) product of two vectors **x** and **y** in $\mathbf{R}^n$ denoted by $\mathbf{x}^T\mathbf{y}$ is a scalar determined as $\mathbf{x}^T\mathbf{y} = \sum_{i=1}^n x_i y_i$. Two vectors are orthogonal if their inner product is equal to zero.

The Euclidean norm of **x** in $\mathbf{R}^n$, denoted by $\|\mathbf{x}\|$, is defined by $\|\mathbf{x}\| = \sqrt{\mathbf{x}^T\mathbf{y}} = (\sum_{i=1}^n (x_i)^2)^{1/2}$.

A matrix is a rectangular array of numbers. An $m \times n$ matrix has m rows and n columns. Matrices are denoted by capital letters, say R, A, P, and so on. The entry of the matrix positioned in the ith row and the jth column is denoted by r_{ij}. We also use a notation such as $R = [r_{ij}]$ that explicitly identifies the corresponding entries of the matrix. The jth column of R is denoted by $\mathbf{r}_j$.

Let R and S be two $m \times n$ matrices. The sum of R and S, denoted by $R + S$, is the matrix whose ijth entry is $r_{ij} + s_{ij}$. The product of a matrix R by a scalar α is the matrix whose ijth entry is αr_{ij}. If R is a $m \times n$ matrix and S is a $n \times p$ matrix, then he product RS is an $m \times p$-dimensional matrix C whose ijth entry c_{ij} is equal to

$$c_{ij} = \sum_{k=1}^n r_{ik} s_{kj}, \quad \text{for} \quad i = 1, \ldots, m \quad \text{and} \quad j = 1, \ldots, p$$

Let R be an $m \times n$ matrix. The transpose R^T of R is a $n \times m$ matrix whose ijth entry is equal to r_{ji}. We say that a square matrix R is symmetric if $R = R^T$. The determinant of R, denoted det$[R]$, is defined iteratively as follows:

$$\det[R] = \det[R] = \sum_{i=1}^n r_{i1} R_{i1}$$

where R_{i1} is the i1-cofactor of R. The cofactor R_{i1} of R is defined as $(-1)^{i+1}$ times the determinant of the sub-matrix formed by deleting the ith row and the first column of R. The determinant of a scalar is the scalar itself.

A square matrix R is nonsingular if there exists a matrix R^{-1}, called the inverse matrix, such that $RR^{-1} = I$, where I is the identity matrix, whose entries on the main diagonal are equal to 1 and all other entries are equal to zero. The inverse of a matrix, when it does exist, is unique. A matrix has an inverse if and only if its determinant is not zero.

Let R be a $m \times n$ matrix. The rank of R is the maximum number of linearly independent rows or, equivalently, the maximum number of linearly independent columns of R. If the rank of R is the minimum of $\{m, n\}$, then R is of full rank.

If R is a $m \times n$ matrix, a scalar λ and a nonzero vector **x** satisfying the equation $R\mathbf{x} = \lambda \mathbf{x}$ are called, respectively, an eingenvalue and an eingenvector of R. To compute eigenvalues of R, we must solve the equation $\det[R - \lambda I] = 0$. If R is symmetric, then its eingenvalues are real numbers. Eingenvalues of a symmetric matrix associated with distinct eigenvalues are orthogonal.

Let R be a $n \times n$ symmetric matrix. We say that R is positive definite if the scalar quantity $\mathbf{x}^T R \mathbf{x} > 0$ for all nonzero $\mathbf{x} \in \mathbf{R}^n$. R is positive semidefinite if $\mathbf{x}^T R \mathbf{x} \geq 0$. Likewise, R is negative definite if $\mathbf{x}^T R \mathbf{x} < 0$ for all nonzero $\mathbf{x} \in \mathbf{R}^n$ Similarly, R is negative semidefinite if $\mathbf{x}^T R \mathbf{x} \leq 0$. A matrix R is positive definite, positive semidefinite, negative definite, and negative semidefinite if and only if its eingevalues are positive, nonnegative, negative, nonpositive, respectively.

A $m \times n$ matrix R is orthonormal if $RR^T = I_m$, where I_m is the $m \times m$ identity matrix. This means that the columns $\mathbf{r}_j$ of matrix R are mutually orthogonal vectors of unit length, that is, $\mathbf{r}_j^T \mathbf{r}_j = 1$.

Let R be a $m \times n$ matrix, where $m \geq n$. Then R can be decomposed as $R = UWV^T$, where U is a $m \times n$ orthonormal matrix and V is $n \times n$ orthonormal matrix. W is a $n \times n$ diagonal matrix, $W = [w_{ij}]$, $w_{ij} = w_i$ if $i = j$, and $w_{ij} = 0$ if $i \neq j$. The diagonal elements w_i, $i = 1, \ldots, n$, are called the singular values. The singular value decomposition always exists and is unique up to same permutations in columns of U, W, and V, and linear combinations for columns of U and V with equal singular values (Golub and Van Loan, 1989). Given the linear equation $R\mathbf{x} = \mathbf{b}$, the solution $\mathbf{x}^{\text{opt}} = V\hat{W}U^T\mathbf{b}$ is least squares, namely, $\mathbf{x}^{\text{opt}} = \arg\min_{\mathbf{x}} \| R\mathbf{x} - \mathbf{b} \|^2$ where $\hat{W}$ is a diagonal matrix whose elements are $\hat{w}_i$ with

$$\hat{w}_i = \begin{cases} \dfrac{1}{w_i}, & \text{if } w_i > 0 \\ 0, & \text{if } w_i = 0 \end{cases}$$

Sets

In a very descriptive manner, we can view a set as a collection of objects (elements). A set can be specified by listing its elements or by specifying the properties that such elements must satisfy.

The set of all possible elements of concern in a particular context or application is called an universe or, alternatively, universe of discourse, space, domain, and space. Universes are denoted by boldface capital letters such as **X**, **Y**, and **Z**.

Let a and b be two real numbers. The closed interval $[a, b]$ denotes all real numbers such that $a \leq x \leq b$. Real numbers satisfying $a \leq x < b$ are represented by $[a, b)$ whereas those satisfying the relationship $a < x \leq b$ are denoted by $(a, b]$. The set of all real numbers satisfying $a < x < b$ is represented by the open interval (a, b).

Let A be a set of real numbers. Then the greatest lower bound, or infimum of A is the largest possible scalar α such that $\alpha \leq x$ for each $x \in A$. The infimum is denoted by $\inf\{x : x \in A\}$ or, alternatively, $\inf(A)$. The least upper bound, or the supremum of A is the smallest possible scalar α such that $\alpha \geq x$ for each $x \in A$. The supremum is denoted by $\sup\{x : x \in A\}$ or, alternatively, $\sup(A)$.

Given a point $\mathbf{x} \in \mathbf{R}^n$ and some $\varepsilon > 0$, The set $N_\varepsilon(\mathbf{x}) = \{\mathbf{y} | \| \mathbf{y} - \mathbf{x} \| \leq \varepsilon\}$ is called an ε-neighborhood of $\mathbf{x}$.

A set A in $\mathbf{R}^n$ is compact if it is closed and bounded. The set A is closed if it is equal to its closure. The closure of a set A of $\mathbf{R}^n$ is the set of all points that are arbitrarily close to A.

A set A in $\mathbf{R}^n$ is said to be convex if for every $\mathbf{x}_1$ and $\mathbf{x}_2 \in A$ and every real number $0 \leq \lambda \leq 1$, the convex combination belongs to A, that is, the point $\lambda \mathbf{x}_1 + (1 - \lambda)\mathbf{x}_2 \in A$.

Functions

A real valued function f defined on a subset S of $\mathbf{R}^n$ associates with each point $\mathbf{x} \in S$ a real number $f(\ )$. The notation $f : S \to \mathbf{R}$ means that the domain of f is S and that the codomain (alternatively, the range) of f is a subset of $\mathbf{R}$. If f is defined everywhere on $\mathbf{R}^n$, or if the domain is irrelevant, we use the notation $f : \mathbf{R}^n \to \mathbf{R}$. A collection of real valued functions $f_1, f_2, \ldots, f_m$ can be viewed as a single vector function $\mathbf{f} = [f_1, f_2, \ldots, f_n]^T$.

A function $f : S \to \mathbf{R}$ is continuous at $\bar{\mathbf{x}} \in S$ if, for any $\varepsilon > 0$, if there exists a $\delta > 0$ such that $\mathbf{x} \in S$ and $\| \mathbf{x} - \bar{\mathbf{x}} \| < \delta$ imply that $|f(\mathbf{x}) - f(\bar{\mathbf{x}})| < \varepsilon$. A vector-valued function is continuous at $\bar{\mathbf{x}}$ if each of its components is continuous at $\bar{\mathbf{x}}$.

Let S be a nonempty set in $\mathbf{R}^n$. A function $f : S \to \mathbf{R}$ is upper semicontinuous at $\bar{\mathbf{x}} \in S$ if for each $\varepsilon > 0$ there is a $\delta > 0$ such that $\mathbf{x} \in S$ and $\| \mathbf{x} - \bar{\mathbf{x}} \| < \delta$ imply that $f(\mathbf{x}) - f(\bar{\mathbf{x}}) < \varepsilon$. Similarly, $f : S \to \mathbf{R}$ is lower semicontinuous at $\bar{\mathbf{x}} \in S$ if for each $\varepsilon > 0$ there is a $\delta > 0$ such that $\mathbf{x} \in S$ and $\| \mathbf{x} - \bar{\mathbf{x}} \| < \delta$ imply that $f(\mathbf{x}) - f(\bar{\mathbf{x}}) > -\varepsilon$. A vector-valued function is upper or lower semicontinuous if each of its components is upper or lower semicontinuous, respectively.

Let S be a nonempty set in $\mathbf{R}^n$ and assume that $f : \mathbf{R}^n \to \mathbf{R}$. If f is lower semicontinuous, then it has a minimum over S. A point $\mathbf{x}^* \in S$ is a minimum of f if $f(\mathbf{x}) \geq f(\mathbf{x}^*)$ for all $\mathbf{x} \in S$. Likewise, if f is upper semicontinuous, it has a maximum over S, that is, there exists a $\mathbf{x}^* \in S$ such that $f(\mathbf{x}) \leq f(\mathbf{x}^*)$ for all $\mathbf{x} \in S$. If x^* is a minimum of f, then it is also a maximum of $-f$. Continuous functions are both upper and lower semicontinuous. Therefore, they achieve both a minimum and a maximum over any compact set.

A point $\mathbf{x}^* \in S$ is a local minimum over S if there is an $\varepsilon > 0$ such that $f(\mathbf{x}) \geq f(\mathbf{x}^*)$ for all $\mathbf{x} \in S$ within a neighborhood of x^*. If $f(\mathbf{x}) > f(\mathbf{x}^*)$, then x^* is a strict local minimum. Similarly, $\mathbf{x}^* \in S$ is a local maximum over S if there is an $\varepsilon > 0$ such that $f(\mathbf{x}) \leq f(\mathbf{x}^*)$ for all $\mathbf{x} \in S$ within a neighborhood of x^*. If $f(\mathbf{x}) < f(\mathbf{x}^*)$, then x^* is a strict local maximum.

Consider a nonempty open set S in $\mathbf{R}^n$, $\bar{\mathbf{x}} \in S$, and $f : S \to \mathbf{R}$. Then f is differentiable at $\bar{\mathbf{x}}$ if there exists a vector $\nabla f(\bar{\mathbf{x}})$ in $\mathbf{R}^n$-called the gradient of f at $\bar{\mathbf{x}}$ and a function $\beta(\bar{\mathbf{x}}, \mathbf{x})$ satisfying $\beta(\bar{\mathbf{x}}, \mathbf{x}) \to 0$ as $\mathbf{x} \to \bar{\mathbf{x}}$ such that, for all $\mathbf{x} \in S$,

$$f(\mathbf{x}) = f(\bar{\mathbf{x}}) + \nabla f(\bar{\mathbf{x}})^T (\mathbf{x} - \bar{\mathbf{x}}) + \| \mathbf{x} - \bar{\mathbf{x}} \| \beta(\bar{\mathbf{x}}, \mathbf{x})$$

The gradient is a n-dimensional vector consisting of partial derivatives of f taken with respect to each component of the vector $\mathbf{x}$, that is,

$$\nabla f(\bar{\mathbf{x}}) = \left(\frac{\partial f(\bar{\mathbf{x}})}{\partial x_1}, \frac{\partial f(\bar{\mathbf{x}})}{\partial x_2}, \ldots, \frac{\partial f(\bar{\mathbf{x}})}{\partial x_n} \right)^T$$

A function $f(\mathbf{x})$ is smooth as it is continuous and differentiable at $\mathbf{x}$; otherwise it is nonsmooth.

A function f is twice differentiable at $\bar{\mathbf{x}}$ if, in addition to the gradient vector, there exists a $n \times n$ symmetric matrix $H(\bar{\mathbf{x}})$, called the Hessian matrix of f at $\bar{\mathbf{x}}$, and a function $\beta(\bar{\mathbf{x}}, \mathbf{x})$ satisfying $\beta(\bar{\mathbf{x}}, \mathbf{x}) \to 0$ as $\mathbf{x} \to \bar{\mathbf{x}}$ such that, for all $\mathbf{x} \in S$,

$$f(\mathbf{x}) = f(\bar{\mathbf{x}}) + \nabla f(\bar{\mathbf{x}})^T (\mathbf{x} - \bar{\mathbf{x}}) + \frac{1}{2}(\mathbf{x} - \bar{\mathbf{x}})^T H(\bar{\mathbf{x}})(\mathbf{x} - \bar{\mathbf{x}}) + \| \mathbf{x} - \bar{\mathbf{x}} \| \beta(\bar{\mathbf{x}}, \mathbf{x})$$

The Hessian matrix of f at $\bar{\mathbf{x}}$ is defined as

$$H(\bar{\mathbf{x}}) = \begin{bmatrix} \frac{\partial^2 f(\bar{\mathbf{x}})}{\partial x_1^2} & \cdots & \frac{\partial^2 f(\bar{\mathbf{x}})}{\partial x_1 \partial x_n} \\ \vdots & \ddots & \vdots \\ \frac{\partial^2 f(\bar{\mathbf{x}})}{\partial x_n \partial x_1} & \cdots & \frac{\partial^2 f(\bar{\mathbf{x}})}{\partial x_n^2} \end{bmatrix}$$

Let S be a nonempty convex set in $\mathbf{R}^n$. The function $f : S \to \mathbf{R}$ is convex if, for ach $\mathbf{x}_1, \mathbf{x}_2 \in S$ and $\lambda \in [0, 1]$ we have

$$f(\lambda \mathbf{x}_1 + (1 - \lambda)\mathbf{x}_2) \le \lambda f(\mathbf{x}_1) + (1 - \lambda)f(\mathbf{x}_2)$$

The function is strictly convex if the inequality holds strictly. The function f is quasiconvex if the following inequality is satisfied

$$f(\lambda \mathbf{x}_1 + (1 - \lambda)\mathbf{x}_2) \le \max\{f(\mathbf{x}_1), f(\mathbf{x}_2)\}$$

The function is strictly quasiconvex if inequality holds strictly and $f(\mathbf{x}_1) \ne f(\mathbf{x}_2)$ and is strongly quasiconvex if strict inequality holds and $\mathbf{x}_1 \ne \mathbf{x}_2$.

The notions of convex and quasiconvex extend to the concept of concavity when we replace f by $-f$.

Nonlinear Optimization

Consider the problem of finding a minimum of a smooth function $f : \mathbf{R}^n \to \mathbf{R}$,

$$\min_{\mathbf{x}} f(\mathbf{x})$$

We assume that f is nonlinear. Linear optimization is covered in Bertsimas and Tsitsiklis (1997). This optimization problem is called an unconstrained minimization problem. A solution $\mathbf{x}^*$ is a stationary point of f if $\nabla f(\mathbf{x}^*) = 0$, where 0 is the zero vector, that is, a n-dimensional vector whose components are all equal to zero. Stationary points can be minimum maximum or saddle points. To investigate which one is the case we need to look at second order information using the Hessian matrix at $\mathbf{x}^*$.

The Hessian matrix of a smooth function f is positive (negative) semi-definite at every unconstrained local minimum (maximum). This is a necessary condition for a local minimum (maximum). A stationary point is an unconstrained local minimum (maximum) if the Hessian matrix is positive (negative) definite. This is a sufficient condition for a local minimum (maximum). Every stationary point of a smooth convex (concave) function is an unconstrained global minimum (maximum).

In practice, local minimum is found using gradient search algorithms. In the simplest one, we follow a direction opposite to the gradient of f. The gradient of

f at $\mathbf{x}$, $\nabla f(\mathbf{x})$, points at the steepest behavior of the function in a neighborhood of $\mathbf{x}$. Therefore, if we look for the maximum, this is the right direction to follow. If our objective is to find a minimum, then we should proceed along the steepest descent direction, which is positioned at the opposite direction of the gradient, that is, $-\nabla f(\mathbf{x})$. This search strategy translates into the following iterative optimization algorithm.

procedure GRADIENT-SEARCH (f) **returns** a solution
input: objective function: f
local: step size: α
         tolerance: ε

         Choose $\mathbf{x}^0$
         $t \leftarrow 0$
         repeat
           compute $\nabla f(\mathbf{x}^t)$
           $\mathbf{x}^{t+1} \leftarrow \mathbf{x}^t - \alpha \nabla f(\mathbf{x}^t)$
           $t \leftarrow t + 1$
         until $\nabla f(\mathbf{x}^t) \leq \varepsilon$
         return $\mathbf{x}^t$

The gradient procedure shown above finds local maximum if we update x in the direction of $\nabla f(\mathbf{x})$ instead of $-\nabla f(\mathbf{x})$. The step size α should be chosen carefully to guarantee convergence of the algorithm. Too large values of α provoke divergence whereas its values that are too small slow down convergence speed. Appropriate value of α can be found by performing line search. In other words, find a value of α at each iteration such that, given $\mathbf{x}^t$ and $\nabla f(\mathbf{x}^t)$, it minimizes the expression $f(\mathbf{x}^t - \alpha \nabla f(\mathbf{x}^t))$.

There are several variations of the basic gradient search algorithm. The underlying idea is to find more efficient search directions. Among them we note conjugate directions (Bazaraa et al., 2006; Luenberger, 2003; Bertsekas, 1999) and quasi-Newton methods such as the method of Broyden, Fletcher, Goldfarb, and Shanno—BFGS (Rardin, 1998).

Suppose we are interested in finding a solution $\mathbf{x}^*$ to the following optimization problem:

$$\min_{\mathbf{x}} f(\mathbf{x})$$
$$\text{s.t.} \quad h_j(\mathbf{x}) \leq 0, \quad j = 1, \ldots, p$$

with f and h_j being two smooth functions. The term "s.t" abbreviates the expression "subject to."

We can determine a minimum of $f(\mathbf{x})$ as follows: Given the set of constraints expressed by functions $h_j(\mathbf{x})$, we form the Lagrangian function $L(\mathbf{x},\mathbf{v})$

$$L(\mathbf{x}, \mathbf{v}) = f(\mathbf{x}) + \sum_{j=1}^{p} v_j h_j(\mathbf{x})$$

where v_j is a scalar called Lagrange multiplier. $\mathbf{v} = [v_1, v_2, \ldots, v_p]^T$. Using the Lagrangian we convert the constrained optimization problem into a unconstrained problem with variables **x** and **v**. From previous discussion, a solution **x*** comes as a stationary point of $L(\mathbf{x}, \mathbf{v})$ and must satisfy the following condition:

$$\frac{\partial L(\mathbf{x}, \mathbf{v})}{\partial \mathbf{x}} = \frac{\partial f(\mathbf{x})}{\partial \mathbf{x}} + \sum_{j=1}^{p} v_j \frac{\partial h_j(\mathbf{x})}{\partial \mathbf{x}} = 0$$

Equivalently we write this down as

$$\nabla_{\mathbf{x}} L(\mathbf{x}, \mathbf{v}) = \nabla f(\mathbf{x}) + \mathbf{v}^T \nabla h(\mathbf{x})$$

where $\nabla h(\mathbf{x})$ is the Jacobian, a matrix whose jth column is the gradient of h_j with resptect to **x**, that is,

$$\nabla h(\mathbf{x}) = \begin{bmatrix} \frac{\partial h_1(\mathbf{x})}{\partial x_1} & \cdots & \frac{\partial h_p(\mathbf{x})}{\partial x_1} \\ \vdots & \ddots & \vdots \\ \frac{\partial h_1(\mathbf{x})}{\partial x_n} & \cdots & \frac{\partial h_p(\mathbf{x})}{\partial x_n} \end{bmatrix}$$

The solution is obtained solving the resulting equations for **x** and **v** that leads to the solutions denoted as **x*** and **v***.

Consider now the more general problem of finding a minimum of a smooth function $f: \mathbf{R}^n \to \mathbf{R}$. In contrast to the previous optimization problems, we require that solutions **x*** must satisfy some constraint set Φ. Often, the constraint set is specified in terms of functions $g_i(\mathbf{x}), i = 1, \ldots, m$ and $h_j(\mathbf{x}), j = 1, \ldots, p$ where g_i, h_j: $\mathbf{R}^n \to \mathbf{R}$, and a open set $S \subseteq \mathbf{R}^n$, that is, $\Phi = \{\mathbf{x} \in S | g_i(\mathbf{x}) \leq 0, i = 1, \ldots, m$ and $h_j(\mathbf{x}) = 0, j = 1, \ldots, p\}$. We call this case a constrained optimization problem and express it in the form

$$\min_{\mathbf{x}} f(\mathbf{x})$$
$$\text{s.t. } g_i(\mathbf{x}) \leq 0, i = 1, \ldots, m$$
$$h_j(\mathbf{x}) = 0, j = 1, \ldots, p$$
$$x \in S$$

The necessary optimality conditions for constrained minimization problems are given by the Karush-Kuhn-Tucker (KKT) theorem. This theorem states that if **x*** is a local solution, then, under Kuhn-Tucker constraint qualification, there exists a pair of vectors $(\mathbf{u}, \mathbf{v})$ such that

$$\nabla f(\mathbf{x}^*) + \sum_{i=1}^{m} u_i \nabla g_i(\mathbf{x}^*) + \sum_{j=1}^{p} v_i \nabla h_i(\mathbf{x}^*) = 0$$
$$u_i g_i(\mathbf{x}^*) = 0, \ i = 1, \ldots, m$$
$$u_i \geq 0, \quad i \in I$$

where $I = \{i | g_i(\mathbf{x}^*) = 0\}$ is called the set of active constraints. Under suitable convexity conditions, the KKT conditions are also sufficient (Bazaraa et al., 2006).

Similar conditions apply to the maximization problems once we recall that max $f = \min -f$. The KKT conditions provide existence results only, however, they do not suggest algorithms to find the constrained minimum. In practice, we must rely on iterative gradient-like algorithms. These algorithms depend on the form of the objective function "f," and constraints g_i and h_j. Detailed coverage of this subject is given in Bazaraa et al. (2006) and Bertsekas (1999).

REFERENCES

Bazaraa, M., Sherali, H., Shetty, C. *Nonlinear Programming: Theory and Algorithms*, Wiley Interscience, Hoboken, NJ, 2006.

Bertsekas, D. *Nonlinear Programming*, Athena Scientific, 2nd ed., Belmont, MA, 1999.

Bertsimas, D., Tsitsiklis, J. *Introduction to Linear Optimization*, Athena Scientific, 2nd ed., Belmont, MA, 1997.

Golub, G., Van Loan, C. *Matrix Computations*, 2nd ed., John Hopkins University Press, Baltimore, MD, 1989.

Luenberger, D. *Introduction to Linear and Nonlinear Programming*, Kluwer, Norwell, MA, 2003.

Rardin, R. *Optimization in Operations Research*, Prentice Hall, Upper Saddle River, NJ, 1998.

Appendix B

NEUROCOMPUTING

Neurocomputing is a comprehensive computational paradigm inspired by mechanisms of neural sciences and brain functioning that is rooted in learning instead of preprogrammed behavior. In this sense, neurocomputing becomes fundamentally different from the paradigm of programmed, instruction-based models of optimization. Artificial neural networks (neural networks, for short) exhibit some characteristics of biological neural networks in the sense the constructed networks include some components of distributed representation and processing as well as rely on various schemes of learning during their construction. The generalization capabilities of neural networks form one of their most outstanding features. The ability of the neural networks to generalize, namely, develop solutions that are meaningful beyond the scope of the learning data is commonly exploited in various applications.

From the architectural standpoint, a neural network consists of a collection of simple nonlinear processing components called neurons, which are combined together via a net of adjustable numeric connections. The development of a neural network is realized through learning. This means to choose an appropriate network structure and a learning procedure to achieve the goals of the application intended. Neural networks have been successfully applied to a variety of problems in pattern recognition, signal prediction, optimization, control, and image processing.

Here, we summarize the most essential architectural and development aspects of neurocomputing.

Computational Model of Neurons

A typical mathematical model of a single neuron (Anthony and Barlet, 1999) comes in the form of an n-input single-output nonlinear mapping (Fig. B1) described as follows:

$$y = f\left(\sum_{i=1}^{n} w_i x_i\right) \tag{B1}$$

Fuzzy Systems Engineering: Toward Human-Centric Computing, by Witold Pedrycz and Fernando Gomide
Copyright © 2007 John Wiley & Sons, Inc.

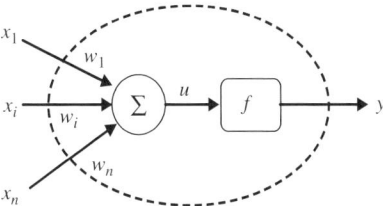

Figure B1 A topology of a neuron; note a two-phase processing.

where $x_1, x_2, \ldots, x_n$ are the inputs of the neuron and $w_1, w_2, \ldots, w_n$ are the associated adjustable connections (weights). The nonlinear nondecreasing mapping f brings a component of nonlinear processing to the functioning of the neuron.

Positive values of weights correspond to excitatory synapses of the neuron whereas negative weights model inhibitory synapses. The adjustable character of the connections makes the neuron (and the neural network as a whole) highly plastic and facilitates all learning faculties. Quite commonly, the neuron is equipped with a bias meaning that we admit a constant input as a part of the topology of the neuron. Expression (B1) is modified as reads $y = f(\sum_{i=1}^{n} w_i x_i + w_0)$. With the acceptance of the vector notation, $\mathbf{x} = [x_1, x_2, \ldots, x_n, 1]^T$ and $\mathbf{w} = [w_1 \; w_2 \; \ldots \; w_n \; w_0]^T$, we make the description of the neuron more concise. Now the output of the neuron reads as $y = f(\mathbf{w}^T \mathbf{x})$

The nonlinear activation function (f) may assume different forms as shown in Figure B2. The sigmoid and Gaussian activation functions are among the most frequently used.

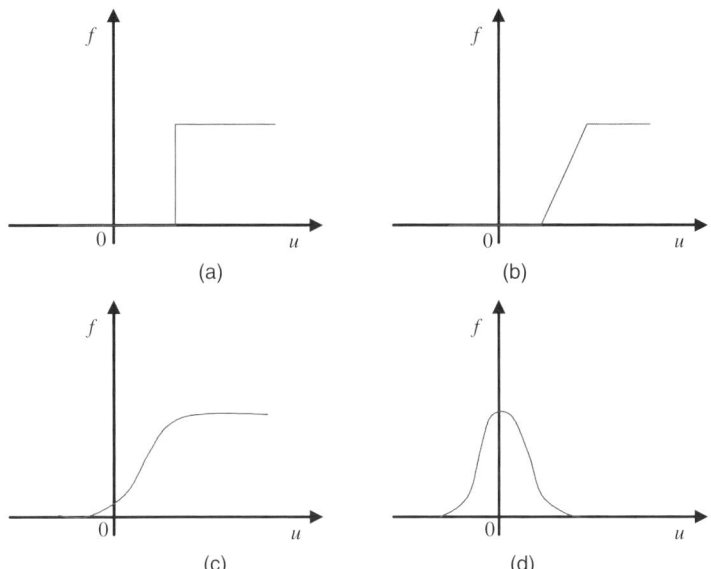

Figure B2 Examples of activation functions: (a) threshold, (b) piecewise linear, (c) sigmoid, and (d) Gaussian.

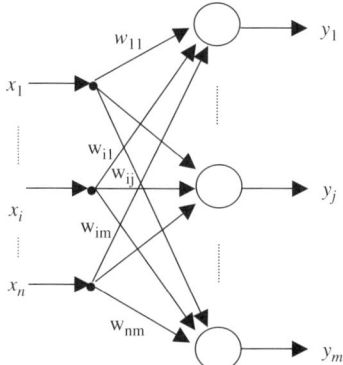

Figure B3 Two-layer feedforward neural network.

Architectures of Neural Networks

Neural network structures are characterized by the connection patterns that link the neurons arranged in layers. There are two generic topologies of neural networks, namely feedforward and recurrent (feedback) networks. Feedforward neural networks can exhibit a single layer of neurons or could come as a multilayer structure. An example of a two layer network is illustrated in Figure B3 whereas Figure B4 shows a three-layer network. In general, we may envision multilayer topologies, say, an L-layer neural network.

In feedforward networks, input signals (inputs) are processed by the units of the first layer whose outputs become inputs for the next layer and so on for the rest of the network. Typically, the neurons of each layer have as their inputs the outputs coming from the preceding layer only. Feedforward neural networks produce static nonlinear input-output mappings. Intermediate layers between input nodes and output layer are called hidden layers (as they are not affected directly by the input signals). We say that the neural network is fully connected if every node in each layer is connected to

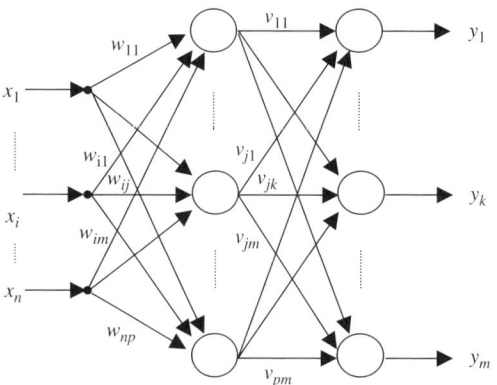

Figure B4 Three-layer feedforward neural network.

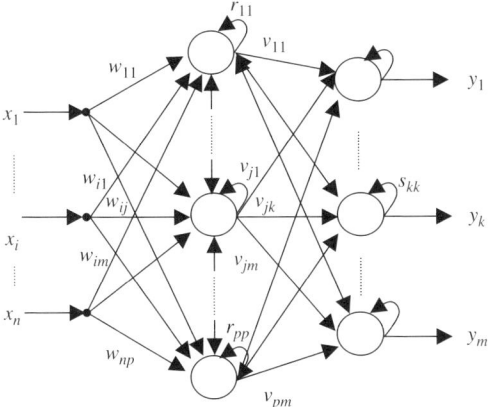

Figure B5 Recurrent neural network.

every other node in the adjacent forward layer. If this does not hold, we refer to partially connected neural network. The set of localized neurons partially connected to a hidden layer neuron constitute the receptive field of the neuron.

Recurrent neural networks distinguish themselves from feedforward networks by admitting feedback loops (Fig. B5). These networks may or may not have hidden layers. Feedback can be local if there exist self-feedback loops only, namely, if the outputs of neurons are fed back to its own input. Recurrent neural networks can exhibit full or partial feedback, depending on how the feedback loops have been structured.

Feedback loops has a profound impact on the input-output behavior of recurrent neural networks and its learning capabilities. Feedback involves the use of q-delay elements, denoted by z^{-q}. These are units whose output at step t is the same as the input at step $t - q$. For instance, unit delay element z^{-1} outputs at t the value of its input at $t - 1$, that is, the input occurring at the previous time instant. The overall neural network produces a nonlinear dynamic input–output behavior (owing to the feedback loops and the nonlinear nature of processing of the neurons themselves).

Neural Networks as Universal Approximators

Neural networks implicitly encode in its structure a function that maps inputs in outputs. The sort of functions that can be encoded depends on the structure of the network. Currently, there is no definite result that indicates which type of networks describe corresponding classes of functions. However, existence results do exist. They are summarized below.

As far as the representation capabilities of neural networks are concerned, they are expressed in the form of a so-called theorem of universal approximation. This theorem states that a feedforward network with a single hidden layer (where the neurons in this layer are equipped with sigmoid type of transfer function) and an output layer composed of linear neurons (viz. with linear transfer functions) is a universal approximator (Cybenko, 1989; Hornik et al., 1989). In other words, there exists a neural network of such topology that can approximate any given bounded continuous function $\mathbf{R}^n \rightarrow \mathbf{R}$ to any arbitrarily *small* approximation error.

A wide class of continuous functions can be approximated by a weighted sum of Gaussians or any bell-shaped functions, as those encountered in RBF neural networks (Baldi, 1991). More generally, any continuous function can be approximated to arbitrary approximation accuracy by a feedforward neural network with two hidden units of sigmoid type of neurons and a linear output layer. In fact, this result also holds when neurons have nonpolynomial activation functions in single hidden layer networks (Hornik, 1993; Leshno et al., 1993).

Fully connected recurrent neural networks can encode states of arbitrary continuous and discrete time dynamic systems with any required accuracy. Therefore, recurrent neural networks approximate continuous trajectories in finite time intervals (Jin et al., 1996).

The theorem about universal approximation of neural networks has to be put in a certain context. It is definitely an important and fundamental finding because it assures us about their representation capabilities (if the function satisfies some continuity assumptions, we are confident that there is a neural network that approximates it to any desired accuracy). This finding is a typical existence theorem because it does not offer any constructive clue on how such a neural network could be constructed. For an interesting discussion of linkages between theoretical results and resulting algorithms refer to Scarselli and Tsoi (1998). Theoretical foundations on approximation of functions completed in any metric space as well as algorithmic issues are addressed in Courrieu (2005).

Learning Mechanisms in Neural Networks

There are three main learning strategies: (a) supervised, (b) unsupervised, and (c) reinforcement learning. In supervised learning, the network is provided with a training set, pairs of input and the corresponding output samples. Weights are adjusted in such a way that we construct the network to produce outputs that are as close as possible to the known outputs (targets) of the training set. Unsupervised learning does not require any outputs associated with the corresponding input. The objective of this learning is to reveal the underlying structure existing in the data (e.g., correlations or associations between patterns in data leading to emergence of their possible categories). Reinforcement learning concerns learning processes in which the network receives only high-level guidance as the correctness of its behavior (for instance, we offer a numeric assessment of performance of the network over a collection of some temporal data rather than each data point individually).

Hybrid learning combines supervised and unsupervised learning; here a subset of weights is updated using supervised learning, whereas the remaining ones are formed through unsupervised learning.

Learning regarded as an optimization process exhibits two facets: parametric and structural learning. Parametric learning concerns adjustments of the numeric values of the connections. Structural learning involves an optimization of the structure (topology) of the network.

Supervised Learning

Supervised learning use four basic types of learning rules: error-correction, stochastic, correlation, and competitive learning. Here we briefly summarize the error-correction rule, which is one of the most commonly used. Refer to Haykin (1998) for a comprehensive discussion of the remaining rules.

In supervised learning, the network takes advantage of desired outputs (targets) for each input to correct (adjust) the values of weights. The underlying principle of error correction rules is to use error signals, the differences between the desired network outputs and actual outputs during the learning process, to update connections weights, and to gradually reduce the approximation errors. The most common learning algorithm in multilayer neural network uses a gradient search technique to find the values of the network, connections so that the error becomes minimized. Typically, this criterion (approximation error) is expressed as a sum of squared errors. For instance, for the neural network depicted in Figure B4 we have

$$Q(\mathbf{w}, \mathbf{v}) = \sum_{i=1}^{N} Q_i(\mathbf{w}, \mathbf{v})$$

where

$$Q_i(\mathbf{w}, \mathbf{v}) = \frac{1}{2} \sum_{k=1}^{m} (y_k(\mathbf{x}^q) - d_k(\mathbf{x}^q))^2$$

where $\mathbf{w} = [w_{11}, \ldots, w_{np}]^T$ and $\mathbf{v} = [v_{11}, \ldots, v_{pm}]^T$ collect the connections of the network, $\mathbf{x} = [x_1, x_2, \ldots, x_n]^T$ is a vector of inputs, $y_k(\mathbf{x}^q)$ the kth output of the network corresponding to the qth input $\mathbf{x}^q$, $d_k(\mathbf{x}^q)$ the kth desired network output for this input, $q = 1, \ldots, N$, and N is the number of training samples.

From now on, let us denote by w_{ij} the weight associated with a link connecting the output of the ith neuron of a layer ℓ with an input of the jth neuron positioned in the successive adjacent layer of the network, layer $\ell + 1$. Let a_i be the ith input and u_j denote the weighted sum of the inputs of unit j, namely

$$u_j = \sum_i w_{ij} a_i \qquad (B2)$$

Then from the output of neuron j of the next adjacent layer is $y_j = a_j$ with

$$a_j = f(u_j) = f\left(\sum_i w_{ij} a_i\right)$$

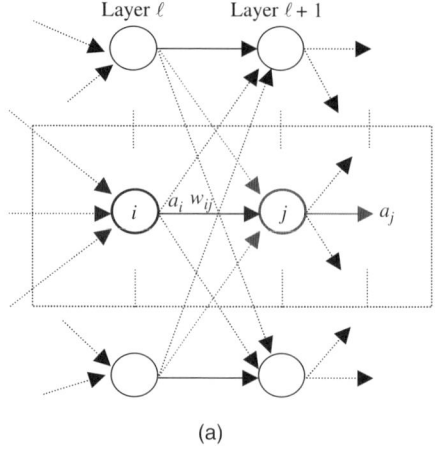

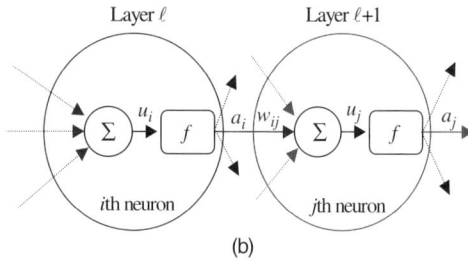

Figure B6 General multilayer network (a) and detail of connection between the ith neuron of layer ℓ with the jth neuron of layer $\ell + 1$.

as shown in Figure B6.

For instance, for the jth neuron of the second layer, we obtain $a_i = x_i$, $i = 1, \ldots, n$. The error between the desired and actual output of neuron j of the output layer is defined as follows

$$e_j = y_j - d_j \tag{B3}$$

Let $f'(u) = \partial f(u)/\partial u$ denote the derivative of the activation function computed with respect to its argument. The gradient search procedure to determine the connections of the network that minimize the sum of squared errors is summarized as follows (Hush and Horn, 1993).

procedure GRADIENT-SUPERVISED-LEARNING (S) **returns** a network
input: training data $S = \{(\mathbf{x}^q, \mathbf{d}(\mathbf{x}^q)), q = 1, \ldots, N\}$
local: network structure with L layers
tolerance: ε
learning rate: α

INITIALIZE-NETWORK-WEIGHTS
repeat
   randomly order training data
   for $\ell = 2: L$ **do**
     $u_j \leftarrow \Sigma_i w_{ij} a_i$
     $a_j \leftarrow f(u_j)$
   for each neuron j in the output layer L **do**
     $\delta j \leftarrow f'(u_j) e_j$
   for $\lambda = L - 1: 1$ **do**
     for each neuron i in layer ℓ **do**
       $\delta i \leftarrow f'(u_i) \Sigma_i w_{ij} \delta_j$
       for each neuron j in layer $\ell + 1$ **do**
         $w_{ij} \leftarrow w_{ij} + \alpha a_i \delta_j$
until sum squared error $\leq \varepsilon$
return weights w_{ij}

Approximations of the gradient supervised learning can be developed to handle recurrent neural networks. An example is the real-time recurrent learning algorithm (Williams and Zipser, 1989). This algorithm attempts to match the outputs of the neurons in a processing layer with the desired values (target values) at specific instants of time. Alternative approaches are addressed in Atiya and Parlos (2000).

Radial Basis Function Neural Networks

Radial basis function (RBF) networks are feedforward neural structures that combine a weighted collection of Gaussian kernels for function approximation and classification. Typically, RBF are three layer networks with Gaussian activation functions in the neurons of the second layer and linear neurons in the output layer. Linear neurons are neurons with linear activation functions, namely, $f(u) = u$. Therefore they compute their outputs as follows:

$$y = u = \sum_{i=1}^{n} w_i x_i = \mathbf{w}^T \mathbf{x}$$

There are two phases of the learning process. First, unsupervised learning takes place. Through clustering, we determine the number and modal values of the Gaussian receptive fields. The number of these fields is the number of clusters to be formed in the data set whereas the modal values of the Gaussians are the prototypes (centers) of the clusters themselves. During the second phase, we determine the weights (connections) of the Gaussians using either a gradient or a least squares algorithm (Haykin, 1998; Ellacott and Bose, 1996). The name radial basis functions comes from the fact that the Gaussian functions are radially symmetric, that is, each neuron produces an identical output for inputs that lie within the same radius from the modal value of the Gaussian function.

There are several items to consider when designing feedforward neural networks:

1. number of layers needed for an application
2. number of neurons per layer
3. generalization capability of the neural network
4. size of training data to achieve the desired generalization level

Despite current advances in learning theory (Müller et al., 2001; Morejon and Principe, 2004), the applications of feedforward neural networks still require careful experimentation and a prudent use of engineering judgment to make them successful.

Unsupervised Learning

Unsupervised learning involves four basic categories of learning rules, namely, error-correction, stochastic learning, correlation learning, and competitive learning. Here we discuss only the scheme of competitive learning.

Competitive learning is a learning paradigm in which output neurons compete among themselves for gaining activation. As a result, only a single output unit is active at a given instant and weight update proceeds in its own and weights of neighbor neurons. When the neighborhood involves only a single winning neuron, the update formula is called *winner-take-all* rule. Competitive learning often clusters input data in the sense that neighboring inputs activate neighboring outputs neurons. Often it has a recurrent structure organized in one-dimensional array such as the one illustrated in Figure B3(a) and called the learning vector quantizer. It is a two-dimensional array of linear neurons (Fig. B7) called self-organized maps (Hassoun, 1995).

Let $\mathbf{x} = [x_1, x_2, \ldots, x_n]^T$ be a vector of inputs and $\mathbf{w}_j = [w_{1j}, w_{2j}, \ldots, w_{nj}]^T$ be the vector of weights associated with the jth neuron, $j = 1, \ldots, M$. $\{\mathbf{x}^q\}$, $q = 1, \ldots, N$ denotes a set of training data.

The competitive self-organizing learning algorithm is described in the following way:

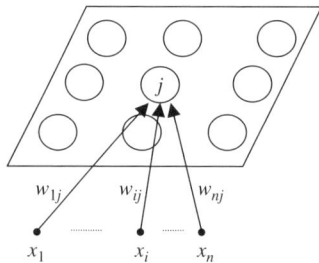

Figure B7 Competitive self-organizing neural networks.

procedure COMPETITIVE_UNSUPERVISED_LEARNING (S) **returns** a network
input: training data $S = \{\mathbf{x}^q, q = 1, \ldots, N\}$
local: network structure with M neurons
 neighborhood of neuron j at step t : $N_j(t)$
 threshold: ε
 learning rate: α

  INITIALIZE-NETWORK-WEIGHTS
  INITIALIZE-NEIGHBORHOODS
  $t \leftarrow 1$
  repeat
    randomly order training data
    for $q = 1 : N$
      select j^* such that $j^* = \arg\min_j \| \mathbf{x}^q - \mathbf{w}_j \|$
      if $j \in N_{j^*}(t)$ **then** $\mathbf{w}_j(t+1) \leftarrow \mathbf{w}_j(t) + \alpha(t)(\mathbf{x}(t) - \mathbf{w}_j(t))$
        else $\mathbf{w}_j(t+1) \leftarrow \mathbf{w}_j(t)$
    decrease $\alpha(t)$
    decrease $N_{j^*}(t)$
    $t \leftarrow t + 1$
  until changes in weight values $\leq \varepsilon$
  return weights $\mathbf{w}_j$

Self-organizing maps (SOMs) are useful in the representation of multidimensional data, probability density approximation, and data clustering and categorization. Design components include such crucial parameters as the dimensionality of the array of neurons, the number of neurons in each dimension, the form of the neighborhood function, and the mechanisms implementing the decrease of the learning rate and shrinkage of neighborhood.

REFERENCES

Anthony, M., Bartlet, P. L. Neural *Network Learning: Theoretical Foundations*, Cambridge University Press, Cambridge, 1999.

Atiya, A., Parlos, A. New results on recurrent network training: unifying the algorithms and accelerating convergence, *IEEE T. Neural Networ.*, **11**(3), 2000, 697–709.

Baldi, P. Computing with arrays of bell-shaped and sigmoid functions, in: R. Lippmann, J. Moody, D. Touretzky (eds.), *Neural Information Processing Systems*, Morgan Kaufmann, San Mateo, CA, 1991, pp. 735–742.

Courrieu, P. Function approximation in non-Euclidean spaces, *Neural Networks* **18**, 2005, 91–102.

Cybenko, G. Approximation by superposition of a sigmoidal function, *Math. Control Signal.*, **2**, 1989, 303–314.

Ellacott, S., Bose, D. *Neural Networks: Deterministic Methods of Analysis*, Thomson Computer Press, London, 1996.

Hassoun, M. *Fundamentals of Artificial Neural Networks*, MIT Press, Cambridge, MA, 1995.

Haykin, S. *Neural Networks: A Comprehensive Foundation*, 2nd ed., Prentice Hall, Upper Saddle River, NJ, 1998.

Hornik, K., Stinchcombe, M., White, H. Multilayer feedforward networks are universal approximators, *Neural Networks* **2**, 1989, 359–366.

Hornik, K. Some new results on neural network approximation, *Neural Networks* **6**, 1993, 1069–1071.

Hush, D., Horne, B. Progress in supervised neural networks: What's new after Lippmann? *IEEE Signal Proc. Mag.* **10**(1), 1993, 8–39.

Jin, L., Gupta, M., Nikiforuk, P. Approximation capabilities of feedforward and recurrent neural networks, in: M. Gupta, N. Sinha (eds.), *Intelligent Control Systems: Theory and Applications*, IEEE Press, Piscataway, NJ, 1996, pp. 234–264.

Leshno, M., Lin, Y., Pinkus, A., Schocken, S. Multilayer feedforward networks with a nonpolynomial activation function can approximate any function, *Neural Networks* **6**, 1993, 861–867.

Morejon, R., Principe, J. Advanced search algorithms for information-theoretic learning with kernel-based estimators, *IEEE T. Neural Networ.* **15**(4), 2004, 874–884.

Müller, K., Mika, S., Rätsch, R., Tsuda, K., Schölkopf, B. An introduction to kernel-based learnng algorithms, *IEEE T. Neural Networ.* **12**(2), 2001, 181–201.

Scarselli, F., Tsoi, A. Universal approximation using feedforward neural networks: A survey of some existing results and some new results, *Neural Networks* **11**, 1998, 15–37.

Williams, R., Zipser, D. A learning algorithm for continually running fully recurrent neural networks, *Neural Comput.*, **1**, 1989, 270–280.

Appendix C

BIOLOGICALLY INSPIRED OPTIMIZATION

T#o fully benefit from the potential of fuzzy sets and information granules as well as all constructs emerging there, there is a genuine need for effective mechanisms of *global* optimization. It is equally important that such an optimization framework comes with substantial capabilities of *structural* optimization of fuzzy systems. It is highly advantageous to have systems whose structure could be seamlessly modified to fully exploit the capabilities of the constructs of fuzzy sets. It would be highly desirable to consider constructs whose scalability can be easily realized. Biologically inspired optimization offers a wealth of optimization mechanisms that tend to fulfill these essential needs. The underlying principles of these algorithms relate to the biologically motivated schemes of system emergence, survival, and refinement. Quite commonly, we refer to the suite of these techniques as Evolutionary Computing to directly emphasize the inspiring role of various mechanisms encountered in the Nature that are also considered as pillars of the methodology and algorithms. The most visible feature of most, if not all, such algorithms is that in their optimization pursuits they rely on a collection of individuals that interact between themselves in the synchronization of joint activities of finding solutions. They communicate between themselves by exchanging their local findings. They are also influenced by each other.

Evolutionary Optimization

Evolutionary optimization offers a comprehensive optimization environment in which we encounter a stochastic search that mimics natural phenomena of genetic inheritance and Darwinian strife for survival. The objective of evolutionary optimization is to find a maximum of a certain objective function f defined in some search space **E**. Ideally, we are interested in the determination of a global maximum of f.

Fuzzy Systems Engineering: Toward Human-Centric Computing, by Witold Pedrycz and Fernando Gomide
Copyright © 2007 John Wiley & Sons, Inc.

A Population-Based Optimization Principle of Evolutionary Computing

The crux of the evolutionary optimization process lies in the use of a finite population of N individuals (represented as elements of the search space $\mathbf{E}$) whose evolution in the search space leads to an optimal solution. The population-based optimization is an outstanding feature of the evolutionary optimization and is practically present in all its variants we can encounter today. The population is initialized randomly (at the beginning of the search process, say, $t = 0$). For each individual we compute its fitness value. This fitness is related to the maximized objective function. The higher the value of the fitness, the more suitable is the corresponding individual as a potential solution to the problem. The population of individuals in $\mathbf{E}$ undergoes a series of generations in which we apply some evolutionary operators and through them improve the fitness of the individuals. Those of the highest fitness become more profoundly visible by increasing chances to survive and occur in the next generation.

In a very schematic and abstract way, a computing skeleton of evolutionary optimization can be described as follows:

procedure EVOLUTIONARY-OPTIMIZATION (f) **returns** a solution
input: fitness function f
local: evolutionary operators rates
        population: set of individuals
        INITIALIZE (population)
        evaluate population
        repeat
           select individuals for reproduction
           apply evolutionary operators
           evaluate offsprings
           replace some old individuals by offsprings
        until termination condition is true
        return a best individual

Let us briefly elaborate on the main components of the evolutionary computing. Evaluation concerns a determination of the fitness of individuals in the population. The ones with high values of fitness have chances to survive and appear in the consecutive populations (generations of the evolutionary optimization). The selection of individuals to generate offsprings is based on the values of the fitness function. Depending on the selection criterion (which could be stochastic or deterministic), some individuals could produce several copies of themselves (clones). The stopping criterion may involve the number of generations (which could be set up in advance, say 200 generations), which is perhaps the simplest alternative. One could also involve the statistics of the fitness of the population; say, no significant changes in the average values of fitness may trigger the termination of the optimization process. There are two essential evolutionary operators whose role is to carry out the search process in $\mathbf{E}$ and make sure that it secures its effectiveness. The operators are applied to the current individuals. Typically, these operators are of stochastic nature, and their

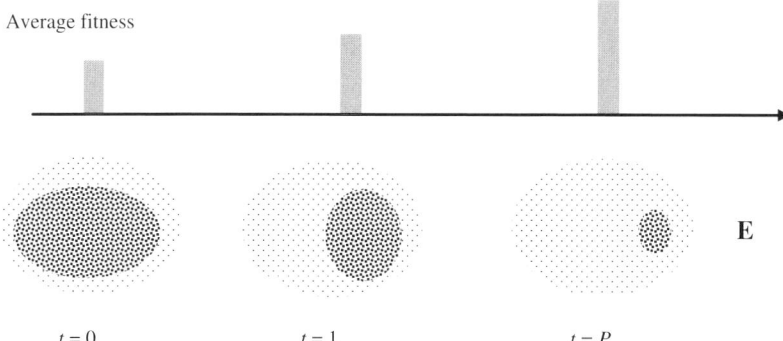

Figure C1 A schematic view at evolutionary optimization; note a more focused population of individuals over the course of evolution and the increase in fitness values of the individuals and average fitness of the entire population.

intensity depends on the assumed probabilities. There are two groups of operators. Crossover (recombination) operators involve two or more individuals and give rise to one or more offsprings. In most cases, the crossover operator concerns two parents and leads to two offsprings. Formally, we can view such a crossover operator as a mapping of the form $\mathbf{E} \times \mathbf{E} \to \mathbf{E} \times \mathbf{E}$. The objective of crossover is to assure that the optimization exploit new regions of the search space as the offsprings vary from the parents. The mutation operator affects a single individual by randomly affecting one or several elements of the vector: In essence, it forms a mapping from $\mathbf{E}$ to itself, $\mathbf{E} \to \mathbf{E}$.

The evolutionary optimization process is transparent: We start with some initial population of individuals and evolve the population by using some evolutionary operators. An illustration of evolutionary optimization is illustrated in Figure C1.

Observe that in successive populations, they start to be more "focused," producing individuals (solutions) of higher fitness. Typically, an average fitness of the population could fluctuate, however, on average; it exhibits higher values over the course of evolution. The best individual (viz. the one with the highest fitness) is retained from population to population, so we do not loose the best solution produced so far. This retention of the best individual in the population is referred to as an *elitist* strategy.

The Main Categories of Evolutionary Optimization

There are four major categories of evolutionary optimization. While they share the underlying principles, they differ in terms of the representation issues and computational aspects.

Evolution strategies (ES) (Schwefel, 1995) are predominantly focused on parametric optimization. In essence, a population consists only of a single individual, that is, a vector of real numbers. This individual undergoes a Gaussian mutation in which we add a zero mean Gauusian variable of some standard deviation, $N(0, \sigma)$. The fittest from the parent and the offspring becomes the next parent. The value of the standard deviation is adjusted over the course of evolution. The main operator is mutation. One can also encounter population-based versions of ES, known as $(\mu + \lambda)$–ES in which μ parents generate λ offsprings.

Evolutionary programming (Fogel et al., 1966) originally focused on evolving finite state machines was focused on the phenotype space. Similar to ES, there is no initial selection and every individual generates one offspring. Mutation is the evolution operator. The best individuals among parents and offsprings become the parent of the next generation.

Genetic Algorithms (GAs) (Holland, 1975; Goldberg, 1989; Michalewicz, 1996) are one of the most visible branches of evolutionary optimization. In its standard format, GAs exploit a binary genotype space $\{0,1\}^n$. The phenotype could be any space as long as its elements could be encoded into binary strings (bitstrings, for short). The selection scheme is proportional selection, known as the roulette wheel selection. A number of random choices is made in the whole population, which implies that the individuals are selected with probability that is proportional to its fitness. The crossover operation replaces a segment of bits in the first parent by the corresponding string of the second parent. The mutation concerns a random flipping of the bits. In the replacement, offsprings replace all parents.

Genetic Programming (GP) (Kinnear, 1994; Koza, 1994) originated as a vehicle to evolve computer programs, and algebraic and logic expressions, in particular. The predominant structures in GP are trees. These are typically implemented in the form of LISP expressions (S-expressions). This realization helped define crossover operation as "swapping to subtrees between two S-expressions is still a valid S-expression."

Knowledge Representation: from Phenotype to Genotype Space

A suitable problem representation in evolutionary optimization becomes a key issue that predetermines success of the optimization process and implies quality of the produced solution. Let us note that evolutionary optimization is carried out in the *genotype* space **E**, which is a result of a transformation of the problem from the original space, a so-called *phenotype* space **P**, realized with the use of some encoding and decoding procedures; refer to Figure C2.

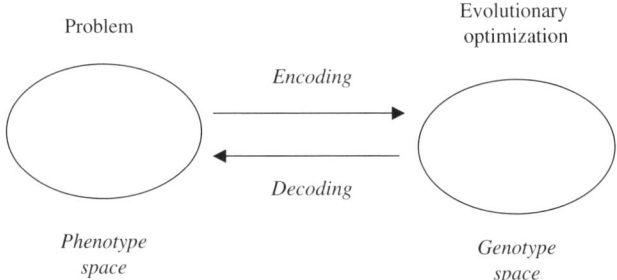

Figure C2 From phenotype space to genotype space: links between optimization problem and its representation in evolutionary optimization.

In a more descriptive way, we could view representation issues as being central to the nature of the underlying optimization problem. Knowledge representation is a truly multifaceted problem, and as such one has to proceed with prudence realizing that the effectiveness of this scheme implies the quality of evolutionary solution.

In what follows, several examples of encoding and decoding serve as an illustration of the diversity of possible ways of knowledge representation.

1. *Binary encoding and decoding*: Any parameter assuming real values can be represented in the form of the corresponding binary number. This binary coding is used quite commonly in GAs. The strings of bit are then subject to evolutionary operations. The result is decoded into the corresponding decimal equivalent. More formally, the genotype space, $\mathbf{E} = \{0, 1\}^m$, hypercube where m stands for the dimensionality of the space and depends on the number of parameters encoded in this way and a resolution (number of bits) used to complete the encoding.

2. *Floating point (real) encoding and decoding*: Here we represent values of parameters of the system under optimization using real numbers. Typically, to avoid occurrence of numbers in different ranges, all of them are scaled (e.g., linearly) to the unit intervals, so in effect the genotype space is a unit hypercube, $\mathbf{E} = [0, 1]^p$ with p denoting the number of parameters. The resulting string of real numbers is retransformed into the original ranges of the parameters.

3. *Representation of structure of fuzzy logic network*: Fuzzy logic network exhibits a diversity of topologies. In particular, this variety becomes visible in the development of networks involving referential neurons. Given four types of referential neurons, that is, similarity, difference, inclusion, and dominance, we can consider several ways of representation of the structure in the genotype space: (a) one can view a binary encoding where we use two bits with the following assignment: 00—similarity, 01—difference, 10—inclusion, and 11—dominance, (b) alternatively, we can consider a real coding and in this case, we can accept the decoding that takes into consideration ranges of values in the unit interval, say, [0.00 – 0.25]—similarity, [0.25, 0.50]—difference, [0.50, 0.75)—inclusion, and [0.75, 1.00]—dominance. The dimensionality of the genotype space depends on the number of the referential neurons used in the network. An example of the binary encoding for the fuzzy logic network with five referential neurons is illustrated in Figure C3.

4. *Structure representation of subsets of variables*: In many cases, in order to reduce problem dimensionality, we might consider a problem of selecting a subset of input variables. For instance, when dealing with hundreds of variables, practically we can envision the use of a handful of them, say 10 or so, in the development of the fuzzy system (say, a rule-based system). Given these 10 variables, we develop a network and assess its performance. This performance index could be regarded as a suitable fitness function to be used in evolutionary optimization. Let us also note that the practicality of a

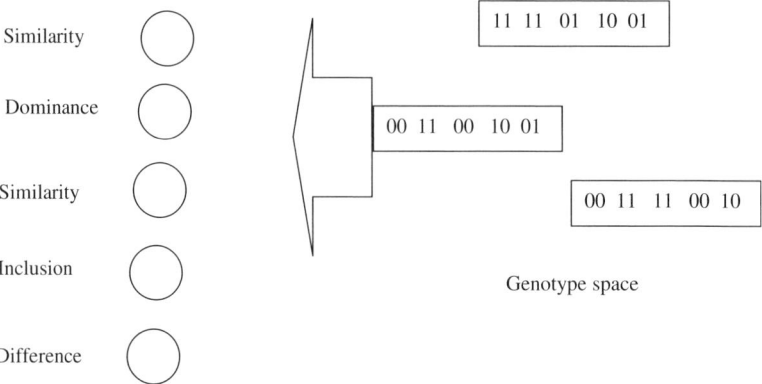

Figure C3 Binary encoding of the fuzzy logic network.

plain enumeration of combinations of such variables is out of question; say, choosing 10 variables out of 200 variables leads to $\binom{200}{10}$ possible combinations. Here the representation of the structure can be realized by forming 200-dimensional strings of real numbers, that is, $\mathbf{E} = [0, 1]^{200}$. To decode the result, we rank the entries of the vector and pick the first 10 entries of the vector. For 100 variables and 10 variables to be selected, we end up with 1.731×10^{13} possible alternatives.

An example of this representation of the genotype space is illustrated in Figure C4. Note that the plain selection of the entries decoded with the use of the intervals of the unit interval (say, [0–1/200]—variable #1, [1/200–2/200)—variable #2, ...) of the vector will not work as we could quite easily encounter duplicates of the same variable. This could be particularly visible in the case of the large number of variables.

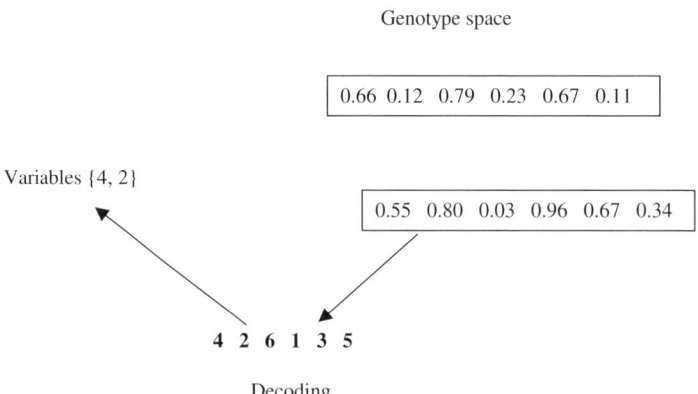

Figure C4 Variable selection through ranking the entries of the vectors of the genotype space E; here the total number of variables under consideration is five and we are concerned about choosing two variables.

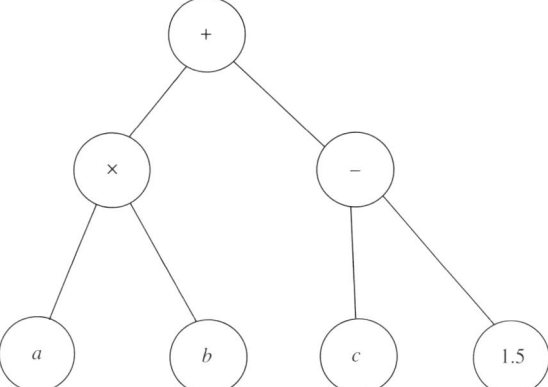

Figure C5 Tree representation of the genotype space.

5. *Tree representation of the genotype space*: This form of knowledge representation is commonly encountered in genetic programming. Trees such as shown in Figure C5 are used to encode algebraic expressions. For instance, the first tree in this figure (i.e., a) encodes the expression whereas the second one reads as $(a \times b) + (c - 1.5)$.

Depending on the representation of the genotype space, the evolutionary operators come with different realizations. As an example, consider a mutation operator. In the case of binary encoding and decoding, mutation is realized by a simple flipping of the value of the specific entry of the vector of bits. In the real number encoding and decoding, we may use the complement operator, namely, replacing a certain value in [0,1] by its complement, say $1 - v$, with v being the original component of the vector.

Some Practical Design and Implementation Guidelines

Evolutionary optimization offers a number of evident advantages over some other categories of optimization mechanisms. They are general and their conceptual transparency is definitely very much appealing. The population-based style of optimization offers a possibility of a comprehensive exploration of the search space and provides solid assurance of finding a global maximum of the problem. To take full advantage of the potential of evolutionary optimization, one has to exercise prudence in setting up the computing environment. This concerns a number of crucial parameters of the algorithm that concern evolutionary operators, size of population, and stopping criterion, to name the most essential ones. Moreover, what is even more fundamental concerns a representation of the problem in the genotype space. Here a designer has to exercise his/her ingenuity and fully capture the essence of domain knowledge about the problem. There is no direct solution to the decoding problem. We could come up with a number of alternatives. In many cases, it is not obvious up front what could work the best, namely, lead to the fastest convergence of the algorithm to the global solution, prevent from premature convergence, and help avoid forming an excessively huge genotype space. The scalability aspect of coding has to be taken into consideration as well. Several examples presented below help emphasize the importance of the development of a suitable genotype space.

The use of evolutionary optimization in the development of fuzzy systems or neurofuzzy systems can be exercised in many different ways. As we do not envision any direct limitations, we should exercise some caution and make sure that we really take full advantage of these optimization techniques while not being affected by their limitations.

(a) Structural optimization of fuzzy systems provided by evolutionary optimization is definitely more profitable than the use of evolutionary methods for their *parametric* optimization. Unless there are clear recommendations with this regard, we could be better-off considering gradient-based methods or exercising particle swarm optimization rather than relying on evolutionary optimization. Another alternative would be to envision a hybrid approach in which we combine evolutionary optimization regarded as a preliminary phase of optimization that becomes helpful in forming some initial and promising solution and then refine it with the aid of gradient-based learning.

(b) The choice of the genotype space is critical to the success of evolutionary optimization; this, however, becomes a matter of a prudent and comprehensive use of the existing domain knowledge. Once the specific genotype space has been formed, we need to be cognizant of the nature and role of specific evolutionary operators in the search process. It might not be clear how efficient they could be in the optimization process.

(c) The choice of the fitness function must fully capture the nature of the problem. While in evolutionary optimization we do not require that such function be differentiable with respect to the optimized component (which is a must in case of gradient-based techniques), it is imperative, though, that the requirements of the optimization problem be reflected in the form of the fitness function. In many cases, we encounter a multiobjective problem, and a caution must be exercised so that all the objectives are carefully addressed. In other words, a construction of the suitable fitness function is an essential component of the evolutionary optimization.

One may note that reciprocally, the technology of fuzzy sets could be helpful in structuring domain knowledge that could effectively be used in the organization of evolutionary optimization. This could result in a series of rule (or metarules, to be specific) that may pertain to the optimization. For instance, we could link the values of the parameters of the evolutionary operators with the performance of the process. For instance, " if there is *high* variation of the values of the average fitness of the population, a *substantial* reduction of mutation rate is advised."

Particle Swarm Optimization

The biologically inspired optimization technique, Particle Swarm Optimization (PSO), is an example of the modern search heuristics that belongs to the category of so-called Swarm Intelligence methods (Eberhart and Shi, 2001; Kennedy and Eberhat, 2001; Parsopoulos and Vrahatis, 2004). The underlying principle of PSO deals with a population-based search in which individuals representing possible solutions carry out collective search by exchanging their individual findings while taking into consideration their own experience and evaluating their own performance. In this sense, we encounter two fundamental aspects of the search strategy. The one deals with a *social* facet of the search; according to this, individuals ignore their own experience and adjust their behavior according to the successful beliefs of individuals occurring in their neighborhood. The *cognition* aspect of the search underlines the importance of the individual experience where the element of population is focused on its own history of performance and makes adjustments accordingly. In essence, PSO dwells its search by using a combination of these two mechanisms. Some applications of PSO are presented in Abido (2002), Gaing (2004), Ozcan and Mohan (1998), Robinson and Rahmat-Samii (2004), and Wang et al. (2004).

The vectors of the variables (particles) positioned in the *n*-dimensional search space are denoted by $x_1, x_2, \ldots, x_N$. In the search, there are N particles involved, leading to the concept of a swarm. The performance of each particle is described by some objective function referred to as a fitness (or objective) function.

The PSO is conceptually simple, easy to implement, and computationally efficient. Unlike the other heuristic techniques, PSO has a flexible and well-balanced mechanism to enhance the global and local exploration abilities. As in the case of evolutionary optimization, the generic elements of the PSO technique involve

1. *Performance (fitness)*: Each particle is characterized by some value of the underlying performance (objective) index or fitness. This is a tangible indicator stating how well the particle is doing in the search process. The fitness is reflective of the nature of the problem for which an optimal solution is being looked for. Depending upon the nature of the problem at hand, the fitness function can be either minimized or maximized.

2. *Best particles*: As a particle wonders through the search space, we compare its fitness at the current position with the best fitness value it has ever attained so far. This is done for each element in the swarm. The location of the *i*th particle at which it has attained the best fitness is denoted by **x_best**$_i$. Similarly, by **x_best** we denote the best location attained among all the **x_best**$_i$.

3. *Velocity*: The particle is moving in the search space with some velocity that plays a pivotal role in the search process. Denote the velocity of the *i*th particle by $\mathbf{v}_i$. From iteration to iteration, the velocity of the particle is governed by the following expression:

$$\mathbf{v}_i = w\mathbf{v}_i + c_1 r_1(\mathbf{x\_best}_i - \mathbf{x}_i) + c_2 r_2(\mathbf{x\_best} - \mathbf{x}_i)$$

or equivalently

$$v_{ik} = wv_{ik} + c_1 r_1 (x\_best_{ik} - x_{ik}) + c_2 r_2 (x\_best_k - x_{ik}) \qquad (C1)$$

$i = 1, 2, \ldots, N, k = 1, 2, \ldots, n$, where r_1 and r_2 are two random values located in [0 1], and c_1 and c_2 are positive constants, called the acceleration constants. They are referred to as cognitive and social parameters, respectively. As the above expression shows, c_1 and c_2 reflect on the weighting of the stochastic acceleration terms that pull the ith particle toward **x_best**$_i$ and **x_best** positions, respectively. Low values allow particles to roam far from the target regions before being tugged back. High values of c_1 and c_2 result in abrupt movement toward, or past, target regions. Typically, the values of these constants are kept close to 2.0. The inertia factor "w" is a control parameter that is used to establish the impact of the previous velocity on the current velocity. Hence, it influences the trade-off between the global and local exploration abilities of the particles. For the initial phase of the search process, large values enhancing the global exploration of the space are recommended. As the search progresses, the values of "w" are gradually reduced to achieve better exploration at the local level.

As the PSO is an iterative search strategy, we proceed with it until the point there is no substantial improvement of the fitness or we have exhausted the number of iterations allowed in this search.

Overall, the algorithm can be outlined as the following sequence of steps:

procedure PARTICLE-SWARM-OPTIMIZATION (f) **returns** best solution
input: objective function f
local: inertia weights
             swarm: population of particles

> Generate randomly N particles, $\mathbf{x}_i$, and their velocities $\mathbf{v}_i$. Each particle in the initial swarm (population) is evaluated using the given objective function. For each particle, set **x_best**$_i = \mathbf{x}_i$ and search for the best value of the objective function. Set the particle associated with as the global best, **x_best**.
>
> **repeat**
>
> **adjust weight**: the value of the inertia weight w. Typically, its values decrease linearly over the time of search. We start with $w_{max} = 0.9$ at the beginning of the search and move down to $w_{min} = 0.4$ at the end of the iterative process,
>
> $$w(\text{iter} + 1) = w_{max} - \frac{w_{max} - w_{min}}{\text{iter}_{max}} \text{iter} \qquad (C2)$$
>
> where iter_{max} denotes the maximum number of iterations of the search and "iter" stands for the current index of the iteration.
>
> **adjust velocity**: Given the current values of **x_best** and **x_best**$_i$, the velocity of the ith particle is adjusted following (C1). If required, we

clip the values making sure that they are positioned within the required region.

adjust position: Based on the updated velocities, each particle changes its position following the expression

$$x_{ik} = v_{ik} + x_{ik} \qquad (C3)$$

Furthermore, we need to keep the particle within the boundaries of the search space, meaning that the values of x_{ik} have be confined to it following the expression $x_k^{\min} \leq x_{ik} \leq x_k^{\max}$, where the ith coordinate of the space assumes the values in $[x^{\min}, x^{\max}]$

move particles: move the particles in the search space and evaluate their fitness both in terms of **x_best**$_i$ and **x_best**.

until termination criterion is met.
return x_best

REFERENCES

Abido, M. Optimal design of power-system stabilizers using particle swarm optimization, *IEEE T. Energ. Convers.* **17**(3), 2002, 406–413.

Eberhart, R., Shi, Y. Particle swarm optimization: developments, applications and resources, *Proceedings of the IEEE International Conference on Evolutionary Computation*, IEEE Press, 2001, pp. 81–86.

Fogel, L., Owens, A., Walsh, M. *Artificial Intelligence Through Simulated Evolution*, John Wiley & Sons, Inc., New York, NY, 1966.

Gaing, Z. A Particle swarm optimization approach for optimum design of PID controller in AVR system, *IEEE T. Energ. Convers.* **19**(2), 2004, 384–391.

Goldberg, D. *Genetic Algorithms in Search, Optimization and Machine Learning*, Addison Wesley, Reading, MA, 1989.

Holland, J. *Adaptation in Natural and Artificial Systems*, University of Michigan Press, Ann Arbor, MI, 1975.

Kennedy, J., Eberhart, R. C. *Swarm Intelligence*, Morgan Kaufmann, San Francisco, CA, 2001.

Kinnear K. (ed.), *Advances in Genetic Programming*, MIT Press, Cambridge, MA, 1994.

Koza, J. *Genetic Programming: On the Programming of Computers by Means of Natural Evolution*, MIT Press, Cambridge, MA, 1994.

Michalewicz, Z. *Genetic Algorithms + Data Structures = Evolution Programs*, Springer-Verlag, Berlin, 1996.

Ozcan, E., Mohan, C. Analysis of a simple particle swarm optimization system, *Proceedings of the Conference on Intel. Eng. Syst. Through Artificial Neural Networks*, 1998, pp. 253–258.

Parsopoulos, K., Vrahatis, M. On the computation of all global minimizers through particle swarm optimization, *IEEE T. Evol. Comput.* **8**(3), 2004, 211–224.

Robinson, J., Rahmat-Samii, Y. Particle swarm optimization in electromagnetics, *IEEE T. Antennas Propagation* **52**(2), 2004, 397–407.

Schwefel, H. *Numerical Optimization of Computer Models*, 2nd ed., John Wiley & Sons, Inc., New York, NY, 1995.

Wang, Z., Durst, G., Eberhart, R., Boyd, D., Miled, Z. Particle swarm optimization and neural network application for QSAR, *Proceedings of the 18th International Parallel and Distributed Processing Symposium*, 2004, 194–201.

Index

Agent, 467
Aggregation operations, 119
Aggregative neurons, 335
Approximation, 316
Averaging operations, 120
Alfa cut (α cut), 47
Associative memories, 178

Cardinality, 49
Cartesian product, 144
Characteristic function, 29
Characterization of fuzzy sets, 45
Clustering, 86, 418, 421, 437, 441, 444, 448
Compatibility relations, 154
Compensatory operations, 128
Complement, 102
Completeness, 325
Composition of fuzzy relations, 160
Computational intelligence, 17, 382
Consistency, 325
Construction of fuzzy sets, 94
Constructors of t-norms, 107
Constructors of t-conorms, 114
Coverage, 59
Convexity, 48
Cylindrical extension, 145

Database, 296
Decoding, 228
Degree of inclusion, 51
Distributed control, 480
Dichotomy, 27

Encoding, 227
Energy measure, 52
Entropy measure, 54
Equality, 50

Equalization, 92
Equivalence relation, 152
Estimation of membership function, 75, 77
Estimation problem, 169
Evolving fuzzy systems, 405, 407
Extension principle, 156

Feedback loops, 351
Focus of attention, 61
Frame of cognition, 59
Functional fuzzy models, 310, 406
Fuzzy arithmetic, 180
Fuzzy codebooks, 234
Fuzzy conjunction, 282
Fuzzy decision trees, 259
Fuzzy disjunction, 284
Fuzzy events, 212
Fuzzy implication, 285
Fuzzy inference, 297
Fuzzy integral, 131
Fuzzy intervals, 39
Fuzzy measure, 129
Fuzzy numbers, 180
Fuzzy quantization, 230
Fuzzy relational equations, 167
Fuzzy relations, 140
Fuzzy rough sets, 196
Fuzzy set, 30
Fuzzy sets higher order, 194
Fuzzy models, 252, 254, 255, 257
Fuzzy neural networks, 259, 352
Fuzzy numbers, 39
Fuzzy rules, 276

Genetic fuzzy systems, 394, 471
Gradual fuzzy models, 315
Granulation, 57, 61, 80

Fuzzy Systems Engineering: Toward Human-Centric Computing, by Witold Pedrycz and Fernando Gomide
Copyright © 2007 John Wiley & Sons, Inc.

525

Granular mappings, 372
Granular neuron, 425

Hierarchical genetic fuzzy system, 394

Inclusion, 50
Information granules, 369
Information hiding, 62
Information retrieval, 462
Interfacing, 225, 364
Intersection, 102
Interpretation, of fuzzy sets, 31, 56
Interpretation of fuzzy neural networks, 358
Interval arithmetic, 182
Interval-valued fuzzy sets, 199
Inverse problem, 174

Learning, 354, 363, 387
Linearization 69
Linguistic approximation, 94
Linguistic fuzzy models, 302
Linguistic variables, 40
Logic networks, 347
Logic processor, 349
Logical operations, 117

Membership function, 30, 33
Model validation, 268
Model verification, 264
Multiagents, 467

Necessity measure, 237
Negations, 133
Normality, 46
Nullnorms, 122

Operations on fuzzy relations, 143
Ordered weighted operations, 122
Ordinal sums, 110, 117, 126

Parameter estimation, 322, 409
Participatory learning, 407, 448
Possibility measure, 237
Projection of fuzzy relations, 144
Proximity relations, 154

Reconstruction of fuzzy relations, 148
Recurrent neurofuzzy network, 384
Referential neurons, 340
Relational ontology, 459
Relations, 138
Representation theorem, 220
Rough fuzzy set, 196
Rule base, 282, 289
Rule-based fuzzy systems, 256, 275, 279, 301, 317

Semantic soundness, 60
Similarity relation, 152
Shadowed sets, 203
Solvability conditions, 177
Specificity, 55, 61
Standard operations, 100
Support, 47
Symmetric sums, 127

Term set, 41
Transparency, 268
Transitive closure, 151
Triangular norms, 104
Triangular conorms, 111
Type-2 fuzzy sets, 200

Uninorms, 122
Unineurons, 343, 344
Union, 102